THE GEOLOGY OF THE SUDAN REPUBLIC

Jebel Kassala, Basement Complex Gneiss, eastern Sudan Air view looking east towards Ethiopia.

THE GEOLOGY OF THE SUDAN REPUBLIC

BY

A. J. WHITEMAN

CLARENDON PRESS · OXFORD

1971

Oxford University Press, Ely House, London W. 1

GLASGOW NEW YORK TORONTO MELBOURNE WELLINGTON
CAPE TOWN SALISBURY IBADAN NAIROBI DAR ES SALAAM LUSAKA ADDIS ABABA
BOMBAY CALCUTTA MADRAS KARACHI LAHORE DACCA
KUALA LUMPUR SINGAPORE HONG KONG TOKYO

PRINTED IN GREAT BRITAIN
BY WILLIAM CLOWES & SONS LIMITED
LONDON, COLCHESTER AND BECCLES

Preface

AN attempt is made here for the first time to provide a comprehensive account of the geology of the Sudan Republic.

In a work of this type one naturally has to draw heavily on the observations and conclusions of those who have gone before, and I trust that due acknowledgement has been made in the text and bibliography throughout.

The availability so far of no more than a standard geological map at the scale of only 1:4 000 000 (third edition, 1963), a few colour-printed maps at 1:250 000 scale, and a few published memoirs has given rise to the view that little is known of Sudan geology, and to many African geologists this part of 'Bilad es Sudan' is very much an unknown land. In some ways this is untrue, for considerable amounts of detailed information exist in unpublished files and unpublished maps of the Sudan Geological Survey (Fig. 5), the Sudan Topographic Survey, the Land Use and Rural Development Department (now incorporated in the Rural Water and Development Corporation), government archives, international aid agencies, consultant survey companies, etc. Data from these sources have been assessed and assembled, together with new information collected since 1960 on surveys and expeditions made by staff and students of the Department of Geology, University of Khartoum.

Field geology in a predominantly desert and semi-desert country like the Sudan with its extremely high temperatures, lack of water, difficulties of survival, and lack of roads can be undertaken only on an expedition basis, and much of the new information presented in this book was obtained by the joint efforts of colleagues and students, often on safari under arduous and trying conditions for men, animals, and vehicles. I should therefore like to acknowledge especially the assistance of my colleagues Dr. D. C. Almond, Mr. S. V. Bell, Professor L. Berry, Dr. I. M. Boushi, Dr. A. R. Gindy, Dr. I. R. Qureshi, Dr. S. M. Rabaa, and Dr. J. Sestini.

The late Professor S. E. Hollingworth of University College London was a frequent visitor to the Sudan as external examiner in geology and his clarity of vision and depth of knowledge have helped us on many occasions in understanding problems of Sudanese geology. He will long be remembered as a friend and surveyor in the Sudan.

I must mention also the contribution of my graduate students: Ustasen Yassin Abdl Salaam Hugaz, Kheiralla Mahgoub Kheiralla, Farouk Ahmed, Abdl Ati Ahmed, Badr el Din Khalil, Ahmed Suleiman Daoud, Mohammed Zein Shaddad, and Omer Badri. They have all contributed to this work, either directly by providing material from their M.Sc. theses for the chapters on stratigraphy, ground-water geology, and mineral resources or in collecting information from files and archives, especially M. Z. Shaddad who collected much of the data on gold, chrome, gypsum, and manganese.

Much of the field work was made possible by the generous financial assistance of the Research Committee of the University of Khartoum, the Ford Foundation, and the

Rockefeller Foundation; by the use of R.V. *Malakal* of the University of Khartoum Hydrobiology Unit; and by the use of Land Rovers and safari equipment of the University of Khartoum Arid Zone Research Unit. For all this help I am indeed grateful. I should like to thank Lt.-Col. H. W. Dixon, M.B.E., ex-General Manager of Sudan Portland Cement Company Works, Atbara for information on the cement industry and mining lore about the gold miners of the 'Savage Sudan'.

Also I should like to record my thanks to all those Sudan Government officials scattered throughout the country who helped us in our work. If I mention specifically Sayed Mohmoud Ahmed Abdulla, Director of Sudan Geological Survey; Sayed Charly Antoun, ex-Director of Sudan Topographic Survey; Sayed Abdl Rahim Bayoumi, Director of Land Use and Rural Development Department; Sayed Kamil Shawgi, Director-General, Rural Water and Development Corporation; and Sayed Hassan Dafalla, ex-Commissioner Wadi Halfa Resettlement, it is not to detract from my gratitude to others unnamed who helped us on our way.

Finally I should like to mention especially Mr. R. J. D. Oliver, draughtsman, Department of Geology, who with great skill and patience drew the maps and diagrams for this book; his assistant Miss L. F. Boulis; Sayda Atiat Gamal el Din of the University of Khartoum, Sudan, and Messrs. S. O. Aiyewa and W. O. Otovoh of the University of Ibadan, Nigeria, who typed part of the manuscript.

Credit for photographs, sources used in constructing maps and diagrams, etc. is given in the captions to the illustrations.

Professor Kevin Burke, University of Ibadan, kindly read and helped correct the manuscript.

University of Ibadan
Nigeria, 1968

A. J. WHITEMAN

Contents

List of Plates

I

Introduction

THE Republic of the Sudan occupies 967 498 square miles (2 496 138 km²) and is the largest country in Africa. It is about one-third the size of Australia and two-and-a-half times that of France. The boundaries of the Sudan are shown in Fig. 1. Geologically it is among the most poorly mapped of the African states, although considerable local detail is available in the unpublished reports and files, and on maps of the Sudan Geological Survey concerning areas of economic interest (Fig. 5).

1. PHYSICAL FEATURES

The dominant features of Sudan topography are

(i) the Nile Valley, a drainage system of considerable antiquity;
(ii) the great erosional scarp that bounds the Red Sea Hills or Jubal el Bahr el Ahmer to the east, overlooking the Red Sea (Whiteman 1965*a*);
(iii) the series of great pediplains and prominent inselbergs covering thousands of square miles and ranging back in age into the Tertiary, Mesozoic, and beyond;
(iv) the vast depositional plains and basins;
(v) the volcanic uplands of Jebel Marra and Meidob in Darfur;
(vi) the southern highlands bordering Kenya, Uganda, and Congo; and
(vii) the foothills of the Ethiopian volcanic plateau.

(The names used in this book are mainly those printed on Sudan Topographic Survey 1:250 000 maps.)

High ground is restricted to the Red Sea Hills in the north-east (Jebel Asotriba, 7272 ft), Jebel Marra (9676 ft) in the west, and the jebels of the southern Sudan such as the Imatongs and the Dongatona Mountains (Mt. Kinyeti, 10 198 ft) along the Uganda frontier.

Less than 2 per cent lies lower than 960 ft (300 m) above sea-level; about 45 per cent lies between 960 and 1600 ft (300 m and 500 m); and a further 50 per cent lies below 3840 ft (1200 m) (Barbour 1961).

The Sudan occupies a major part of the Nile basin but small parts of the country drain into the Chad basin, for example the Wadi Azum and its tributaries, which drain the eastern slopes of Jebel Marra in Darfur. A small area in the south-east drains into Lake Rudolf.

The Nile and its tributaries are the main rivers. From north to south they include the Atbara, which joins the Nile near the edge of the desert; the Bahr el Azrak or Blue Nile rising near Lake Tana, and its tributaries the Rahad and the Dinder; Bahr el Abiod or White Nile with its main tributaries the Sobat, Bahr el Jebel, and its continuation in the Albert and Victoria Niles, Bahr el Arab, Bahr el Ghazal, and Bahr el Zeraf. The Gash or Mareb descends from the Eritrean highlands near Asmara and loses itself north-west of Kassala in a large desert delta.

There are many desert wadis that flow occasionally, but few reach either the Nile or the Red Sea. The largest of those draining to the Nile are the Wadi Howar and the Wadi el Milk, both of which were important watercourses during Pleistocene and Recent times. The Khor Arbaat and Khor Langeb–Baraka drain to the Red Sea.

Numbers of wadis occur in the Red Sea Hills, for example the Wadi el Arab which drains towards the Atbara and the Wadi Gabgaba which drains towards the Nile.

2. CLIMATE

The climate of the Sudan is wholly tropical and varies from complete desert north of 18° N through regions of semi-desert with rainfall of varying intensity and duration, passing southwards into a continental equatorial type of climate with a considerable dry season, even in the extreme south (Barbour 1961, Lebon 1965).

Rainfall is clearly seasonal and for much of the country it is related to the position of the

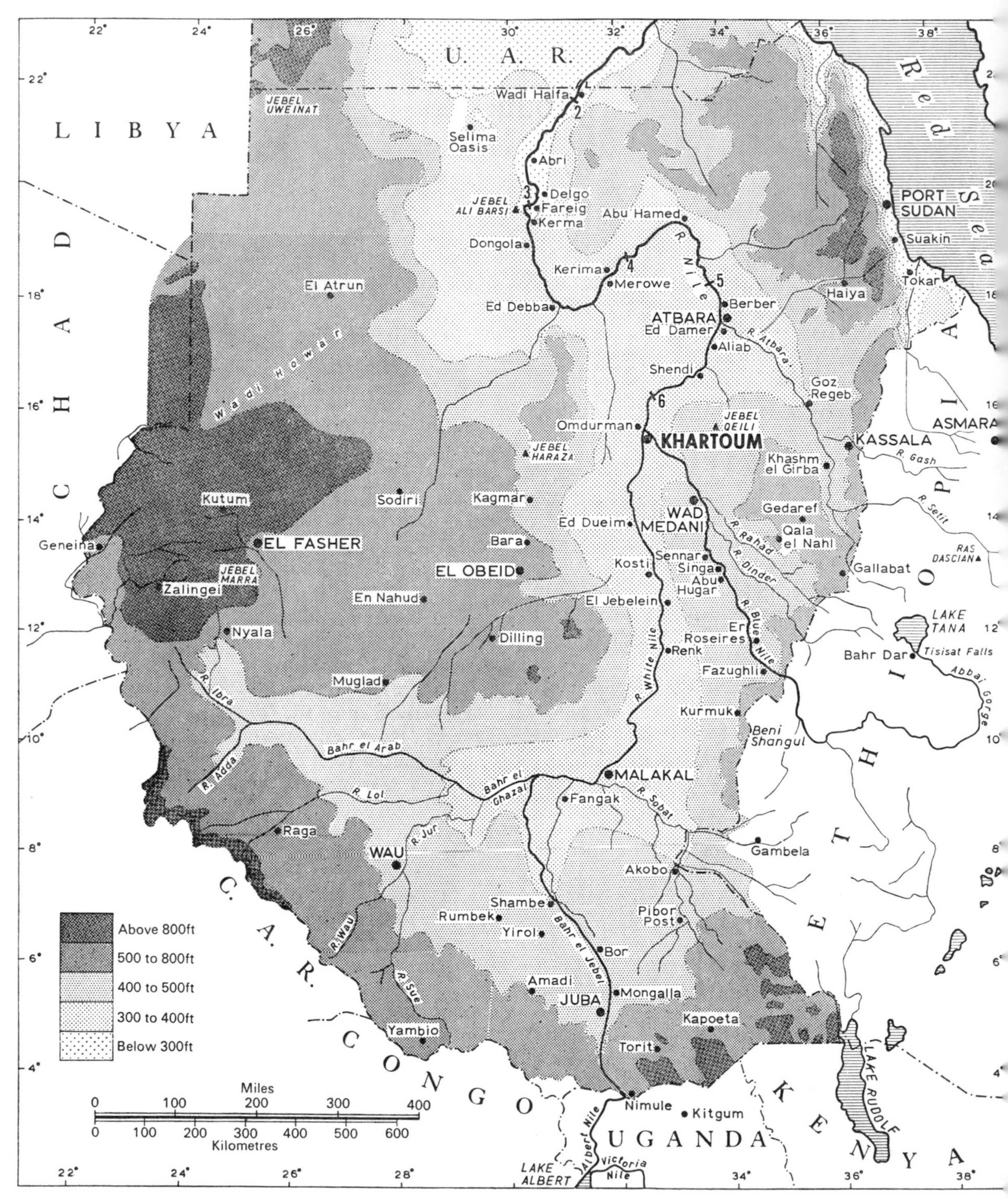

FIG. 1. Physical and general locality map, Sudan Republic (Based on Sudan Topographic Survey Maps, 1:2 000 000 and 1:250 000 scales) Nos. 2–6 refer to traditional cataracts

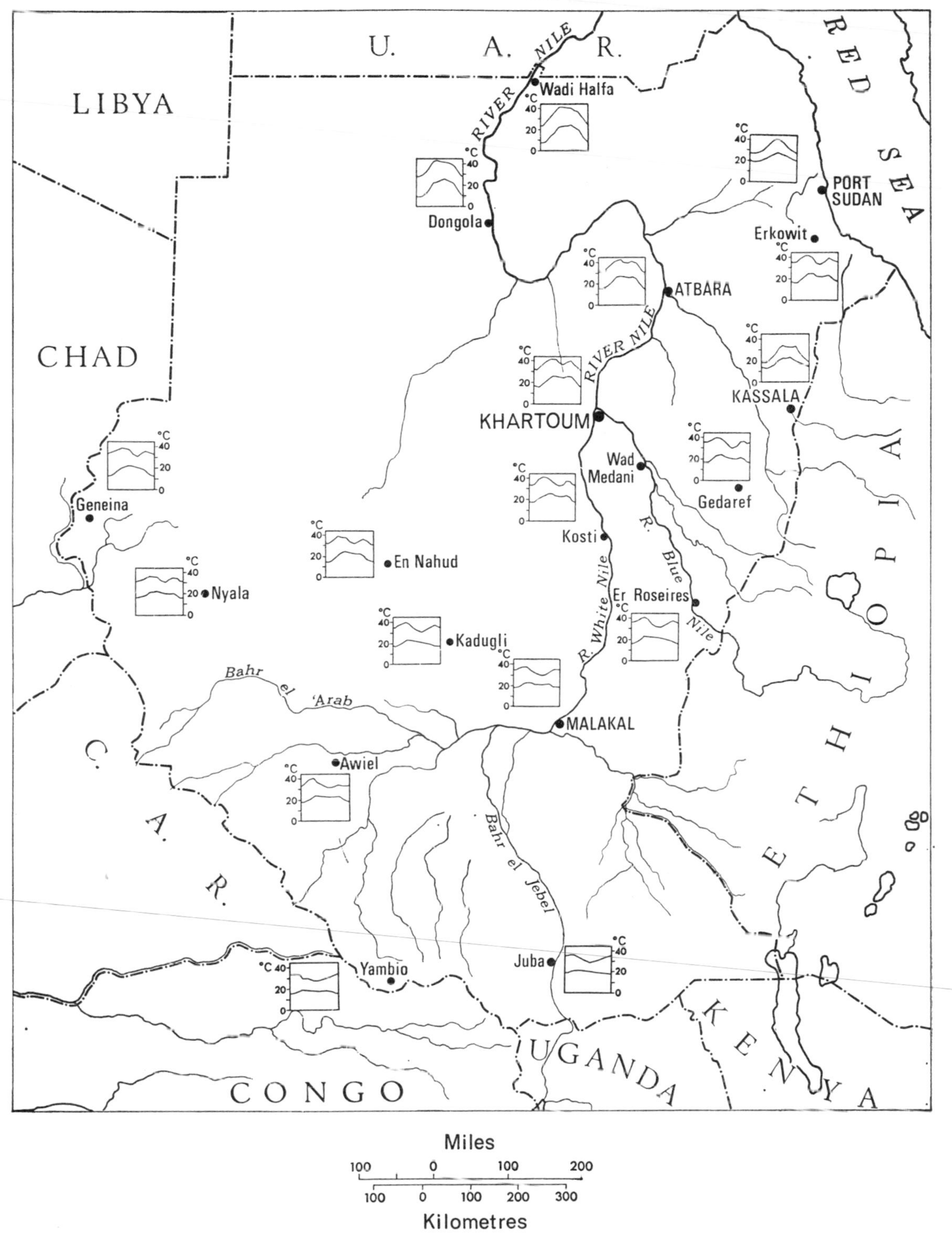

FIG. 2. Climatic map—mean daily temperature (Based on Sudan Topographic Survey Map and Meteorological Department data. Selected information)

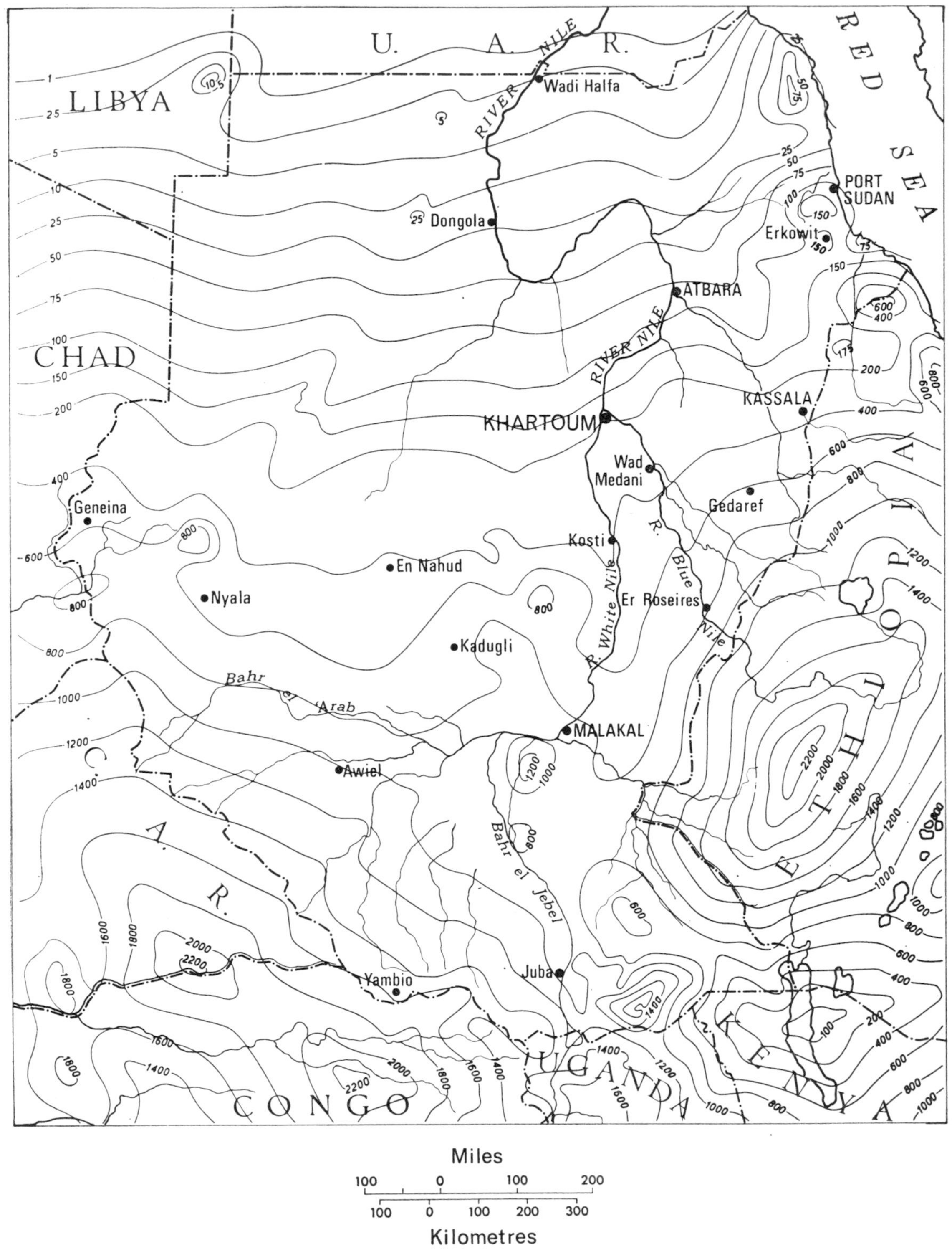

FIG. 3. Climatic map—mean monthly rainfall map in mm (Based on Sudan Topographic Survey Map and Meteorological Department data. Selected information)

inter-tropical convergence zone and the descending easterly jet stream.

Temperatures are high, with a mean daily winter temperature of 60·8° F in the north and 84° F in the extreme south. The diurnal range in the desert in the north is often as much as 40° F. During the summer the highest mean daily summer temperature at Atbara on the desert edge is 109° F, while at Wadi Halfa in the desert temperatures of 126° F are common. At Khartoum 116° F is frequently experienced in the months preceding the kharif (rainy season). Rock temperatures are exceedingly high throughout the year, 182·5° F having been recorded by Berry and Cloudsley Thompson (1960) in the Red Sea Hills. Rock decay is at a maximum in the northern and central Sudan along the savannah-desert edge because of the combination of high humidity and temperature for part of the year and because of the subsequent desiccation which produces intense salt-weathering in places.

3. GEOLOGICAL SEQUENCE

The most extensive formations in the Sudan, excluding the superficial deposits, are those of the Basement Complex. They are assumed to be mainly of Pre-Cambrian age, and consist of igneous, metamorphic, and sedimentary rocks that occupy more than 49 per cent of the surface area of the country. Together with the flat-lying Umm Ruwaba Formation (Tertiary to Quaternary fluviatile and lacustrine clays, sands, etc.) and the Nubian Sandstone Formation (predominantly sandstones and grits of Lower to Upper Cretaceous age), the Basement Complex occupies 97 per cent of the total surface area (Table 1).

A generalized geological column for the Sudan is given in Table 1. The approximate percentage surface areas of each formation or group of formations are based on the 1949 second and the 1963 third editions of the Sudan Geological Survey 1:4 000 000 Map.

The Sudan Geological Survey did not contribute to the *Lexique Stratigraphique International* and there is no official published list of formations. Delany (1966), who was an officer of the survey, independently provided a list of formations for the Sudan in this series. Whiteman (1970) has appealed for the recognition of the 'Nubian Sandstone' as the Nubian Group. 'Nubian Sandstone Formation' is here used throughout, however, because this work was completed before Whiteman (1970) was published.

TABLE 1

General geological column, Sudan Republic

Name of unit Sudan Geological Survey 1:4 000 000 maps, 2nd and 3rd edns	Name of unit (this work)	Age (this work)	Surface area (%)
Superficial deposits	Superficial deposits	Quaternary–Tertiary	Not differentiated
Coastal deposits (Red Sea margin)	Continental–marine formations Red Sea Littoral	Quaternary–Tertiary–Mesozoic	<1
Umm Ruwaba Series	Gezira Umm Ruwaba and El Atshan Formations	Quaternary–Tertiary	19
Lavas, Tertiary	Volcanic and intrusive rocks, mainly lavas (undifferentiated)	Quaternary–Tertiary	2
Hudi Chert, Middle Tertiary	Hudi Chert Formation	Tertiary? (Oligocene?)	<1
Nubian Series, Mesozoic	Nubian Sandstone Formation	L. Cretaceous?	28
Yirol beds	Yirol Formation (?equivalent to Nubian Formation)	L. Cretaceous?	<0·1
Not differentiated	Gedaref Sandstone Formation	Jurassic (pre-Oxfordian)	Included in Nubian estimate
Nawa Series	Nawa Formation	Palaeozoic?	<1
Palaeozoic sandstones	Palaeozoic Formations (undifferentiated)	Palaeozoic?	<0·5
Basement complex	Basement Complex Group	Pre-Cambrian mainly	49

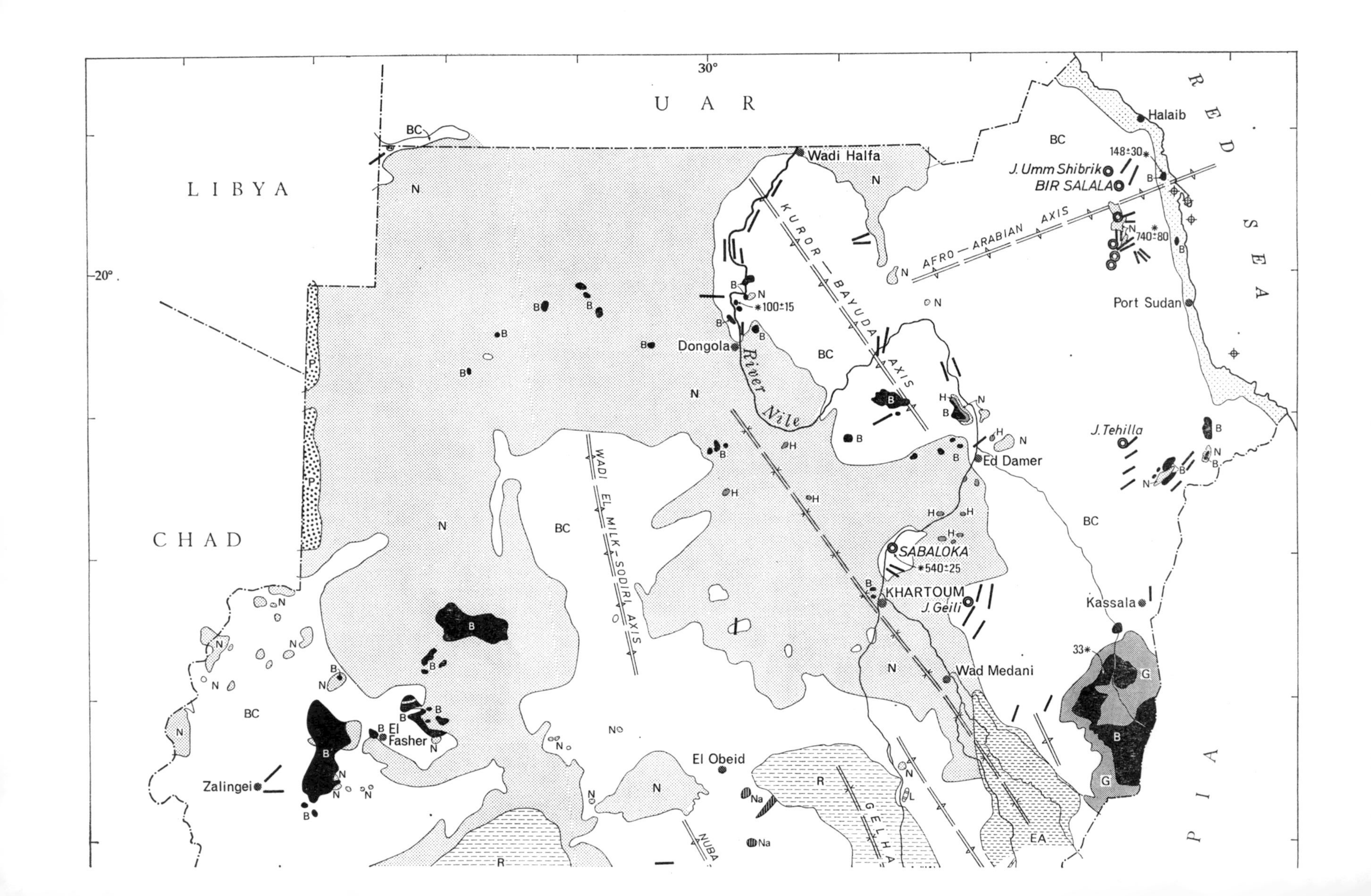
UAR
LIBYA
CHAD
RED SEA
PIA
30°
20°
Halaib
148±30*
J. Umm Shibrik
BIR SALALA
AFRO-ARABIAN AXIS
740±80
Port Sudan
Wadi Halfa
KUROR-BAYUDA AXIS
*100±15
Dongola
River Nile
J. Tehilla
Ed Damer
SABALOKA
*540±25
KHARTOUM
J. Geili
Kassala
33*
Wad Medani
WADI EL MILK-SODIRI AXIS
El Obeid
NUBA
GELHA
El Fasher
Zalingei
BC
N
B
H
G
EA
R
P
Na
L

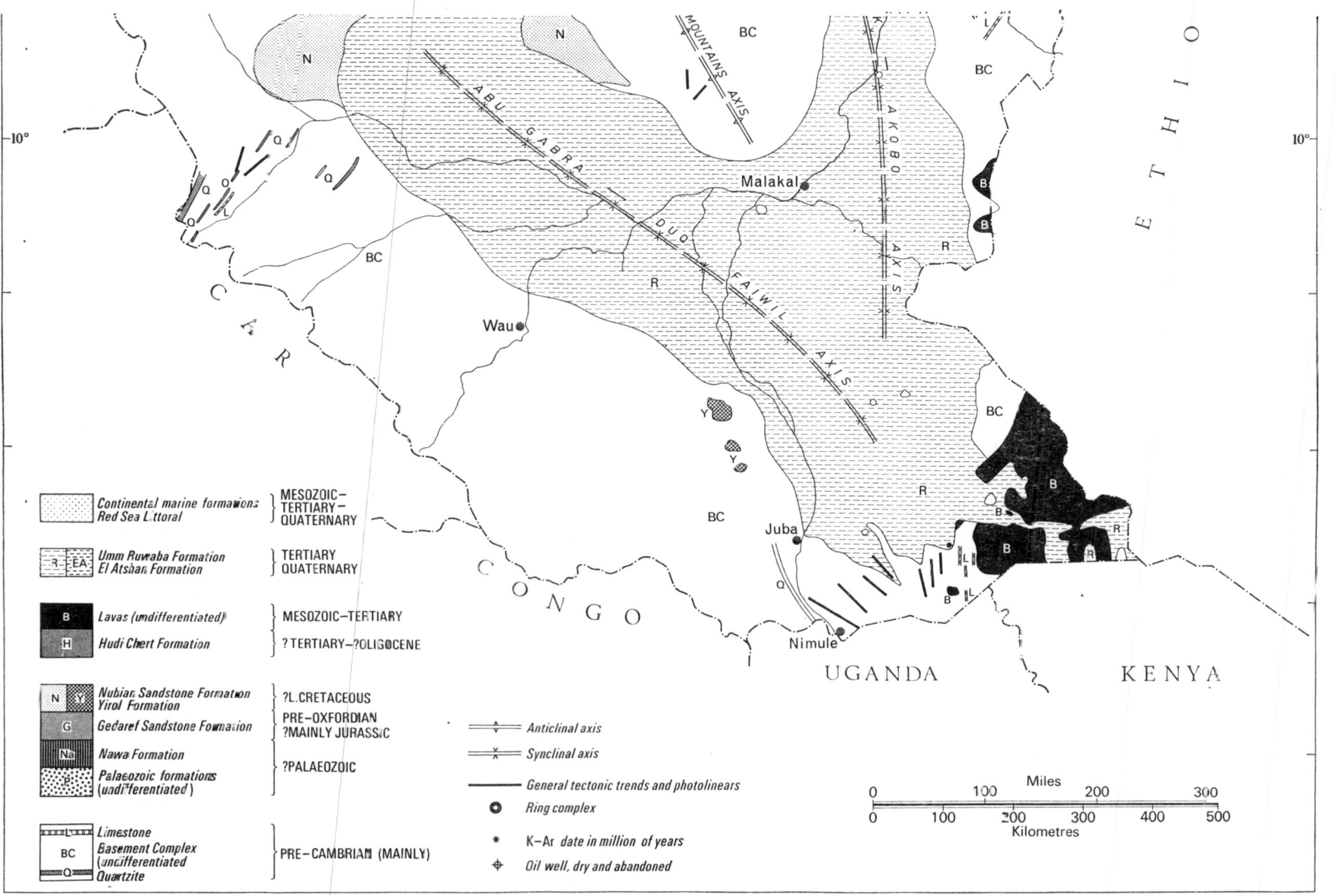

FIG. 4. General geological map, Sudan Republic (Based on Sudan Geological Survey General Geological Map 1947, 1949, and 1964 editions, 1:4 000 000; 1955 edition, 1:2 000 000, southern Sudan. New data University of Khartoum, Geology Department; potassium–argon dates (Geochron Inc.); structural interpretations)

II

History of Geological Research

DURING historical times gold, copper, and iron ores were mined. Recent finds at Buhen, Wadi Halfa recorded by Professor W. B. Emery indicate that copper was smelted there in Early Dynastic times. This probably implies that it was mined near by, although the source has not been located. The gold of Kush has influenced Egyptian–Sudanese relations for generations and copper was mined at Hofrat en Nahas in Darfur before the Condominium (see Chapter VIII).

Iron ore was mined and smelted on an extensive scale in ancient times at Bagarawiya Meroe, during the ultimate centuries of the pre-Christian era and later. It has been said that from this 'Birmingham of ancient Africa' geologically-minded smiths carried the knowledge of iron-smelting southwards into eastern and southern Africa (Davidson 1959). Archaeological evidence does not support this, however (Shinnie 1967).

Geological information bearing on routes, water supply, etc. was accumulated during the campaigns that resulted in the formation of the Condominium of the Anglo-Egyptian Sudan in 1899, and following in the wake of the soldiers came engineers, miners, geologists, and fortune-hunters seeking gold and other minerals. Numerous short references to the geology of the Sudan were made by the nineteenth-century explorers and 'Nile seekers'. Russegger (1837, etc.), for instance, produced a general account of the northern Sudan and the country beyond Sennar extending into the Ethiopian foothills, gained on his searches for gold and iron on behalf of Mohammed Ali Pasha, ruler of Egypt. Lyons (1894, 1897, etc.) published papers on the geology of the Libyan and Nubian deserts. These and other early papers are listed by Andrew (1945) and are included in the bibliography. Llewellyn (1903) and Dunn (1911) reported on gold prospecting in the Sudan and in 1905 the Sudan Geological Survey was formed with T. Barron as government geologist.

Few memoirs were published by the Survey, mainly apparently because of financial stringency, but up to 1945 the unpublished reports of the Geological Survey numbered 282. They were compiled by Barron (1904–1906), S. C. Dunn (1906–1918), G. W. Grabham (1907–1939). (He published also a very short account (Grabham 1935)), G. Y. Karkanis (1918–1962), G. T. Madigan (1920–1921), G. V. Colchester (1922–1932), J. W. Evans (1930), J. M. Edmonds (1934–1939), and G. Andrew (1939–1955). The reports cover water supplies and storage, haffir (reservoir) construction, dams, wells, irrigation, coal and oil, gold and lead, non-metallic minerals such as salt, guano, limestone, building stone, pottery and brick clays, foundations, etc. Some 132 of these reports are the work of the late G. W. Grabham.

Grabham, who was Director of the Survey for many years, unfortunately did not publish a comprehensive account of Sudan geology and it was not until 1948 that Andrew (*in* Tothill 1948*b*) produced a general account. This was accompanied by regional maps of the solid rocks and superficial deposits at a scale of 1:10 000 000. Hume (1934) made numerous references to the geology of the Sudan, especially along the river and in the northern Red Sea Hills.

Maps at the scale of 1:4 000 000 (drift and solid editions) were published in 1947 (first edition) and in 1949 (second edition) but accompanying memoirs did not appear. Lack of structural information makes these maps difficult to interpret. More detailed maps were published in the 1950s. These include the Khartoum Geological Map 1:1 000 000 scale, published in 1952, and compiled by Delany in 1951. Eastern Khartoum Province 1:250 000, compiled by Delany in 1949–1951 and published in 1953; and the Derudeb sheet 46I, 1:250 000 also surveyed by Delany in 1951–1955 and published in 1956. A geological map of the southern Sudan 1:2 000 000 based on data by G. Andrew, J. B. Auden, and E. P. Wright was produced in 1955. The date of publication was not given. A map showing the geology of the Mohammed Qôl sheet 36M, scale 1:250 000 and an accompanying memoir were published in 1962, based on work by Gass, Ruxton, and Kabesh. The final compilation and presentation was made by Kabesh (1962).

Annual Reports of the Sudan Geological Survey from 1950 to 1952 and from 1952 to 1953 dealt with the reinvestigation of the Hofrat en Nahas copper deposits, undertaken by two commercial organizations, geophysical surveys (gravity and resistivity) undertaken by Dr. L. Z. Makowiecki of the Overseas Geological Surveys, Great Britain, who was then a member of the Sudan Geological Survey; the borehole sunk for coal in 1949 at Wad Kabu in the Gedaref area; comments on the state of gold mining in the Sudan; beginning of reconnaissance mapping of Dungunab sheet 36I and Mohammed Qôl sheet 36M; the discovery of the Fodikwan iron ore deposit by Dr. I. G. Gass; etc. The highlights of the Annual Reports for 1953–1955 include: the continuation of gravity and resistivity surveys on the Red Sea coast; reopening of interest in the petroleum prospects of the Sudan; and the report for 1957 includes a policy statement by the Director of the Geological Survey. Thirteen memoirs and bulletins of the Geological Survey, published since independence in 1956, are listed in the bibliography (up to 1968).

Delany (1955, 1958, etc.) described the ring complexes of the Sudan, adding to our knowledge of the petrological history of the Sabaloka and Jebel Geili rings.

The third edition of the Sudan Geological Survey Map (1:4 000 000) was published in 1964. Bad copying of boundaries shown on the second edition (whole outcrops are omitted and boundaries misplaced) and the addition of new data and modification of some boundaries make this map difficult to interpret. Again a memoir was not issued to accompany the map.

Berry (1964) and Whiteman (1964) published bibliographies and summaries of research dealing respectively with geography and geology in the Sudan. A general article dealing with geology and scenery in the Sudan was produced by Whiteman (1966) and a bibliography of Sudanese geology is in course of publication (Whiteman, *Sudan Notes and Records*, in press).

Berry and Whiteman (1968) described the evolution of the Nile drainage in the Sudan and adjacent areas; and Whiteman (1968) has described the origin of the Red Sea depression referring to the Sudan Red Sea Littoral.

Four geological sheets at 1:250 000 covering part of Kordofan were published in 1964. These were produced by Hunting Technical Services for the Department of Land Use and Rural Water Development, United Nations Special Fund and Food and Agriculture Organization. New formational limits for the Umm Ruwaba Formation are shown on these maps together with new boundaries for the Basement Complex (Fig. 9).

The first of the Sudan Geological Survey Regional Memoirs, sheets 66A Rashad and 66E Talodi appeared in 1966. The memoir was written by Mansour and Sammuel and is given a publication date of 1957.

The progress of mapping by the Sudan Geological Survey up to 1967 is shown in Fig. 5.

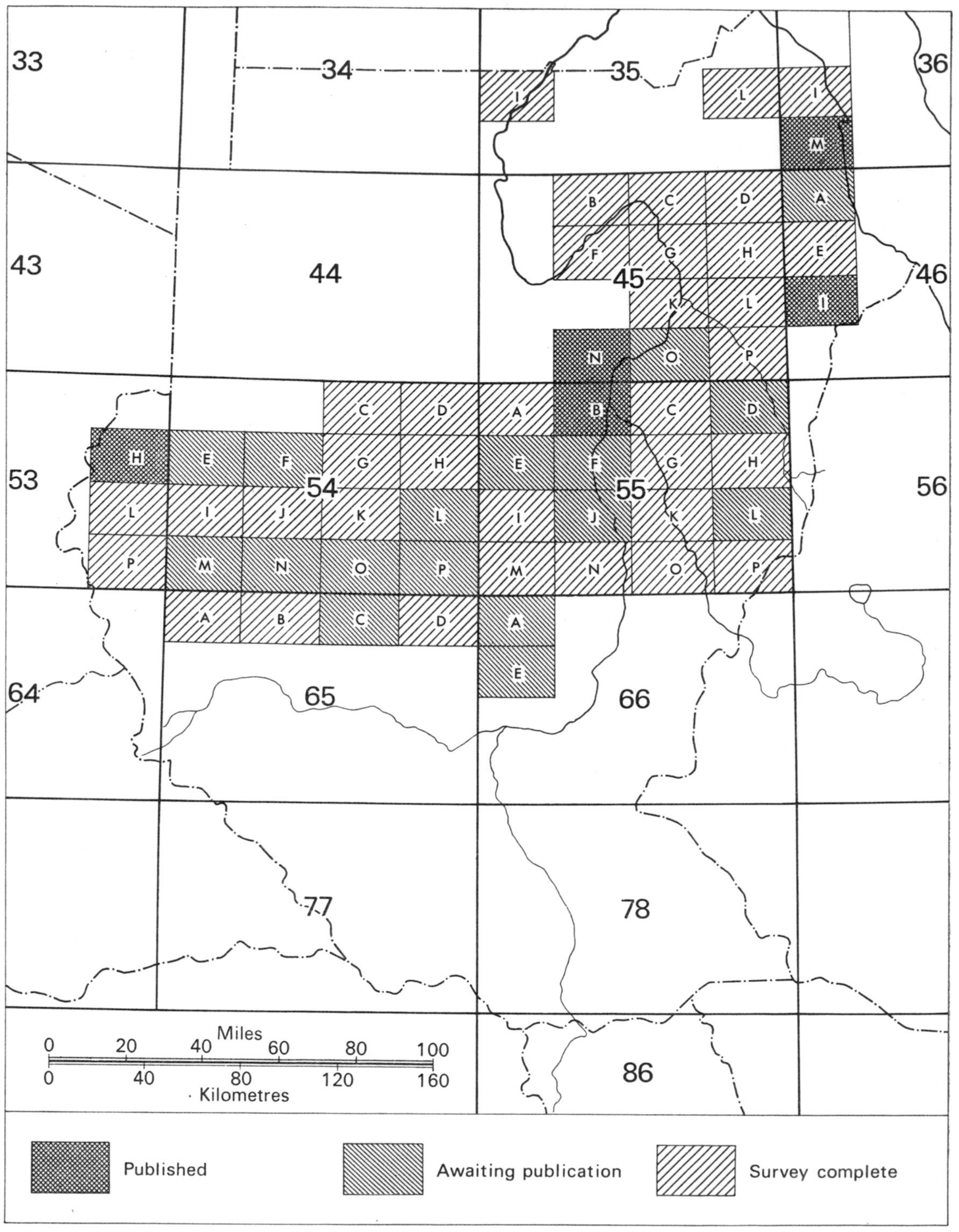

FIG. 5. Sudan Geological Survey, Progress of mapping 1:250 000 sheets up to 1967. Data from Sudan Geological Survey Compiled by S. M. Rabaa for Erkowit Conference, September 1967)

III

Stratigraphy

1. BASEMENT COMPLEX

(a) General

THE Basement Complex of the Sudan has been described by many travellers and geologists, for example Rüssegger (1837, etc.), Lyons (1894, 1897), Blanckenhorn (1902), Linck (1903), Lynes and Smith (1921), Tyler and Wilcockson (1933), Hume (1937), Andrew (1948, etc.), Delany (1952, etc.), Ruxton (1956), Gabert, Ruxton, and Venzlaff (1960), Kabesh (1962), Lotfi and Kabesh (1964), and in government reports by Barron, Dunn, Grabham, etc. Various formations have been distinguished, but up to the present a satisfactory comprehensive classification has not been established, largely because great variations in metamorphic grade occur, and because of lack of isotopic dates, our limited knowledge of many areas, and the isolated nature of many jebel masses. Very little modern structural mapping has yet been done.

The name 'Basement Complex' is used below to include those igneous, metamorphic, and sedimentary rocks that are overlain by horizontal and sub-horizontal Palaeozoic or Mesozoic sedimentary or igneous rocks. They are assumed to be mainly of Pre-Cambrian age although some formations may be younger, for example the cooling age for the Sabaloka ring complex appears to be Cambrian according to a potassium–argon date recently obtained by the author. Delany (1966, pp. 77–9), however, assigned the whole of the Basement Complex to the Pre-Cambrian. In the author's view the matter must be left open until more isotopic dates are available, as other ring intrusions similar to the Sabaloka occurrence are known.

In Table 2 the classifications of Andrew (1948), Wilkinson and Tyler (1933), Ruxton (MS 1956), Delany (1952, etc.), Ruxton (1956), Gabert, Ruxton, and Venzlaff (1960), Kabesh (1962), and Lotfi and Kabesh (1964) are summarized and a general attempt is made at correlation. This is discussed at the end of this chapter.

(b) Sudan, south of latitude 6° N

At present the rocks of the southern Sudan south of latitude 6° N cannot be correlated with those of the northern Sudan. Obviously they are related to those of Uganda and the Congo but a correlation has not been established, for very little geological work has been done along the frontier zone. Some general data exist in the files of the Sudan Geological Survey and a geological map of the Southern Sudan (1:2 000 000) has been published (Andrew, Auden, and Wright 1955). Unfortunately it was not accompanied by a memoir.

The Basement Complex of southern Equatoria consists in part of a group of granoblastic, foliated basic-to-acid gneisses with hypersthene and feldspar, intruded into foliated paraschist and paragneisses. The hypersthene-bearing rocks, which are related to the charnockitic group mapped in Acholi and Karamoja in Uganda, are cut by non-charnockitic intrusions. According to G. Andrew (personal communication) charnockitic rocks are known only from areas east of the White Nile. Strike is constant over wide areas and in parts of Equatoria varies from 180° to 337° (Figs. 4 and 6).

The Madi quartzites of the West Nile Province of Uganda continue into the Sudan, and the course of the Nile for many miles north of Nimule is determined by the Madi structures that are a continuation of the Assua Mylonite zone. The Madi structures appear to run as far as Kajo Kaji and then leave the river, which follows a northerly course determined by major joints (Berry and Whiteman 1968). They cut the Juba–Yei Road about 30 miles from Juba and the zone may continue as far as Kidi.

Andrew (1948) recognized the following rock types in the southern Sudan:

Group 8	Charnockitic orthogneiss similar to the charnockitic rocks of northern Uganda
Group 9	Foliated granites and granodiorites
Group 10	Feldspathoidal soda syenites (?post-Palaeozoic)

TABLE 2

Correlation chart for Basement Complex formations, Sudan Republic

Sudan south of lat.6° north Andrew 1948	Qala En Nahl Wilcockson and Tyler 1933	Gedaref and Sennar Ruxton Ms. 1956?	Sheet 55 Andrew in Ruxton MS. 1956? and Delany 1952	Sudan north of lat. 60° north Andrew 1948	S.Red Sea Hills Derudeb sheet Delany 1956
Feldspathoidal soda syenite possibly post-Palaeozoic	Newer intrusions	Soda granite — Jebel Ummat Liha	Sabaloka-series — Riebeckite-granite 540 ± 25m.y. Whiteman-Geo-chron sample, volcanic rocks	Soda granites	Soda granites unfoliated granite of Tehilla ring complex
Non-charnockitic foliated granite and granodiorites	Ban Balos granite quartz reefs silicification of serpentines	Unfoliated granite — Ban Balos group	Unfoliated granite	Unfoliated granite	Foliated and porphyritic granites
				Folding of non-Metamorphic rocks	
				Unmetamorphosed greywackes and lavas, predominantly andesitic	
Intense regional Metamorphism				*Folding and regional metamorphism*	
		Unfoliated gabbros and andesites			
	Foliation Beila granite	Foliated granites	Unfoliated and foliated granite and granodiorite	Plutonic intrusions and granite	Gabbro
	Ultrabasic rocks and older gabbros	Foliated basic rocks and ultrabasic, serpentines and and talc-magnesites	Gabbros and ultrabasic rocks	Ultrabasic rocks, serpentines, gabbros,and norite,all more or less foliated	
				Folding and regional metamorphism	
		Basal Schists and Phyllites: Green series — Chlorite-epidote-schist	Para-schists: Butana green series		
	Basal Schists	Basal Schists and Phyllites: Carbon Schists — Quartzites carbonaceous mudstones, limestones	Para-schists: Butana group	Para-schists — Regionally metamorphosed but with slightly metamorphosed bedded rocks including lavas	Odi Schists — Limestone, quartzi lava, graphite. schist undifferentiated
		Basal Schists and Phyllites: Basal Schists — Quartz sericite phyllites spilites marbles			
				Folding and regional metamorphism	
				Folding and regional metamorphism	
Charnockitic orthogneiss parametamorphic succession			Gneiss — Foliated and banded at Sabaloka and El Obeid	Orthogneisses	Older gneissose Pre-Cambrian formation
1	2	3	4	5	6

Northern Red Sea Hills and Mohd. Qol sheet Ruxton 1956	Deraheib Dunganab Mohd. Qol sheets Gabert Ruxton and Venzlaff 1960	N.Red Sea Hills Kabesh			Dunganab sheet Gass 1955	
Acidic intermediate and basic dykes 740 ± 80 m.y. Jebel Hama-Shaweib Bir Salala Whiteman-Geochron sample Ruxton 8 Red granite	Younger granites: Soda granites and syenites Red granite	Dykes equivalent to Ruxton 8 Post-geosynclinal intrusive rocks: Alkalines, granites acid intrusive rocks including granites, diorites, and syenites		Cambrian	Younger intrusives: Granites, granophyres, quartz-syenites forming bosses, ring dykes, stocks, and some intrusions of post-Nubian age	Lower Tertiary & Upper Jurassic
Injection granite *large-scale metasomatism*		*Mild orogeny*				
Awat series: Acid volcanic rocks, silty mudstones, acid volcanic rocks, conglomerates, silty mudstones	Awat series: Acid volcanic rocks silty mudstone	Third geosynclinal deposition: Acid and intermediate volcanic rocks micro-conglomerates, greywackes, and mudstones				
Uplift and erosion		*Unconformity*			*Unconformity*	
Basic intrusive rocks	Basic intrusions including norites, gabbros, and serpentines	Dykes equivalent to Ruxton 8 Post-geosynclinical deposition: Serpentines, gabbros, and troctolites	Basement Complex	Pre-Cambrian	Basic intrusive rocks: Gabbros, troctolites, pyroxenites forming dyke-like intrusions	Basement Complex
Large-scale folding	*Large-scale folding and regional metamorphism*	*Orogeny*				
	Batholithic microcline granite, hornblende gneiss	Batholithic granite				
Nafirdeib series: Limestone, quartzites and pelites, intermediate volcanic rocks and greywackes, basic and intermediate volcanic rocks and greywackes	Nafirdeib series: Limestone quartzites and pelites intermediate volcanic rocks and greywackes basic and intermediate volcanic rocks and greywackes	Second geosynclinal deposition: Limestones and quartzites conglomerates hornblendic rocks greywackes slates volcanics conglomerates			Oyo series: Pelites, quartzites, marbles, conglomerates, acid volcanic rocks interbedded	
Uplift and erosion		*Unconformity and orogeny*				
Basic dyke swarm		First geosynclinal deposition: Calcite-schists conglomerates greywackes slates and volcanic rocks				
Large-scale folding and regional metamorphism	*Folding and metamorphism*	*Unconformity and orogeny*			*Unconformity*	
Granite pegmatites, acid gneiss, schist and slates, hornblende-schist and gneiss	Kashebib series: Mixed and banded gneiss and chlorite-gneiss	Para and orthogneisses			Granites, schists, gneisses, acid volcanic rocks found in the basal conglomerate of Oyo series	
7	8	9			10	

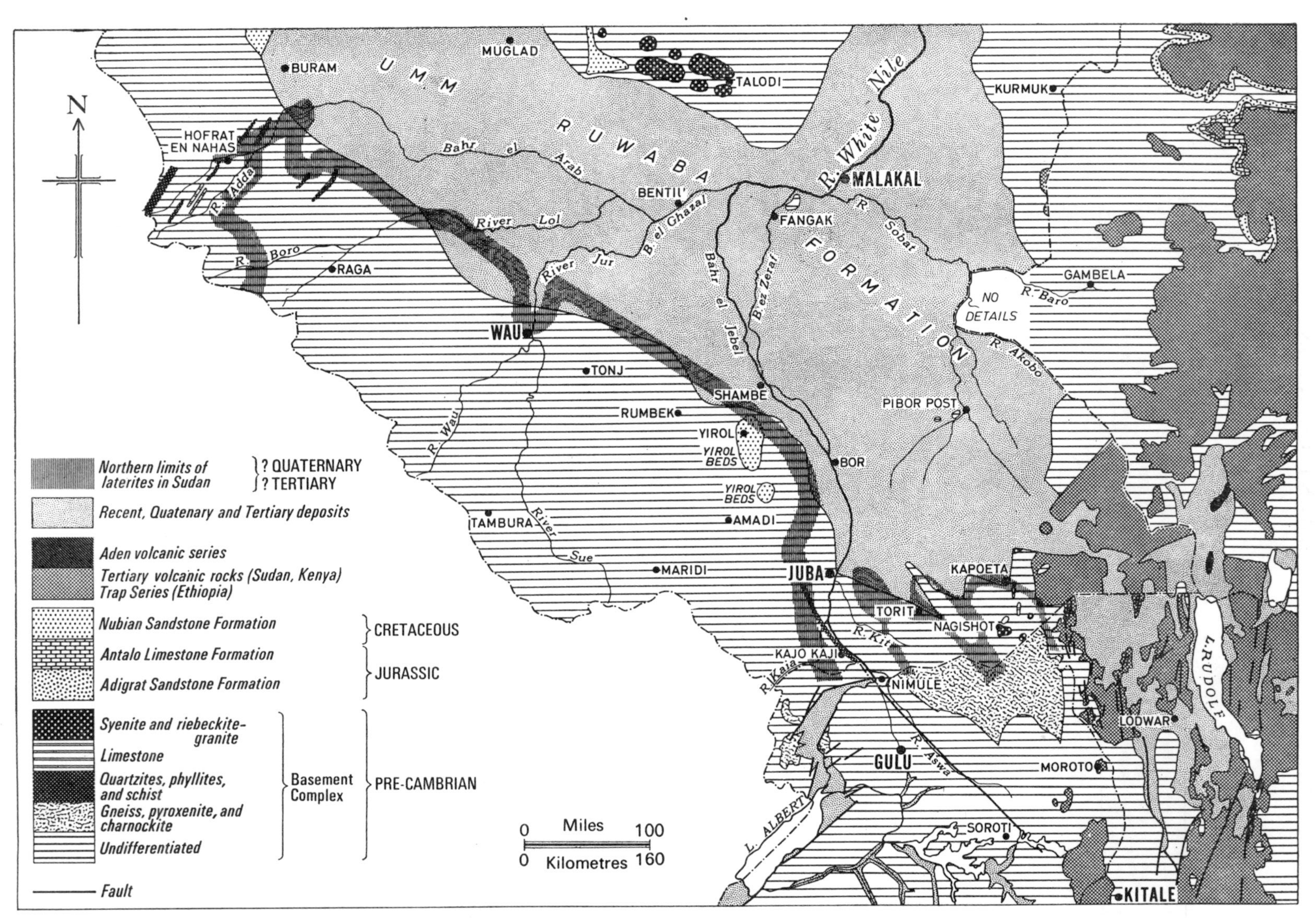

FIG. 6. Southern Sudan and adjacent areas, general geological map (Based on Sudan Geological Survey Map 1:2 000 000 ASGA-UNESCO geological map of Africa 1:5 000 000, and Uganda Geological Survey Map 1:1 250 000)

No order of superposition is implied in Andrew's classification (Table 2).

In the absence of isotopic dates we can only speculate about the relative ages of these rocks, using evidence from adjacent areas. Cahen and Snelling (1966, Figs. 17, 4) have shown the adjacent rocks in the Congo Republic and Uganda as part of the 'north Congo shield' and the 'Uganda Basement', consisting of rocks folded and metamorphosed 2500–2600 × 10⁶ years ago or earlier. The Albert Rift area in both the Congo and Uganda is shown with a distinctly separate trend as a belt folded and metamorphosed 730–600 × 10⁶ years ago (Katangan cycle) with post-tectonic events in these belts as young as 450 × 10⁶ years. Likewise the Mozambiquian belt of the Katanga cycle is shown as extending through Turkana as far as the Sudan–Ethiopian frontier. On the International Tectonic Map of Africa (ITMA 1969) a different version is shown (Whiteman 1969).

(c) Gedaref–Sennar region

About 1000 square miles of country were surveyed in the Qala en Nahl area by W. H. Tyler between 1930 and 1932 (Tyler 1932; Wilcockson and Tyler 1933); and Ruxton surveyed some 25 000 square miles between Gedaref and Sennar, in part covering the talc–magnesite–serpentine–chromite area of Qala en Nahl (Ruxton MS 1956). Officers of the Sudan Geological Survey have worked in the area in recent years, especially in the talc–magnesite area, but their results have not so far been published.

The account that follows is based mainly on the above-mentioned published works and a short visit made to the Qala en Nahl area by the author (Fig. 7 (a) and (b)).

Wilcockson and Tyler (1933, p. 308) recognized the following succession in the Qala en Nahl region:

Deposition of 'cotton soil'
Newer intrusions (diorites and gabbros)
Intrusion of Ban Balos granites accompanied by quartz reefs and silicification of serpentines
Foliation
Intrusion of Beila granite with serpentinization followed by carbonation of the ultrabasic rocks
Intrusion of ultrabasic rocks and older gabbros
Deposition of Basal schists

(i) *Basal schists*

These are the oldest rocks exposed in the area and consist of quartz–chlorite schists with occasional

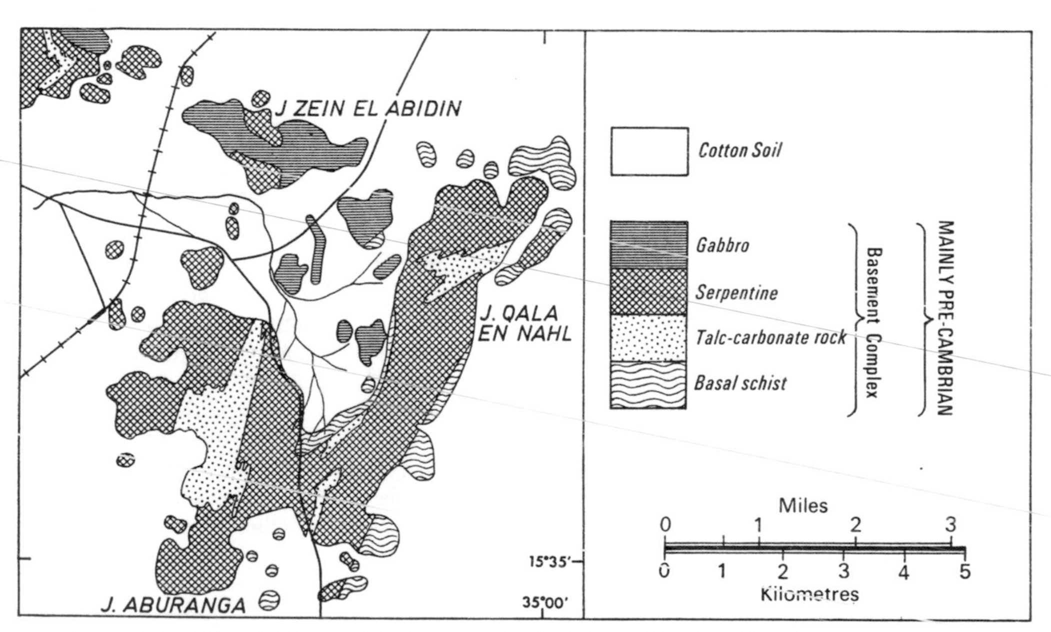

FIG. 7(a). Qala en Nahl, central Sudan, general geological map (Based on Wilcockson and Tyler 1933)

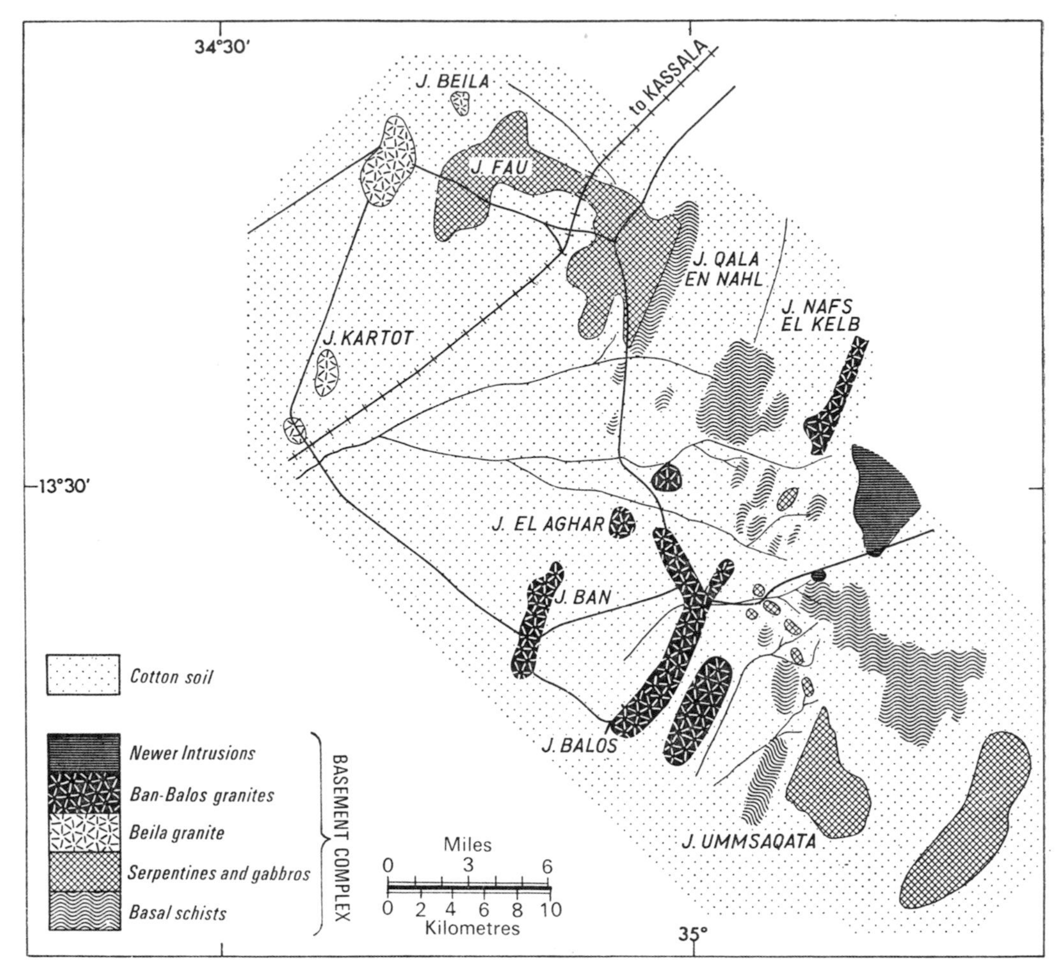

FIG. 7(b). Qala en Nahl, central Sudan, geological map (Based on Wilcockson and Tyler 1933)

epidote, calcite, and cordierite. Well-foliated quartz–sericite schists with haematite or magnetite occur also.

Limestones (marbles) have been recorded at Jebel Ghanam, where a thin dark band is exposed on the north-west slopes; at the north-east end of the Qala en Nahl Hills; and at Nafs el Kelb, where there is a large mass of coarsely crystalline calcite infolded among schists.

Hornblende schists, coarsely crystalline and sodic, occur between Qala en Nahl and Beila.

(ii) *Ultrabasic rocks and older gabbros*

The Basal schists are intruded by gabbros, serpentines, talc–carbonate, and quartz–carbonate rocks in the Qala en Nahl region.

The serpentine is a fresh-looking massive green rock interfingering marginally with talc–carbonate bodies. It is probably an altered dunite and exhibits antigorite habit. Magnetite and chromite are common in the serpentine, and chrysotile asbestos occurs at several localities. Carbonates occur in almost all specimens of serpentine and there are large outcrops of talc–carbonate. According to Wilcockson and Tyler (1933) the largest known mass is situated west of the Qala en Nahl jebels and covers about $2 \times \frac{3}{4}$ miles. Most of the bodies are smaller, however.

Two analyses of the carbonate rock made by W. H. Tyler and the mineralogical composition of these rocks suggested by Wilkinson and Tyler (1933) are given in Tables 3 and 4 respectively.

TABLE 3

	%	%
SiO_2	20·16	31·43
TiO_2	0	0
Al_2O_3	2·27	2·53
Cr_2O_3	0·89	0·37
FeO	6·85	4·50
CaO	0·30	0·37
MgO	36·35	35·44
Alkalis	n.d.	n.d.
Loss on ignition	33·60	24·60
Total	100·52	99·24

TABLE 4

	%	%
Talc	33·00	49·00
Magnesite	55·00	42·00
Other constituents	12·00	9·00

Another feature of the talc–carbonate rock is that the bodies have a core of quartz–carbonate rock along their centres. The cores pinch and swell and vary from a few inches to 20 ft. The existence of quartz filigree suggests that the quartz was introduced last.

The rocks of the Qala en Nahl, Jebel Ghanam, and Jebel Fau areas are also silicified, a complete replacement of the silicate minerals by quartz and chalcedony having taken place. Magnesia appears to have been removed totally during silicification.

Associated with the serpentine masses are highly altered gabbros called 'older intrusions' by Wilcockson and Tyler (1933). Their boundaries are obscure and the age relations uncertain.

(iii) *Beila and Ban Balos granites*

These rocks constitute the main granitic intrusions in the area. The Beila granite is foliated but the Ban Balos granite is not; and on this evidence Tyler and Wilcockson (1933) consider the Beila granite to be the older. Both are younger than the serpentine.

The Ban Balos granite is composed essentially of microcline with subsidiary albite–oligoclase feldspar and contains little or no orthoclase. Biotite is the most common ferromagnesian mineral; there is some muscovite and hornblende.

The Beila granites are granophyric, coarse-grained, and pegmatitic. The feldspar is mainly plagioclase and the quartz-feldspar intergrowths are developed on a large scale. Biotite is the main ferromagnesian mineral.

Although 'cotton soils' obscure the contacts of these granites with the country-rock, the surrounding schists appear to have been thermally altered.

(iv) *Newer intrusions*

These consist of gabbros and diorite. Fresh gabbros occur at Jebel Qala et Takarir.

Other igneous rocks such as dykes of basalt and dolerite occur in the region but their age is uncertain.

(v) *Sequence proposed by Ruxton* (1956)

The larger size of the area studied by Ruxton (MS 1956) enabled him to elaborate and develop the succession established by Wilcockson and Tyler (1933) for the Qala en Nahl region. The sequence presented by Ruxton is as follows:

6 Soda granite of Jebel Ummat Liha
5 Unfoliated granite of the Ban Balos Group
4 Unfoliated gabbros, and andesites
3 Foliated granites
2 Foliated basic and ultrabasic rocks, serpentines, and talc–magnesites

1 Basal schists and phyllites	Green series	Chlorite–epidote schists
	Carbonaceous schists	Quartzites, carbonaceous mudstones, limestones
	Basal schists	Quartz–sericite phyllites, spilites, marbles

(vi) *Basal schists and phyllites*

Like Wilcockson and Tyler (1933), Ruxton recognized the Basal Schist Group as the oldest rocks exposed in the region. Ruxton divided these rocks into three formations (Table 2).

Basal schists. Typically these rocks are quartz–sericite phyllites with pyrites. Quartzose slaty beds occur and the unit becomes more chloritic and calcareous towards the north. Spilites, graphitic schists, and andalusite hornfels have been recognized at some localities.

Carbonaceous schists. These constitute the second subdivision of Ruxton's Basal schists and phyllites

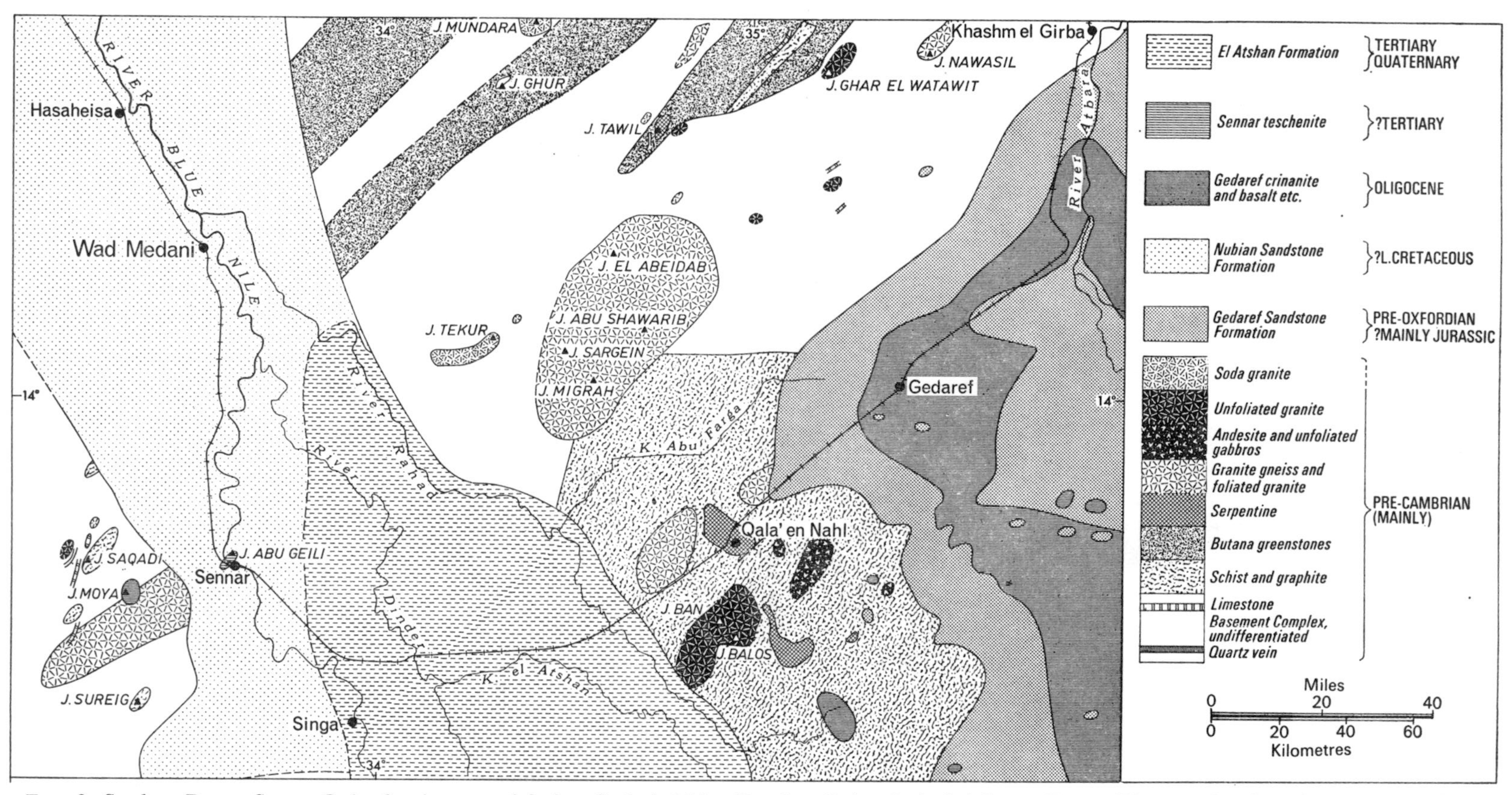

FIG. 8. Southern Butana–Sennar–Gedaref region, central Sudan, Geological Map (Based on Sudan Geological Survey Eastern Khartoum Province sheet 1:250 000, Ruxton (MS 1956) Geological Map, and additional field observations)

and consist of quartzites, black carbonaceous mudstones, with occasional limestones striking at 022°. North of the Gedaref region these rocks are referred to as the Butana Group (Andrew MS 1943, University of Khartoum Geology Library). They extend to the White Nile near Kosti and apparently pass beneath the Umm Ruwaba depression towards Rahad in the Nuba Mountains. Similar rocks also occur in the Ingessana Hills. The carbonaceous schists carry kyanite in a dark fine-grained mass in the southern Gezira.

Green Series. They include some epidiorites (derived from dolerites and basalts), hornblende–epidote schist, and biotite–hornblende feldspar schists. The predominant strike of these rocks is 022°.

(vii) *Foliated basic and ultrabasic rocks*

Ruxton (1956) described the two main areas of these rocks at Qala en Nahl, and Umm Sagata some 30 miles to the south-east. He added little, however, to the petrological descriptions of Tyler and Wilcockson (1933).

(viii) *Unfoliated intrusions*

Ruxton (1956) described the Ban Balos and Beila granites and a dyke-like mass of aegirine-granite at Jebel Liha.

Ruxton's Basal schists and phyllites are probably equivalent to the paraschists of Andrew (1948, p. 95). The underlying orthogneisses are not exposed in the Gedaref and Sennar regions, however (Table 2). The serpentines and ultrabasic rocks shown in Table 2 are the Qala en Nahl occurrences referred to by Ruxton.

In much of the central Sudan foliation and bedding appear to be generally parallel, and the strike is often constant over wide areas.

(d) Kordofan

Rodis *et al.* (1964, p. 19) described the Basement Complex of Kordofan as being composed predominantly of granite, gneiss, and schist with quartzites and crystalline limestones. Unfortunately no information about foliation is given. In the Salara Hills, near Dilling, trends and photo-linear trends are roughly east–west but in the Rashad and Talodi areas of the Nuba Mountains north-east–south-west trends have been noted (Mansour and Samuel 1957).

The geologists of Hunting Technical Surveys (1964) recognized the following succession for the Basement Complex Group of Kordofan in the El Obeid area (Fig. 9):

7 Younger intrusive masses
6 Quartz veins
5 Acid volcanic rocks with associated mudstones
4 Basic volcanic rocks
3 Greywacke Series with or without associated volcanic rocks
2 Metasediments
1 Acid gneiss group (metamorphosed igneous rocks) and augen gneiss

Unfortunately the account describing the formations is not available for publication (Fig. 9). The strike of the metamorphic rocks is predominantly north to north-north-east but some east–west trends were recorded.

Jebel Ed Dair, a prominent jebel south of Er Rahad, is shown as a younger intrusive mass as is Jebel Kareir. Jebel Dumbeir, north-east of Jebel ed Dair is composed of soda granite and has xenoliths of essexite-type syenite. Over much of the area mapped the Basement Complex is shown as 'undifferentiated' because of the thick superficial cover. In October 1966 the Jebel Dumbeir area was shaken by a strong earthquake (Qureshi and Sadig 1967).

Mansour and Samuel (1957) recognized the following succession in the Talodi and Rashad areas:

Granites and syenite
Foliated hornblende 'quasi-granitic gneiss'
Banded metasedimentary series
Banded gneisses and schists

(1) *Banded gneiss and schists.* These are mainly hornblende-bearing rocks occasionally interbedded with mica-rich bands. The formation covers much of the area and is the oldest division of the Basement Complex exposed. The predominant regional trend is north-north-east–south-south-west.

(2) *Banded metasedimentary series.* These consist of mica-schists, graphitic schists, slates, phyllites, quartzites, and marbles. The graphitic schists and slates are most common, especially in the eastern part of the Rashad sheet. Greasy graphite-rich bands occur at Jebel Tirmi, east of Rashad, and north of Abu Gubeiha crystalline marbles occur (Mansour and Samuel 1957).

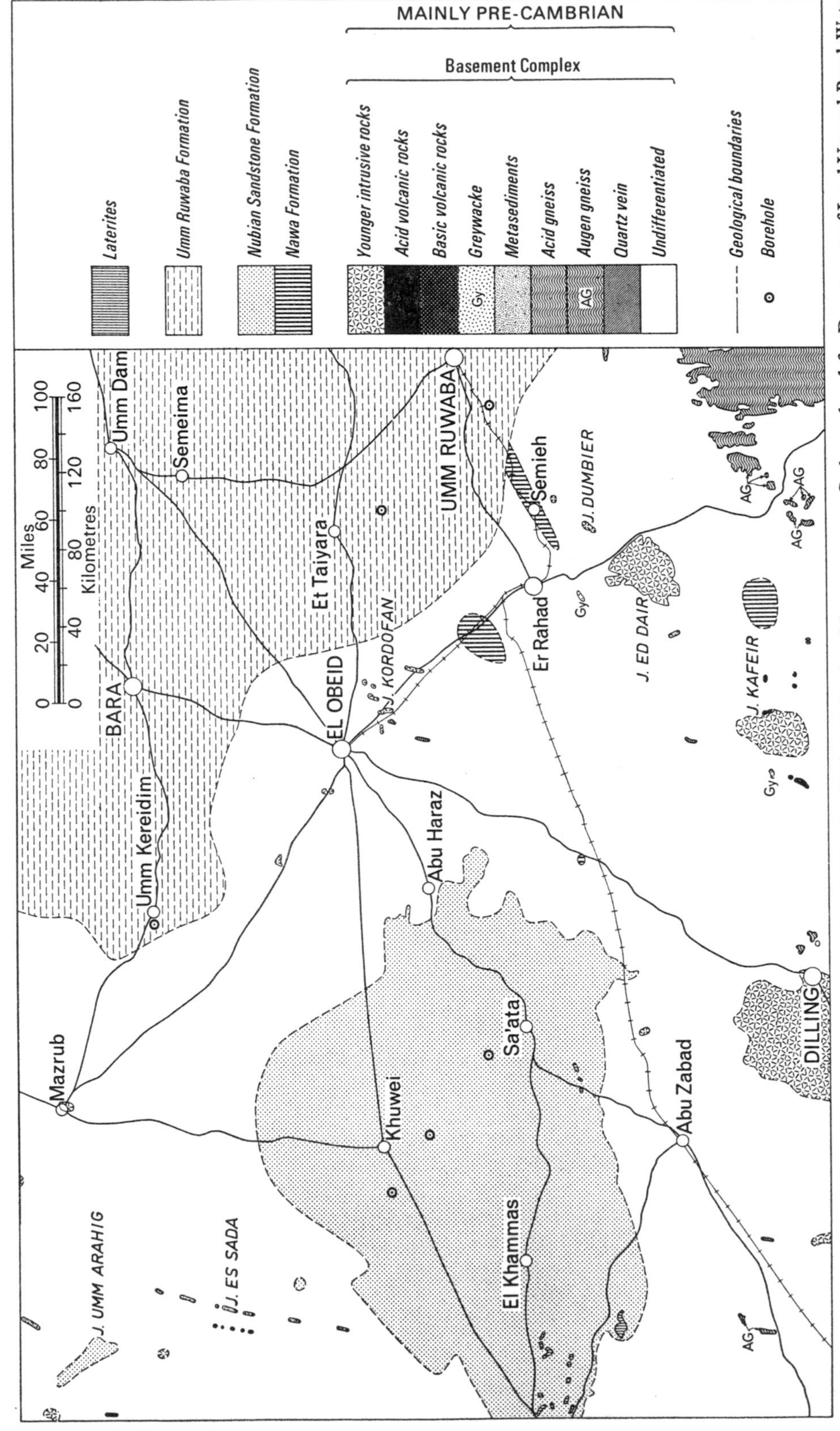

FIG. 9. El Obeid–Dilling–Bara region, central Sudan (Based on Hunting Technical Services Maps 1:250 000 Scale produced for Department of Land Use and Rural Water Development, UN Special Fund and Food and Agricultural Organization)

Near Abu Gubeiha there is a wide shear zone trending north-north-east–south-south-west.

(3) *Foliated hornblende 'quasi-gneiss'*. Mansour and Samuel (1957) described these rocks as granoblastic gneisses 'of a hornblende granitic to granodioritic composition'. Lineation parallel to the foliation is well developed.

(4) *The granites and syenite.* The granites of the western half of the Rashad sheet are fine- to medium-grained, somewhat porphyritic, grey hornblende-granites and are slightly foliated. North of Rashad the granites are of fine-grained biotitic type with foliation trending at 180–200°. They are greyish to pinkish-grey in colour. Biotite-granites occur in the Masahin Qisar Hills and Jebel Korongo and at Jebel Liri; and hornblende-granite occurs at Jebel Gedir, Jebel Kologi, Jebel Morung, and J. Werri. The Limon Hills and Jebel Talodi are composed of medium- to coarse-grained syenite.

Gold occurs in quartz veins at Jebel Tira Mandi, Jebel Sheibun, and Jebel Umm Gabrallah (see Chapter VIII).

(e) Ingessana Hills, south-eastern Sudan

This area was mapped by Kabesh (1961) and consists of calcitic and dolomitic marbles, meta-quartzites, crystalline schists, and gneisses. Metamorphism reached the amphibolite grade. These rocks seem to be conformable and are clearly the

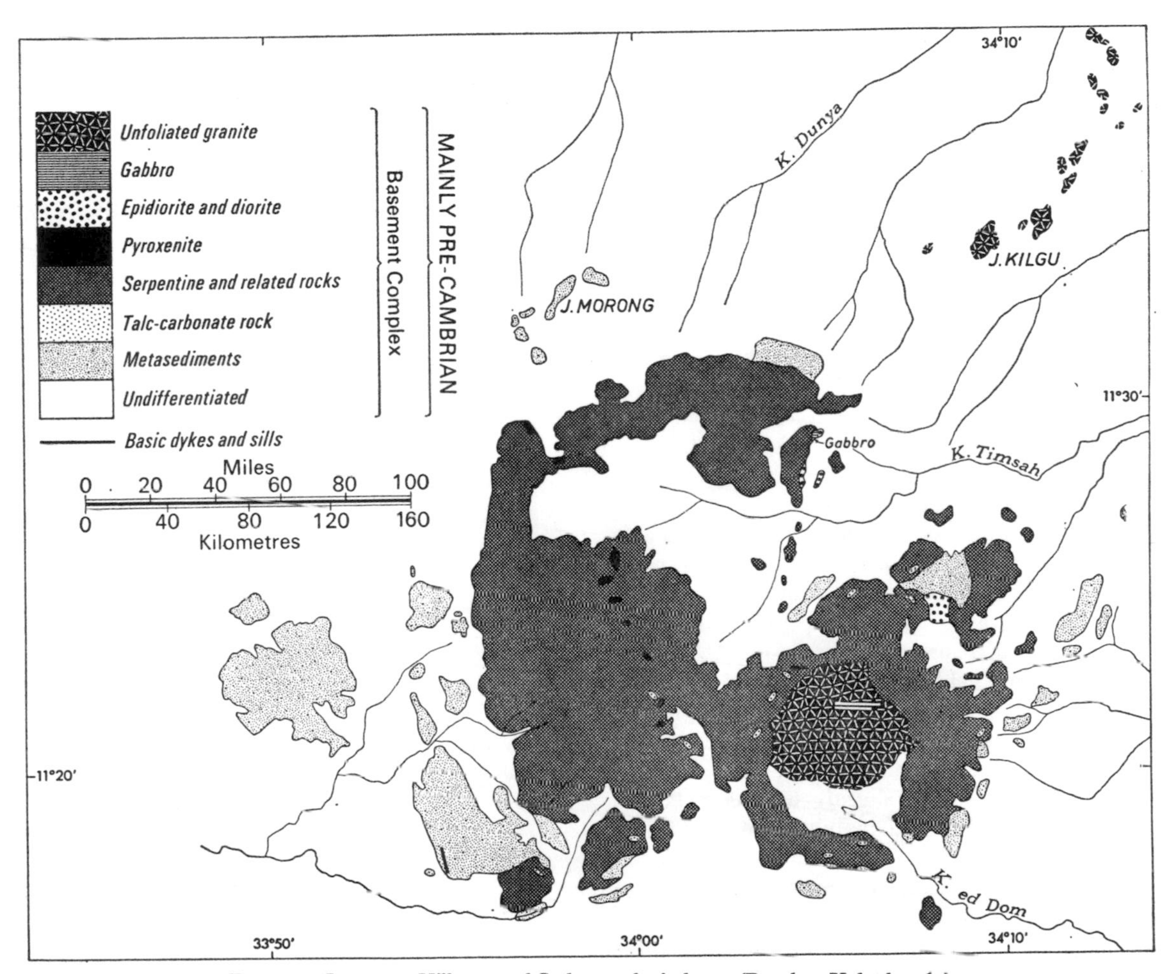

FIG. 10. Ingessana Hills, central Sudan, geological map (Based on Kabesh 1961)

oldest rocks exposed in the region. Ultramafic bodies, which were subsequently serpentinized and heavily carbonated, were then intruded and in many areas they have been replaced by talc–carbonate rocks. Fold axes trend at 022°, 337°, and 090°. The main regional trend is said to be 022°.

Porphyritic and non-porphyritic biotite-granites were emplaced on a large scale in the area, and acidic and basic dykes that cut the serpentines and granites were finally intruded (Fig. 10). Chromite is present in the serpentines and talc carbonates and chromite are mined on a small scale (see Chapter VIII).

Kabesh (1961) did not attempt to correlate the Ingessana rocks with those of other areas in the central Sudan, but the sequence of metasediments, intruded by ultrabasic rocks, etc., is best correlated with the paraschist sequence and ultrabasic rocks (Table 2).

(f) Central Sudan sheet 55 (1952 edn), Sabaloka, and Jebel Qeili

The following rock-types were recognized in the Sudan north of 6° N by Andrew (1948).

Group 1 Schists and gneisses, crystalline, highly foliated, and recrystallized and apparently interbedded with Group 2.

Group 2 Schistose, parametamorphic rocks including limestones and marbles, graphite-slates and schists, quartzites, arkoses, rhyolites, and other volcanic rocks.

Group 3 Pelites, calc-pelites, slightly metamorphosed, interbedded with graphite slate, quartzite, and limestone.

Group 4 Acid gneiss, medium to coarse, highly crystalline, foliated or foliated-granoblastic, granitic to dioritic in composition.

Group 5 Intrusive igneous rocks, acid, intermediate basic and ultrabasic types ranging from gabbro and norite (with or without olivine) to hornblende plagioclase gneiss, dunite, pyroxenite, etc., to serpentine, actinolite, and talc schist.

Group 6 Greywackes, slightly or non-metamorphosed, found mainly in the north-east Sudan associated with volcanic rocks, predominantly andesitic. These continue into Eritrea.

Group 7 Non-foliated granitic intrusions of which soda granites are the latest example.

As with the classification of the rocks of the southern Sudan, no order of superposition is implied because the rock-types are arranged mainly according to metamorphic grade.

On sheet 55, 1:1 000 000 scale, Delany (1952) presented a lithological sequence in the map margin but unfortunately few formational names were given and the map was not accompanied by a memoir.

Andrew, according to Ruxton (MS 1956), grouped the Basement Complex rocks of sheet 55 as shown in Table 2. Delany (1966) in the *Lexique Stratigraphique International* described the Basement Complex, the Green Series, Odi Schists, and Sabaloka Series.

Ages have not been determined for the paraschists and gneiss (Table 2, column 5) but two potassium–argon dates are available for the ring-dyke complex exposed in the Sabaloka inlier some 50 miles north of Khartoum. One of these analyses, made using an amphibole concentrate from the porphyritic Jebel Geri granite, showed argon loss and gave a date of $400 \pm 40 \times 10^6$ years. This would make the Geri and associated younger granites younger than the ring-dyke complex, which is clearly impossible from the field evidence. A biotite concentrate from the Sabaloka ring-dyke, however, gave a date of $540 \pm 25 \times 10^6$ years, so placing the cooling of the intrusion in the Lower Cambrian (or earlier) on the Holmes–Kulp revised time-scale (Pan-African Orogeny of Kennedy 1965, p. 48). Unfortunately an isotopic age is not available for the country-rock.

In the central Sudan a fivefold division of the Basement Complex may be recognized:

5 Intrusion of the soda granites—'younger granites', and the formation of the Sabaloka ring complex and Jebel Qeili igneous complex.

FOLDING

4 Deposition of unmetamorphosed greywackes and lavas

FOLDING AND REGIONAL METAMORPHISM

3 Intrusion of basic and ultrabasic rocks

FOLDING AND REGIONAL METAMORPHISM

2 Formation of paraschists of Andrew (1948) and Basal schists and phyllites of Ruxton (MS 1956)

FOLDING AND REGIONAL METAMORPHISM

1 Formation of older gneiss and foliated granites

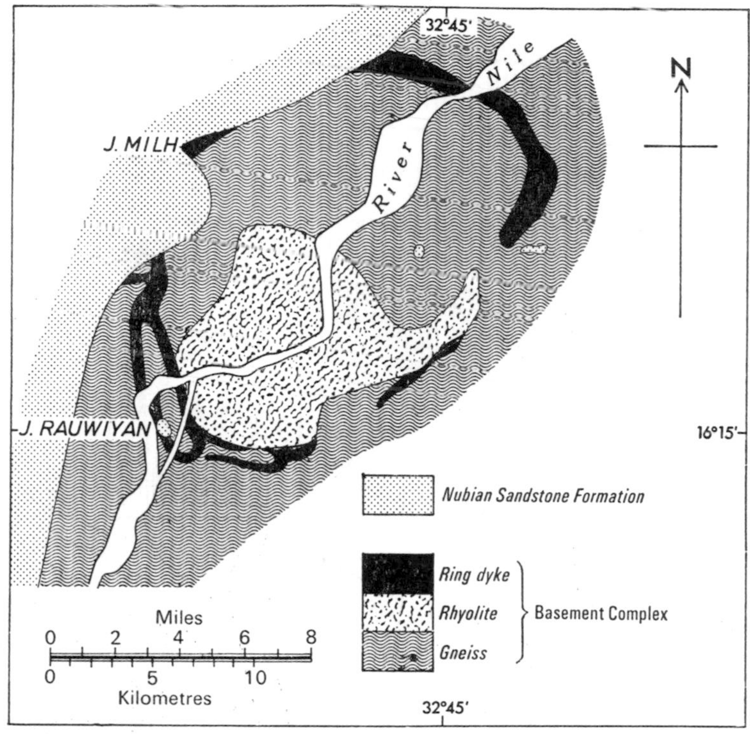

FIG. 11. Sabaloka inlier, 50 miles north of Khartoum, General geological map (Based on Delany 1955)

The best-known region in the central Sudan is at Sabaloka, where the ring complex and older Basement Complex is exposed in a series of impressive jebels and in the famous gorge of the Nile in the Sixth Cataract area (Plates 1 and 8, Figs. 11 and 12).

The succession at Sabaloka, based on Andrew (1948), Delany (1955, etc.), Almond (1964, 1967), and personal observations is as follows:

PLEISTOCENE–RECENT
Nile silts, alluvial fans, pediments, hamada, blown sand, terraces, breaching of rock sill that dammed back Pleistocene and recent lakes of Nile Valley

UNCONFORMITY

TERTIARY
? Silicification of Nubian Sandstone Formation at Jebel Rauwiyan

Superimposition of Nile drainage from a Hudi Chert or Nubian Sandstone surface on to the buried hill that forms the Sabaloka inlier

UNCONFORMITY

MESOZOIC (? Jurassic–Lower Cretaceous)
Nubian Sandstone Formation composed of sandstones, grits, conglomerates, siltstones, mudstones, frequently silicified or ferruginized, abundant fossil wood (*Dadoxylon*) found scattered throughout the formation

UNCONFORMITY

BASEMENT COMPLEX—CAMBRIAN (?)
9 Felsite dykes and sheets and other minor intrusions
8 Ring-dyke porphyritic microgranite. A potassium-argon determination on the ring-dyke microgranite gave an age of $540 \pm 25 \times 10^6$ years. Tin and wolfram stockwork in ring-dyke near Abu Dom
7 Intrusion of mica-granite (west bank of Nile) with associated tin and wolfram mineralization
6 Ignimbrite tuffs
5 Upper rhyolites
4 Agglomerates, breccias, airborne tuffs
3 Lower rhyolites
2 Trachybasalt
1 Subvolcanic metamorphosed shales, quartzites, and flaggy sandstones

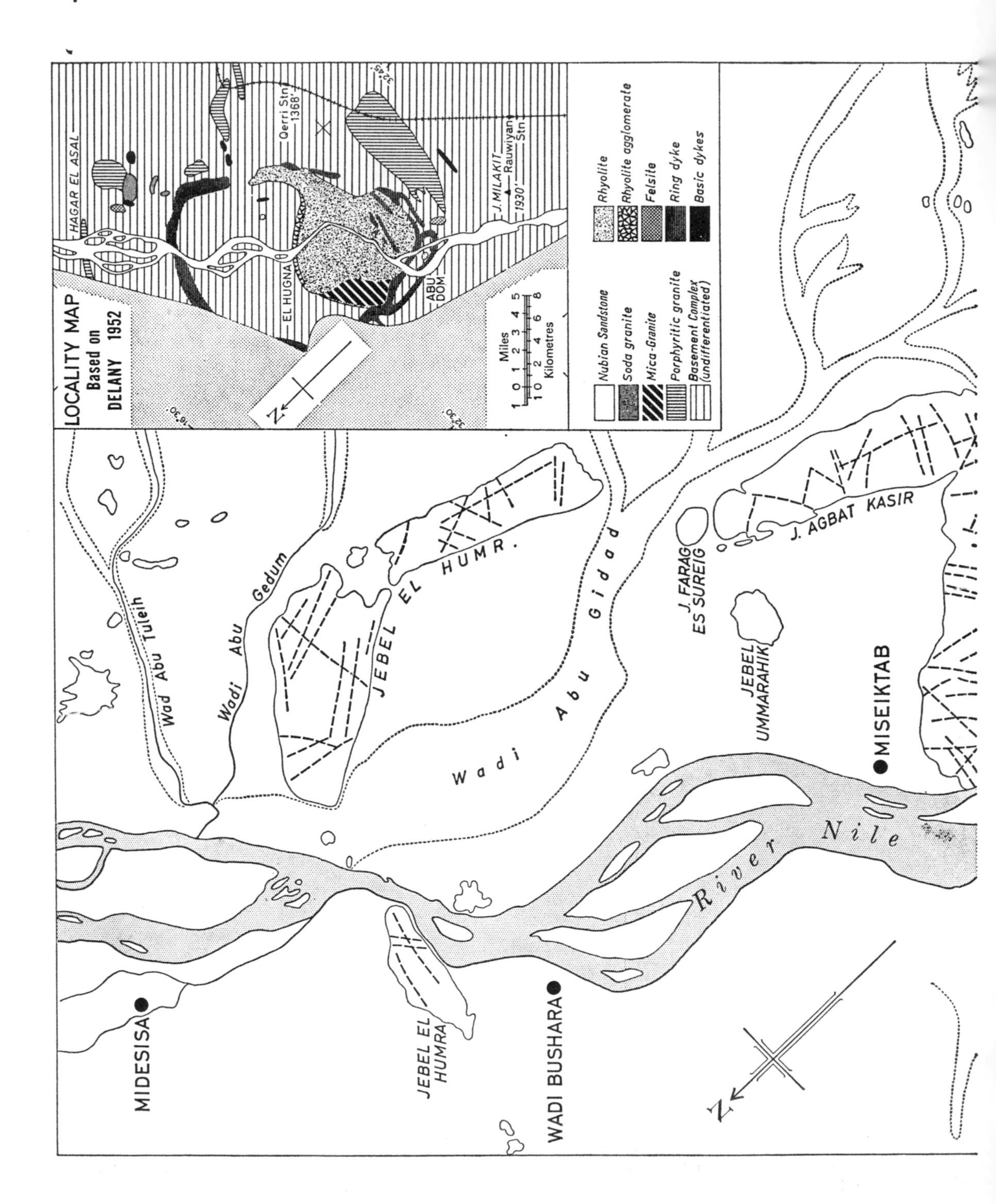
LOCALITY MAP
Based on
DELANY 1952
HAGAR EL ASAL
Qerri Stn
1368'
J. MILAKIT
Rauwiyan Stn
1930'
EL HUGNA
ABU DOM
Miles
0 1 2 3 4 5
0 2 4 6 8
Kilometres
Nubian Sandstone
Soda granite
Mica-Granite
Porphyritic granite
Basement Complex (undifferentiated)
Rhyolite
Rhyolite agglomerate
Felsite
Ring dyke
Basic dykes
Wad Abu Tuleih
Wadi Abu Gedum
JEBEL EL HUMR
Wadi Abu Gidad
J. FARAG ES SUREIG
J. AGBAT KASIR
JEBEL UMMARAHIK
MISEIKTAB
River Nile
MIDESISA
JEBEL EL HUMRA
WADI BUSHARA
N

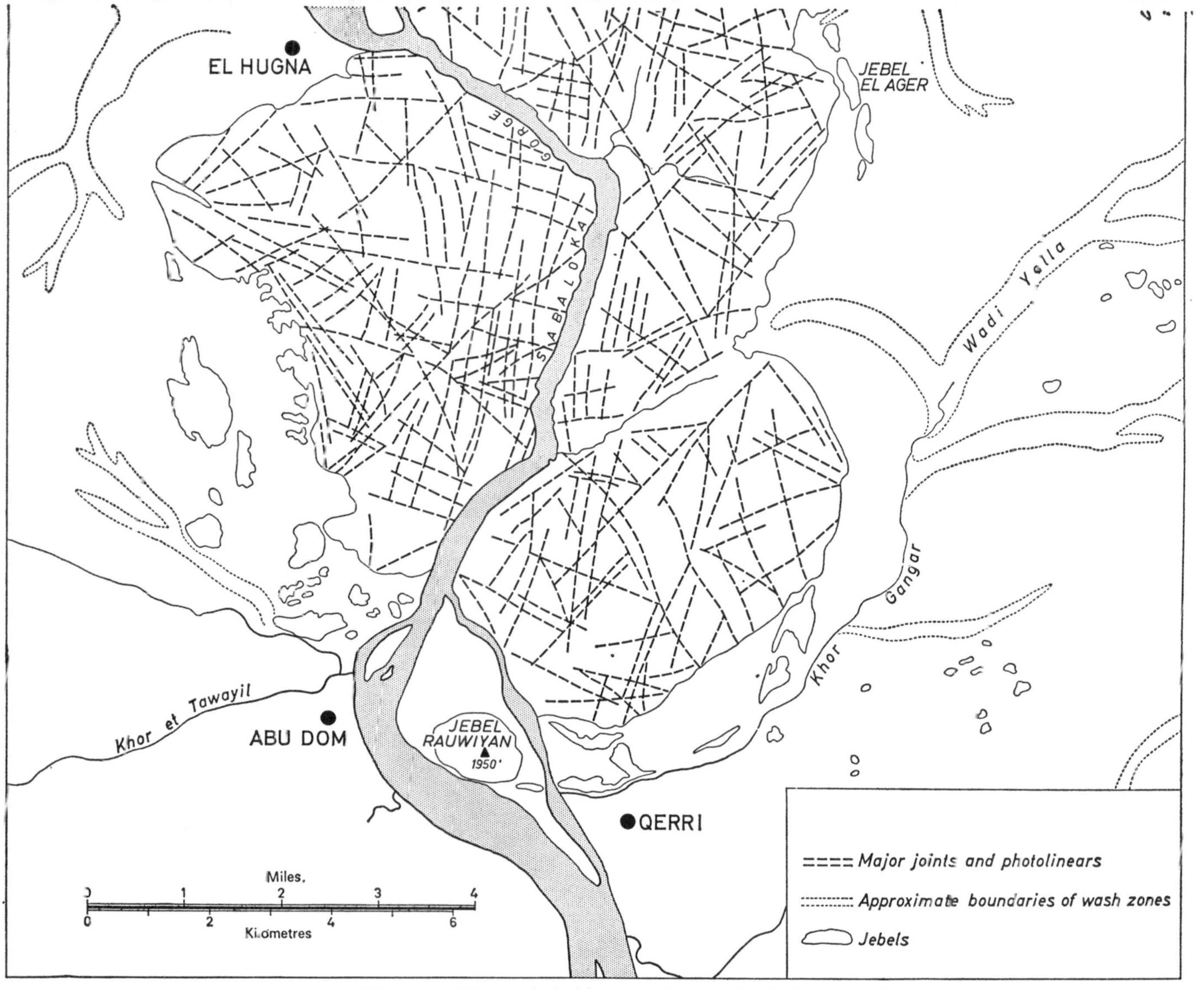

FIG. 12. Sabaloka inlier, 50 miles north of Khartoum. Photogeological interpretation showing fault control of the 'superimposed' course of the Nile, and inset map of Saboloka ring complex (Photointerpretation by Dr. D. C. Almond 1965)

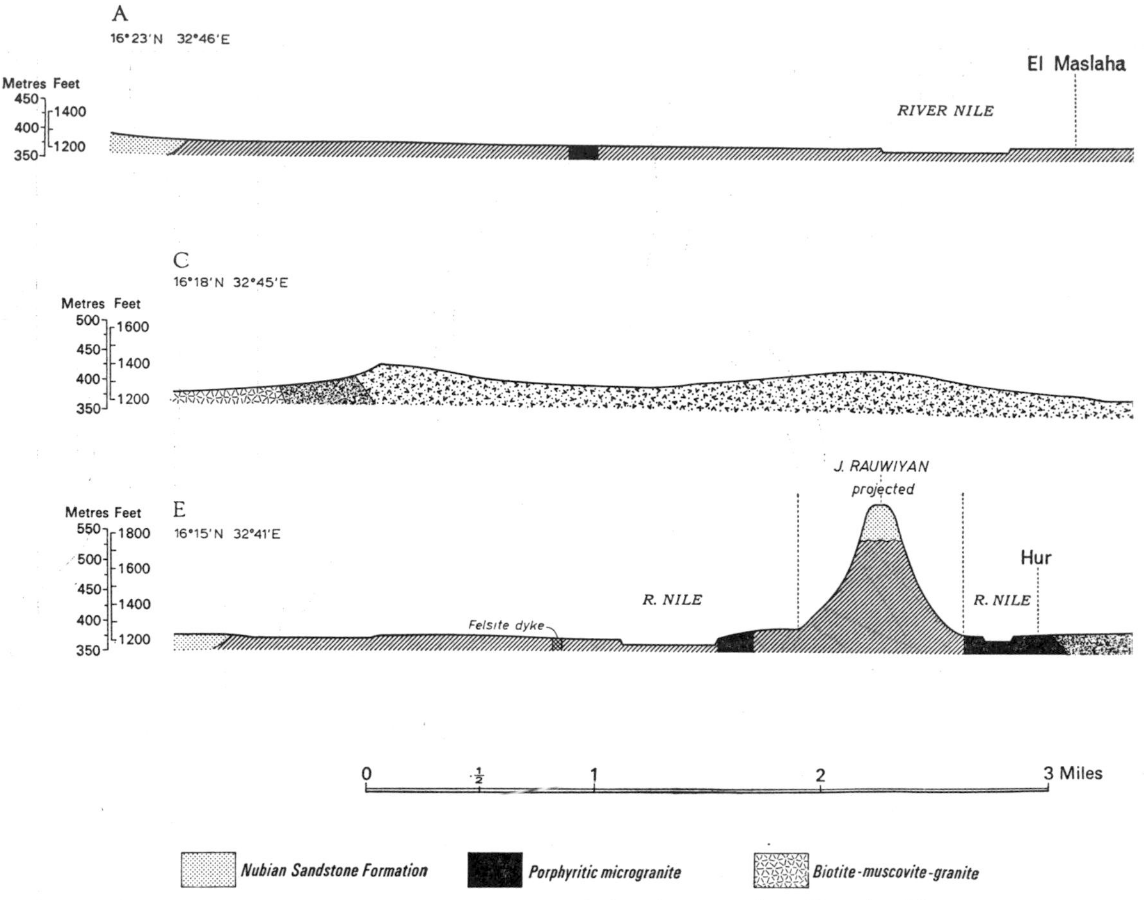

FIG. 13. Sabaloka inlier, 50 miles north of Khartoum, geological sections. (Based on Almond 1965). Topographic Data by A. A. Sadig (MS Thesis, University of Khatoum, in preparation)

UNCONFORMITY
METAMORPHISM

BASEMENT COMPLEX—PRE-CAMBRIAN
High-grade metamorphic gneisses, garnetiferous in places, granites, and amphibolites

---

The details of the Basement Complex described below are based on a 1:12 000 survey of the Sabaloka region conducted by Dr. D. C. Almond and students of the Department of Geology, University of Khartoum, Delany (1955), Almond (1967), Berry and Whiteman (1968), and personal observations.

The central hill mass at Sabaloka, cut through by the Nile in a gorge some 320 ft deep and in places only 1300 ft wide, is composed of a 'cake' of rhyolitic lavas and ignimbrite tuffs, agglomerates, etc. resting on metamorphosed Basement Complex gneisses, granites, and amphibolites, and in places on metamorphosed sandstones and mudstones. It is partially enclosed by a ring-dyke consisting of granophyric granite-porphyry derived from rhyolitic magma and intruded into the zone of fracture bordering the central zone of collapse (Fig. 11).

The oldest rocks exposed in the region are quartzo-feldspathic gneisses with some garnetiferous gneiss, granitic gneiss, hornblende gneiss, amphibolites, and quartz dykes. The foliation of these rocks generally trends between 292° and 315° and dips at about 60°. The gneisses are mainly of the almandine-

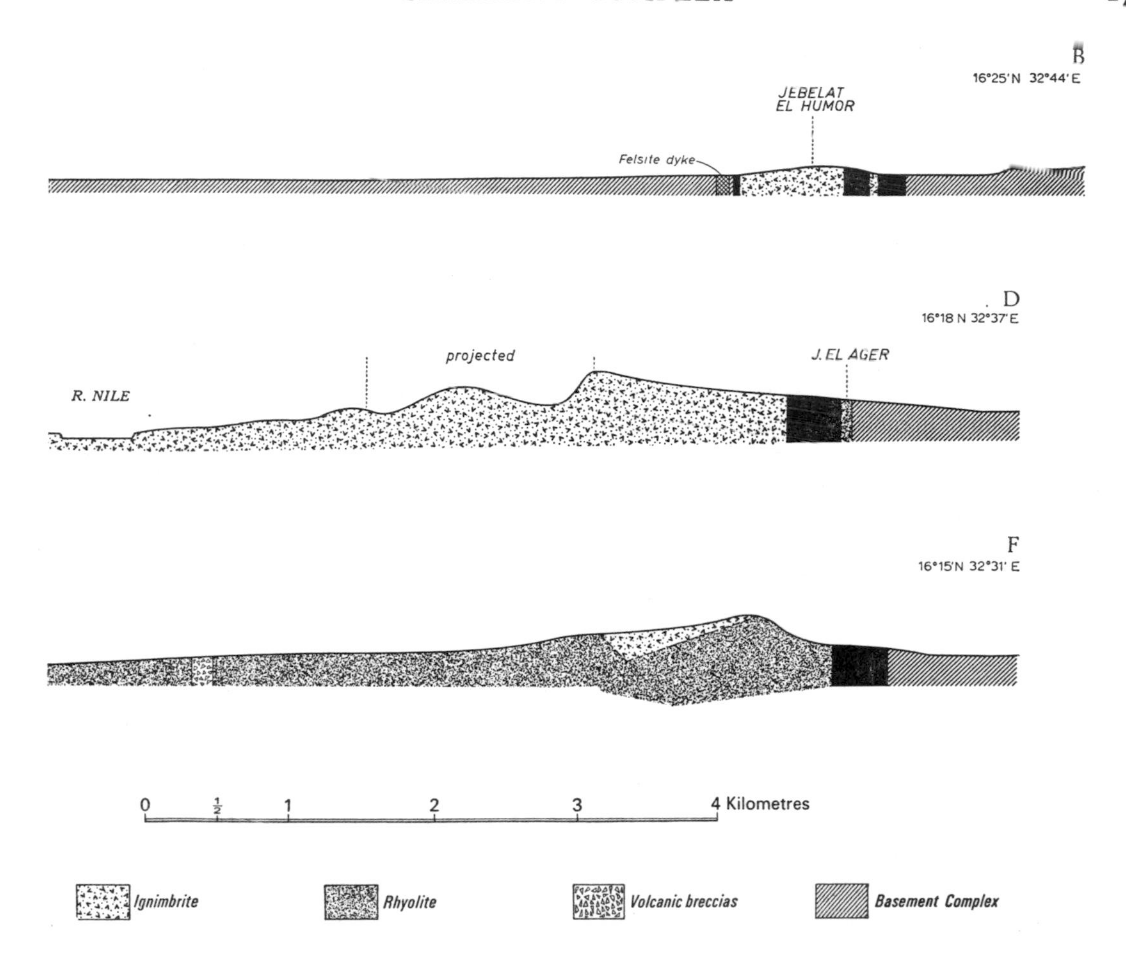

amphibolite facies but small patches of granulite facies occur.

These highly metamorphosed rocks are succeeded by 'subvolcanic sediments' consisting of metamorphosed basal quartzites and flaggy sandstones, followed by metamudstones on the west bank of the Nile only (Amond 1967).

The first signs of volcanic activity were marked by the extrusion of thin trachybasalt. Flow-banded, columnar Lower Rhyolites were extruded next, followed by a complex of agglomerates (absent on the southern part of the plateau), breccias, airborne, and ignimbrite tuffs. The ignimbrite tuffs are extensive and form the greater part of the Sabaloka plateau. There are minor non-sequences but two general successions have been made out (Almond 1967):

NORTH SIDE OF PLATEAU	SOUTH SIDE OF PLATEAU
Ignimbrite tuffs	Ignimbrite tuffs
Upper rhyolites	
Agglomerates, etc.	Lower rhyolites
Trachybasalt	Trachybasalt

Mica-granite, consisting of sodic and potassic feldspar with abundant quartz and biotite, occurs on the west bank of the Nile and metamorphoses the rhyolite in the cataract zone. A tin–wolfram–molybdenite mineralization is associated with this body and the ring-dyke (Almond 1967).

The ring-dyke, consisting of porphyritic microgranite, was intruded next. It is distinctive in the field because it weathers dark red, contrasting with the brighter red of the central core.

According to Delany (1955) the ring-dyke 'is slightly oval in shape with diameters of 20 km by 15 km. It is divided by the Nile gorge which follows the larger axis. Ridges of the red granite-porphyry, weathering to large dark red boulders, encompass entirely the rhyolite hills. . . .The ring is slightly excentric to the north, where a considerable area of granitic gneiss lies inside it'.

The ring-dyke is discontinuous (Fig. 11) and sometimes bifurcates enclosing screens of gneiss, as in the Rauwiyan Island area. The contact between the gneiss in part of this area is vertical and 'the dyke is chilled to a fine-grained compact green rock and the first 0·1 mm of hornblende-gneiss at the contact has been altered to a fine hornfels' (Delany 1955). Elsewhere, in the few places where the contact has been seen, it dips towards the rhyolite hills at angles of 60°.

Within the ring complex, dykes of red granite-porphyry, generally trending east–west, cut the rhyolites of the plateau and are clearly closely connected.

The intrusion of the ring-dyke appears to have been brought about because of the collapse of the central part of the volcanic complex. No doubt at the surface a large caldera was formed, and it is interesting to note that the dimensions of the collapse as determined by the size of the ring are similar to those of the present-day Ngorongoro caldera in Tanzania.

The porphyritic microgranite was intruded into the marginal fracture zone, which had no doubt been widened by 'fluidization', and the central ash and lava 'cake' were dropped into the Basement Complex in early Cambrian time. Felsite dykes and sheets and minor intrusions, one of which is composed of soda granite, mark the last igneous activity in the region. Ashes and lavas do not occur outside the ring fracture zone at present in the Sabaloka area and therefore it is not possible to determine the extent of the volcanic complex. A. A. Sadig has recently completed a gravity survey of the Sabaloka ring but details of his M.Sc. Thesis, University of Khartoum, were not available at time of going to press.

The most that can be said from the field evidence about the age of the Sabaloka ring complex is that it is pre-Nubian. The Nubian formation contains occasional pebbles of rhyolite and is unmetamorphosed. It rests unconformably on the complex, as is visible at Jebel Rauwiyan. The small outlier of Nubian Formation just north of the rhyolite plateau has been faulted in. Because the field evidence is inconclusive the author submitted a specimen of porphyritic microgranite from the ring-dyke for age-determination by the potassium–argon method. This gave an age of $540 \pm 25 \times 10^6$ years, i.e. Lower Cambrian or older. It is probably a cooling age.

The broader relationship between the Sabaloka sequence and the Paraschists, etc. of Andrew (1948) cannot be determined directly but the occurrence of rhyolites at Jebel Abu Sha'afa and of soda granite and rhyolite at Jebel Mundara in the Butana Greenstone belt to the south-east indicate that probably the Sabaloka sequence is younger than the Paraschists of Andrew.

Another well-developed igneous complex occurs some 85 miles east of Khartoum at Jebel Qeili (15°30′N, 33°45′E), in Kassala Province (Figs. 14 and 15). It was described by Delany (1955), who studied the area with J. B. Auden. The account that follows is based on this work, on a report by Ustaz Farouk Ahmed, who studied the area in detail, and on observations made by the author. Jebel Qeili consists of a series of syenitic, granitic, and gabbroic intrusions derived from an alkali-rich magma.

The complex is composed of a central plug (Qalat Qeili syenite) and three associated ring-shaped bodies intruded into the quartzites, hornblende schists, and altered andesitic lavas of the Butana Green schist. As a result of differential erosion the syenitic and granitic rocks now form a group of hills that rise about 260 ft above the surrounding clay plain of the Butana (Fig. 15).

The emplacement of the ring complex was preceded by the extrusion of rhyolitic lavas. These lavas are compact, fine grained, blue–black rhyolites that dip vertically and trend between 000° and 015°. They are clearly younger than the andesitic lavas, which have been completely saussuritized. The rhyolitic lavas are now preserved as a narrow strip in contact with the outermost ring (El Rom granite bodies).

The main body of the ring complex is composed entirely of syenites and granites intruded in the following order: outermost ring (El Rom granite bodies), outer ring (Qalat Busta syenite), inner ring (El Hawya syenite), and the central plug (Qalat Qeili syenite). According to Farouk Ahmed (MS 1968) the field evidence indicates that we are dealing

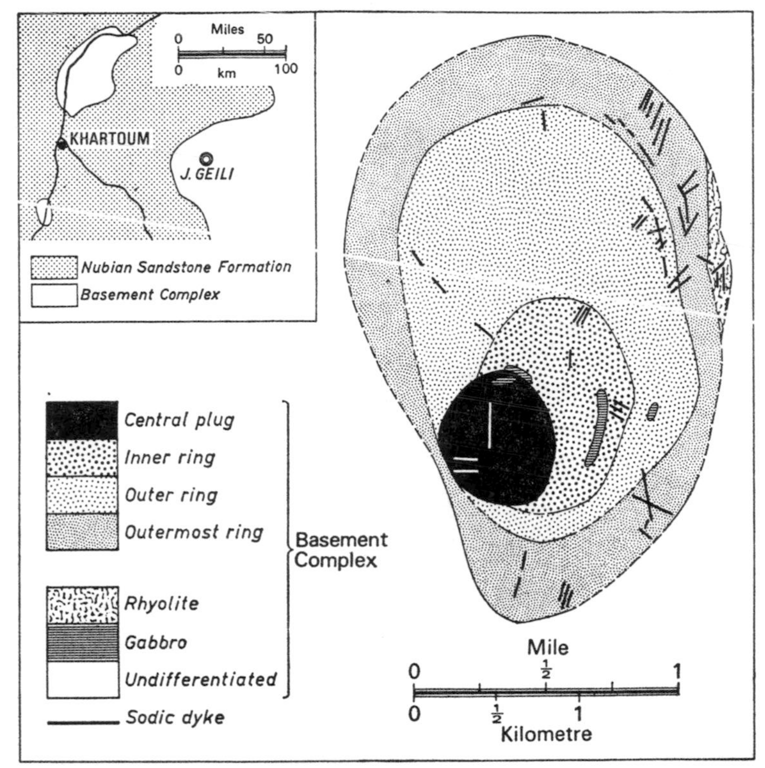

FIG. 14. Jebel Qeili, 85 miles east of Khartoum, geological map (Based on Delany 1955 and Farouk Ahmed, M.Sc. Thesis, University of Khartoum, in preparation)

with plugs. The chilled margin of Delany (El Rom granites) forms the outermost ring and has a vertical contact with the rhyolite. The contact between the chilled margin and the syenite in places dips east but in general it is vertical.

The outer ring (Qalat Busta syenite) is composed of grey–brown syenite that appears to have been more contaminated than the inner syenite. Jointing is well developed and a conspicuous set of joints dips to the east at 9°–15°. On the west the joints are vertical.

According to Farouk Ahmed (MS 1968) the following sequence can be recognized at Jebel Qeili:

JEBEL QEILI

8 Minor intrusions	(mainly dykes)
7 Qalat Qeili syenite	(= central plug of Delany)
6 El Hawya syenite	(= inner ring of Delany)
5 Qalat Busta syenite	(= outer ring of Delany)
4 El Rom granites	(= chilled margin of Delany)
3 Alkali rhyolites	
2 Gabbros and dolerites	(= essexite of Delany)
~~ UNCONFORMITY ~~~~~~	
1 Butana Green Schists	

The inner ring is composed of syenite, and although petrographically similar to the outer ring it is a separate intrusion.

The central mass of Jebel Qeili is a vertical plug of pegmatitic to even-grained, white, riebeckite–quartz–syenite. Its eccentric position with regard to the other syenite bodies indicates that it is the youngest intrusion. Arcuate dykes of microsyenite cut the outer syenite and the chilled margin, and vein the older rhyolites. The latest minor intrusions include syenite–pegmatite, trachytes, and micro-syenite dykes.

Farouk Ahmed (MS 1968) described the syenites as composed of potassic feldspars forming more than 80 per cent of the rock; dark minerals are mainly alkali amphiboles, pyroxenes, biotites. Some crystals of fayalite occur in the outer ring only. Quartz is abundant in the outer ring but completely absent

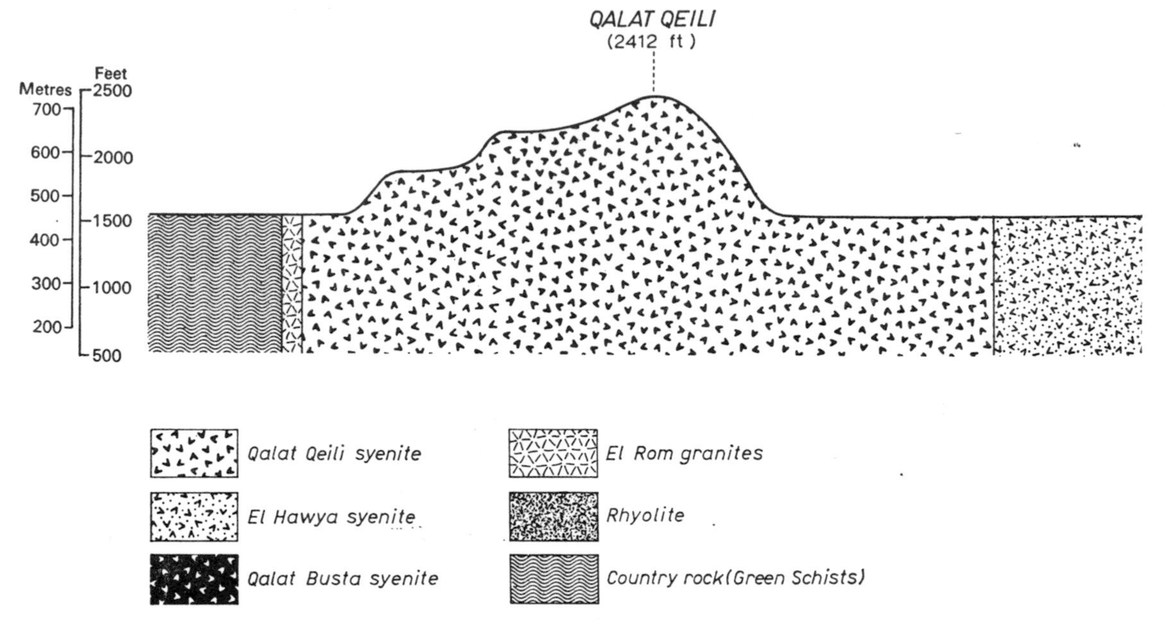

FIG. 15. Jebel Qeili, 85 miles east of Khartoum, geological sections (Based on Farouk Ahmed, M.Sc. Thesis, University of Khartoum, in preparation)

in the other rings. Apatite, iron oxides, zircon and carbonates are the common accessories.

Scattered information is available for several parts of the central Sudan in Andrew (MS 1943). Limestones, quartzites, and graphitic slates, striking at 022° were recorded east of the Gezira. Exact positions were not given. Presumably these rocks are similar to the Basal schists and phyllites of Ruxton mentioned above.

In the southern part of the Gezira Andrew (1948), Iskander (MS 1957), Abdl Salaam (1966), and the author have recorded marbles, quartzites, graphitic slates, pelitic schists, and gneisses. The metasediments are intruded by granites such as those exposed at Jebel Moya and Jebel Saqadi. According to Iskander (MS 1957) the intervening ground between these two jebels is occupied by metasediments which crop out in low dark hills known locally as 'galas'. These metasediments strike predominantly east-north-east–west-north-west and dip 70°E.

Contorted graphite schists occur mainly at Galaa el Agoos, and green micaceous schists are interbedded at some localities.

Acidic gneiss occurs south of Jebel Moya at Jebel Awar, Jebel Abu Sabun, and Jebel Humeil. Red, well-jointed, porphyritic granite is injected along foliation planes in the metasediments of Jebel Saqadi.

Jebel Moya is composed of three main rock-types according to Iskander (MS 1957); medium grained, black, acidic rocks forming the country rock, charnockitic quartz diorites, and pink granites. The charnockitic quartz-diorites form most of Jebel Moya. They are dark in colour and contain blue quartz, dark plagioclase, and hypersthene.

Jebel Moya stone is fresh and well-jointed and is quarried for dimension stone. The stone is widely used by Sudan Railways. It is an excellent building stone and could be used to a much greater extent than at present. The pink granites are coarse and pegmatitic and are intrusive into the hypersthene diorites.

There are several outcrops of Basement Complex rocks along the western margin of the Gezira. These include Jebel Huweida, Jebel Helbi, Jebel Bereima, Jebel Harashkol, etc. A felsite dyke intruded into mica schists occurs at Jebel Huweida and Jebel Bereima is a highly brecciated felsite dyke. Both these dykes and associated rocks strike north–south and dip steeply east. Jebel Helbi occurs west of Jebel Huweida and is composed of sheared micaceous gneiss.

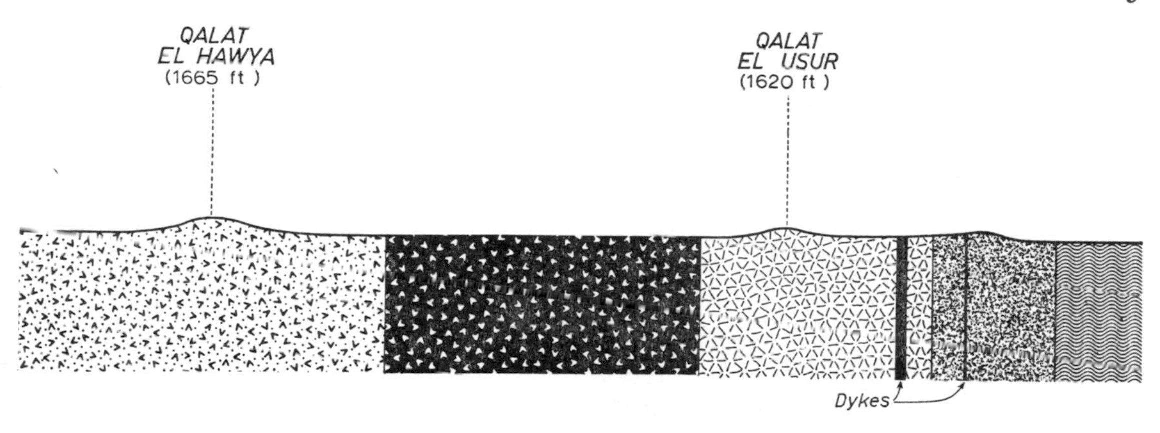

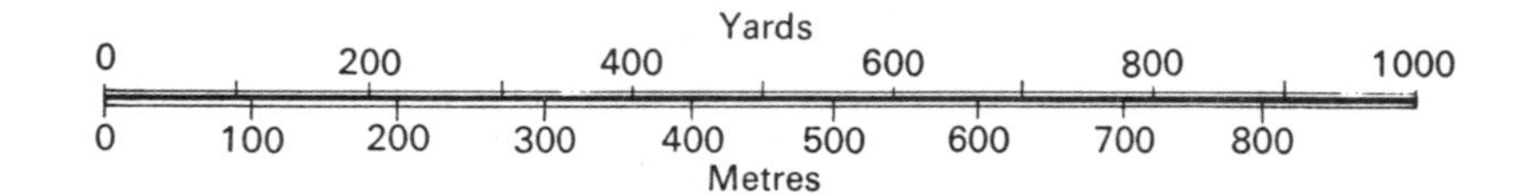

South-east of the Gezira in the Roseires area, limestone intruded by foliated granites crops out in a broad belt along the Blue Nile. Details of the engineering geology of this area are given by Jones (1954), who described the Roseires Dam site.

In the Beni Shangul foothills, south of Roseires, several basic and ultrabasic intrusions are known and a limestone, trending slightly east of north, has been recorded in the middle reach of the Dabus (or Yabus) valley.

The Beni Shangul Hills are said to granitic. Primary gold lodes and placers occur in this area and alluvial deposits on either side of the frontier are gold-bearing as far as the Blue Nile (see Chapter VIII).

Jebel Mirr (north-east of Kosti), Jebel Dali (south-west of Sennar), and Jebel Guli (east of Renk) are isolated hills of soda granites (Andrew MS 1943).

(g) Southern Jubal el Bahr el Amer, Derudeb sheet 46I, and Kassala district

Delany (1956) recognized an older gneissose Basement Complex followed by the Odi Schist Group (Table 2). Most probably these schists are equivalent to the paraschist of sheet 55. Like the Butana rocks, the Odi Schist Group contains many rock-types such as limestones and quartzites, altered andesitic lavas, and chlorite–epidote schists. North of Jebel Tehilla the country is formed mainly of banded mica-gneiss, which passes northwards into a migmatite.

The main feature of the Derudeb district is the presence of a large ring complex, the granite walls of which encircle a plain roughly 5 miles north–south and 7 miles east–west (Fig. 17; Plates 4A and 7).

In the Tehilla district Delany (1955, pp. 142–7) described the following sequence:

Tisibrahimit dykes, a subparallel swarm of aplites of uncertain but pre-Nubian Sandstone age.

Intrusion of soda granite.

Unfoliated granite of the Tehilla ring complex; this granite invaded and assimilated older rocks but had little effect on the gabbro. (The age of the gabbro is, however, uncertain.)

(?) Emplacement of gabbro near the contact with the Odi Schist and the older gneissose Basement Complex.

The Tehilla ring is only one of nine ring-structures that have been recorded in the Sudan (Fig. 4). The

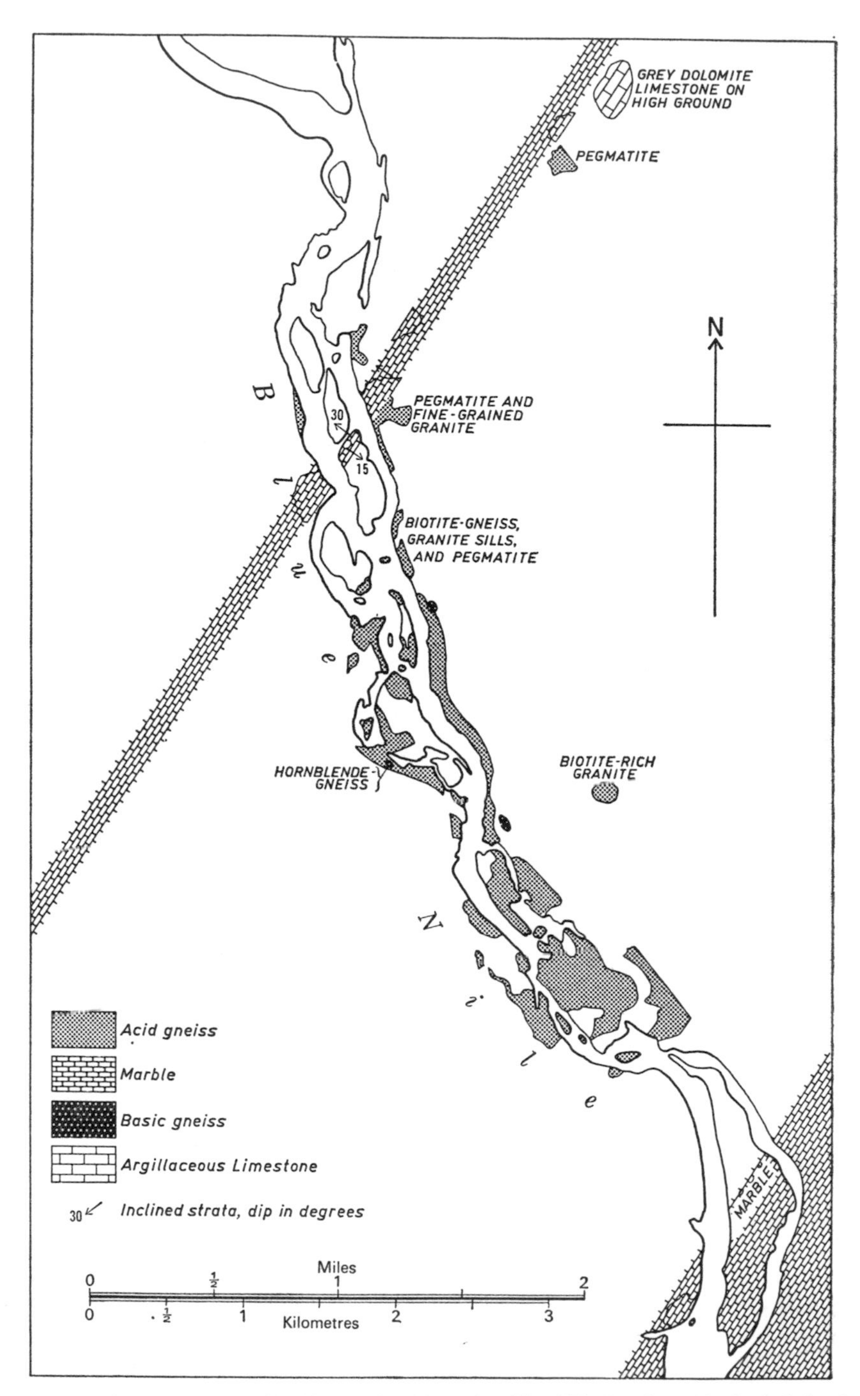

FIG. 16. Damazeen rapids, Er Roseires Dam site, Blue Nile Province (Geological sketch map. Based on Jones 1954)

Sabaloka, Tehilla, and Qeili rings were described by Delany and are the latest acid bodies intruded into the Basement Complex. They crystallized from soda-rich magma; riebeckite and aegirine are common minerals. According to Delany (1955) all the rings are approximately contemporaneous and are to be correlated with the rhyolites associated with some Nigritian rings in the western Sahara, although she pointed out that isotopic dates are required to confirm this.

In the Kassala district, Jebel Kassala, perhaps the most striking jebel in the Sudan (Plate 10), is composed of grey gneiss that trends roughly north-south with conspicuous east-west cross jointing. A line of limestone outcrops runs east of Kassala but very little is known of the area. It may well be the same limestone that crops out at Roseires (Andrew, unpublished MS).

(h) Northern Jubal el Bahr el Ahmer, Mohammed Qôl sheet 36M and Dungunab sheet 36I

This area constitutes perhaps the best-known part of the Basement Complex of the Sudan, having been studied recently by Gass (1955), Ruxton (1956, pp. 314–30), Gabert, Ruxton, and Venzlaff (1960, pp. 241–69), Kabesh (1962, pp. 1–61), Lotfi and Kabesh (1964), and by the author (1962, 1967). A composite sequence is given in Table 2. The subdivisions given below are those of Ruxton (1955) and those adopted largely by Kabesh (1962). Considerable discrepancies exist, however, between the maps of Gabert *et al.* (1960) and Kabesh (1962) (Fig. 18).

(i) *Primitive 'System' or Group*

Some uncertainty exists about the order of the formations making up the 'Primitive Group', but

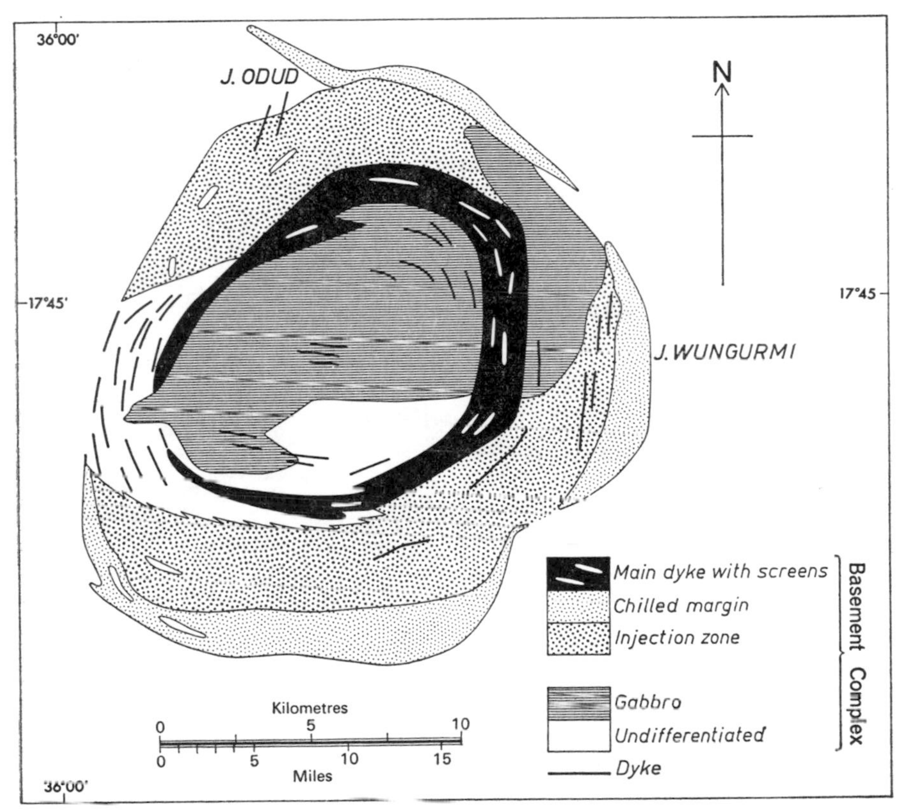

FIG. 17. Tehilla ring complex, southern Jubal el Bahr el Ahmer (Based on Delany 1955 and Sudan Geological Survey, Derudeb sheet)

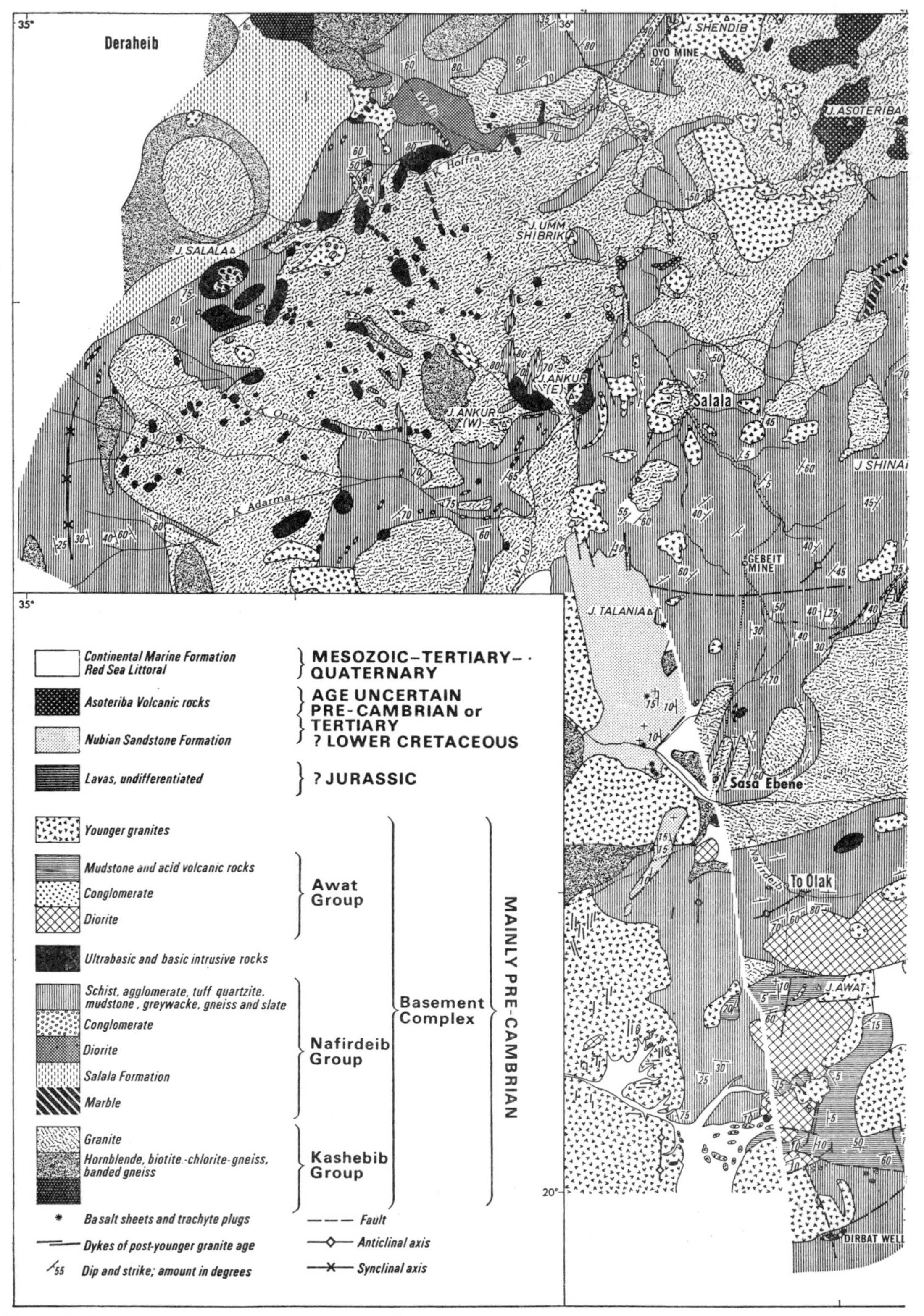

FIG. 18. Deraheib, Dungunab, and Mohammed Qôl areas, northern Jubal el Bahr el Ahmer, composite geological map (Based on Gass 1955, Ruxton 1956, Gabert, Ruxton, and Venzlaff 1960, Kabesh 1962, and personal observations in 1962 and 1967)
Note: Map does not join in Sasa Ebene–Dirbat Well area.

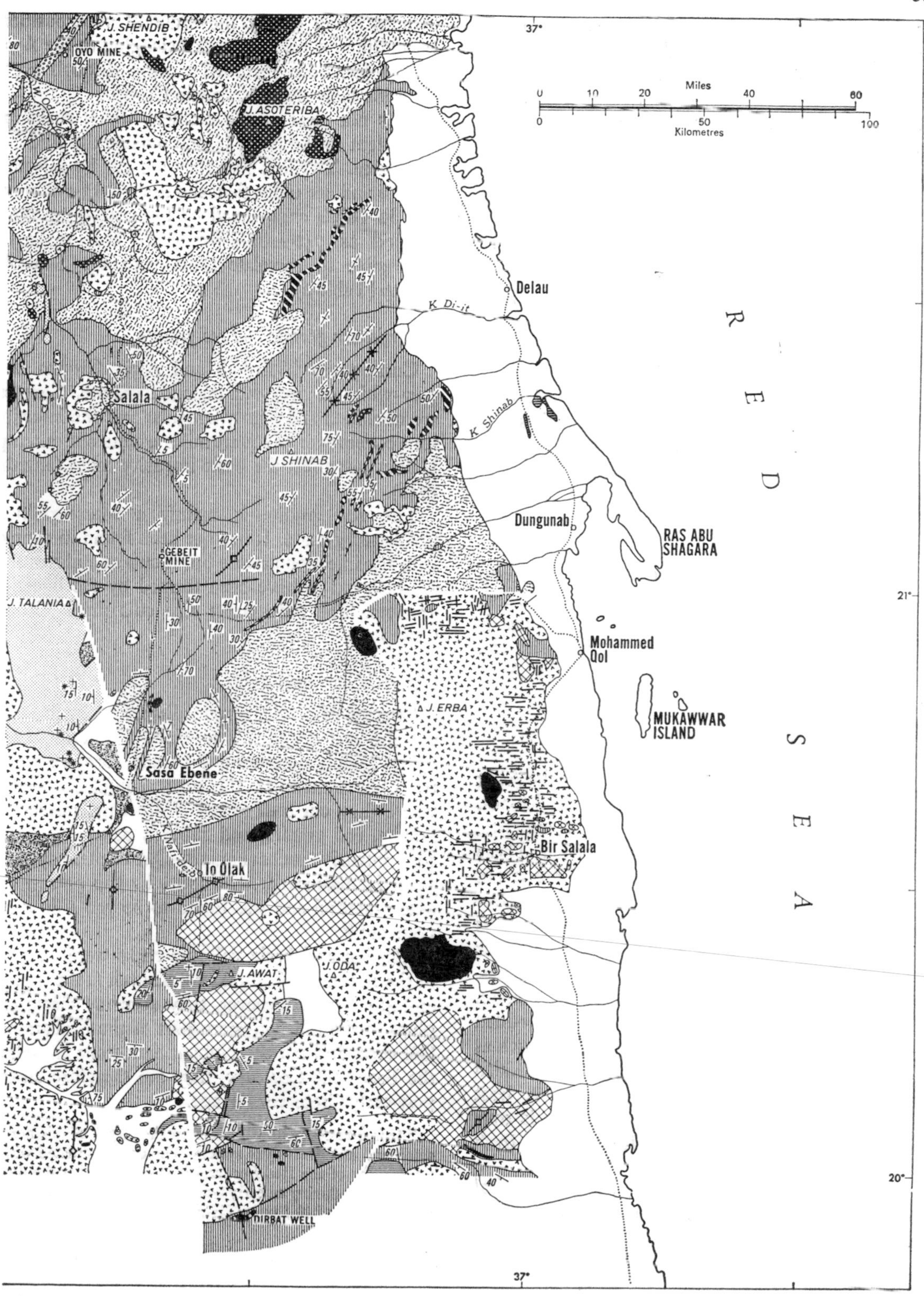
J. SHENDIB
OYO MINE
J. ASOTERIBA
Miles
Kilometres
Delau
K. Di-it
R E D S E A
Salala
K. Shinab
J. SHINAB
GEBEIT MINE
Dungunab
RAS ABU SHAGARA
J. TALANIA
Mohammed Qol
J. ERBA
MUKAWWAR ISLAND
Sasa Ebene
Bir Salala
To Olak
J. ODA
J. AWAT
DIRBAT WELL
37°
21°
20°

4

according to Ruxton a period of large-scale folding and regional metamorphism followed by the intrusion of basic dyke-swarms separates the Primitive Group from the succeeding Nafirdeib Group. The Primitive Group consists of acid gneiss, hornblende schist, schists, and slates.

(ii) *Nafirdeib 'Series' or Group*

The Nafirdeib Group may be correlated with the Odi schist and the paraschist group of sheet 55 (Table 2). In the To–Olak area of the Red Sea Hills the Nafirdeib Group is estimated to be more than 10 000 ft thick. It is folded in a series of open folds with axes plunging at a low angle in a direction 030°, whereas the regional strike of foliation ranges from 090° through 045° to 000° in the Primitive Group. The Nafirdeib Group is composed of schists and slates, andesitic tuffs, agglomerates, meta-andesites with minor quartzites, limestones, metadacites, and greywackes.

(iii) *Basic intrusive rocks*

Large gabbroic masses cut the Primitive Group and norites cut calcareous members of the Nafirdeib Group. A skarn zone extending over 4 miles is associated with the norite of Dirbat Well (Fig. 18). The basic intrusive rocks do not cut the Awat Series.

(iv) *Awat 'Series' or Group*

According to Ruxton (1956, pp. 322–4) the Awat Group consists of grey, massive silty mudstones, with some microconglomerates. A few andesitic lavas occur at the bottom of the succession. These beds are more than 2000 ft thick (the base is not exposed). They are followed by conglomerates over 400 ft thick. The matrix of the conglomerate consists of silty mudstone. Pebbles up to 2 ft in diameter occur and are composed of silty mudstones, meta-andesites, tuffs and agglomerates, porphyries, and granite. The conglomerates pass upwards into an interbedded sequence of silty mudstones, acid volcanics such as dacites, rhyo-dacites, and rhyolites with interbedded tuffs and agglomerates. These beds are at least 2500 ft thick. Altogether the Awat group is over 5000 ft thick. Its most striking feature is its fresh and unaltered state.

Flat-lying Awat volcanic rocks, overlying vertical Nafirdeib Group, occur at some localities, so demonstrating clearly the order of superposition. In addition, a volcanic neck belonging to the Awat Group has been recorded cutting through green andesitic beds of the Nafirdeib Group.

It is not clear at the moment which rocks are equivalent to the Awat Group outside the Red Sea Hills area. The fact that they are cut by injection granite and are older than the basic dyke-system of the Bir Salala area, dated by the potassium–argon method as $740 \pm 80 \times 10^6$ years old, makes it impossible to correlate them with the Sabaloka Group (Table 2).

(v) *Injection granite and red granite*

Cutting the Nafirdeib and Awat Groups is a grey-white or pink-red injection granite. Contacts are unchilled and highly irregular and the injection zone covers over 40 square miles around one outcrop. It does not affect the basic intrusive rocks (Ruxton 1956).

The red granite is mainly a coarse-grained, unfoliated granite. It has not been found in contact with the injection granite but it cuts complexes that are believed to have been formed by the injection granites. The red granite also cuts the Nafirdeib Group.

(vi) *Acid and basic dykes*

Swarms of acidic, intermediate, and basic dykes cut the red granite. The black-weathering basic dykes, mainly dolerites and basalts, form a striking reticulate system in the area around Bir Salala where they cut buff to red weathering, microcline granite, red granite, and gabbros belonging to the basic intrusive rocks (Fig. 18). One of these dykes at Jebel Hamashaweib was dated by the potassium–argon method and was found to be $740 \pm 80 \times 10^6$ years old.*

Alkaline and basic volcanic rocks were assigned to the Basement Complex by Kabesh (1962, pp. 14 and 48). Since, however, they occur as sheets and plugs in the north-western part of the Mohammed Qôl sheet and cut the Nubian Sandstone as well as the Basement Complex, they are therefore clearly post-early Cretaceous in age.

* It is interesting to speculate on the origin of this north–south and east–west dyke system. The area is situated within a few miles of the Red Sea, and it is possible that the dykes were formed during an early tension phase in the development of the Red Sea Rift System (Whiteman 1965, 1968).

(vii) *Dungunab sheet 36I*

Part of the region described by Ruxton (1956) was mapped by Gass and submitted in an M.Sc. thesis to the University of Leeds (1955). Neither Ruxton (1956) nor Kabesh (1962) made reference to this work, although Kabesh cited the unpublished Sudan Geological Survey Report on which the thesis is based. Some fundamental differences exist between the successions reported by Ruxton, Kabesh, and Gass (Table 2). The succession reported by Gass (1955) for the Dungunab sheet is as follows:

BASEMENT COMPLEX

DUNGUNAB SHEET 36 I

RECENT
Blown sands, outwash fans, and coral limestones.

UNCONFORMITY

PLIO-PLEISTOCENE
Coastal plain sediments and volcanic rocks, marls, chalks, evaporites, and conglomerates interbedded with olivine basalts.

UNCONFORMITY

TERTIARY
Asotriba volcanic rocks, rhyolitic lavas, tuffs, agglomerates, vent breccias, and volcanic necks.

UNCONFORMITY

LOWER TERTIARY–UPPER JURASSIC
Nubian Series, current-bedded sandstones, conglomerates, and shales.

UNCONFORMITY

Uplift associated with the intrusion of younger granites.

LOWER TERTIARY–UPPER JURASSIC
Younger granites, granophyres, quartz-syenites forming bosses, stocks, ring dykes, and some intrusions of post-Nubian age.

UNCONFORMITY

ARCHAEAN OR BASEMENT COMPLEX
Basic intrusives including gabbros, troctolites, and pyroxenites, forming dyke-like intrusions, probably plutonic, representative of dolerite dykes of some age.

Batholitic granite, mainly biotite–hornblende–granite.

Assimilation granite, biotite–hornblende–granodiorite.

UNCONFORMITY

Oyo Series, pelites, quartzites, marbles, conglomerates, interbedded with acid volcanic rocks.

UNCONFORMITY

Granites, gneisses, schists, acid volcanic rocks found only in the Basal conglomerate of the Oyo Series.

Gass (1955) stated that the greater part of the area is granitic. He recognized three main types of granite, the assimilation, the batholitic, and the younger granite masses. The rocks of the sedimentary–volcanic Oyo Series are the oldest in the area. It is postulated that the remnants of this series are simply large roof pendants, and the whole area is founded on granite that exists at no great depth.

(i) Jebel Uweinat, Northern Province

The Basement Complex of the Jebel Uweinat area, Northern Province was described in a general way by the Royal Dutch–Shell and British Petroleum Survey Party, led by A. F. Hottinger, who traversed from the Nile at Dongola to Uweinat in 1959. The Basement was studied in the Uweinat Group, and at Jebel Tageru, Jebel Abyad, Jebel Rahib, Laquiya Arbain, etc. It is said to consist of magmatic rocks and metasediments of meta- and cata-metamorphic grade including phyllites and biotite schists.

Hottinger *et al.* (1959) noted that plutonic rocks predominate, except in the Jebel Rahib area, and consist of large bodies of granite, syenite, and diorite. Gneisses, quartzites, marbles, and mica-schists were recorded also.

Despite the great distance involved, and the limited knowledge of the area, the Basement rocks were correlated with the Suggarian and Pharusian groups of the Ahaggar. Such a correlation can have little validity in the absence of isotopic dates.

Said (1962, pp. 48–50) also described the Basement Complex of Jebel Uweinat basing his account on Hassanein (1924), Hussein (1928), Menchikoff (1927, 1926), Beadnell (1931), Clayton (1933), Desio (1933), and Sandford (1935).

The country-rock is composed of highly-folded metamorphics, which form subdued topography, intruded by undeformed plutonic masses that form the main jebel groups. Subsequently hypabyssal rocks were emplaced. Burollet (1963) and Sandford (1935) briefly described the Basement Complex of the Uweinat area (Figs. 19–21).

(j) Rubatab area, Northern Province

Basement Complex rocks have been described from the Rubatab mica-mining area of the northern Sudan but although (Kabesh 1960) gave detailed

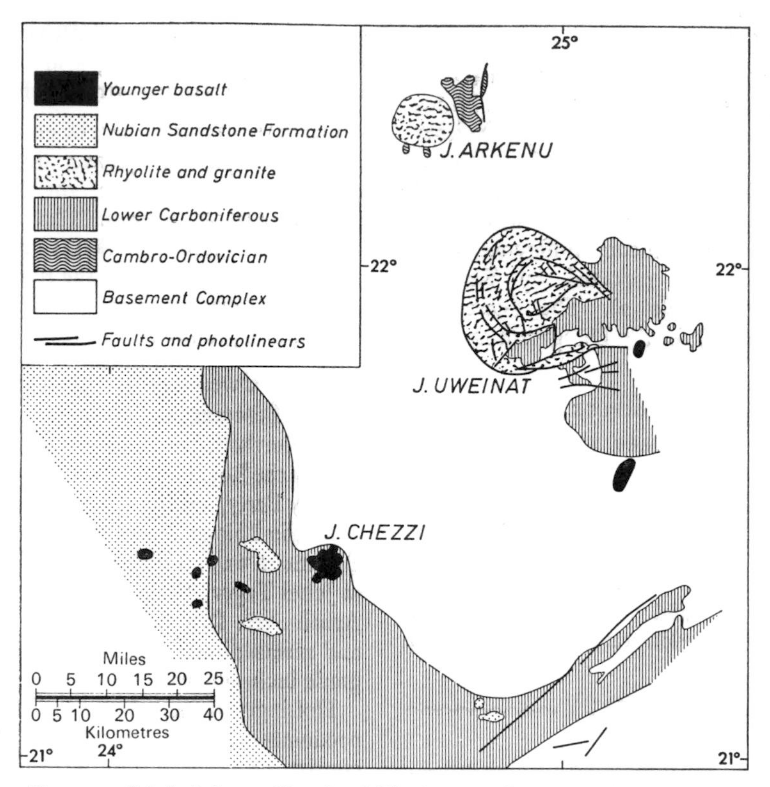

FIG. 19. Jebels Arkenu, Chezzi and Uweinat, south-eastern Libya and north-western Sudan, general geological map (Based on Hottinger *et. al.* 1959 and Burrollet 1963)

lithological descriptions, little attempt was made to work out the sequence or unravel the structure. Large areas of Kabesh's map are structurally incorrect and from the map it is not possible to estimate the thickness of mica-bearing rocks involved. The original rocks were calcareous, arenaceous, and argillaceous in various admixtures. They are cut by tourmaline–garnet–beryl- and apatite-bearing pegmatites. Basic dykes and sills were formed during the last phase of igneous activity in the Rubatab area.

Kabesh (1960) did not attempt to correlate the Rubatab sequence of metasediments but probably it is equivalent to Odi–Nafirdeib paraschist sequence (Table 2).

(k) Atbara area, Sudan Portland Cement Company quarry

Highly foliated crystalline gneiss occurs on the west bank of the Nile, and about 20 km west of the river there are outcrops of coarse-grained, vari-coloured marble bands interbedded with bands of paragneiss. These outcrops extend over several square miles. The rocks strike at 070° and dip on average at 35° to the east-south-east. They are overlain to the west by the Nubian Sandstone Formation.

The marbles are 960–1280 ft wide and extend considerably in depth, although paragneiss was visible in some of the pitching folds seen on the working faces of November 1963 and April 1967. About 1½ miles north of this quarry face the pitch causes the paragneiss bands to disappear. The marble outcrop is over 320 ft wide and 2240 ft long. Folding is complicated, the marble bands having acted as incompetent beds.

An iron ore deposit occurs about 1½ miles slightly north-east of the quarry face. The ore is haematite and the reserves down to 80 ft must be of the order of 1–2 million tons for an outcrop area of 160 × 960 ft.

The foregoing observations are based on a report by F. L. Smith and Co. presented by Lt.-Col. H. W. Dixon, General Manager of the Sudan Portland Cement Company; a magnetometer survey made by Dr. I. R. Qureshi, University of Khartoum; and on personal observations by the author.

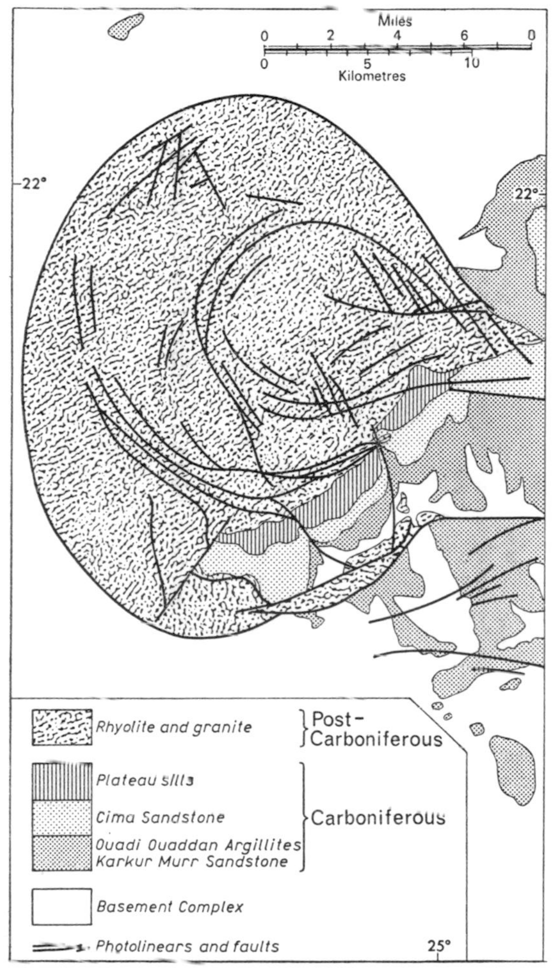

FIG. 20. Jebel Uweinat ring structure (Based on Burollet 1963)

The ore replaces limestone and was formed most probably by percolation of iron-rich solutions derived from the iron-rich Nubian Formation, which has now been stripped off. Outcrops of Nubian Formation occur to the south and west of the marble quarry.

(1) The bend of the Nile and the region from the Second to Fifth Cataracts

Basement Complex is exposed along the bend of the Nile, and in the region east of it, between Wadi Halfa, Akasha, Dongola, Merowe, Abu Hamed, and Atbara. It is difficult, however, to build up a comprehensive picture from the information at present available.

Details of sections exposed along the river, especially in the cataract zones, are given by Hume (1925, 1934, 1935, 1937). The account that follows is based on Hume's data and on personal observations made by the author on a reconnaissance expedition from Khartoum to Wadi Halfa, via Dongola and Abu Hamed in 1961 (Fig. 22).

A variety of schists, gneisses, granites, and to a lesser extent marbles, dolerites, and epidiorites crop out in this area. Structural trends are predominantly north–south for long distances along the Nile but there are marked deviations, often in the cataract areas (Berry and Whiteman 1968).

Llewellyn (1903) described the gold-bearing country east of the Wadi Halfa Railway between 20° to 22° N, and 32° to 34° 30′ E. The rocks are grey granites merging into gneisses, overlain unconformably by an extensive series of micaceous, talc, and hornblendic schists and 'shales', clay slates, arkoses, and crystalline limestones interbanded with gneiss. There are broad areas of intrusive igneous rocks, and some basic (?) extrusive igneous rocks (Plate 5).

Hume correlated the grey granite with the 'Fundamental Gneiss of the Cataracts', and the schists, etc. with the Schist Group of the Nile Valley between Semna and Wadi Halfa.

The Second Cataract area (Plate 5) is composed essentially of diorites in the north and of dolerites and gabbros in the south. The older diorites are penetrated by younger diorite dykes and by felsitic rocks. Their relationships with the schists south of the Second Cataract are uncertain. To the west the diorites are bounded by granite but to the north and north-east the whole complex passes beneath the Nubian Formation.

Fine-grained mica-schists cut by dolerite dykes occur along the banks of the Nile from Gemmai to Sarras. The beds are often vertical and in places are clearly overturned. In the prominent bend of the Nile at Murshid, where the river swings through

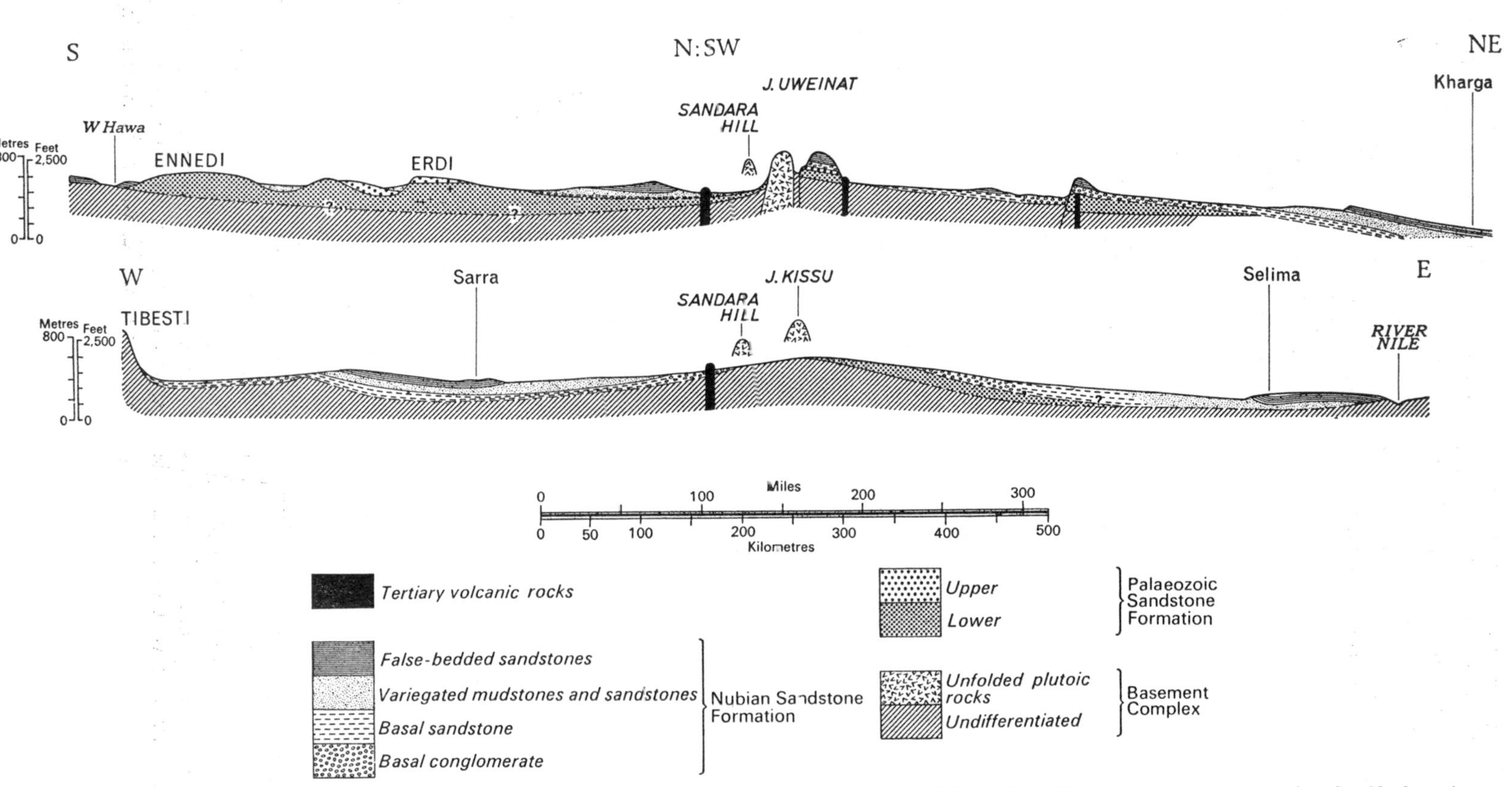

FIG. 21. Ennedi, Erdi, Jebel Uweinat, Kharga Oasis geological section and Tibesti, Sarra, Jebel Kissu, Selima, River Nile geological section (Based on Sandford 1935)

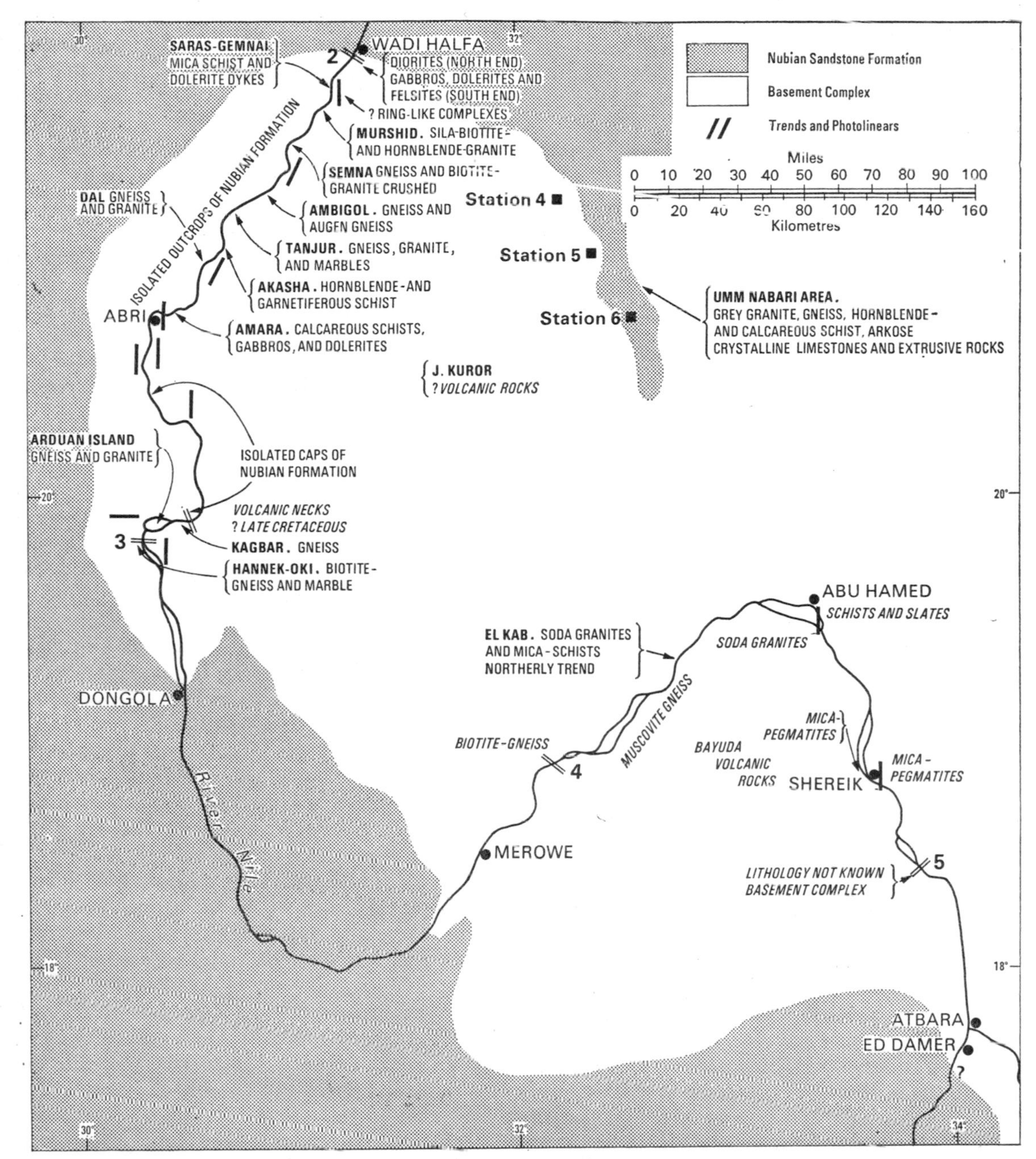

FIG. 22. Second to Fifth Cataract Region, Nile Valley, reconnaissance geologic map. Numbers refer to the traditional cataract areas (Based on data collected by the author on Khartoum–Wadi Halfa reconnaissance expedition 1961 and Hume 1925–37)

almost a right angle, the Sila granite occurs. This is a biotite–hornblende granite that is locally highly foliated. The joints in the granite have enabled the river to cut an east–west course, whereas north of Murshid the meridional trend of the schists controls the course of the river (Berry and Whiteman 1968).

Prominent ring-like structures occur in the desert, south-east of Murshid. These are clearly visible from the air and appear to be as large, if not larger, than the ring complex at Sabaloka, near Khartoum. According to G. Andrew (personal communication) many of these structures when inspected on the ground are granitic intrusions that dip outwards from the centre. Some, however, appeared to the author from the air to be volcanic features.

South of Sarras the road swings into the desert and passes through a sequence of schists and gneisses folded into a broad but complex anticlinal structure.

The Semna Cataract, famous for the Nilometer found by Lepsius in 1842–3, is formed by a bar of hard red and grey gneiss and crushed biotite-granite, which confines the Nile, here about 1280 ft wide, into a channel only 128 ft wide.

The Ambigol Cataract is composed of gneiss, and augen gneiss occurs in the Atiri area. Gold has been mined in the Dowshat area south-east of Atiri and elsewhere (see Chapter VIII). The workings were abandoned because, among other things, of the rise in the Nile consequent on the construction of the Sayed Ali Dam at Aswan. South and east of the Cataract the strike is highly variable (Plate 3).

The Tanjur Cataract, between Akasha and Ambigol, is composed of southward-dipping foliated gneisses. The whole mass is veined with granite, and to the east highly metamorphosed marbles occur interbanded with gneiss. West of the river there are hornblende gneisses.

At Akasha hornblendic schists crop out and there are garnetiferous schists with quartz veins. The general strike is north–south, apparently dipping towards the river. There are also hot springs at Akasha but little is known about their origin. Lepsius (trans. Homer 1853) recorded a temperature of 122° F and noted that the springs were sulphurous.

The Dal Cataract is formed of granitoid gneiss in the south and porphyritic granite and gneiss in the north. The hills between Ferka and Dal are composed of schist and gneiss, easily eroded by the river.

The Nile Valley from Ferka to Kosha is formed of mica-schists but these extend only a short distance eastwards and disappear south of the Dal Cataract. The banks are then formed of red and grey gneiss.

At the Amara Cataract both banks of the Nile are composed of highly calcareous schists, and the east–west bend of the river cuts across schists striking north–south. Gabbros and dolerites also follow the strike, giving rise to rapids, and numerous steep watercourses have been cut in the soft schists that dip towards the Nile at 25°–45°.

The Kagbar or Kaibar Cataract occurs between Arduan Island and Delgo where the Nile again makes a prominent bend at Fareig. Navigation is barred in the spring and early summer by a strip of gneiss that rises from beneath the Nubian Sandstone banks. The strike is apparently 335° but swings nearer to north in the Delgo area.

Arduan Island is composed of gneiss and granite and in the Jebel Barga area granite jebels and pediments are spectacularly developed (Plate 9).

The Third or Hannek Cataract is situated north of Kerma and is formed of biotite gneiss, locally associated with marble bands to the west. The gneiss strikes almost east–west across the river, and is traversed by compact granitic bodies which, because of their superior resistance, form ridges at right angles to the river giving rise to rapids. There are local variations of strike, as on Musul Island where the trend changes to 022° (Plate 9).

South of Abu Fatima's Tomb, near Dongola, the Nile flows over Nubian Sandstone, which occupies the bend of the Nile as far as a point just above the confluence with the Wadi Muqaddam, below the Fourth Cataract (Plate 12).

The Fourth Cataract is composed of biotite gneiss, generally very friable and seamed by masses of pegmatite and quartz veins. The strike of the biotite gneiss at Jebel Us varies between 000° and 030°.

Between El Kab and the Fourth Cataract the rock is a highly foliated muscovite gneiss and mica-schist, and in the Abu Hamed district there are white and pink marbles interbedded with the schists and gneisses.

North of Abu Hamed, near Umm Nabari near Station 6, the tectonic trend is mainly 090°, but varies to 045°, and the schists dip off a calcareous group of schists underlain by rhyolites (Plate 2).

At Abu Hamed, the main trend of the schists and

slates is 000°, while at Shereik, Rubatab area, it is variable but predominantly 000°.

The Fifth Cataract, near Karaba is composed of Basement Complex rocks. No details are available.

North of the Bayuda district, forming the core of the bend of the Nile, there are soda granites intruded into northerly-trending gneisses and schists.

Generalized tectonic trends are shown for the Basement Complex. These indicate that in the northern Sudan, west of the Nile, the predominant trends are meridional, swinging to east-north-east near the Egyptian frontier in the Red Sea Hills (Figs. 4 and 22).

(m) Hofrat en Nahas, Darfur Province

The copper mines of Hofrat en Nahas (09°48′N, 24°05′E), in south-western Darfur have been known for many years (Fig. 23). They appear to have been worked by the Kreish natives but the mines were abandoned before 1899.

Rüssegger reported on the area between 1835 and 1839, having been commissioned by Mohammed Ali Pasha to do so. Purdy's expedition visited the area in 1876, and in 1903 Colonel Sparkes reported that the mine had been abandoned.

A concession was granted in 1920 to the Nile–Congo Divide Syndicate, a subsidiary of Tanganyika Concessions, and until 1925 prospecting for copper and other minerals such as gold was carried on at Hofrat en Nahas and to the south-west.

T. D. Guernsey and P. E. Fairbairn made a reconnaissance expedition in the area in 1948 on behalf of Anglo-American Corporation of South Africa, Ltd. and an investigation was made by geologists of the German Democratic Republic in 1956 in the course of their work in Bahr el Ghazal and Darfur provinces.

The Sudan Geological Survey started its own investigation in 1957 and continued into 1960. The results of these and previous investigations are embodied in a memoir written by Afia and Widatalla (1961). This account is based on their work. Latterly both Italian and Japanese interests have worked in the area. In 1967–9 UNDP, at the request of the Sudan Government, investigated the area under the direction of A. M. Quennell.

According to Afia and Widatalla (1961) the country-rock is composed of metasediments and associated intrusions of presumed Pre-Cambrian age. The rock-types present in the mine area include chlorite schist, sericite schist, acid gneisses, amphibolites, and talc schists. Subsequently, the rocks were altered and sheared and mineralization took place along the shear planes. The ore is mainly chalco-

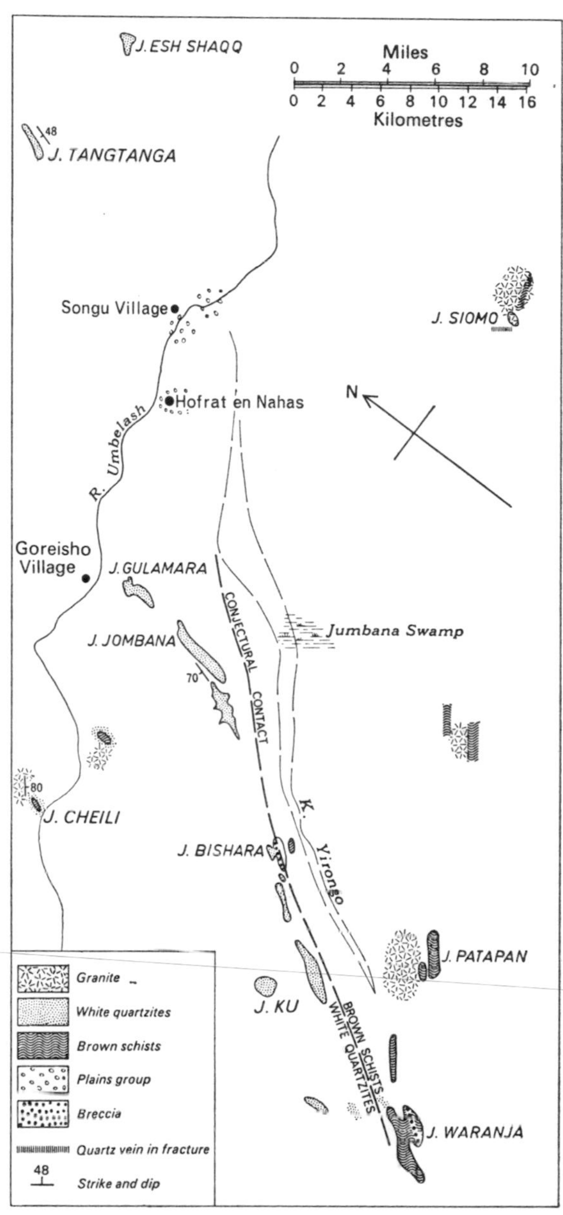

FIG. 23. Hofrat en Nahas copper area, south-western Sudan, geological map (Based on Nile–Congo Divide Syndicate data and Afia and Widlalla 1961)

pyrite and pyrite in a gangue of quartz and calcite. Intense tourmalinization has taken place. Gold and uranium minerals also occur in the area. A. M. Quennell, after a preliminary reconnaissance of the area in early 1968, concluded that the deposit may be of the porphyry type (Quennell, personal communication, March 1968).

Guernsey and Fairbairn (Afia and Widatalla 1961, pp. 25–33) recorded the following lithological subdivisions:

3 Intrusive rocks — mainly granite dykes; hornblende syenite; tourmalinized, limited extent.

2 The Younger Group — consisting of feldspathic quartzites and other quartzose rocks; brown schists probably of volcanic origin.

1 The Older Plains Group — consisting of granites, hornblende, and chlorite schists of igneous origin; basic intrusive rocks; crystalline quartzites; grey schists.

Older Plains Group. Outcrops are not common on the plains in Hofrat en Nahas area. Foliation strikes at 083° and dips northwards at 30–80° in general. In the grey schists local variations occur and the foliation strikes from 015° to 054° and dips from 30° to 65° to the north-west.

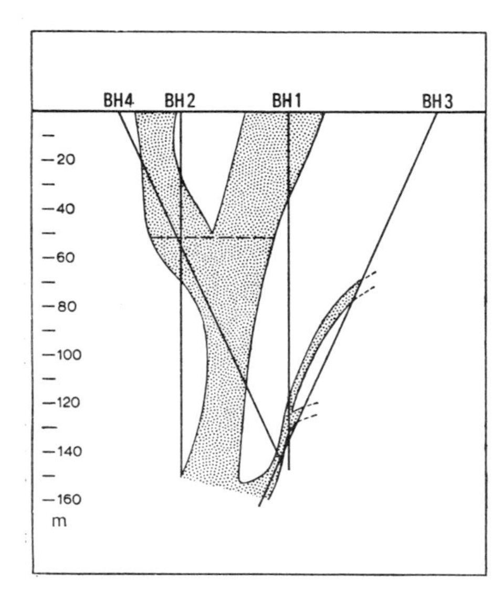

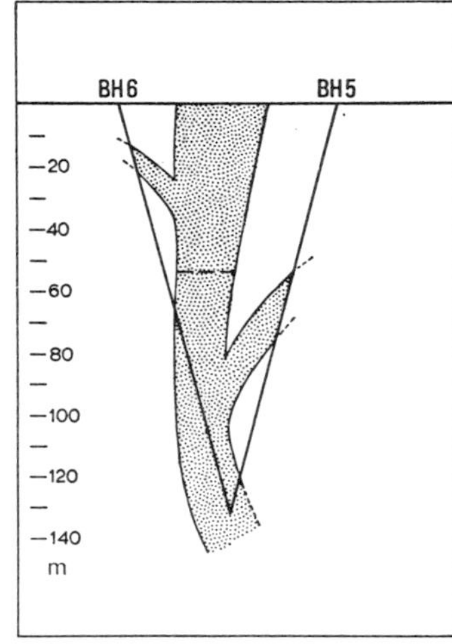

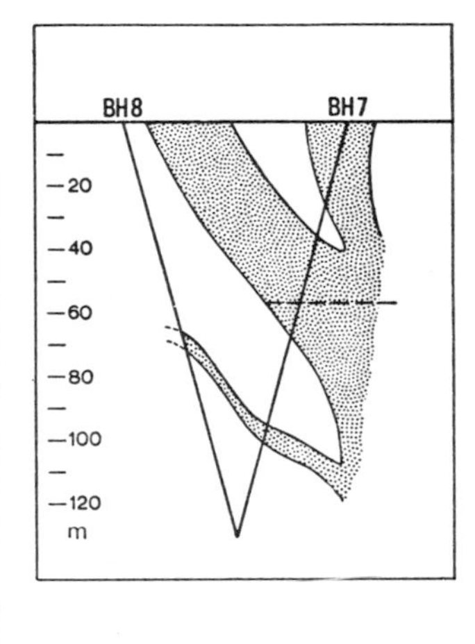

Vertical and Horizontal Scales Equal

Pecked line indicates beginning of sulphides

Depths are given in metres from ground level

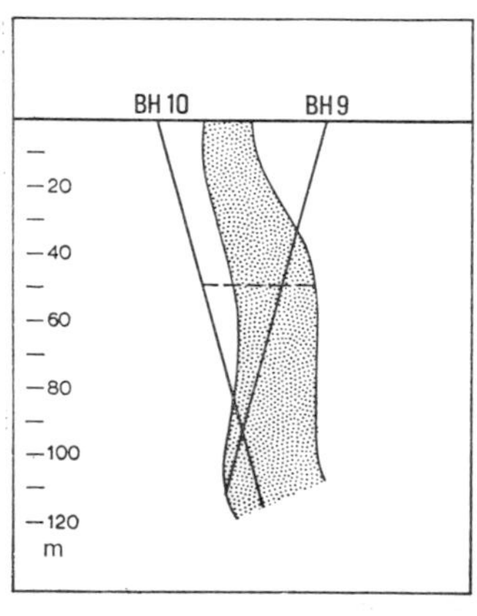

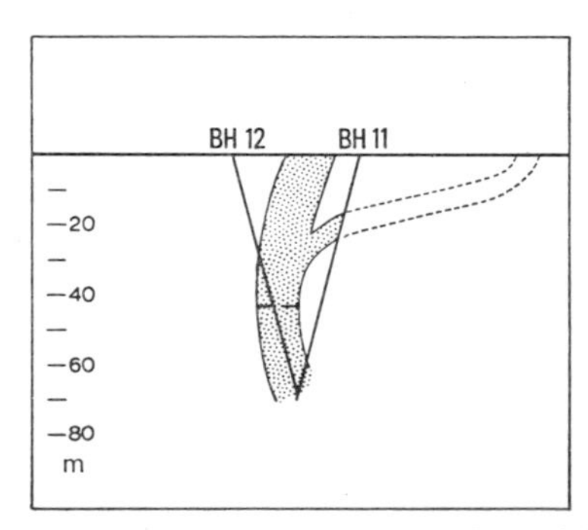

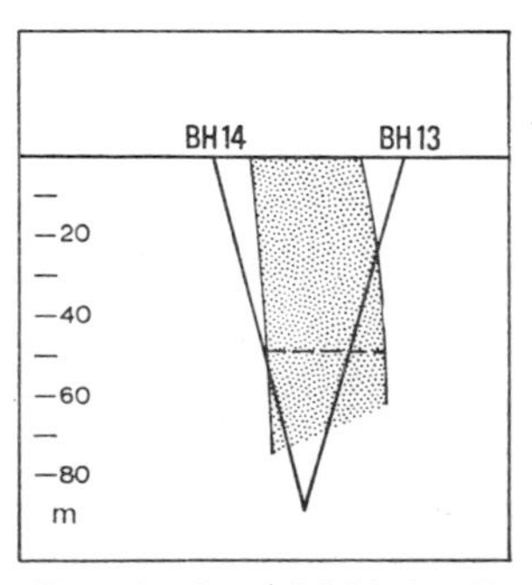

FIG. 24. Hofrat en Nahas, geological sections showing extent of mineralization mainly proved by Nile–Congo Divide Syndicate (Based on Afia and Widatalla 1961)

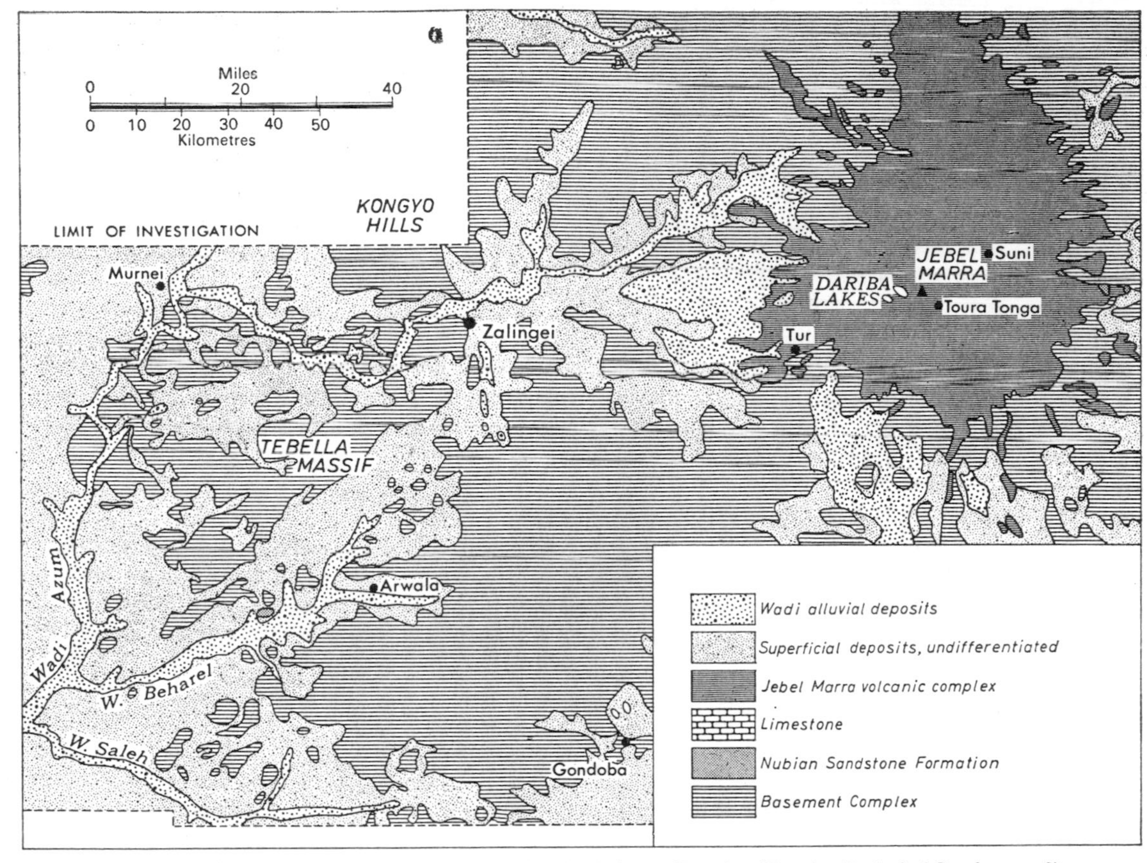

FIG. 25. Jebel Marra, western Sudan, general geological map (Based on Hunting Technical Services 1958)

In the 'Younger Group' the quartzites strike from 355° to 040° with a concentration in the direction of 020°. Dips are generally steep and to the west. Multiple fracture transverses to the strike are common in the harder quartzites.

(n) Jebel Marra region

The only published account dealing with Basement Complex of this region was written by the geologists of Hunting Technical Services (1958, pp. 29 32) (Fig. 25, Plate 17).

The general subdivision of the Basement Complex given by Andrew (1958) was adopted for the region but information so far available is insufficient to enable correlation to be made outside the area.

Hunting Technical Services (1958) recorded the following sequence:

ZALINGEI–NYERTE ROAD

3	Coarse-grained biotite injection gneiss with pink *lit-par-lit* granite	Structurally highest strata
2	Fine grained biotite–feldspar schist with occasional *lit-par-lit* veins of granite and cross-cutting aplitic veins	
1	Flaggy quartz–feldspathic gneisses with *lit-par-lit* granite veins	Structurally lowest strata

This section is exposed for approximately 5 miles across the strike. The general foliation direction in this section is 090° with northerly dips, swinging round to 045° with north-westerly dips at the eastern end of the section. These rocks are assigned on lithological grounds to Andrew's Group I Crystalline Schists and Gneisses.

The Paraschist and Gneiss Group includes calc-silicate gneisses, fine-grained garnetiferous biotite granulites, graphitic rocks, garnetiferous mica-schists and staurolite schists. Most of these rocks are relatively resistant in weathering and form hills, in contrast to the Crystalline Schists and Gneiss Group, which generally occupies low ground.

In places these beds are folded isoclinically and the geologists of Hunting Technical Services (1958) are of the opinion that a case could be made for separating these Basement rocks into two groups on degree of metamorphism and folding.

Both foliated and unfoliated igneous rocks occur in the region. The former are mainly granitic in composition but there are some basic and ultrabasic rocks forming minor intrusions. The unfoliated rocks include acid and basic types. Some of the basic rocks may be of Tertiary age, however. A large diorite mass occurs near Guldo and there are extensive areas of granitic rocks near Kalokitting. Pegmatite and aplite dykes are numerous.

(o) Classification of the Basement Complex formations of the Sudan

As mentioned above, the name 'Basement Complex' is used to include those igneous metamorphic and sedimentary rocks that are overlain by horizontal and subhorizontal Palaeozoic or Mesozoic sedimentary rocks. They are assumed to be mainly of Pre-Cambrian age although some formations are known to be younger.

In Table 2 the classifications of Basement Complex rocks proposed by Andrew (1948) for the southern Sudan; Wilcockson and Tyler (1933) and Ruxton (MS 1956) for the Gedaref and Sennar area; Andrew (in Ruxton MS 1956), Delany (1952), and others for sheet 55; Andrew (1948) for the Sudan north of latitude 6° N; Delany (1956) for the southern Red Sea Hills, Derudeb sheet; Gass (1955), Ruxton (1956), Gabert, Ruxton, and Venzlaff (1960), Kabesh (1962), and Lotfi and Kabesh (1964) are summarized.

No order of superposition is implied for the southern Sudan, south of 6° N, in Andrew's classification. Because of our lack of knowledge, the discontinuity of outcrops, the extensive superficial cover, absence of isotopic dates, etc., this sequence must stand alone.

The rock types are similar to those encountered in the Congo, Ethiopia, Kenya, and Uganda but so little work has been done in the frontier zone that it is impossible to make correlations, except for distinctive formations such as the Madi quartzite that cut across the frontier. The contact between the dioritic and charnockitic gneisses shown on the Uganda Geological Survey 1:125 000 map has not been followed into the Sudan, nor have any of the formations mapped by Hepworth (1964) in West Nile Province in Uganda.

Columns 2–5 of Table 2 all more or less apply to the same area—the north central Sudan. Broadly speaking, a fivefold division of the Basement Complex may be recognized. Again much more work needs to be done, as vast areas are mapped simply as 'Gneiss, Undifferentiated' or as 'Green Series'. Equally vast areas are covered by thick residual deposits of Butana clay. On lithological grounds the Odi schists of the Derudeb sheet may be correlated with the paraschists of Andrew (1948) and the gabbros that crop out in the central part of the Tehilla ring complex (Delany 1955) may well have been intruded about the same time as the Qala en Nahl gabbros and ultrabasic rocks. Whether the unfoliated granites and the soda granites of the two areas are synchronous is uncertain at present, but they are shown as such in Table 2.

Four sequences have been proposed for the northern Red Sea Hills for adjacent areas on the Deraheib, Dungunab, Mohammed Qôl, and Port Sudan sheets (Table 2). Considerable differences exist between them; the main ones are between the sequence and dating proposed by Gass (1955), who mapped the Dungunab sheet, and between Delany (1955), Ruxton (1956), Gabert, Ruxton, and Venzlaff (1960), Kabesh (1962), and Lotfi and Kabesh (1964), who mapped adjacent areas.

Gass (1955) assigned the younger granites, granophyres, and quartz–syenites forming bosses, stocks and ring-dykes of the Dungunab sheet to the Upper Jurassic–Lower Tertiary interval, citing as evidence that the Nubian Sandstone (to which he assigns a Jurassic–Upper Cretaceous age) has been affected by an iron mineralization intricately associated with the younger intrusive rocks of Bir Salala and Umm Shibrik areas (Fig. 18). Additional support for this is said to be gained from 'the fact that aegirine trachytes associated with the same soda-rich igneous episode are found intruding through rocks of the Nubian Series. This is especially well seen at Jebel Gainbal where a plug of trachyte is intruded into the Nubian sandstones and shales.'

Delany (1955), Ruxton (1956), and others, however, consider that all the ring structures in the Red Sea Hills belong to the Basement Complex and are probably Pre-Cambrian in age. Another difference is that Gass (1956) assigned the Asotriba volcanic rocks to the Tertiary but Gabert, Ruxton, and Venzlaff (1960) placed them in the Pre-Cambrian Awat Formation.

Lotfi and Kabesh (1964, pp. 93–9) have proposed 'A new classification of the Basement Rocks of the Red Sea Hills, Sudan'. No new field evidence was presented in this paper, however, to justify the changes from the sequence proposed by Ruxton (1965), and indeed from the sequence proposed by one of the authors (Kabesh 1962). Lotfi and Kabesh (1964) claimed that 'three geosynclinal phases' can be recognized 'separated by orogenies and unconformities' in the northern Red Sea Hills and place the Awat Group above Ruxton's Phase 8 'acid and basic sills and dykes' (Table 2, Column 7). These dykes form part of the regional north–south to north-north-east–south-south-west trending dyke swarm, which apparently marks the last major igneous Pre-Cambrian event in the area.

According to the classification proposed by Lotfi and Kabesh (1964) the dykes belong to two groups: post and pre-Awat. Certainly in the Jebel Awat area such a separation cannot be effected; the author takes the view that Ruxton's statement that 'At 20°22′N, 36°23′E horizontal acid volcanics (Awat) overlie the Nafirdeib Series which is striking east–west, with a vertical dip' is correct, and that the Phase 8 and dykes form a single group as Ruxton proposed.

Another point that arises concerning the Lotfi–Kabesh classification is the basis on which they have separated the formation grouped as 'first geosynclinal deposition' and 'second geosynclinal deposition'. An unconformity is said to be well displayed in some localities of the areas examined, as in the extreme northern part of the Mohammed Qôl sheet. Curiously enough Kabesh (1962, p. 25) in the *Mohammed Qôl Memoir* has written: 'Unfortunately, no internal succession has been worked out for the Nafirdeib Series, by the present author', and certainly an unconformity in this position is not shown on the published map.

In the author's view acceptance of the classification proposed by Lotfi and Kabesh (1964) should be withheld until more modern structural mapping is done in the northern Red Sea Hills of the Sudan. No doubt we are dealing with a thick series of geosynclinal deposits that have had a long history of folding, intrusion, and metamorphism, but the evolution of the geosyncline, etc. cannot be worked out clearly until we have much more field evidence and a network of isotopic dates. These last remarks apply equally well to erecting a correlation of the Pre-Cambrian rocks of the Sudan Republic as a whole, and to correlation with adjacent countries.

2. PALAEOZOIC FORMATIONS

Palaeozoic rocks have been recorded from three main areas in the Sudan (Fig. 4):

(a) the continuation of the Erdi–Ennedi outcrops of Chad and Libya, between 16°N and 20°N along the western frontier;

(b) the Jebel Uweinat area extending into Libya and Egypt; and

(c) the outcrops of the Nawa Formation, rocks of doubtful Palaeozoic age in Kordofan.

In all these areas, the Palaeozoic Formations rest unconformably on the Basement Complex.

(a) The continuation of the Erdi–Ennedi outcrops

In Chad and the Sudan, Palaeozoic sandstones and other sediments extend from Libya around the south side of Tibesti into the Erdi and Ennedi regions, where Sandford (1935, Plate 36) labelled them 'Palaeozoic sandstones and sandstones of uncertain age (pre-Nubian)'. Sandford (1935, p. 335) rejected the idea presented by Lynes and Campbell-Smith (1921, p. 208) that comparative weathering had a chronological value and could be used to correlate sandstones of the jebels of Erdi (Chad) with those of Nahud and El Fasher (Sudan). Sandford believed that the rocks described by Lynes and Campbell-Smith from the Lugud–Abiad district are Nubian, and not Palaeozoic, and that Palaeozoic sandstones occur beneath the Nubian Formation of Darfur and Kordofan. However, in the author's opinion, thickness considerations make this unlikely in Kordofan and it is probably true only to a limited extent in Darfur.

For the Murdi Depression, the region that forms the boundary between Erdi and Ennedi, Sandford (1935, p. 336) gave the following succession:

MESOZOIC	8 Buff sandstone (Nubian Series) 9 Variegated white and purple shales and mudstones (Nubian Series)
UPPER PALAEOZOIC	6 Grey limestone 5 Soft, light grey sandstones 4 Black sandstones with ripple marks 3 White sandstones 2 Hard black shales
LOWER PALAEOZOIC SANDSTONES	1 Massive sandstones of the southern side of Murdi, Guroguro, Ennedi, and central Erdi showing fantastic weathering

In Figs. 21 and 26 the Palaeozoic sandstones are shown pinching out eastwards into the Sudan. Elsewhere this pattern has been confirmed by the Royal Dutch–Shell–BP geological party working in the north-western Sudan in 1959. As only a few non-diagnostic fossils have been found in the Erdi–Ennedi area the age is uncertain.

On the 1963 edition of the 1:5 000 000 Geological Map of Africa, however, these outcrops are shown as Devonian and Ordovician, but there is little evidence for this and the rocks should have been shown simply as Palaeozoic undifferentiated.

(b) Jebel Uweinat area

Lower Carboniferous sandstones were recorded by Frittel and Carrier (1924) on the eastern side of Ennedi; and Menchikoff (1926, 1927), as a member of the Hassanein Bey expedition of 1924–5, recorded the following section at Uweinat:

JEBEL UWEINAT

	Thickness (ft)
4 Plant-bearing sandstones	9·6
3 Decayed rhyolite	16
2 Sandstone resting on an eroded surface	6·4

UNCONFORMITY

1 Schist, gneisses, granites, etc.

Sandford (1935, p. 339), combining his personal observations with those of Menchikoff, gave the following succession:

JEBEL UWEINAT

5 Nubian Series, buff, brown, false-bedded sandstone
Conglomerate with fragments of all underlying beds
4 Lower Carboniferous plant-bearing sandstones
3 Rhyolite
2 Lower Palaeozoic sandstone

UNCONFORMITY

1 Pre-Cambrian schists, granite, etc.

All the divisions are apparently separated by unconformities and the only fossils that were recorded are *Archaeosigillaria vanuxemi*, mentioned by Fritel, and a lepidodendron flora, identified by W. N. Edwards of the British Museum (Natural History) (see Chapter III, Palaeontology).

The most up-to-date information available on the Jebel Uweinat area is that given by Burollet (1963, pp. 219–27) and by Vittemberga and Cardello (1963, pp. 228–40).

At Jebel Archenu, 25 miles (40 km) north-west of Jebel Uweinat, some 640 ft (200 m) of Cambro–Ordovician sandstones, attributed to the Hassaouna Formation, were recognized resting on the crystalline Basement Complex. Tigillitic sandstones form the lower 320 ft (100 m) of the Hassaouna Formation; the upper 320 ft (100 m) are coarse-grained and conglomeratic. Apparently *Tigillites* is taken to indicate a Cambro–Ordovician age by analogy with French practice in Algeria (Table 5).

The Memouniat Formation (said to be upper Ordovician) unconformably overlies the Hassaouna Formation. It is over 480 ft (150 m) thick and consists of alternations of fine white and coarse conglomeratic sandstones. These formations are said to have been uplifted on the flanks of Jebel Archenu by the post-Carboniferous intrusion that occupies the Uweinat area. The Hassaouna and Memouniat formations are equated with the Hauaisc Sandstone of Jebel Hauaisc, north-west of Kufra oasis (Table 5).

Burollet (1963, pp. 221–5) recognized the following Palaeozoic succession in the Jebel Uweinat area, 4920 ft (1540 m) thick.

JEBEL UWEINAT

Upper sandstones (? Lower Cretaceous)

10 Nubian Formation	Chieun limestones (Wealden) Lower sandstones (Jurassic)
9 Jebel Uweinat ring complex Post-Carboniferous	Granitic and syenitic rocks showing little contact metamorphism; intruded into all the formations mentioned below; said to be similar to the Sabaloka ring complex near Khartoum

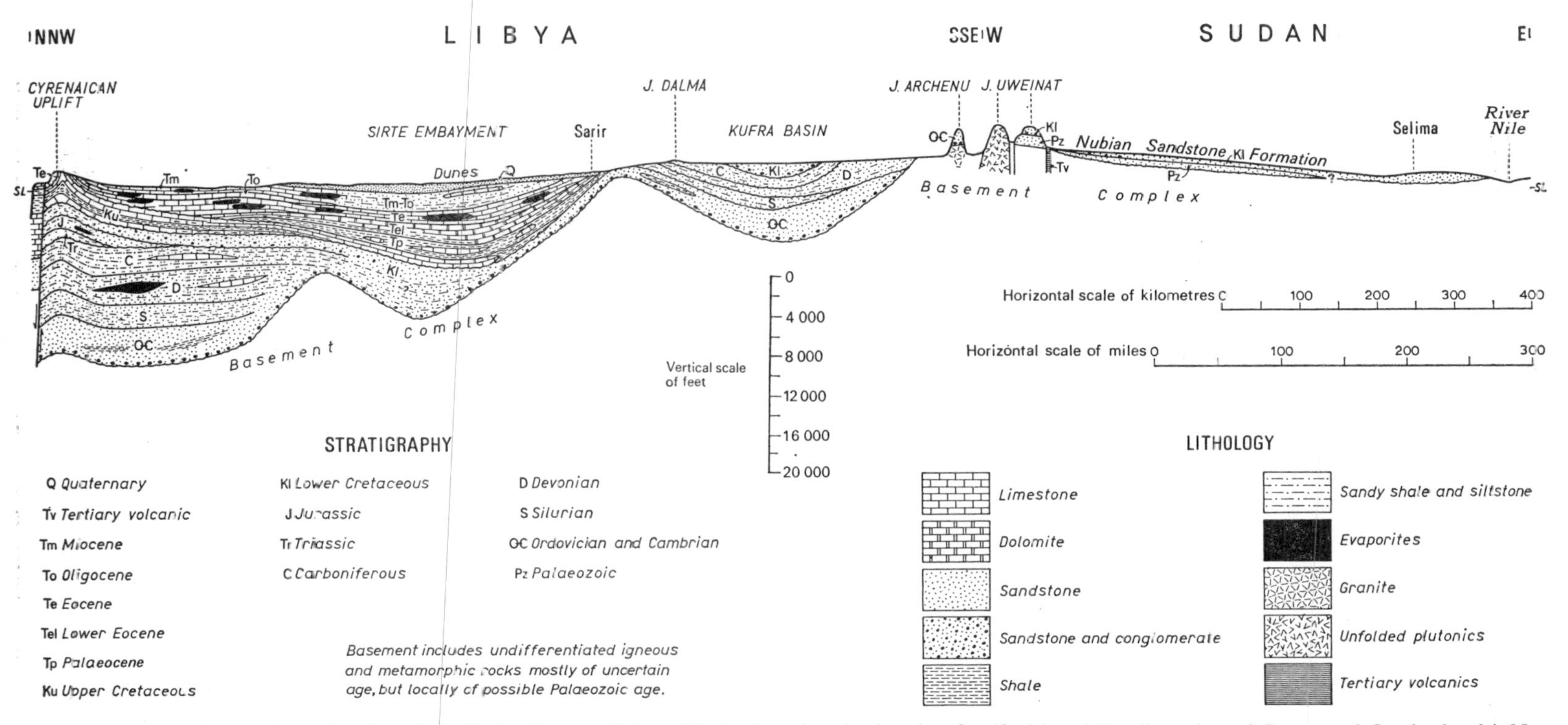

FIG. 26. Libya to Sudan. Cyrenaica–Sirte–Sarir–Kufra–Uweinat–Selima–Nile (geological section based on Sandford (1935) Burollet 1963, and Conant and Goudarzi 1967). Note the prominent Archenu–Uweinat high and the thin Palaeozoic–Mesozoic sequence in Sudan which makes it unlikely that much ground water migrates from the central Sudan to Egypt at the present day

TABLE 5

Correlation chart for Palaeozoic and Mesozoic formations in Ethiopia, the Sudan, and Egypt

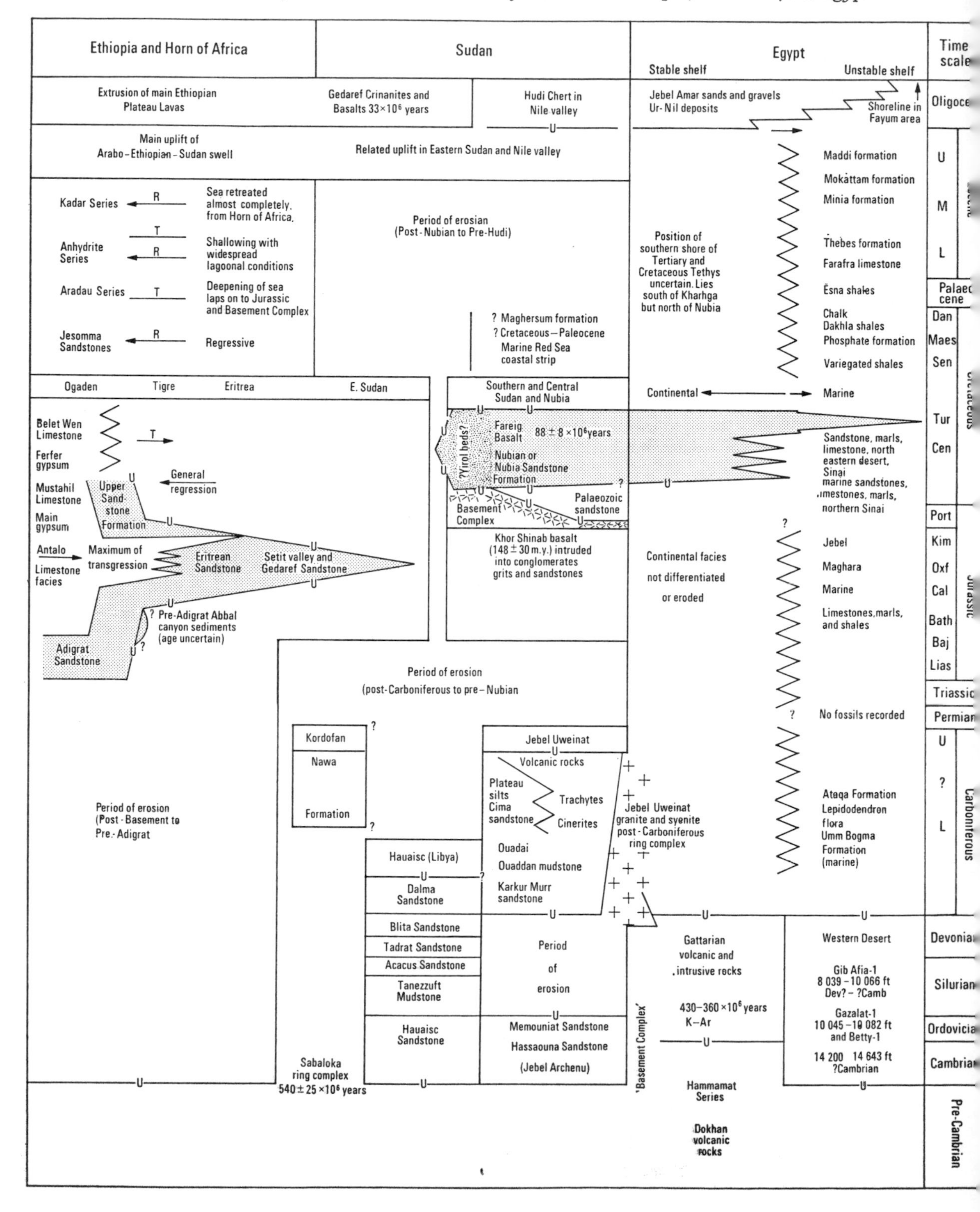

8 Post-Carboniferous extrusive rocks	Trachytes (more or less altered) ranging to rhyolites, and cinerites with some volcanic breccias; interbedded with some sands and gravels. Thickness approximately 1600 ft (500 m)
7 Plateau siltstones	Black silts and mudstones, interbedded with sandy layers intercalated with beds of trachyte. Thickness approximately 192 ft (60 m)? Carboniferous
6 Cima sandstones	Quartzitic sandstones, massive, cliff forming, affected by contact metamorphism. Thickness 1184 ft (370 m)?
5 Argillites Ouadai Ouaddan	Alternations of siltstone, argillite, and sandstone. Thickness 896 ft (280 m)? Carboniferous
4 Karkur Murr sandstones	Sandstone, conglomeratic, and quartzitic, with siltstones and fine sandstones, white to rose, micaceous, plant-bearing with *Archaeosigillaria* cf. *vanuxemi*. Thickness 256 ft (80 m) Lower Carboniferous and Upper 256 ft Devonian
PERIOD OF EROSION	
3 Memouniat Formation	Alternation of sandstone, fine, white, and conglomeratic sandstones, thickness 450 ft (150 m) Upper Ordovician (?) Jebel Archenu
PERIOD OF EROSION	
2 Hassaouna Formation	Sandstone, coarse-grained, conglomeratic. Thickness 320 ft (100 m)
PERIOD OF EROSION	
1 Basement Complex	Assumed to be mainly pre-Cambrian in age

North-westwards in Libya, north of Kufra Oasis, Cambro–Ordovician, Silurian, and Devonian conglomerates, sandstones, and shales are present (Fig. 26 and Table 5) and are greater than 1280 ft (400 m) thick on the south-west flank of the Kufra Basin.

The Palaeozoic sediments in the Kufra–Uweinat region are mainly detrital and consist of conglomerates, sandstones, mostly coarse-grained, siltstones, and shales. Limestones and marine fossils are rare and the sequence is thought to have been deposited in shallow water probably less than 320 ft (100 m) deep (Burollet 1963).

How far the Palaeozoic seas extended into the Sudan is uncertain because of erosion. Southwards and eastwards from Jebel Uweinat the Nubian Formation rests directly on the Basement Complex, and Hottinger *et al.* (1959) have stated that the marine Palaeozoic transgressions recorded in Libya did not reach the Sudan. This is borne out by observations by Burollet (1963), Vittemberga and Cardello (1963), and Freulon (1963, p. 164).

Hottinger *et al.* (1959) discussed the pre-Nubian Sandstone sequence of the Uweinat and Mourdi areas and stated that these rocks are at least 3200 ft (1000 m) thick and that they decrease in grain size from south to north, indicating a source area probably south of Uweinat. Eventually they pinch out in the Sudan, and at Jebel Rahib (18° N, 270° E) and many localities further south the Nubian Formation rests directly on Basement Complex.

The thin sedimentary column and the fact that the erosion has cut through to the Basement Complex make this region a poor prospect for oil. Furthermore, if the Carboniferous age for the intrusion of the Jebel Uweinat ring complex given by Burollet (1963) is correct, then the petroleum prospects are reduced even further because of increased geothermal gradient in post-Carboniferous times (Chapter VII).

(c) Nawa Formation

This formation consists of gently dipping sediments ranging from mudstones to arkosic grits; purple to green in colour, abundantly micaceous, and containing up to 25 per cent undecayed feldspar. Abundant fresh biotite, epidote, apatite, sphene, and detrital chlorite have been recognized in thin section. Some of the mudstones are false bedded, and all the rocks are well compacted, unmetamorphosed, and uncleaved. This is in contrast to the rocks of the Basement which are in advanced state of metamorphism (Andrew 1945, 1948, and unpublished MS 1943; personal observations).

The Nawa Formation is known mainly from boreholes in Er Rahad District of Kordofan taking its name from the small village of Nawa in the Khor Abu Habl basin. Its maximum proved thickness is

443 ft (135 m) from surface and subsurface data (Rodis 1954 *et al.* p. 27). Relationships with the Basement Complex were not proved in the boreholes but the unmetamorphosed state of the formation indicates that it is younger than the crystalline rocks. At Sameih Station, boreholes proved a faulted boundary against granite (Andrew op. cit.). The freshness of the Nawa deposits probably indicates that they are locally derived. Field relationships show that they have been preserved mainly because of faulting.

Rodis *et al.* (1964, pp. 18, 19, and 27), presumably basing their opinion on the presence of limestones in the Nawa sequence (mentioned for the first time in their paper but not described), postulated that after 'a period of prolonged erosion which apparently lasted through most of Palaeozoic time, shallow seas invaded parts of the region in the late Palaeozoic and deposited the sediments of the Nawa Series. Before the close of Palaeozoic time, however, the region was uplifted and most of the Nawa sediments were removed by erosion. Only a few isolated remnants of the Nawa Series are now left in Kordofan as evidence of this once-extensive geologic unit.'

As no evidence of the marine nature of the limestones is given in the paper, judgement must be reserved on the validity of this sweeping palaeogeographical statement. The author still tends to favour a local and continental origin for the Nawa Formation, as did Andrew (op. cit.). The limestones may be of freshwater origin.

In the absence of fossil evidence the Nawa Formation is best classed as Palaeozoic–Mesozoic undifferentiated. However, on the Geological Map of Africa, 1963 edition, it is labelled Karroo, irrespective of the fact that the nearest fossiliferous Karroo Beds lie some 12° of latitude to the south, and that there is little lithological similarity (Whiteman 1965).

Palaeozoic sediments are therefore absent from most of the Sudan, as they are from much of Egypt, Ethiopia, east Africa, and Saudi Arabia. This is perhaps one of the most striking features of African stratigraphy and is not readily explicable. Whether this vast area remained land during the Palaeozoic; or whether Palaeozoic sediments were deposited and subsequently eroded away; or whether Palaeozoic rocks (especially Lower Palaeozoic) are represented among metamorphic and igneous rocks that make up the Basement Complex are questions that cannot be answered at present.

The first explanation is unlikely, because if such a vast area had remained land for millions of years then some continental deposits would have been laid down and preserved, as the Karroo sediments are preserved to the south, for instance. If the second assumption is accepted then we should expect to find structurally preserved remnants of Palaeozoic rocks scattered throughout the region, but this is not the case. The last possibility can be proved only as more isotopic dates become available.

3. MESOZOIC FORMATIONS

The Mesozoic Formations of the Sudan include (Fig. 4 and Table 1):

(a) outcrops of Nubian Sandstone Formation in the north central Sudan, continuous with those of Libya and Egypt and with tongues extending as far south as Suq el Gamal, near Nahud; and Sennar and Singa in the east;
(b) Gedaref Sandstone Formation, continuous with the Adigrat Sandstone Formation of Ethiopia;
(c) Yirol Beds west of the Bahr el Jebel, Bahr el Ghazal Province;
(d) Fareig basalts; and
(e) Khor Shinab basalts and the Cretaceous sediments of the Red Sea coast.

(a) Nubian Sandstone Formation

(i) *Evidence bearing on age and definition*

The term *Nübischer Sandstein* was introduced into Egyptian stratigraphy by Rüssegger (1837). Nubian Sandstone or Nubia Sandstone are the English terms in common use. Nubian Sandstone, in Egypt and north-east Africa, came to be used for any Palaeozoic or Mesozoic Sandstone of uncertain age. The Palaeozoic sandstones of the Uweinat area and of the Gulf of Suez taphrogeosyncline were once referred to as Nubian Sandstone.

According to Said (1962, p. 129) the term 'Nubia Sandstone' or Nubian Sandstone is nowadays given to 'the brownish, highly dissected, and almost horizontal sandstone beds that are widely distributed over the southern parts of Egypt and Nubia in particular. . . . The true Nubia Sandstone, however, is known to be of Upper Cretaceous age. On lithological grounds, the term "Nubia Sandstone" should be retained as a formational name.'

In this definition, Said followed Hume (1906), Shukri and Said (1944), Shukri (1945), etc. who described the formation in Egypt.

Such a concept cannot be applied farther west in the Sahara, where both Nubian Sandstone and *Grès de Nubie* are used to include beds of Palaeozoic as well as Mesozoic age. Nor can it be applied in the Sudan where the Nubian Sandstone Formation has yielded only plant fossils of little chronological value, and where it is bounded above and below by unconformities of uncertain time value. The Nubian Formation rests on Basement Complex rocks of assumed Pre-Cambrian age over most of its outcrop. In a few places in the Sudan it is overlain unconformably by the Hudi Chert Formation of Lower Tertiary (?Oligocene) age, but one must travel at least 120 miles into Egypt before the Nubian Formation is succeeded by the Upper Cretaceous as at Kurkur Oasis, or farther north as Kharga and Dakhla oases, Aswan and Esna. A valuable discussion of the varieties of meaning that 'Nubian Sandstone' has acquired was given by Pomeyrol (1968). Whiteman (1970) and others oppose Pomeyrol's suggestion to abandon the term 'Nubian Sandstone' and Whiteman (1970) has proposed that the Nubian Sandstone Formation should be designated the Nubian Sandstone Group (abbreviated to Nubian Group). The type definition is given below and quoted in Whiteman (1970).

The Nubian Sandstone outcrops of Aswan have been described by many geologists (e.g. Ball 1907, Newton 1909, Barthoux 1922, Shukri and Ayouty 1953, etc.) and are divisible into three conformable units.

ASWAN SECTION

		Thickness (ft)
3	Shales, variegated, sandy shales, sandstones and quartzites, frequently false-bedded	65–275
2	Sandstones, sandy shales, ferruginous sandstones. Oolitic ironstone band at base succeeded by layers of refractory clay and a sandstone bed rich in oolitic iron; most of the fossils collected from Nubian Sandstone of Aswan are from this unit	30–70
1	Conglomeratic, kaolinized sandstone, and sandstones of varying grain size; concretions of barytes	130–175
	Total	225–470

~~~~~~~~ UNCONFORMITY ~~~~~~~~

Basement Complex

Among the fossils collected from unit 2 and identified by Newton (1909) and Cox (1956) are:

Inoceramus balli *Isocardia aegypticum* *Cyprina humei*	Marine forms
Unio jowikolensis *U. attiai* *U. nubianus*	Non-marine forms

Newton (1909) also records *Unio crosthwaitei* and *Mutela mycetopoides*; worm-casts *Galeolaria filiformis*.

Fossil plants occur in the Nubian Formation, and Edwards (*in* Attia 1955), and Barthoux and Frittel (1925) identified *Weichselia* sp., *Magnolia barthouxi*. Rüssegger obtained *Cyclas fara*. The plant fossils are taken to indicate an early Late Cretaceous (Cenomanian) age (Said 1962, p. 90).

The Nubian Sandstone Formation is exposed in the Western Desert and the following apparently conformable sections at Dakhla and Kharga Oases have been recorded (Said 1962):

DAKHLA OASIS

	Thickness (ft)
Dakhla Shale Formation	
Phosphate beds with *Inoceramus regularis, Trigonarca multidentata, Naera aegyptica, Plicatula aschersoni; Lamna libyca, Otodus biauiculatus, Corax pristodontus*, and *Stephanodus splendens*. Age according to Said (1962, p. 69) is Maestrichtian	67
Variegated shales	230
Nubian Sandstone Formation. Unfossiliferous brown sandstone, forming the floor of the oasis	384
Total	681

Provisionally the Nubian Sandstone Formation here is assigned to the Maestrichtian by Said (1963) but clearly it could be older.

KHARGA OASIS

	Thickness (ft)
Dakhla Shale Formation. The lower part of this formation is said to be Maestrichtian in age and the upper part Danian (Hassan 1956)	240
Phosphate formation. Phosphate rock shales with coprolites. No age given	22
Purple and variegated shale. Some fossil fruits but no other fossils are listed	112–160
Nubian Sandstone Formation. Brown unfossiliferous, sandstone proved in Baris well	1907
Total	2329
~~~~~~~~

To summarize, then, for southern Egypt, according to Said (1962) the palaeontological evidence indicates that the Nubian Formation may be pre-Maestrichtian, Cenomanian, or Upper Cretaceous in age, but elsewhere, when summarizing for all Egypt, Said (Table 1, 1962) assigns the Nubian Sandstone to the Turonian and Santonian.

In Libya in the Djebel Ben Ghnema and Dor el Gussa area the Nubian Sandstone Formation is a conglomeratic, coarse-grained sandstone with silt-stone and silty 'shale' bands. These beds have yielded the fossil plant *Cladophlebis zaccagnai* (Principi), which is supposed to indicate an early Cretaceous or Wealden age (Klitzsch 1963, p. 106), but Plauchut (*in* Klitzsch 1963) recorded Jurassic plants in the lower part of the Nubian Formation.

Farther west in the Murzuk Basin, Libya, the name Nubian Formation is given to the continental strata resting with unconformity on the post-Tassilien or older formations. The flora of the Nubian in this area is taken to indicate a late Jurassic to early Cretaceous age (Wealden).

Conclusive palaeontological evidence bearing on the age of the Nubian Formation in the Sudan is not available at present. The formation has yielded plant fossils and many of these are identifiable only at generic level, although it should be added that very little systematic work has been done on the flora. The most common plant fossils are silicified wood, which in places is strewn in great quantities over the weathered outcrops. Frequently cell structure, medullary rays, and annual rings are visible. No palynological work has been done.

Specimens up to 18 yards long have been recorded and in some localities roots and stumps in position of growth have been found. Pieces showing butts of branches are common also. The majority of specimens have been referred to the genus *Dadoxylon*. Near Jebel Umm Marrahi, 20 miles north of Khartoum, a 'silicified forest' covering a number of acres was found by the author.

Edwards (1926) identified the following fossils from Jebel Dirra, 47 miles east El Fasher, Darfur:

Weichselia reticulata (Strokes and Webb),
Fern fragment (Indet.),
Frenelopsis hoheneggeri (Ettinghausen),
Dadoxylon aegyptiacum Unger.

According to Edwards (1926) *Weichselia* and *Frenelopsis* suggest an early Cretaceous age (Neocomian or Barremian) but as he has pointed out *Weichselia reticulata* has been recorded from the Lower Cenomanian of Egypt, so the Nubian of Jebel Dirra may be of this age also (see Chapter III, Palaeontology).

Similar forms have been taken to indicate a Wealden or Lower Cretaceous age in the Sahara (de Lapparent 1954). Clearly then the definition given by Said (1962, p. 129) cannot be applied generally and perhaps the simplest way of dealing with the problem is to consider the Nubian Formation of the Sudan and southern Egypt as the continental facies of marine Mesozoic rocks that occur in northern Egypt and wedge out and rest unconformably on Palaeozoic or Basement Complex rocks towards the Nile–Congo divide.

The position of the shore line varied from time to time across the Unstable Shelf and Stable Shelf areas of Egypt. In Cenomanian times the shore line must have lain south of Aswan but north of the Sudanese frontier. Estimation of earlier positions are very much a matter for conjecture because the Nubian disappears northwards under younger rocks and has been penetrated in relatively few boreholes and wells in Egypt. If the interpretations placed on these boreholes in the Western Desert are accepted (Said 1962, Appendix 1) then in Jurassic and Triassic times the shore line must have lain north of Aswan, probably in the Unstable Shelf area.

In the Sudan, Beadnell (1909, p. 48) first used the term Nubian Series; Sandford (1935, p. 342) adopted this name and applied it to rocks in the north-western Sudan which he considered to be of post-Palaeozoic and pre-Tertiary age.

The term 'Nubian Series' was also used by Andrew *et al.* (1948) on the second edition of the Sudan Geological Survey Map 1:4 000 000; Kleinsorge and Zscheked (1958); Rodis, Hassan, and Wahadan (1964), Karkanis (1965), and others. 'Nubian Series' is in common use among the geologists of the Sudan Geological Survey.

In this account the term 'Nubian Series' is abandoned because of the incorrect use of the term 'series'; this is a time-stratigraphical term and should not be used in a lithostratigraphical sense. The unit is here redefined as the Nubian Sandstone Formation (abbreviated to Nubian Formation) and is applied to those bedded and usually flat-lying conglomerates, grits, sandstones, sandy mudstones, and mudstones that rest unconformably on the

Basement Complex and Palaeozoic Sandstone Formation and are older than the Hudi Chert Formation (early Tertiary) and the 'early Tertiary' lavas. Its distribution is shown in Fig. 4.

Since this was written Whiteman (1970) has proposed that the Nubian Formation should be designated the Nubian Sandstone Group (abbr.: Nubian Group). The description and definition given here is the same as that given in Whiteman (1970). *Formation* is used throughout this work.

(ii) *Lithology*

In the Shendi–Khartoum area Kheiralla (1966) recognized the following lithological types in the Nubian Formation; similar types occur elsewhere:

(1) Pebble conglomerates
(2) Intraformational conglomerates
(3) Merkhiyat sandstones
(4) Quartzose sandstones
(5) Mudstones

(1) *Pebble conglomerates.* According to Kheiralla (1966) the Nubian pebble conglomerates are of mixed origin, being composed of vein quartz, quartzites, and black phyllite pebbles. Rhyolite pebbles, which are locally derived, occur in a conglomeratic bed capping Jebel Rauwiyan, north of Khartoum.

The quartzite pebbles are characterized by banding. The phyllite pebbles consist of a fine quartz mass, shreds of white mica, thin quartzite bands, and lenses. The pebbles are generally well-rounded, smooth, and typically water-worn. Disc-shaped as well as ellipsoidal particles are common. Rounded rod-like particles composed of black phyllite occur.

The rounding of the conglomerate particles may be due either to transport over long distances, or to the pebbles having been derived from older conglomerates. The presence of comparatively soft rounded phyllite pebbles points to the first explanation, because it is unlikely that they would stand recycling.

Concerning texture and fabric, the conglomerates are poorly packed and large voids exist. The particles are poorly sorted and range in size from granules to pebbles and are mixed with a large amount of silt and sand. The pebbles are haphazardly distributed, although some preferred orientation was noted for rod-like pebbles.

The pebble conglomerates are light brown to chocolate brown to black. Dark brown colours are characteristic of conglomerates capping hills, for example Merkhiyat Jebels, north-west of Omdurman. Colour is related to degree of ferruginization as well as to the nature of the iron oxide that is the main cement.

The iron oxide is largely primary but there has been considerable redistribution in places. Its primary nature can be demonstrated from borehole evidence as in a hand-dug well at Naqa, near Shendi, where chocolate-brown iron-cemented bands occur between a white soft ferruginous layer and a yellowish-brown slightly ferruginous layer.

The thickness of the pebble conglomerates varies considerably. The maximum recorded thickness is 35 ft in the Amriya borehole (Sudan Geological Survey, SGS, No. 827). The pebble conglomerates are widely distributed but Kheiralla (1966) reports that they are very rare north of Shendi.

(2) *Intraformational conglomerates.* They are common in the Nubian Formation and, like the pebble conglomerates, are commonly interbedded with the Merkhiyat sandstones. The intraformational conglomerates are mainly made up of mudstone or siltstone fragments mixed with varying quantities of pebbles embedded in a coarse sandy matrix. Some quartz or phyllite pebbles occur but are rare, and in places boulders of grey sandstone have been noted, as at Jebel es Serir at the southern end of the Merkhiyat Jebels.

The sandy matrix is highly ferruginized at some localities and the iron concentration is particularly high in thin bands of conglomerate.

The intraformational conglomerates are very coarse and rough textured. The particles vary in shape from platey fragments with hackly outlines to rounded and pellet-shaped. Some preferred orientation occurs. They are poorly sorted and particles range in size from a fraction of an inch to several feet.

The conglomerates vary in colour from buff to dark brown. The rock fragments vary from blue through purple to grey, and often impart a motley appearance to the rock. Colour again depends largely on the degree of ferruginization.

In the Shendi–Khartoum area studied by Kheiralla (1966) the infraformational conglomerates are usually interbedded with Merkhiyat Sandstone. They commonly occur in lenses but thin bands, 3 to 5 inches thick, also occur and persist for long distances. The contact with underlying strata is usually erosional.

Beds of intraformational conglomerate vary considerably in thickness. Kheiralla (1966) recorded

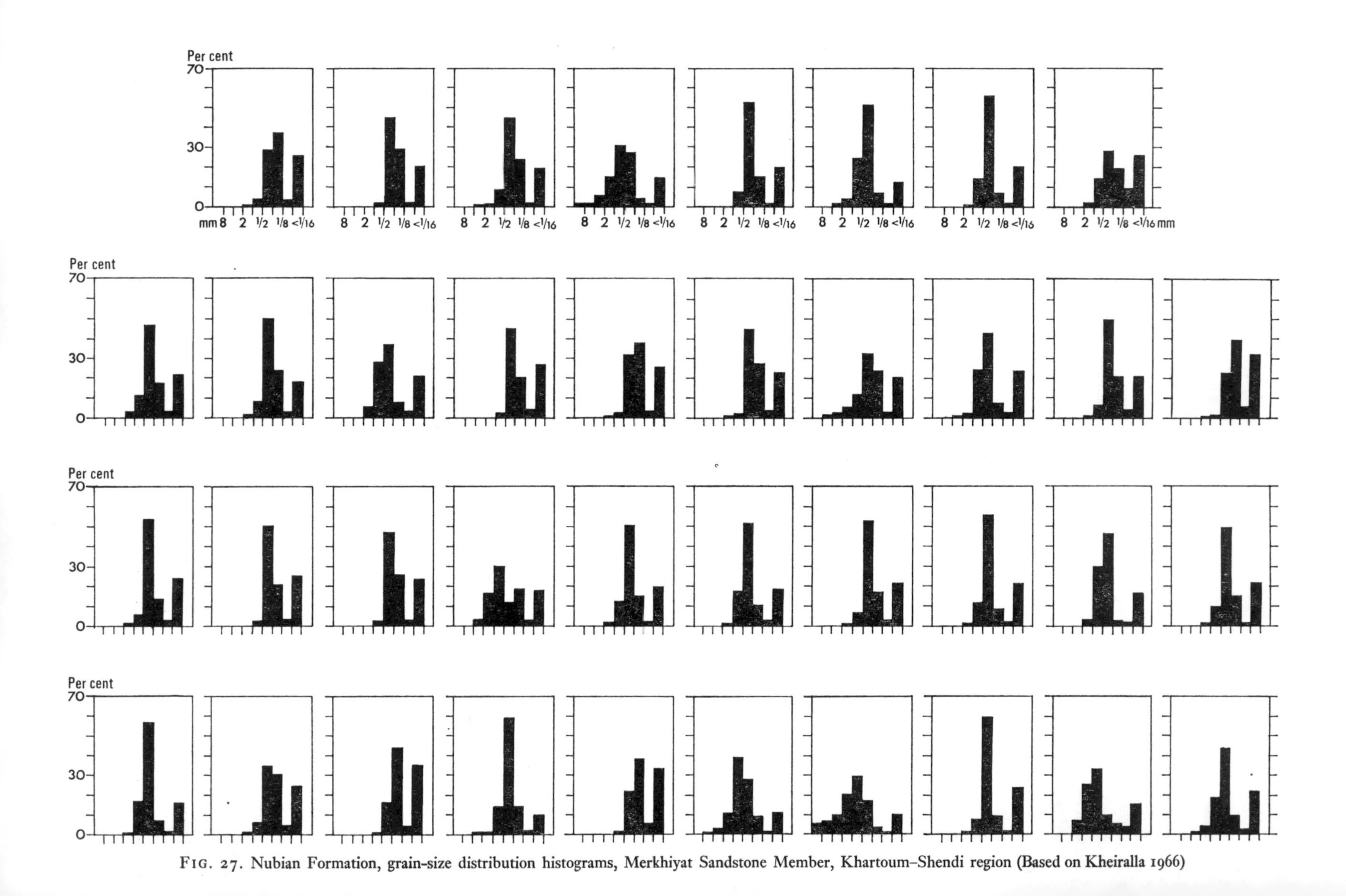

FIG. 27. Nubian Formation, grain-size distribution histograms, Merkhiyat Sandstone Member, Khartoum–Shendi region (Based on Kheiralla 1966)

a maximum thickness of 50 ft in section 26/55B, Jebel el Asfar in the Merkhiyat area. They are widespread in the Khartoum area but again are rare in the Nubian Formation in the Nile Valley, north of Shendi.

(3) *Merkhiyat sandstones*. The bulk of the Nubian Formation consists of sandstones and, as mentioned above, Kheiralla (1966) has recognized two main types.

The characteristics of the Merkhiyat Sandstone Member are (1) high clay-silt content, (2) abundance of white specks (? kaolinized feldspar fragments), (3) pebbly nature, (4) well-developed cross stratification, and (5) abundant fossils. Although designated a member here, according to Whiteman (1970) this should be elevated to Formation and renamed *Omdurman Formation*. The type locality is the Merkhiyat Jebels, Omdurman.

The Merkhiyat sandstones are made up of detrital quartz grains, characterized by undulose extinction. White specks are common in the sandstone and these are thought to be kaolinized feldspar grains. No fresh feldspar grains were noted in the sandstone. Some silica-cemented sandstones contain phyllite fragments up to 4 mm in size. The heavy fraction is small and consists of tourmaline, zircon, rutile, and staurolite. The greatest number of grains are opaque, mainly magnetite.

The sandstones are frequently pebbly. The pebbles are well rounded and range in size up to 30 mm. They consist mainly of quartz but some phyllites occur.

The Merkhiyat sandstones are coarse to medium grained; their average mean diameter in the area investigated by Kheiralla (1966) is 0·31 mm, varying from 0·18 to 0·77 mm. The modal class of the thirty-eight samples analysed falls into the size grades 1 to 0·5 mm (12 per cent) and 0·5 to 0·25 mm (74 per cent). The clay-silt fraction of the Merkhiyat sandstones reaches a maximum of 34·8 per cent by weight and averages 20·6 per cent for the thirty-eight samples analysed.

The Merkhiyat sandstones range from well sorted, with the sorting coefficient ranging from 1·206 to 2·297 with little clay content in the coarse and medium-grained varieties, to poorly sorted, with a sorting coefficient of 6·184 with high clay content in the fine-grained sandstones. The average coefficient of sorting is 1·958 (Fig. 27).

The Merkhiyat sandstones are commonly poorly cemented but locally there are well-cemented bands. Highly silicified sandstones occur at a number of localities, for example Jebel Mudaha; Jebel el Hardan; and Jebel Meagna, near Omdurman. These silicified horizons occur well above the base of the Nubian Formation (about 1200 ft in the Merkhiyat Jebels) as well as near the base of the formation. They do not seem to be restricted to the base of the formation as previously reported (Sandford 1935). The main colour of the Merkhiyat sandstones are yellowish-brown although some friable beds are whitish or grey. Dark-brown to black sandstones commonly cap the jebels.

The thickness of the Merkhiyat sandstones is highly variable. Kheiralla (1966) recorded a maximum thickness of 790 ft in SGS borehole No. 1517 (El Merikh). The Merkhiyat sandstones are widely distributed, but again they were not recognized in the sections measured north of Shendi by Kheiralla (1966). Good sections can be seen in the Masauwarat and Naqa areas; east of the Nile to Abu Duleiq; at Jebels Qisi, Maqrun, Abu Wuleidat, and in many boreholes. In boreholes SGS No. 903, 813, etc. the Merkhiyat sandstones rest directly on Basement Complex according to Kheiralla (1966). A basal conglomerate is not present in these areas.

Cross stratification is common in the Merkhiyat sandstones. The cross strata are of medium scale ranging in length from 1 to 3 ft. Large-scale cross stratification is not typical of the Merkhiyat sandstones but cross units having a length of as much as 42 ft were recorded at Jebel Hardan by Kheiralla (1966). The most common type of cross stratification is the tabular-planar type but trough-planar types were observed, especially in the Ed Datte area, Merowe. The cross strata dips show a wide variation ranging from 6° to 37°. Dips tend to be higher in the coarser-grained sandstones.

Graded bedded is exceptionally well developed in some iron- and silica-cemented sandstones, for instance in the Merkhiyat and Ed Datte areas. Iron concretions are common in the Merkhiyat sandstones; they vary in shape from spheroidal to tube-like. Some concretions are as much as 3 ft in diameter.

A summary of the characteristics of the Merkhiyat and quartzose sandstone types is given in Table 6 (based on Kheiralla, 1966).

(4) *Quartzose sandstones*. They are well bedded, non-pebbly, clean, very well sorted and contain fossil ripple-marks, and rib-and-furrow structures. They consist mainly of quartz, coated with iron oxide, with interstices filled with ferruginous matter. Micaceous sandstones (muscovite) were noted by

TABLE 6

Nubian Formation, a comparison of lithological characteristics of Merkhiyat and quartzose sandstone types in the Gezira–Khartoum–Shendi region (Based on Kheiralla 1966)

Merkhiyat type	Quartzose type
(i) Pebbly, coarse to medium grained.	(i) Non-pebbly, medium to fine grained.
(ii) Poor to well sorted. SO = 1·958.	(ii) Excellently sorted. SO = 1·175.
(iii) Clayey; average clay and silt fraction approximately 20 per cent.	(iii) Clean; average clay and silt fraction 6·3 per cent.
(iv) Interbedded with pebble and intraformational conglomerates.	(iv) Conglomerates not observed in measured sections.
(v) Primary structures; tabular-planar and trough-planar types of cross stratification.	(v) Primary structures; tabular-planar and wedge type of cross stratification; fossil ripple marks, and rib-and-furrow structures.
(vi) Average dip of cross strata 21° and current direction is north and north-westerly.	(vi) Average dip of cross strata 25° and current direction is north and north-easterly.
(vii) Poorly bedded.	(vii) Well bedded.

Kheiralla (1966). They are designated Shendi Formation in Whiteman (1970).

The heavy mineral fraction (0·15 per cent by weight) consists mainly of tourmaline, kyanite, staurolite, rutile, and opaque minerals.

The quartzose sandstones are medium to fine grained. The median diameter ranges from 0·10 to 0·41 mm with an average of 0·23 mm. In 84 per cent of the samples analysed by Kheiralla (1966) the modal class falls into the two size grades of 0·5 and 0·25 mm; and 47 per cent of the samples have their modal class in the coarser grade (Fig. 28). The clay-silt fraction in the quartzose sandstones is very small compared with Merkhiyat sandstones; the average clay-silt content is 6·31 per cent by weight.

The quartzose sandstones are well sorted and have an average sorting coefficient of 1·175 with the maximum value reaching 1·37. They are friable and disintegrate easily. Hard iron-cemented sandstones are found capping hills but this is a secondary feature. Silica-cemented sandstones occur but this appears to be a secondary feature. The common colour of the quartzose sandstones varies from pale pink to yellow brown. Violet- and reddish-coloured sandstones have been noted. Bedding is well developed and in general flat-lying. It is often continuous for long distances and sharply defined.

The maximum recorded thickness of the quartzose sandstone is about 300 ft (S.G.B. No. 1861) at Ed Damer. Normally it ranges between 80 and 150 ft as in the jebels, east of Umm Ali, north of Shendi.

The quartzose sandstones occupy a large area north-east of Kabushiya, near Shendi. This is designated the type area of the Shendi Formation (Whiteman 1970). West of the Nile they are well exposed at Jebel Managarat, Jebel Umm Gerab, Jebel Zaghawi, etc. and at the last two localities they are overlain by the Hudi Chert Formation (Fig. 41). They are distinctive enough to be mappable and are here called the Shendi Formation which has its type locality in the Shendi district.

Fossil ripple-marks are common in the quartzose sandstones. The ripple index falls within the range characteristic of water-formed types and the shape of the ripples indicates that the main current direction was from the south-east.

The tabular-planar type of cross stratification is common in the quartzose sandstones. At some localities crumpled or convolute forset laminae occur between undeformed ones and according to Kheiralla this is probably due to contemporaneous subaqueous slumping of the sediments. The direction of currents indicated by cross stratification is shown in Fig. 29.

Rib-and-furrow structure is well developed in the quartzose sandstones and indicates a current direction from the south-west.

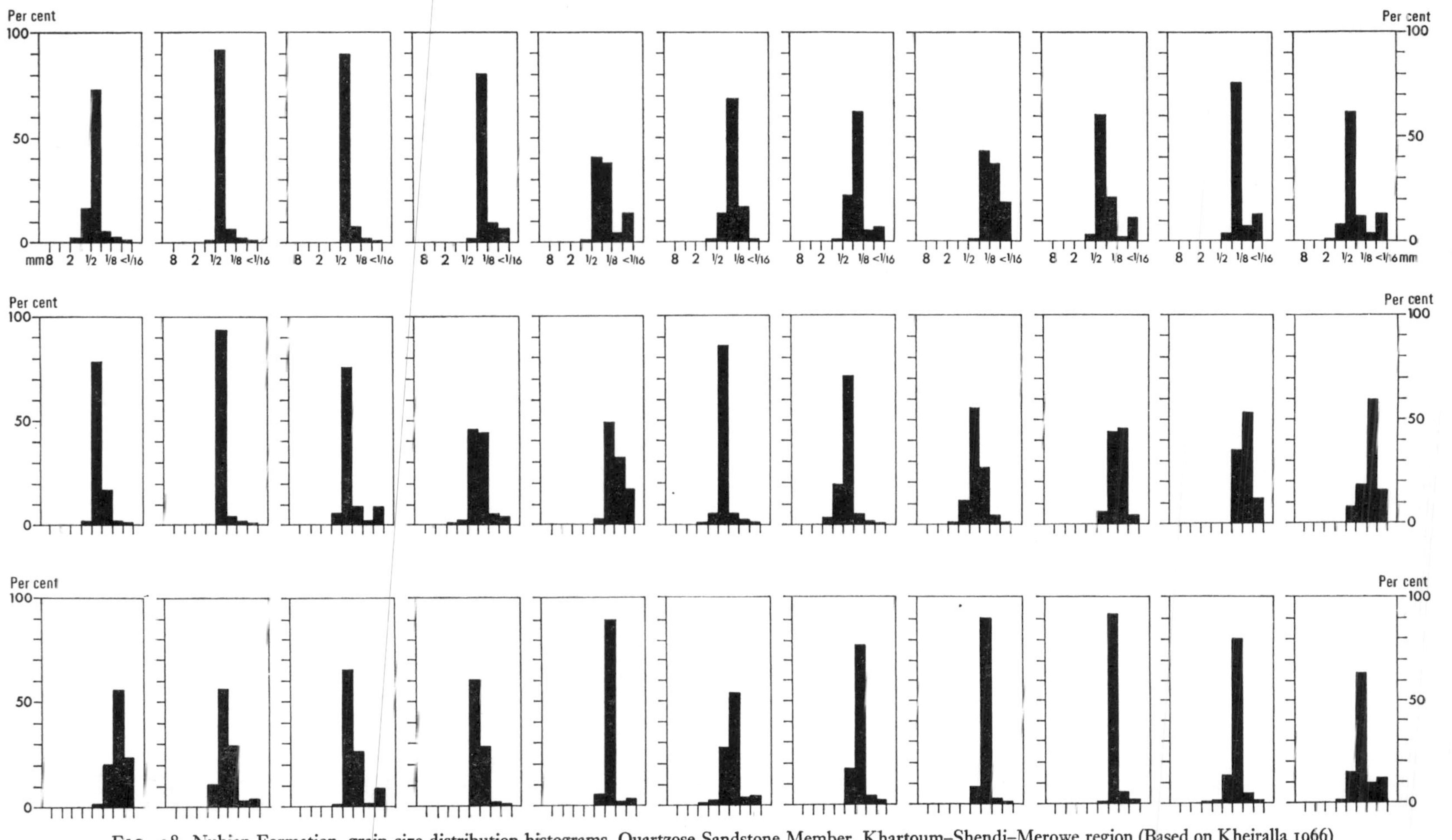

FIG. 28. Nubian Formation, grain-size distribution histograms, Quartzose Sandstone Member, Khartoum–Shendi–Merowe region (Based on Kheiralla 1966)

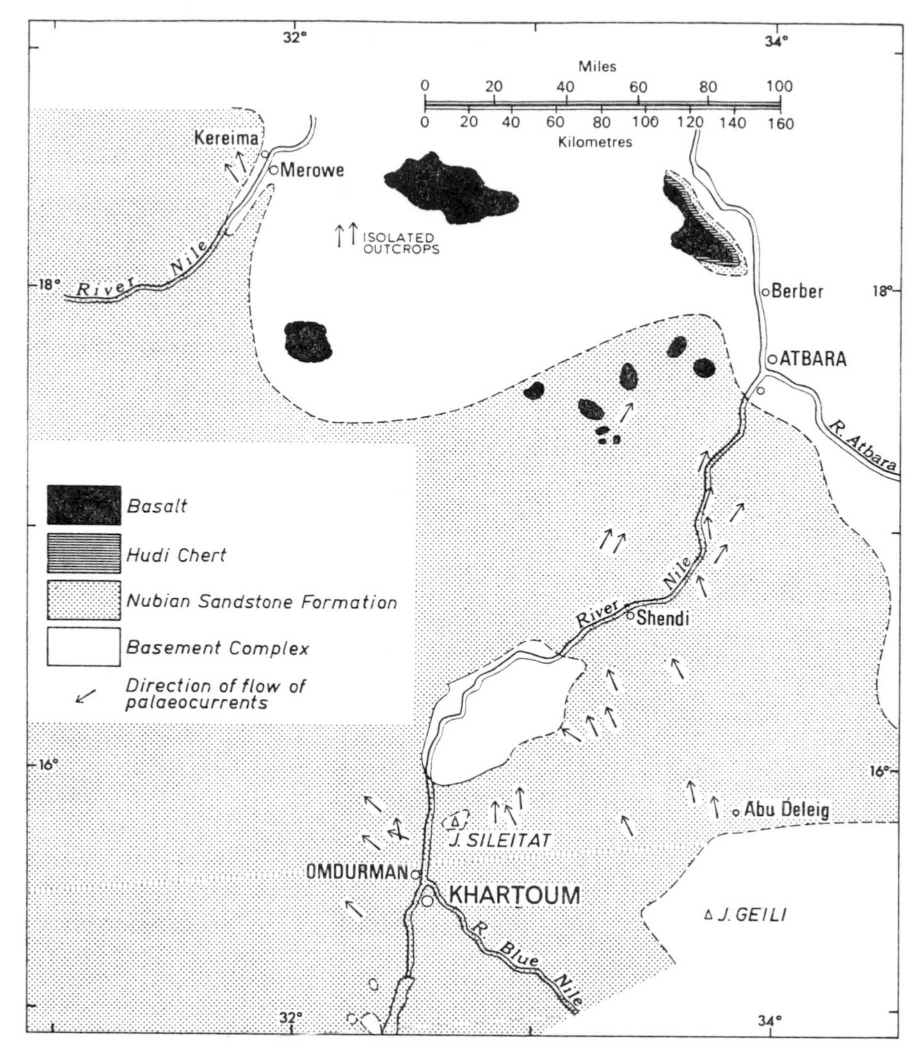

FIG. 29. Nubian Formation, Palaeocurrent direction map, Khartoum, Shendi, Merowe region Based on Kheiralla 1966)

(5) *Mudstones.* The term 'mudstones' is applied to those rocks made up largely of clay particles and having a blocky fracture. They appear to be composed mainly of clay minerals of the kaolinite group together with quartz grains, some mica, and sericite. Occasionally they are pyritous and carbonaceous. Pyrite-bearing mudstones up to 400 ft thick have been recorded. The most common variety is sandy mudstone. Colour varies from grey to purple to yellowish-brown to chocolate and depends on the iron content and degree of weathering.

The maximum thickness of mudstone recorded in the Khartoum–Shendi area studied by Kheiralla (1966) is 540 ft in SGS borehole No. 1728. The shape and thickness of some of the mudstone bodies are shown in Figs. 30 and 31.

Mudstones are widely distributed throughout the Khartoum–Shendi area but because of the ease with which they disintegrate in the highly corrosive climate of the desert–savannah transition zone they are rarely exposed and commonly occupy low ground.

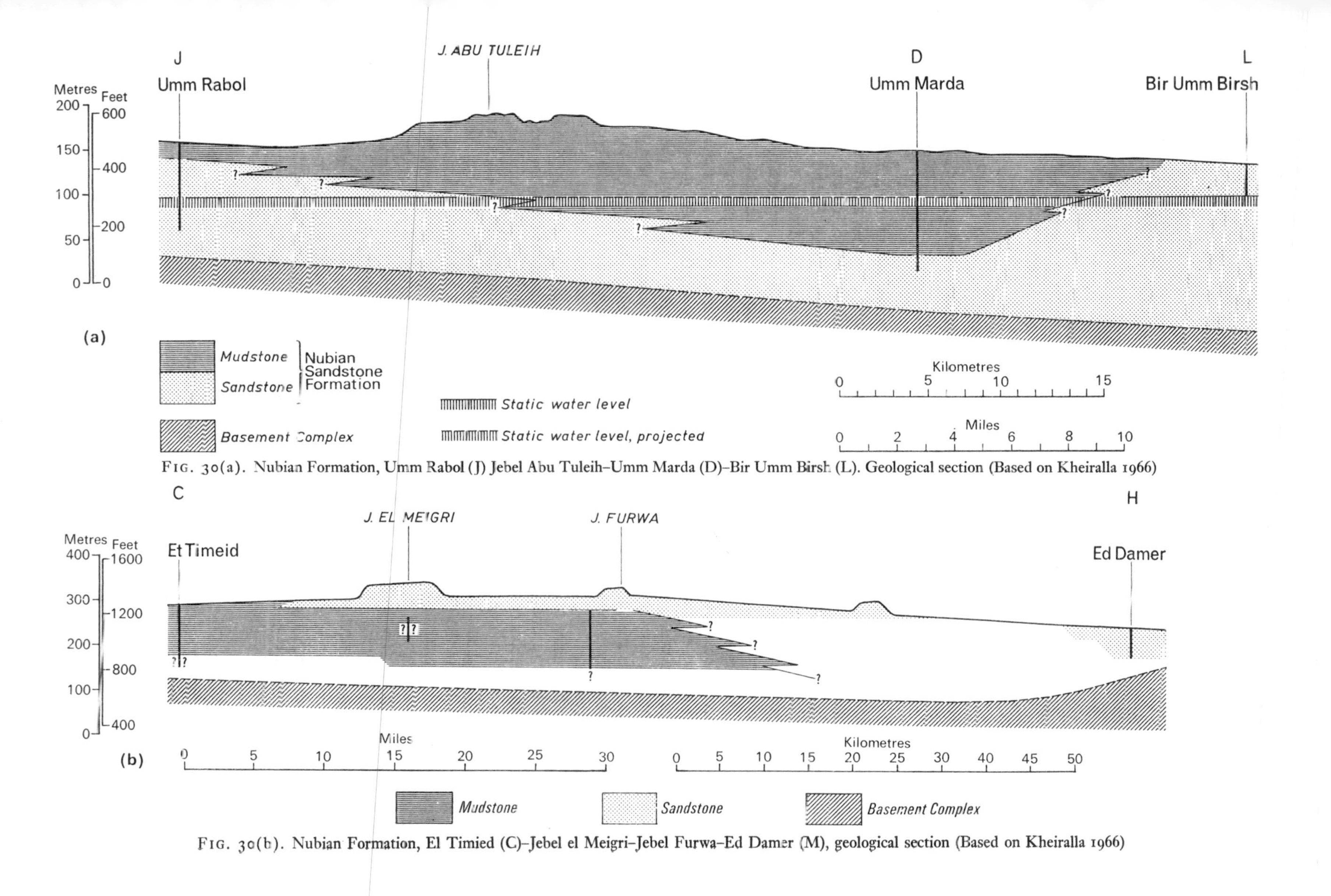

FIG. 30(a). Nubian Formation, Umm Rabol (J) Jebel Abu Tuleih–Umm Marda (D)–Bir Umm Birsh (L). Geological section (Based on Kheiralla 1966)

FIG. 30(b). Nubian Formation, El Timied (C)–Jebel el Meigri–Jebel Furwa–Ed Damer (M), geological section (Based on Kheiralla 1966)

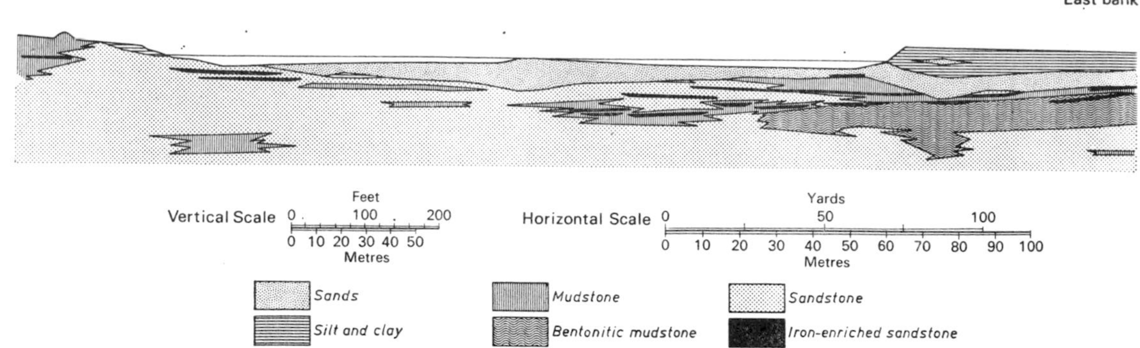

FIG. 31. Nubian Formation, Shambat Bridge section, Main Nile Khartoum, geological section (Based on Sudan Geological Survey engineering geology report)

(6) *General description.* Elsewhere in the Sudan, where geological details such as those obtained by Kheiralla (1966) for the Khartoum–Shendi area are not available, the Nubian Formation may be described in the following general terms.

The Nubian Group consists of conglomerates, intraformational conglomerates, grits, sandstones, sandy mudstones, and mudstones. Colours vary considerably at outcrop dependent on the nature of weathering, as well as the original chemical composition. They range from white, buff, yellow, pink, lavender, purple, brown, foxy-red, grey to black. When fresh, Nubian rocks are often buff or brown or dark-grey. Occasionally they are pyritous and carbonaceous in boreholes.

Often the Nubian Formation is iron-rich and in a number of localities beds of iron ore occur. These are well developed in the Shendi district and were worked in Meroitic and ? later times. They are now of little economic value. The origin of these deposits is not clear but in the Shendi district the main iron nodule and marram-bearing horizons occur only a few metres below a well developed (?) late Cretaceous–early Tertiary erosion-surface and are closely related to Hudi Chert deposits that were laid down within the proto-Nile drainage system. Oolitic iron ore occurs in the Wadi Halfa area in the Nubian Formation but appears to be a metasomatic replacement of limestones intercalated in the sequence. This deposit has little economic value (Andrew MS 1943).

The cement in the Nubian Sandstone is commonly iron oxide or iron carbonate. Calcium carbonate or silica cements also occur. In general the formation is poorly cemented and of medium hardness. Well-cemented silicified beds occur, however.

The suitability of the Nubian Sandstone as a building stone varies considerably. It has been used in the construction of Meroitic temples and pyramids, which are still well preserved, in government buildings, mosques, the Anglican Cathedral, the University of Khartoum, private houses in Khartoum, etc. After exposure for a number of years there is often considerable decay and scaling, especially in exposed places, or if the stone has not been carefully selected or has been laid bed face. This is due to the extremely high rate of chemical decay that takes place during and after the kharif or rainy season (Whiteman 1967).

Silicification of the Nubian Sandstone is a common feature in some areas, for example on the summit of Jebel Rauwiyan, Sabaloka, where the rock is almost a silicrete. The Nubian is often highly silicified below the Hudi Chert or in areas where the Hudi Formation has been stripped off recently. Silicification in these areas is clearly related to downward percolation. Another type of silicification frequently associated with iron enrichment occurs in the Nubian Formation along linear disturbed zones often marked by vertical beds. These zones, which run for hundreds of yards across areas of flat-lying beds, are usually not faults, the beds being unbrecciated or undisturbed in the zones. Most probably these zones were formed when water was expelled along them during compaction and diagenesis; or they may have been formed by dilation consequent on earthquakes. Sometimes they are parallel to the joints and in many cases they weather out as hard bands, being better cemented than the surrounding sandstone.

Bedding is well developed but is of variable nature and extent. Commonly at outcrops major posts are clearly defined but there is often conspicuous cross-bedding in between. Taken together with the wide range of grain size, the presence of ripple-marks, mud-cracked surfaces, mud-flake conglomerates, fossil rain marks, and the great number of silicified trees obviously in position of growth or showing few signs of transportation, this clearly points to a shallow-water terrestrial origin.

(iii) *Origin of the Nubian Sandstone Formation*

Much has been written concerning the origin of the Nubian Formation both in the Sudan and Egypt. Deposition under aeolian, deltaic, estuarine, and marine conditions has been advocated for the Nubian Formation in Egypt. Walther (1888) and Fourtau (1902) considered that the Nubian Formation is a gigantic fossil desert deposit, laid down along the southern margin of the Palaeozoic and Mesozoic Tethys. Barthoux (1922), Cuvillier (1930), and Picard (1943) also believed that it is an aeolian deposit. Hume (1906), on the other hand, favoured a marine origin, basing his opinion on the presence of rounded boulders in the basal conglomerates, the transition of sandstones into mudstones with marine fossils, etc. Ball (1907) thought that the evidence in the Aswan area indicated deposition in shallow water and later assigned a fluvio-marine origin to the Nubian deposits of Sinai (Ball 1916). Shukri and Said (1944), basing their comments mainly on information from Egypt, ascribed a near-shore marine origin to the unfossiliferous sandstone, pointing out that 65 per cent of the pebbles are disc-shaped and that the sands are well sorted and are composed of one main ingredient.

Quite clearly the conditions that prevailed in the Sudan must have been different from the estuarine or fluvio-marine conditions that prevailed along the southern shore of Tethys near Aswan in Cenomanian times and earlier. Most geologists believe that the Nubian Formation in the Sudan was laid down in a subaqueous environment, but opinions differ as to whether it was deposited in a marine or a continental freshwater environment. Sandford (1935) advocated deposition in a continental subaqueous environment for sandstones containing water-worn pebbles and fossil trees, but he believed that mudstones and white sands were mainly of aeolian origin and were probably formed in deserts. Deposition in a continental and/or near-shore environment was advocated by Rodis *et al.* (1964), and B. G. Karkanis (1965) postulated that the highly saline water encountered in the Nubian Formation in the western provinces of the Sudan is connate water. The saline ground-water can, in the author's view, be better explained by infiltration from other formations.

Evidence bearing on the origin of the Nubian Formation falls into two categories: (1) lithological, (2) palaeontological.

(1) *Lithological evidence.* Quite clearly the pebble conglomerates and the intraformational conglomerates mentioned above indicate deposition in water. According to Kheiralla (1966) the composition of the Merkhiyat sandstones is similar to river deposits described by Twenhofel (1961, 219, Fig. 20) and by Krumbein and Sloss (1955, pp. 71 and 201). The high percentage of clay and silt in these sandstones may indicate that they were deposited by currents that were suddenly checked on entering the depositional area. Currents that transport pebbles, gravel, and very coarse sand, as well as clay and silt, usually occur only in narrow channels. The evidence of the Merkhiyat sandstones then favours deposition by fast-flowing streams debouching into lakes on the flood plains of a series small rivers. The tabular type of cross-stratification indicates strong current action and there can be no doubt that the sandstones and conglomerates were laid down in water because of their large grain size.

The impersistent nature of the Merkhiyat sandstones may indicate that they also are river deposits; the graded bedding and intraformational conglomerates certainly indicate subaqueous deposition. The characteristic lens-shaped occurrence of the pebble conglomerate bodies is probably due to the fact that they are channel deposits.

The Shendi Quartzose sandstones indicate a slightly different environment of deposition however. Characteristically they are non-pebbly and are medium to fine grained. They are well sorted and the clay-silt fraction is small. The mechanical composition and the poor degree of rounding of grains may be taken to indicate deposition by weak currents. Ripple-marks and rib-and-furrow structure all indicate deposition in water.

(2) *Palaeontological evidence.* The fossil plants in the Nubian Formation are mostly found *in situ* and

they rarely show any evidence of having been transported long distances.

Silicified wood, mainly parts of tree trunks and branches, are the most common fossils, and at Jebel Marahi, 70 miles north of Khartoum, there is a petrified forest. This evidence certainly indicates a continental origin for the Nubian Formation of the Khartoum area.

The presence of rings on many of the silicified tree stumps and trunks indicates that the climate was seasonal. According to Gotham (1910, p. 11), *Weichselia* and *Frenelopsis*, identified by Edwards (1926) from Jebel Dirra, near El Fasher (see above), are strand-line plants. The reduction of leaves and the assumption of photosynthetic processes by the branches that characterize *Frenelopsis* 'suggests strong insolation and lack of humidity . . .' (Berry 1910, p. 39). However, the occurrence of coal in what is listed as Nubian Formation at Dongola (see Chapter VI) indicates that conditions were wet enough in places for coal swamps to exist.

In the author's view the Nubian Formation in the Sudan was deposited in a great series of fans and playas (rahads) separated by low isolated jebels. This environment extended southwards from the shore line in Egypt for well over a thousand miles, perhaps as far as the Nile–Congo divide. In the initial stages of accumulation vast pediplains must have formed around the Basement Complex jebels and as subsidence progressed the jebels were buried by sands, conglomerates, silts, and muds, which eventually were transformed into conglomerates, grits, coarse and cross-bedded sandstones, fine- and medium-grained sandstones, sandy mudstones, mudstones, etc., which constitute main lithological types of the Nubian Formation. Plant cover was most probably limited to xerophilous plants, as in Cretaceous times grasses had not yet established themselves. The climate was wet enough to flush soils and prevent the formation of saline lakes. Salt deposits have not been recorded in the Nubian Formation in the Sudan.

(iv) *Thickness variations*

The Nubian Sandstone Formation varies considerably in thickness as the regional descriptions given below show. From an estimate of jebel height and borehole depth in the Khartoum area the flat-lying Nubian rocks may be about 1550 ft thick. Thickness is related to erosion and the configuration of the underlying Basement Complex bedrock surface. In areas where the base of the Formation is recorded in boreholes the thickness varies from 120 to 485 ft. In the region north-east of the Sabaloka inlier the preserved thickness of the formation (550 ft) is far less than in areas to the south and south-west around Khartoum where over 1000 ft of Nubian rocks was recorded in Sudan Geological Survey borehole No. 1517 alone.

The configuration of the sub-Nubian surface in the Khartoum–Shendi area is shown in Fig. 32.

(v) *Regional description: Jebel Uweinat–Dongola area*

Sandford (1935) recognized the following succession in the north-western part of the Sudan:

3 Upper buff and brown, cross-stratified sandstone group with silicified wood,
2 Middle mudstone and siliceous sandstone group,
1 Lower conglomerate and siliceous sandstone group.

The succession is applicable to a limited area in the north-west Sudan only.

Hottinger *et al.* (1959) also described the sequence in the Uweinat area:

JEBEL UWEINAT

Upper Nubian Sandstone	Sandy, conglomeratic member, brown to black, ferruginous bands common, fossil wood abundant near base
Middle Nubian Sandstone	Sandstone, brown, red or yellow, coarse to fine grained
Lower Nubian Sandstone	Conglomeratic, hard, siliceous, quartzitic

The thickness is said to vary from 300 to 600 ft and the unit is assigned to the Lower Cretaceous by Hottinger *et al.* (1959) on regional rather than palaeontological evidence.

Between Uweinat, the Nile, Ed Debba, Jebel Rahib, and the Egyptian and Libyan frontiers, the country is occupied mainly by Nubian Sandstone and Andrew (MS 1943) gave the following description:

'The general characteristics of the series are the same as in Kordofan and Darfur. The series consists of interbedded sandstone and subordinate mudstones, white, buff-brown, and red. Bands of ferricrete and concretionary ironstone are fairly common, and these ferruginous layers frequently form the surface outcrops on low ground or cap hills on account of their cohesion. The sandstones are most poorly cemented and friable, but silicified sandstones are most found, particularly near

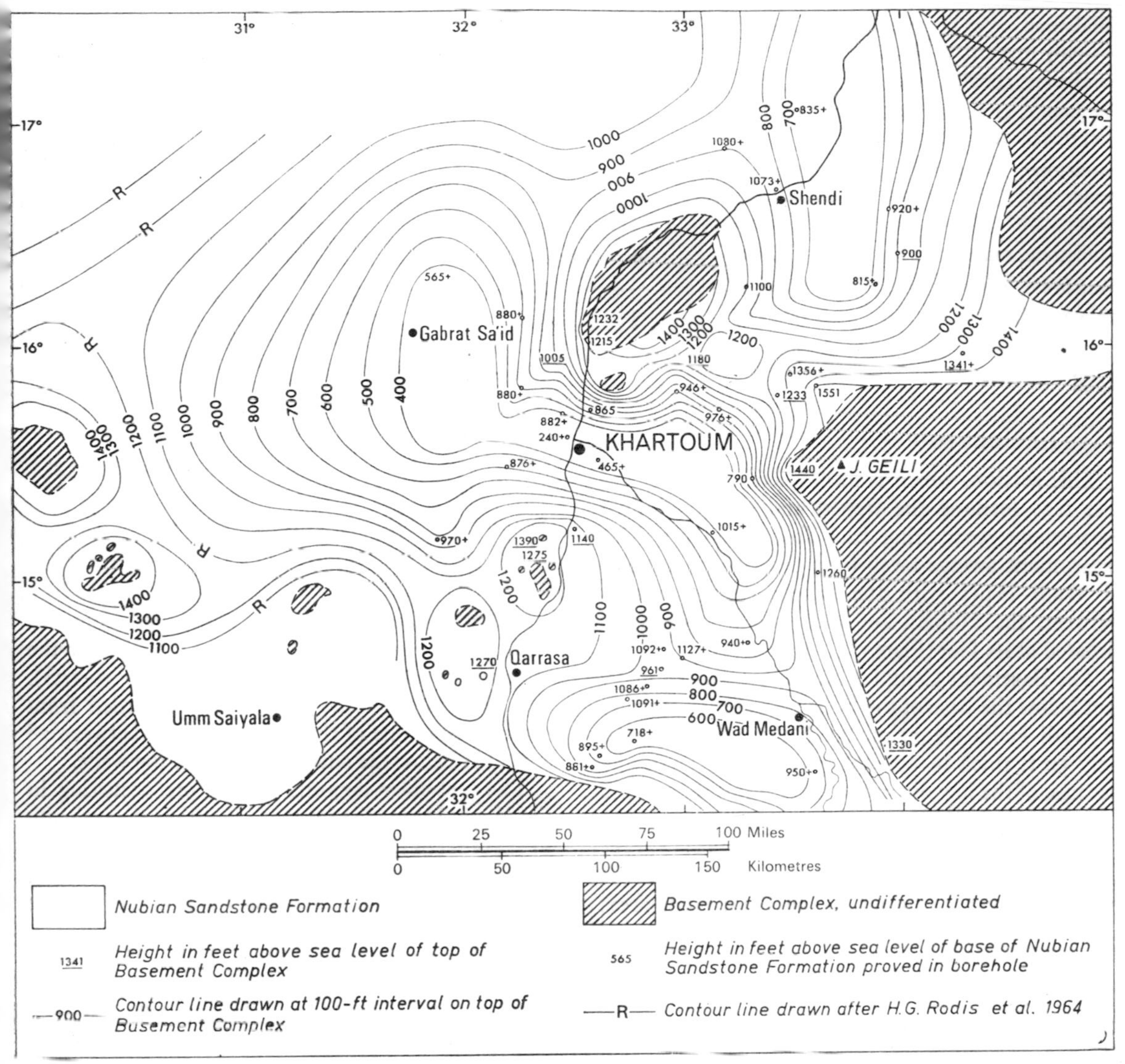

FIG. 32. Nubian Formation, configuration of the Sub-Nubian, Wad Medani–Khartoum–Shendi region (Contours according to Kheiralla 1966)

the base on the series. Silicified wood is very commonly strewn on the surface on or near outcrops of the series, sometimes as large as tree trunks . . . The characteristic of the Nubian Series is the general absence of a basal conglomerate. Conglomeratic beds are rarely at the base of the series, but occur some distance (2–10 m) above the base.'

(vi) *Regional description: Wadi Halfa area, Northern Province*

The following section was measured by the author on a prominent jebel of Nubian Sandstone, east of the Nile at Sidi Amir el Sahaba, 4 miles north of Wadi Halfa, Northern Province:

SIDI AMIR EL SAHABA

	Thickness (ft)
TOP OF SECTION	
Triangulation point sandstone, red-brown, ferruginous	33

Sandstone, cross bedded, ferruginous, bands of sandy mudstone, blue-grey	33
Sandstone, conglomeratic	30
Sandstone, cross bedded, gritty, conglomeratic, micaceous, red-brown	19
Sandstone, micaceous, with bands of mudstone and microconglomerate, abundant mica	33
Sandstone, micaceous, bands of mudstone	34
Sandstone, coarse to medium grained, red-brown with bands of sandy mudstone	5
Talus and wash of Nubian Sandstone	14
Total	211

BASE OF SECTION

This section, which is incomplete, is part of the free face forming the fretted and slightly faulted escarpment weathering back from the Nile. The escarpment swings away from the river south-east of Wadi Halfa with the unconformable contact between the Nubian and the Basement Complex situated at its foot (Plate 5).

The Basement Complex cannot be situated many feet below the base of the El Sahaba section because at Wadi Halfa Degheim, the Nubian Sandstone is only 50 ft thick in borehole 505 (GS/35/1/G), and because of the low dip of the rocks below the Second Cataract where Basement diorites are exposed.

Mansour (open file, Sudan Geological Survey 1960) gave the following general succession for the Nubian Sandstone Group in the Wadi Halfa area:

	Thickness (ft)
TOP OF SECTION	
Sandstone, coarse grained to pebbly-buff, friable, false bedded, rich in kaolinized feldspar; flattened clayey pebbles are common and increase upwards to form thin beds; in places pebbles up to 1 inch form thin lenticular bands	128
Shale, dark red, ferruginous and mostly silicified, hard and cleaved, probably equivalent to the lower oolitic band of haematite	0·75
Sandstone, massive, white-pinkish, pale-grey, calcareous cement, some mica	No thickness given
Total	130

Mansour also gave details of a section situated 3 km west of Jebel Sheitan, 5 km south-east of Wadi Halfa. Two oolitic bands occur in the Nubian Formation and have been of interest as possible sources of iron, as at Aswan, but unfortunately they appear to be uneconomic.

JEBEL SHEITAN

	Thickness (ft)
TOP OF SECTION	
Sandstone, hard ferruginous, with iron oxide in dispersed ooliths	3–13
Oolitic iron ore band B, brick-red, hard siliceous, some small silica grains, oolites diameter 2–3 mm.	1
Sandstone, hard-white	39
Sandstone, white with dispersed ooliths	0·6
Oolitic iron ore band A, fine grained, ooliths 1 mm diameter, yellowish-red colour, seems to be limonitic, occasionally siliceous	0·75
Sandstone, buff, clayey sandstone intercalations	16
Total	70

Oolitic iron ore has also been recorded from two wells in the Argin district of Wadi Halfa, near the Dispensary (Sudan Geological Survey GS/35–1/G). Reddish-brown oolitic ironstone is recorded from the surface in one of the wells and from between 28 and 33 ft below the surface in both of the wells.

The lenticular deposits of iron ore occurring in the Wadi Halfa district were mentioned originally by Lyons (1897) who stated that they are 0·5–1 m thick and extend from 3 to 8 km in the Shemain area. Andrew (MS 1943) stated that the iron ores replace limestones.

North of Station 3 on the Wadi Halfa–Abu Hamed Railway, a second escarpment appears and altogether in this region probably more than 1000 ft of Nubian Sandstone is exposed. The sandstones north of Station 3, and to the south-east, are similar to those recorded near the Nile, except that there are some prominent bands of mudstone and sandy mudstone in the sequence.

The Nubian Sandstone–Basement Complex boundary roughly follows the railway line to Station 8 and then doubles back swinging northwards into Egypt along the west side of Wadi Gabgaba. This long section of outcrop may be fault-controlled in the Sudan (Fig. 22).

The Nubian Sandstone–Basement Complex boundary crosses the Nile a few miles north of the Second Cataract, and, south of the Rock of Abu Sir (Plate 5) follows the west bank of the Nile a few miles from the river as far south as Abu Fatima's Tomb, north of Dongola, where it crosses again to the east bank. In this section there are many jebels (not shown on the Sudan Geological Survey 1:4 000 000 map) capped by the Nubian Formation, such as Jebel Ali Barsi, where there is clear evidence of superimposition of the Nile from a once much more extensive Nubian cover. In some places along the river, for example at the Fareig bend south of Delgo, the Nubian Formation occupies the river banks, forming outliers from the main mass of the Nubian rock to the west. They are preserved either in down-faulted blocks or in synclinal areas.

From Dongola the contact swings in a series of broad lobes into the desert towards Merowe and crosses the river near the town.

(vii) *Regional description: Jebel Nakharu*

This is perhaps one of the most important sedimentary outcrops in Northern Province, for it is one of the few areas in the central Sudan where Tertiary rocks rest on Cretaceous rocks. It is situated on the west bank of the Nile, 8·5 miles north of Berber, and the Nubian Sandstone is well exposed in cliffs forming the bank of the Nile.

The Nubian Formation is overlain by bedded gravels consisting of Hudi Chert pebbles, followed by coarsely-bedded quartz gravels, and capped by basalts. Andrew and Karkanis (1945, p. 4) correlated the cherts, which they described as bedded, with the Hudi 'Series' as recognized at Hudi Station, east of Atbara, and at Zeidab.

The following section was measured on the University of Khartoum Bayuda Expedition 1963 (Fig. 33).

JEBEL NAKHARU

	Thickness (ft)
TOP OF SECTION	
Trigonometrical station within old fortifications, River level approximately 1443·4 ft	
Basalt, dark, blue to black, vesicular bouldery outcrop, subspheroidal to angular	54
Gravels, roughly bedded, consisting mainly of quartz pebbles, diameter 1½ in, well rounded, occasional pebbles of Basement rocks, irregularly iron-stained, pebbles pink-stained surface	60
(N.B. Hudi Chert is not present in this section but occurs between the gravels and the Nubian Group about 0·5 miles to the north where it is bedded and is an approximately 15 ft thick. Hudi chert pebbles are present in talus deposits at Jebel Nakharu)	
Nubian Sandstone, cross bedded, fine grained, micaceous, yellow-brown, bands of secondary iron enrichment, gulls developed opening towards the Nile	90
Talus and gravel	25
Nile alluvium reduced level 1043·5 ft (326·16 m) old datum Khartoum	
Total	229

Andrew and Karkanis (1945) recorded a bed of grit with a clay matrix some 48 ft thick below the pink-stained quartz gravels and a bed of chert 6 ft thick with blocks from 7 to 15 inches thick at Jebel Nakharu. The chert bed is irregularly distributed and is considerably thinner than in the type area to the south. This occurrence is slightly north of Jebel Nakharu and was visited by the University of Khartoum Expedition 1961.

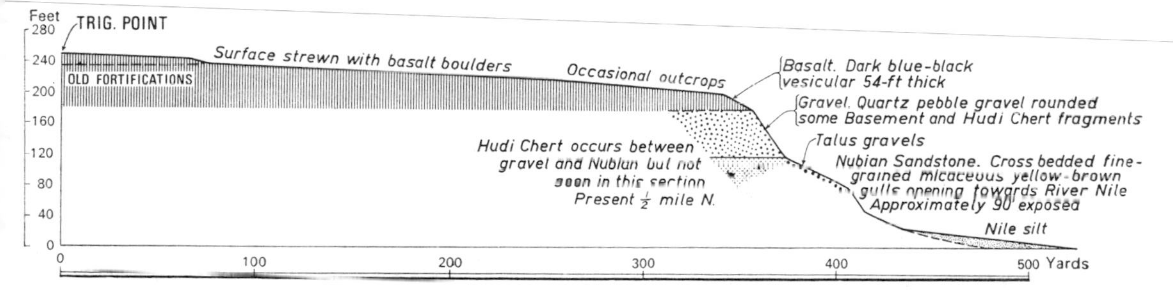

FIG. 33. Nubian and Hudi Chert formations, Jebel Nakharu, Northern Province, geological section

The Nubian Formation in these areas is thin and is perhaps of the order of 150 ft at the most, because a few miles to the north the metasediments of the El Bageir Group crop out. How much erosion of the Nubian Formation took place before the deposition of the Hudi Formation in this region cannot now be estimated.

(viii) *Regional description: Kassala Province*

Isolated outcrops of the Nubian Sandstone Group occur in two areas of Kassala Province (Figs. 34 and 38): (1) Khor Langeb, (2) northern Red Sea Hills.

Khor Langeb area. Delany (1954, pp. 14–15) described the Nubian Formation in the Khor Langeb region, which was first visited by G. H. Thompson in 1916 and later by Barron, who traversed the area. Both geologists recorded rhyolites resting on basalts and noted the relationships of these rocks to the Nubian Formation.

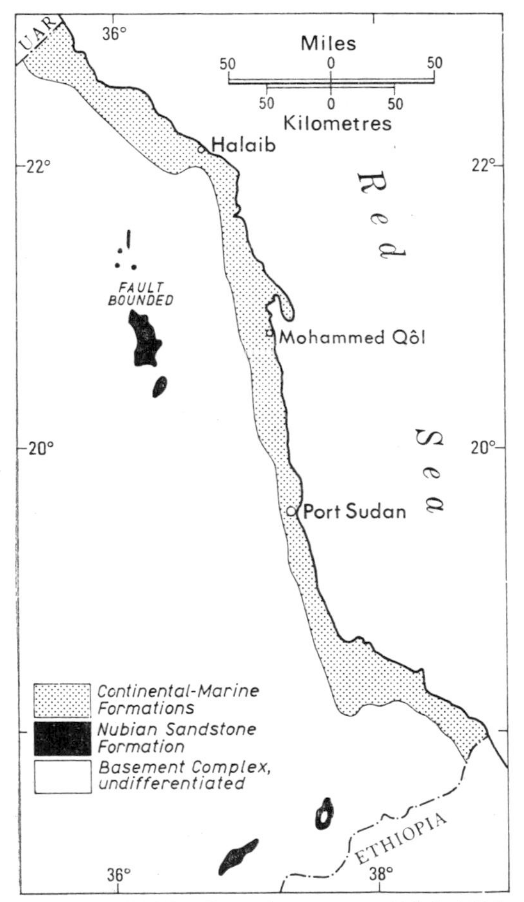

FIG. 34. Nubian Formation outcrops, Jubal el Bahr el Ahmer, geological map

Delany (1954) described the following section:

EL BAB (17°15′N, 36°38′E)

	Thickness (ft)
TOP OF SECTION	
Basalt	
Nubian Sandstone Formation	
Mudstone, white flinty, well stratified, baked siliceous nodules give rock of fossiliferous appearance	80
Sandstone, ferruginous	32
Limestone, blue with chert horizons, overlying sandy marls form transition beds. No fossils found	7
Sandstone, white to red, current bedded, occasional pebble beds	128
Conglomerate, quartz pebbles, subangular loosely cemented	3
Basement Complex	
Mica-schists discordantly overlain by sandstone	
Total (Nubian Sandstone Formation)	250

Although no fossils were found in the El Bab section, Delany (1954) pointed out that the limestone may be correlated on lithological grounds with the Jurassic Antalo Limestone of Ethiopia. If this correlation is correct then the sandstone is perhaps better referred to as Gedaref Sandstone, rather than Nubian.

Sediments occur below the basalts and above the Nubian Formation, but these are dealt with in Chapter III under Tertiary rocks.

Northern Red Sea Hills. Two isolated outcrops occur in the northern Red Sea Hills on the Mohammed Qôl sheet (Kabesh 1962, pp. 45–8) (Fig. 34). They rest unconformably on the Basement Complex, consisting thereabouts of pink red granite, paragneiss, and a schist–mudstone–greywacke series (Nafirdeib Group). The outliers include layers of conglomerate, pebble bands, grits, quartzites, sandstones, clayey sandstones, clays, and shales. The strata are well bedded; the bedding planes generally dip gently (10°) and towards the west. In some places, they are practically horizontal, but near faults the beds dip as much as 20°. The conglomerates and pebble bands generally occur at the base of the series.

Basing his conclusions on the lithological similarity of the Red Sea Hills Nubian Formation and other areas in Egypt and Sudan, Kabesh (1962) concluded that they are of marine shallow-water origin.

Five Nubian outliers have been described from the Dungunab sheet 36–I (Gass 1958, pp. 58–60). They all occur in the south-west of the sheet and are bounded on the east by faults. The largest of these outcrops occupies 100 square miles. Four other small outliers of Nubian Formation occur at Khor Odib (21° 24′ N, 36° 05′), Jebel Rikabeb (21° 22′ N, 36° 03′ E), an unnamed outcrop (21° 17′ N, 36° 00′ E), and Salala (21° 16′ N, 36° 09′ E) (Fig. 34).

The Nubian Formation consists of massive current-bedded sandstone with shale bands. The beds dip to the east at between 5° and 30°, and are cut by faults that were probably formed in Tertiary times. Kabesh noted also that the coarse clastic material is mainly well-rounded vein quartz and only occasionally are fragments of Basement Complex rocks found. The groundmass is made up of small, angular quartz fragments.

(ix) *Regional description: Ed Debba–Khartoum–Sennar lobe*

The Nubian Sandstone extends south-eastwards from Ed Debba on the southernmost part of the bend of the Nile, forming extensive outcrops around Khartoum and extending as far south as Sennar on the Blue Nile. The contact with the Basement Complex then swings westwards and northwards forming an extensive embayment north of Sodiri and south-east of El Atrun. Inliers of Basement Complex rocks occur at Sabaloka, Jebel Seleitat, and Jebel Tabar between Omdurman and Bara, Jebel Bareima and Huweida south of the Jebel Auliya Dam, Jebel Helbi, west of Jebel Auliya, etc. (Fig. 4). In all these areas the Nubian Sandstone rests unconformably on a surface of Basement Complex rocks that had considerable relief. In Fig. 32 an attempt has been made to show the general configuration of the sub-Nubian surface based on scattered borehole data.

There are many good exposures in this area, and the flat-lying Nubian Sandstone forms extensive mesas such as the Elai Jebels (Plate 13), the jebels between Aliab and Abu Duleiq, etc. Along the Nile, erosion has given rise to a series of pediplains and fretted escarpments, which are best developed on the east bank of the Nile between Aliab and Shendi. The escaprments continue southwards, decreasing in height and extending further from the river, east of the Sabaloka inlier. Towards Sennar the Nubian Formation forms part of a flat, almost featureless, plain and outcrops are mainly restricted to the river bed.

For stratigraphical details of the Nubian Formation in part of this area the reader is referred to Kheiralla (1966), who studied the Khartoum–Shendi area (Figs. 35 and 36). The main lithological types recognized by Kheiralla are described above.

The following sections are taken as representative of the Nubian Sandstone in the section along the Nile between Aliab and Wad ben Naga. They constitute the Shendi Formation (Whiteman 1970).

JEBEL ZAGHAWI

West of Aliab, west bank of Nile Northern Province, 17° 16′ N, 33° 42′ E.

	Thickness (ft)
Hudi Chert	
Chert, brownish, horizontal, thin bedded; occasional boulder beds of chert; boulders embedded in a ferruginous coarse-grained silicified sandy matrix	10–40
UNCONFORMITY	
Nubian Sandstone Formation	
Sandstone, fine to medium grained, pink to reddish brown; followed by sandy mudstone, greyish; ferruginous and silica cement at the top of the unit	38
Sandstone, medium to fine, grey to pale pink or brown; thinly bedded; micaceous at top	31
Base of section. Talus	109
Total	208

JEBEL TERABIYA

Section K III, 8 miles north-west of Kabushiya, Northern Province, 16° 54′ N, 33° 46′ E.

	Thickness (ft)
Nubian Sandstone Formation	
Sandstone, fine grained, micaceous laminated, light grey to pale pink; thinly bedded; used for building stone	30

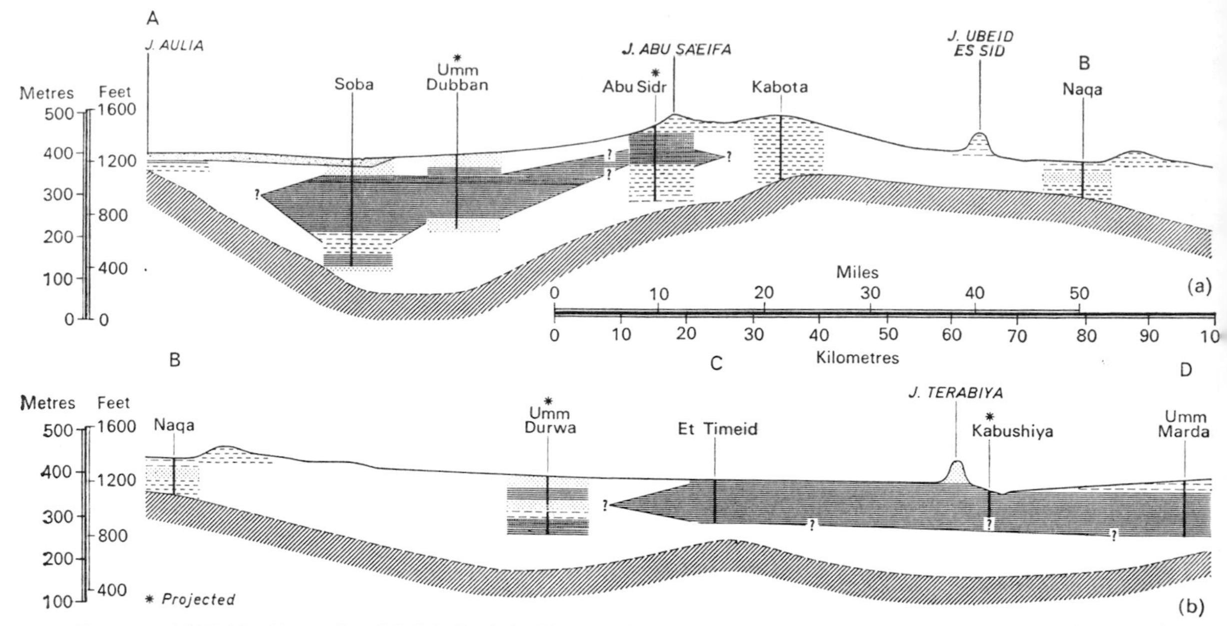

FIG. 35. (a) Nubian Formation, Jebel Aulia–Soba–Umm Dubban–Naqa (B), Northern Province, geological section (Based on Kheiralla 1966. See Fig. 80 for location and Fig. 30 for lithology), (b) Nubian Formation, Naqa (B)–El Timeid–Umm Marda (D), Northern Province, geological section (Based on Kheiralla 1966. See Fig. 78 for location and Fig. 30 for lithology)

Sandstone, coarse to medium grained, cross bedded, coarse laminae, yellowish-reddish or brown	26
Sandstone, medium to fine grained, clayey; pink to white	11
Sandstone, coarse to medium, brown to yellowish; topped by 8-cm ferruginous band; cross bedded	45
Base of section. Talus	
Total	112

JEBEL K IV

Section K IV, Jebel 5 miles south-east of Kabushiya, Northern Province 16° 49′ N, 33° 43′ E.

	Thickness (ft)
Hudi Chert	
Chert, fossiliferous, bedding parallel to underlying Nubian Formation	5
UNCONFORMITY	
Nubian Sandstone Formation	
Sandstone, coarse, gritty, violet, ferruginous and silicified, cross bedded	20
Mudstone, sandy in parts; grey, yellow, brown, to red; ironstone bands up to 1 ft thick are common	32

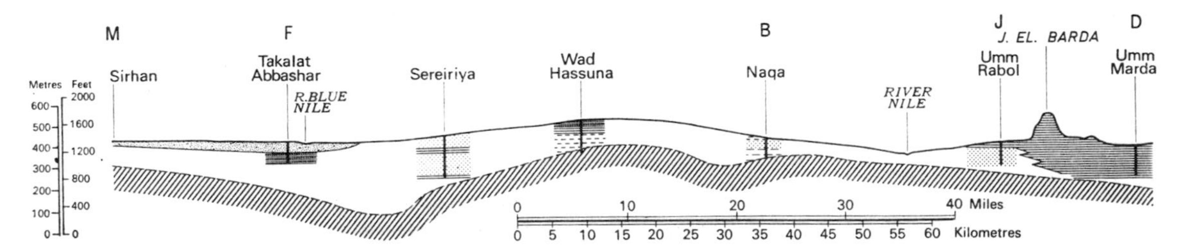

FIG. 36. Nubian Formation, Sirhan (M)–Wad Hassuna–Naqa (B)–Umm Marda (D), Northern Province, geological section (See Fig. 78 for location and Fig. 30 for lithology)

Sandstone, medium to fine grained; reddish-yellow to brown, highly ferruginous in parts with concretions; cross-bedded near the top; thin mudstone lens	47
Total	104

JEBEL NASB ES SAMI

33 miles, 203° Shendi, 16° 15′ N, 33° 14′ E.

	Thickness (ft)
Nubian Sandstone Formation	
Sandstone, coarse to medium grained; brown in colour, limonitic	50
Sandstone, arkosic, clayey, coarse; brown with mudstone fragments, quartz pebbles; cross bedded; overlain by 2 ft of intraformational conglomerate	40
Sandy mudstone, whitish-grey	16
Conglomerate, contains rounded to sub-rounded quartz pebbles, and black, fine grained, metamorphic pebbles (?) phyllite; hard sandy matrix, ferruginous	6
Sandstone, gritty to coarse grained; pebbly in parts; brown; cross bedded	46
Cap	15
Borehole section situated 2 miles west of section across a flat plain continued below	
Sandstone, coarse to medium-grained brown	45
Sandstone, medium grained; light brown; clayey	115
Mudstone; greyish; with quartz pebbles	130
Total (Nubian Formation)	476

UNCONFORMITY

Basement Complex

In these four areas the Nubian Sandstone Formation has yielded only plant fossils referable to the genus *Dadoxylon*. These, unfortunately, indicate a general Cretaceous age only.

In the Khartoum–Omdurman area the Nubian Formation is best exposed in the Merkhiyat Jebels, north-west of Omdurman and in the isolated buttes such as Jebel Kisi, Jebel Abu Wuleidat, etc.

The Nubian Formation was also penetrated in borehole SGS 1156 in the Merkhiyat Jebels described below, and near by at Jebel Abu Umra 243 ft of Nubian Sandstone was recorded. The total thickness of the Nubian Formation proved in wells and in the Merkhiyat area is then about 550 ft. The rocks constitute the Omdurman Formation (Whiteman 1970).

MERKHIYAT JEBELS

Section recorded in SGS 1156 Merkhiyat Jebels, north-west of Omdurman.

	Thickness (ft)
Top of Section	
Nubian Sandstone Formation	
Sandstone, fine to medium, grained, friable, brown, some pebbles and gravel	10
Sandstone, friable, yellow to brown	50
Sandstone, brecciated, quartzitic	10
Sand clay, yellow-brown to yellow	75
Sandstone, friable, light brown	15
Sandy clay, light yellow	70
Sandstone, quartzitic, pebbles present	5
Sands, loose pebbles and gravels, pink	25
Sand, fine with gravel, pale white	45
Total	305

The Nubian Formation was also proved in the Afad School boreholes, e.g. Afad School borehole Omdurman, SGS No. 366.

AFAD SCHOOL, OMDURMAN

	Thickness (ft)
Clay, yellow-brown, limey	110
Clay, dark brown, limey	30
Clay, grey, limey	40
Clay, yellow-brown, limey	50
Clay, dark grey	70
Mudstone, black carbonaceous	142
Mudstone, purple-brown with hard green bands and pebbles	30
Mudstone, grey and silty with sandstone fragments	27
Siltstone and fireclay (?), grey-whitish and soft	4
Mudstone, grey with calcareous nodules	12
Mudstone, silty, grey	4
Sands, fine grey and silty	5
Sand, white and silty grey	15
Siltstone, grey-white with pyrite nodules	5
Siltstone with carbonaceous streaks	5
Clayey silt and sand, ill-sorted	6
Mudstone, grey silty, dirty white with carbonaceous streaks	10
Sand, grey	5

Silt and mudstone, grey	24
Clay brownish to greyish, sandy	9
Sandstone, clayey, red-brown	6
Sandy mudstone, grey-lavender with ochreous fragments	11
Mudstone, grey	5
Sandy clay, pink-brown and grey	5
Sandstone, red ochreous pebbles and fragments	7
Total	636

It is not clear from the record how much mudstone is actually present in this borehole, because of caving and poor circulation of mud, but elsewhere in the Omdurman area sandy mudstone and mudstone lenses have been proved in excavations and boreholes. Extensive lenses of 'bentonitic' mudstone and mudstone were proved in the foundations for the new Shambat bridge across the Main Nile (Fig. 31) and in the Khor Abu Anga bridge (Whiteman MS engineering report). Mudstones occur in the Barok area south-west of Omdurman. Because the sandy mudstones tend to weather more easily than the sandstones they frequently form low ground or often are covered by talus from sandstones higher in the succession, and blown sand, etc., and therefore may be inconspicuous in the field.

Much less mudstone was recorded in borehole No. 757, Wadi el Adara, 20 miles bearing 313° from Omdurman on the road to Gabrat Said, again showing the great variability of the Nubian Sandstone Formation.

At Merikh borehole, Omdurman (SGS No. 1517) 1003 ft of Nubian Sandstone Formation were drilled. Taking into account the height of the jebel and the depth of the boreholes in the Khartoum area, the flat-lying Nubian Formation must be more than 1550 ft thick.

According to Andrew (1943 MS) the Nubian Formation along the Blue Nile consists of 'the usual, brown-reddish, pink or white soft friable sandstone, with silicified beds (silicrete) near the base, often pebbly. . . . Ferricrete or highly ferruginous sandstones are common and are conspicuous. Hill tops are capped by the rock, which is resistant to weathering, and it also forms craggy outcrops in the plain.'

Andrew (MS 1943) believes that these iron-rich sandstones are not superficial alteration products, although there may have been some redistribution of iron at the surface. He cites the occurrence of ironstone as deep as 160 ft in several bores in the Wadi Seidna area as evidence that the iron is original. However, the depth to the present day water-table, some 400–600 ft below the surface in places, and the deep zone of intense oxidation may well point to superficial concentration even for the Wadi Seidna occurrences.

Farther south on the White Nile, the Nubian Sandstone Formation occurs on the east bank, north of Kosti bridge. It is exposed in a small quarry, which has been used for road metal; the sandstone is most probably down-faulted. The beds are broken and folded anticlinally. They are silicified and iron-stained. Although the lithology is atypical these beds are included with the Nubian Sandstone, for they have yielded some small pieces of *Dadoxylon*.

Writing about the Nubian Formation in the Sennar area, Blue Nile, Ruxton (unpublished manuscript 1956) states 'there are very few outcrops of the Nubian Series in the Sennar area and nearly all the evidence is gathered from well and borehole records. There are no records of the depths of the base of the series except near the Basement Complex rocks on the watershed. The series is at least 190 ft thick and the lowest level proved in this area is 1190 ft (374 m) (R.L. Ref. 364·82 m old Khartoum datum) at Sennar Junction but the base is probably well below this. The Nubian Series continues under the Blue Nile Basin. . . .'

The Nubian Formation is characterized by flat dips in the Blue Nile region and overlies the Basement Complex with pronounced unconformity as elsewhere. It is mainly a group of multicoloured unfossiliferous sandstones, sandy mudstones, and mudstones. They are considerably thicker than Ruxton suggests and attain as much as 400 ft at Eritel borehole, south-south-east of Mangeil. There are a few conglomerates with well-rounded vein quartz pebbles. The sandstones are quartzitic and kaolinitic with few detrital feldspars and ferromagnesian minerals. In places there are thick mudstones lenses.

(x) *Regional description: Kordofan and Darfur*

Rodis, Hassan, and Wahadan (1964) described the Nubian Formation of Kordofan. It consists of sandstone and conglomerate, and mudstone with hard ferruginous and siliceous layers. The strata in general are flat-lying or dip gently northwards and

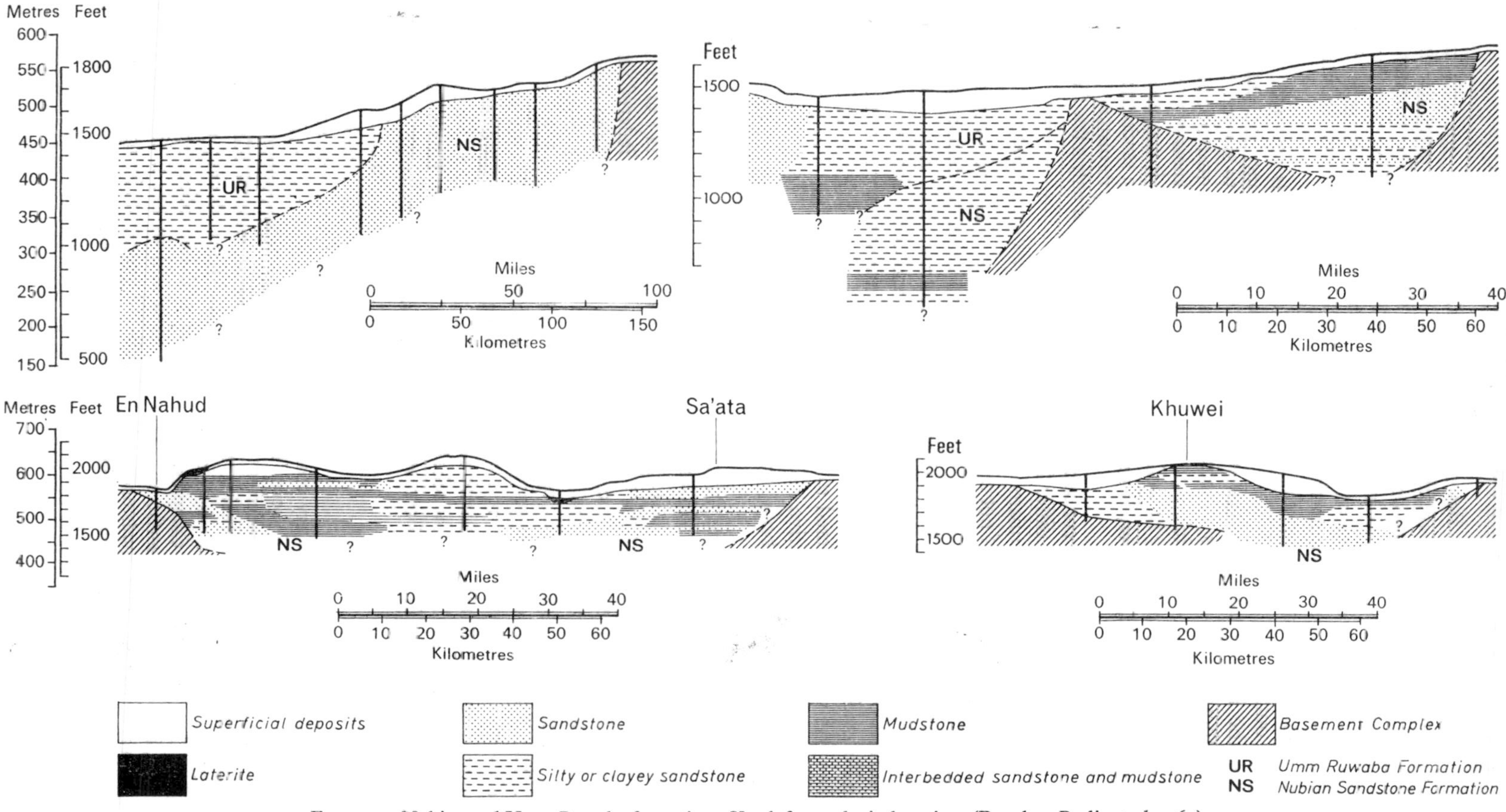

FIG. 37. Nubian and Umm Ruwaba formations, Kordofan geological sections (Based on Rodis *et al.* 1964)

occupy four broad depressions in the Basement Complex. The sandstone grains are sub-angular to rounded and the matrix commonly clayey or silty. In the Nahud area mudstone forms a prominent part of the section and is in places as much as 164 ft thick. The Nubian Formation reaches a maximum thickness of 558 ft in south-western Kordofan and 500 ft in the Nahud outlier.

As a formation the Nubian can be traced over long distances but there are many facies variations between sandstone and mudstone, which can be detected in those areas where the density of wells is high. The bulk of the formation, however, is made up of sandstone (Fig. 37).

The views of Rodis *et al.* (1964, p. 21) concerning the deposition of the Nubian Formation differ from those expressed by the author. They postulated that 'shallow continental seas covered much if not all the province. During this time the clastic sediments of the Nubian Series were laid down in continental and/or near-shore marine environments and over a basement rock surface of considerable relief.'

If the Nubian Formation was deposited in a marine environment then one would have expected to find marine fossils and other features indicative marine sedimentation, as in Egypt. None have been found, however, in the Sudan.

B. G. Karkanis (1965) described the Nubian Sandstone Group in Western Kordofan and Darfur, overlapping the region studied by Rodis *et al.* (1964). The Nubian Formation is said to consist of 'friable to well consolidated sandstone, mudstone, and conglomerate unconformably overlying the basement complex . . . abrupt changes in lithology are common. . . . The Nubian sandstone is generally flat-lying or dips gently to the north; it forms flat or gently undulatively smooth surfaces with isolated flat-topped mountains of relatively low relief.'

Karkanis (1965) divided the Nubian Formation into a basal, middle, and upper 'series'.

The basal series consists of

'an alternation of light brown, coarse- to very coarse-grained silicified sandstone and conglomeratic beds. The basal series is cemented with silica thus rendering the formation very hard and resistant to weathering and erosion; where exposed it forms mesas of basal conglomerate capping hills of Basement Complex. It weathers into rough, iron-stained gravelly surfaces.

The middle series is composed of mudstone and sandstone beds, with mudstone being the dominant lithology. The middle series is multi-coloured and exhibits a variation in grain size from coarse-grained sand to silt and clay. The sandstone is quartzitic; the clay varies in composition with bentonitic clay being most common. The middle series is poorly cemented and friable; it weathers into undulating surfaces with a thick sandy cover.

The upper series is composed of brown and buff, poorly-sorted sandstone. Its grain size ranges from medium to fine sand to silt and clay. The sandstone is poorly cemented, and quartzitic in composition.

The maximum thickness of the Nubian Sandstone in the area is not known; it may exceed 500 ft in places. In the Nubian section the sandstone and the conglomerate are the water-bearing formations, the water being under low artesian head of only a few feet.'

(b) Gedaref Formation

The name 'Gedaref Formation' is proposed to include all those sandstones and sandy mudstones and mudstones that crop out in the area around Gedaref and along the Ethiopian frontier and pass laterally into sandstones of the Setit valley and the Adigrat Sandstone (Fig. 4 and Table 5). Eventually they pass under the westernmost tongues of the Antalo Limestone of Upper Jurassic (Oxfordian) age in Ethiopia. They are therefore probably older than the Nubian Formation as proved in the rest of the Sudan, and it is for this reason that they are classified separately here.

On the Geological Map of Africa, 1963 edition, 1:5 000 000 scale, the Gedaref Sandstone is labelled 'Nubian Sandstone' and the stratigraphical nomenclatural problems that ensue from this wrong description are resolved by adopting an artificial facies change colour system between 'Nubian sandstones' and the '*Continental intercalaire*' colour pattern used for the Nubian Group outcrops between Abu Duleiq, Khartoum, and Wadi Muqaddam. Such a facies change does not exist in the field, however. It is an artificial device to bring in the Nubian Formation, which is finally accomplished along the map join between sheets 2 and 3 of the above-mentioned map (Whiteman 1965). There is therefore no '*Continental intercalaire*' cropping out in the region west of Khartoum as the Geological Map of Africa shows; the investigations by the author and his colleagues between Khartoum and Dongola and in the Elai Jebels (Plate 13), by the Shell–BP geologists in 1959 between Jebel Uweinat and the Nile, and by Sandford (1935, etc.), demonstrate clearly that the rocks extending westwards across the Sudan into Chad and beyond into Ennedi and Mourdi are Nubian Sandstone.

According to Ruxton (MS 1956) the beds now included in the Gedaref Formation consist of conglomerates, sandstones, sandy mudstones, and mudstones and exhibit many of the characteristic features of the Nubian Formation that crops out further west. In many places the sandstones are silicified to such an extent that they are almost quartzites, for example Jebel Matna area. Outcrops are rare, and throughout much of the region the Gedaref sandstones are overlain by crinanite and basaltic intrusive and extrusive rocks.

The total thickness of the formation has not been proved in boreholes but Gedaref town bore 157 (Ruxton MS 1956) proved 640 ft of crinanite, followed by 415 ft of Gedaref Formation. Grits, fine sands, calcareous sands, chert, and brecciated limestones occur in the boreholes. Unfortunately no fossils were recorded and it is impossible to determine whether the limestone is marine or freshwater from the evidence available. The extent of the Jurassic sea in the Ethiopian marches is still very much a matter for speculation and the size and the shape of the Gedaref Basin are also unknown.

Ruxton (MS 1956) also recorded three outliers of the Gedaref Formation: one at Jebel Kasamor, north of Gedaref where the sandstone reaches a maximum level of 2160 ft above river level: another at Jebel Umm Biliel, south of Gedaref where conglomerate and breccias occur dipping north-eastwards at 30°; and a small capping of quartzite probably belonging to the Gedaref Formation which occurs on one of the jebels of the Jebel Salmin Group at about 3144 ft above river level.

The position of the base of the Gedaref Formation in the Gedaref region varies between 800 and 1600 ft above sea-level and between 800 and 2144 ft if the Jebel Salmin occurrence is included. This may be due to variations of pre-Gedaref topography or to post-Gedaref earth movements. There is no doubt that considerable movements have taken place and large-scale warping is suggested by Ruxton along north-north-east–south-south-east axes in the Ethiopia marches. Mohr (1962), however, favours faulting.

(c) Yirol Group

The Yirol Formation, according to Andrew (*in* Tothill 1948), consists of an isolated succession of sandstones and mudstones, with the latter prepominating, occurring in a variety of colours from lavender, purplish grey, brown or red to pink. The formation is micaceous and contains some thin quartz pebble beds. It is known only from wells in the Yirol and Tali Post areas, Equatoria Province, and is 16–20 ft thick at Yirol and 73 ft at Tali.

Andrew (*in* Tothill 1948) believes that the Yirol is older than the ?Tertiary laterite formation and resembles the Nubian Formation in lithology. On the Geological Map of Africa, 1963 edition, the Yirol Formation, like the Nawa Formation, is shown in stripes of Carboniferous and Jurassic colour which give a wrong impression of age. Either Nubian or Mesozoic undifferentiated would have been a better choice for the Yirol Formation (Whiteman 1965).

(d) The Fareig basalts

Between Kerma and Delgo the Nile makes a prominent right-hand bend, extending from Arduan Island, where the river swings east–west, to the Kaibar (or Kagbar) Cataract near Fareig, where it again turns north (Fig. 22, Plate 9).

In this region the Nubian Sandstone caps some of the jebels, for example Jebel Ali Barsi, and in places sweeps down to the banks of the Nile, as at the bend of the Nile near the Kaibar Cataract. The Nubian Formation in the bend is gently folded synclinally but in general it appears to have been downfaulted. However, if the whole of the region northwards from Abu Fatima's Tomb to south of Delgo is considered, the structure may well be part of an eroded buried hill, as at Sabaloka near Khartoum. The Nubian Formation and the Basement Complex in the Fareig area are cut by dykes and volcanic necks at a number of localities. These consist mainly of basalts which show the following features under the microscope. Specimens KG 1641 and 1637 are both olivine-basalts, fine-grained, holocrystalline, and fresh. They are made up almost entirely of forsteritic olivine, labradorite, augite, and opaque iron ore. Plagioclase laths show a crude flow orientation. Deuteric alterations and alterations due to weathering have produced serpentine, green chlorite, and saussurite but on the whole these effects are negligible. On the 1:4 000 000 maps of the Sudan they are classed as Tertiary.

Two potassium–argon determinations were made

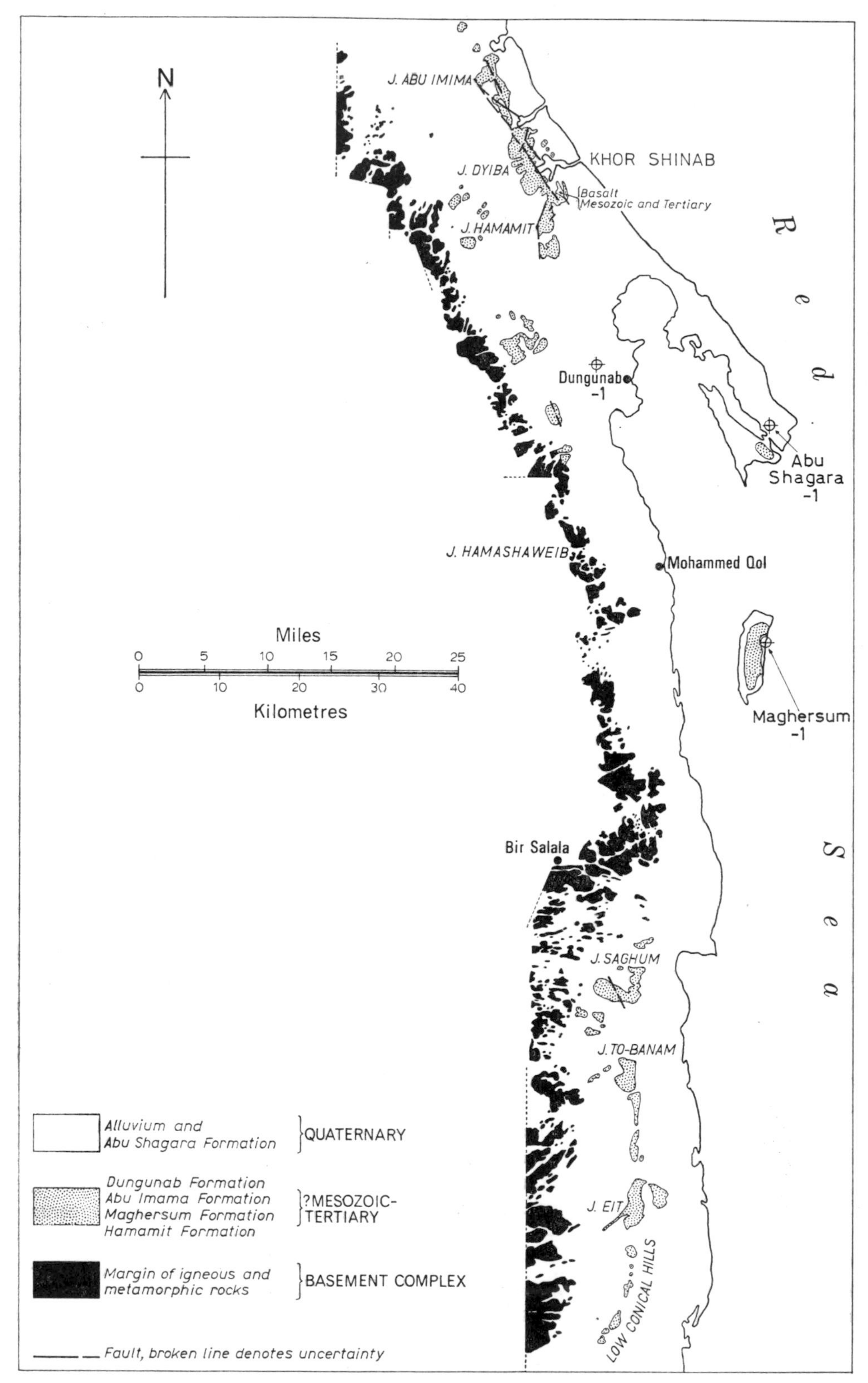

FIG. 38. Main outcrops of Mesozoic, Tertiary, and Quaternary formations, Red Sea Littoral. (Based on Willis *et al.* 1961 and Carella and Scarpa 1962, University of Khartoum geophysics expedition Red Sea Hills 1962)

on the above-mentioned basalts by Geochron Laboratories, Inc.

KG 1637. Whole rock, −40/+80 mesh. Volcanic rock from jebel on east bank of Nile, 2 miles 190° Fareig, Sudan. Age, 100 (± 15) × 10^6 years.

KG 1641. Whole rock, −40/+80 mesh. Volcanic rock from jebel on east bank of Nile, 3 miles 114°S Fareig, Sudan. Age, 88 (± 8) × 10^6 years.

Despite the reasonably large errors in these determinations, owing to the low potassium content, it is reasonable to assume that the necks were intruded in the early part of the late Cretaceous and are of Turonian, or earliest Senonian, age. The age for the Nubian Sandstone Formation in the Sudan is consistent then with the ages deduced for southern Egypt. It is possible that other so-called 'Tertiary basalts' neary by are also of this age.

(e) Khor Shinab and the basalts of the Red Sea Littoral

Several outcrops of basic volcanic rocks have been recorded along the Red Sea coast, interbedded with and intruded into stratified rocks that in the past have been assumed to be of Tertiary age (Fig. 38). Recently, however, a potassium–argon determination indicating a Jurassic or early Cretaceous age was made on a basalt obtained by the author from the south end of the Jebel Dyiba structure near Khor Shinab, in the northern section of the Red Sea Littoral. Lithological similarity and occurrence points towards a similar age for other basic volcanics of the Red Sea coast. Acceptance of this date necessitates revision of current ideas on Red Sea palaeogeography.

Lees and Wyllie (1923–4), while making a reconnaissance survey of the petroleum prospects of the Red Sea coast, recorded a basaltic intrusion some 50 ft thick resting on sandstones and shales and overlain by variegated paper shales in the Khor Shinab area. The shales are 'locally indurated, contorted, and veined with (?) calcitic tongues of basalt and were seen to transgress the bedding, and the basalt margin sometimes enclosed pieces of baked shales'. The anticlinal structures that occur in the region were thought to have been caused by the basaltic intrusions (Lees and Wyllie *in* Abdullah 1955, p. 166). Gass (1955) studied the basaltic rocks of the coastal plain and came to the conclusion that they are flows and are not intrusive into the coastal plain sediments. Kabesh (1962, pp. 52–4) also described the lava flows of the Red Sea coast and plotted their distribution on a map.

The rocks are of doleritic type and are especially well developed near the base of the sedimentary sequence, for example west of Khor Eit Well and near Jebel Saghum. Often the tops of the flows are obscured by talus but the underlying sediments are clearly baked. In the few localities where the tops of the flows have been studied the presence of vesicles and weathered zones and the absence of baking in the overlying sediments indicates clearly that we are dealing with flows.

Kabesh (1962) described the rocks as hard, massive, fine- to medium-grained rocks, greyish-green to dark-brown in colour which show onion-like weathering. The flows in general dip seaward at dips ranging from 4° to 20° and in places are cut by faults.

In thin section the rocks are seen to be porphyritic olivine-basalts. Kabesh (1962) assigned them to the late Tertiary volcanic episode that has been recorded in several parts of Egypt and elsewhere among the sedimentary rocks of the Red Sea coast (Andrew 1937, Rittmann 1954, Sabet 1958). He pointed out, however, that little work has been done on the macrofauna of the Red Sea sediments of the Sudan and that it is not possible to assign a more precise age than Tertiary (possibly Miocene) to these rocks and the flows.

During the University of Khartoum Red Sea Hills geophysics expedition in 1962 the author studied the basalt flows and sills of the Khor Shinab–Jebel Dyiba area, Jebel Hamamit, Jebel Saghum, and Khor Eit. In the Jebel Dyiba–Khor Shinab area the basalts are exposed at the southern end of the faulted Jebel Dyiba anticline. They dip seaward at more than 25° in places and are overlain unconformably by the Abu Imama Formation, which according to Carella and Scarpa (1962) is of medial Miocene age (Tortonian). Both these formations are overlain and are succeeded unconformably by Raised Reef Complex deposits of Monastirian age (Berry, Whiteman, and Bell 1966). Fossils were not found by the author in the Abu Imama Formation immediately above the basalt flows but they appear about half a mile east of the top of the flows.

Apparently Carella and Scarpa (1962) identified the Abu Imama Formation in the Jebel Dyiba area on lithological grounds.

The formation underlying the basalt flows in the Khor Shinab area according to Carella and Scarpa (1962) is the Maghersum Formation. At Khor Shinab it consists of sandstones and shales that have been baked by the basalt flows. These are assigned to the A Member of the Maghersum Formation and classed as Middle Miocene (Helvetian), again it would seem purely on lithological grounds. These beds appear to be faulted against the Abu Imama Formation, which passes upwards into the predominantly gypsiferous and salt-bearing Dungunab Formation. In the type locality, according to Carella and Scarpa (1962), the Dungunab Formation consists of the A Member, massive white rock-salt with minor grey anhydrite 2105 ft (658 m), and the B Member, predominantly gypsum with interbedded clays and sandstones 173 ft (54 m) thick.

In Dungunab–1, some 15 miles to the south, a basalt layer 75 ft thick was encountered between the B Member and the Abu Imama Formation. The age of this formation is also uncertain because no fossils have been found. However, because of its stratigraphical position, it is considered to be post-Middle Miocene. Volcanic tuffs are said to underlie the Maghersum and overlie the granitic Basement Complex.

Basalts were also encountered in Abu Shagara–1 below the predominantly sandstone-conglomerate section of the Maghersum Formation. Carella and Scarpa (1962) assumed that the basalt is closely associated with the Basement Complex although they think that the Hamamit Formation may be present.

From the subsurface evidence it can be shown that tuffs occur below the Maghersum and above the Basement Complex and between the Maghersum Formation B Member and the Abu Imama Formation, both assigned to the Middle Miocene by Carella and Scarpa (1962).

The question now arises about the validity of the dating of the formation mentioned above in the Khor Shinab area. Only the A Member of the Maghersum Formation has yielded diagnostic fossils and these indicate a Helvetian (Middle Miocene) age. The B Member has yielded fossils but these are not diagnostic. In a short search in the Khor Shinab area the author did not find a recognizable macrofauna and, as noted above, it would appear that dating of the Maghersum Formation in the Khor Shinab area was done mainly on lithological grounds by Carella and Scarpa (1962). In fact the pre-Abu Imama Beds can all be older than Miocene, as the potassium–argon evidence presented below may indicate.

According to the analytical data the basalt flow at Khor Shinab forming the south end of the Jebel Dyiba anticline must be of Jurassic to early Cretaceous age. Unfortunately the amounts of potassium and argon in the sample were extremely low but Geochron Laboratories have written: 'In conclusion, I would treat this age determination as follows: I would consider it as analytically correct within the errors given. This means that an age of Jurassic to early Cretaceous must be considered a very real possibility for this sample. I would add however the reservation that the age may be in error because of certain geological reasons beyond the control of this laboratory.'

RSH 143. Basalt, Khor Shinab area, north of Port Sudan, south end of Jebel Dyiba anticline, Kassala Province, Sudan. Age, 148 ($\pm$ 30) $\times 10^6$ years.

Dr. I. G. Gass (personal communication 1965) has, however, obtained an age of $24 \pm 5 \times 10^6$ years for a basalt from the Khor Shinab area. These discrepancies are not easy to explain if the specimens were collected from the same outcrops. The possibilities are: that they were collected from basalts outcrops of different ages; that the Geochron result is too high because of analytical error; or, if the analytical data are correct, that the rock was contaminated at the time of crystallization by variable amounts of radiogenic argon (Whiteman 1968*a*, Gass 1969).

In support of a Jurassic or early Cretaceous age for the Khor Shinab basalts we may cite the evidence put forward by Carella and Scarpa (1962, pp. 16–17) for a Cretaceous or Palaeocene age for the Mukawwar Formation. This formation was recorded in Maghersum–1 and consists of grey to red-brown silty shales interbedded with grey quartzitic sandstones and occasional limestones. The formation has yielded mollusc casts, and a microfauna with echinoid spines, fish teeth, *Bracocythere* cf. *ledaforma* (Israelsky) *Brachycythere* sp., *Isohabrocythere* sp., *Buntonia* sp., *Bairdia* sp., and other microfossils. On the basis of the ostracod assemblage the formation is assigned to the 'Upper Cretaceous transitional to Palaeocene'. It is interesting to note that Karpoff (1957) has recorded marine marls of Maestrichtian age near Jeddah on the Arabian coastal plain and that these outcrops are almost opposite the location of the Maghersum-1.

The evidence given above may indicate that a gulf of Tertiary and late Cretaceous Tethys extended southwards along the Red Sea depression (Whiteman 1968). Information about its ultimate extension must await publication of new exploration data from Eritrea and Arabia. The shore line in Nubian Sandstone times in southern Egypt may well have swung south from the Aswan area across the northern Red Sea Hills, and it is indeed unfortunate that the isolated outcrops of Nubian Formation in the northern Red Sea Hills of the Sudan have not yielded marine fossils. Kabesh (1962, p. 48) states that these outcrops are of marine origin, but his arguments are not convincing, for they are based on lithological comparison with other areas where the marine origin of the Nubian Formation is also in doubt (Whiteman 1968*a*).

4. TERTIARY FORMATIONS

The Tertiary formations of the Sudan include:

(a) marine, lagoonal, and continental formations of the Red Sea Littoral, the islands and the Suakin Archipelago;
(b) Asotriba volcanic rocks of Dungunab sheet 36I;
(c) Umm Ruwaba and el Atshan Formations of south central and southern Sudan;
(d) Hudi Chert Formation of the Nile Valley and Wadi Muqaddum and Wadi el Melik areas;
(e) Volcanic rocks of the Jebel Marra, Kutum, and Melit areas, Berti or Tagabo Hills, Meidob Hills, and isolated outcrops in Darfur;
(f) Bayuda volcanic field, Jebel Nakharu Basalt, isolated outcrops of 'Tertiary lavas' south-west of Laqiya Arbain and north of Dongola, and Jebel Kuror;
(g) 'Infra-basaltic sediments', basalts, and rhyolites of Khor Langeb, Kassala Province;
(h) Gedaref and Gallabat basalts, crinanites, etc. of the Ethiopian Marches;
(i) Sennar teschenite;
(j) Akoba and Boma basalts of the south-eastern Sudan;
(k) Jebel et Toriya and Omdurman basalts;
(l) laterites, lateritic soils, iron caps, silicretes, etc. of the southern and central Sudan.

(a) Tertiary formations, Red Sea Littoral, the islands, and Suakin Archipelago

Lees and Wyllie (1923–4) made a reconnaissance expedition in the Red Sea Littoral and described the petroleum prospects of the region. According to Abdullah (1955, pp. 165–6):

'They determined the age of the Red Sea coastal belt to be Plio-Pleistocene. They recorded very few fossils amongst which was a *Pecten vasseli*, which they determined was of Indian Ocean derivation, and ruled out the existence of any fossils which could have come from the north, hence they concluded that the Miocene sea did not penetrate as far as the Sudan.

In one locality, Mersa Shinab, they described a basaltic intrusion as being injected in the sediments. They described this basaltic intrusion sheet as being 50 ft thick, resting on sands and shales, and overlain by variegated paper shales, the shales being locally indurated, contorted and veined with calcite tongues of basalt were seen by them to transgress their bedding, and the basalt margin sometimes enclosed pieces of baked shales.

They considered the area to have domal or anticlinal structures, caused by the basaltic intrusion. They concluded that the source rocks for oil were lacking and that the Cretaceous and Miocene rocks, from which the Egyptian oil draws its supplies were not represented at the surface, and may never have been deposited or if deposited may have been broken up by denudation and carried into deeper parts of the Red Sea trough, and that the structures observed, did not warrant finding oil in the coastal belt of the Red Sea.'

Grabham (1935) made a reconnaissance tour from Port Sudan to Halaib and noted the relationship between the coastal basaltic intrusions and the sediments, and details of the unconformable relationship of the Tertiary formations and the Basement Complex.

Andrew (*in* Tothill 1948 and unpublished MS 1943) gave a general account of the geology of the Red Sea Littoral and pointed out that the generally accepted view is that the Red Sea was formed in mid-Tertiary, post-Eocene, probably Oligocene times. He reiterated that no sediments older than 'Plio-Pleistocene' by analogy with those of the southern part of the Gulf of Suez in Egypt, have been found along the coastal strip of the Sudan. The 'Plio-Pleistocene' beds are limestones, often shelly or coralline, shales, marls, clays, grits, conglomerates, and gypsum beds. They are overlain by recent raised coral reefs, and near the foothills by thick and very coarse angular river terrace deposits. The characteristic fossils are *Pecten vasseli* and *Laganum depressum* with a small variety of lamellibranchs and gasteropods.

'Dips are generally seawards, less than 30°, and are probably original. Local fold-structures occur and some faulting, with dips of 20° or more near the faults. In several places the series is intruded by basaltic sills or dykes. The thickness of the series, measured on low hills of the coastal plain, is over 100 m (Jebel Eit) lying on the basement complex, but it is fairly certain that the base is very irregular and a greater thickness would be found in bores.

A bore near Port Sudan reached a depth of 788 ft (246·3 m), through a monotonous succession of clayey grits, of unknown age. There is a probability that much of this succession is Miocene but of this there is no direct evidence, and the succession must be regarded vaguely as Upper Tertiary to Pleistocene' (Andrew, unpublished MS 1943).

Concerning the wadi deposits of the coastal plain Andrew wrote: 'The wadis which debouch onto the coastal plain are flanked by several series of terraces built of a very coarse conglomerate or breccia of large boulders, some rounded, some angular, up to half a metre in diameter, in a sandy matrix . . .' (Andrew, unpublished MS 1943).

Abdullah (1955, pp. 166–7) challenged Lees and Wyllie's interpretation of the 'dome-like' structures of the Red Sea Littoral pointing out that the volcanic rocks are in the main concordant with the sediments. At Shinab the paper shales reported by Lees and Wyllie were found to contain fish fossils. The occurrence of black shales with carbonaceous plant fossils on Mukawwar or Maghersum Island was recorded by Abdullah (1955).

Gass (1955) described the sediments and volcanic rocks of the Littoral, and Kabesh (1962, pp. 50–1) gave a brief account of these rocks in the *Mohammed Qôl Memoir*. In the absence of precise palaeontological evidence Kabesh (1962) simply assigned the outcropping sediments to the Tertiary and criticized the Plio-Pleistocene age assigned to them by Gass (1960, unpublished Petroleum Report, General Exploration Company of California).

The geologists of the General Exploration Company of California produced two reports dealing with the coastal strip and the company have kindly made these and other information available to the author. The first report, *Geologic study of the Coastal Province, Sudan between latitudes* 20° 10′ N *and* 21° 52′ N was written by Willis *et al.* (1961). The region studied was divided into two main blocks and said to differ

'both as to stratigraphy and geological structure, and therefore as to oil possibilities. The more westerly consists of the greater part of the mainland area of the coastal plain and will be referred to as the mainland block. The more easterly includes the peninsula known as Ras Rawaya, Mukawwar Island and a strip of coastal waters extending from five to fifteen miles from shore. It is herein designated the marine block. The marked differences between these two can best be accounted for on the justifiable assumption that they lie on opposite sides of one of the larger faults of the Red Sea graben or rift system. Large scale lateral displacement on this fault is indicated, the east block having moved south relative to the west block, thereby bringing into proximity sediments originally laid widely separated localities.

'The topography of the mainland block consists of a flat plain of deposition, broken locally by residual hills of Pliocene–Pleistocene sediments. The plain merges to the west with the mountain area where the basement complex is exposed. There is no evidence of warping or displacement of the topographic surface by recently active folding or faulting. The Plio-Pleistocene sediments lie directly on the Basement Complex at the western margin and dip prevailingly eastward. They are thought to thicken eastward from zero to an estimated maximum of 5500 ft at the coast. The east dip is interrupted along two trends which form sinuous, irregular structures with the appearance of anticlines, namely the Mersa Arakiya and Mersa Shinab structures . . .'

Dealing with the stratigraphy of the region Willis *et al.* (MS 1961) wrote:

'The Plio-Pleistocene sediments of the mainland block, exposed in the residual hills, which rise above the plain, are quite different from those of the marine block as exposed on Mukawwar Island. The former consist predominantly of coarse clastics and evaporites, with smaller percentages of coral limestone, chalk and marl; the latter contain no pure evaporites, but a larger percentage of fine clastics and limestone. Thin bands of oil shale have been noted in them.

These differences strongly suggest deposition in distinct geological provinces with different stratigraphic and structural histories.'

This last comment is not supported by the field evidence, nor are the structural interpretations (Chapter V).

The second report of the General Exploration Company of California was written by Bickel and Peters (MS 1962) and is more detailed concerning the stratigraphy and structure. New palaeontological evidence enabled the section to be dated on Mukawwar Island and the mainland as Miocene and Pliocene. Stratigraphical and palaeontological details are given below.

Berry, Bell, Qureshi, and Whiteman visited the Red Sea Littoral in 1962 and made a reconnaissance gravity and geological survey between Suakin and Khor Shinab.

Carella and Scarpa (1962, pp. 1–23) published some of the geological results of oil exploration in the Sudan by AGIP Mineraria (Sudan) Ltd. Detailed definitions and descriptions of the main surface and subsurface rocks units were given, based on field surveys started in 1959 and information from exploratory wells Durwara–1, Dungunab–1, and Abu Shagara–1. This is the most complete account available to date. The well log-data are scrambled, however.

Sestini (1965) described the Cenozoic stratigraphy and depositional history of the Red Sea coast of the Sudan. Comments upon his age designations are given below by S. V. Bell.

The formational details that follow were abstracted mainly from Carella and Scarpa (1962) (Table 7).

(i) *Mukawwar Formation* (*late Cretaceous or Palaeocene*)

This name was introduced by Carella and Scarpa (1962, pp. 16–17) and has its type locality in Maghersum–1 (20° 39′ N, 37° 00′ E). The formation was drilled between depths of 1560 and 1761 m and is said to be of Upper Cretaceous or Paleocene age. The corrected thickness of the formation is about 190 m (Fig. 39).

The Mukawwar Formation consists of shales, silty, dark grey to red brown, indurated, fine to medium grained, quartzitic sandstone, with subordinate marls and limestone beds. It rests unconformably on a basaltic, tuffaceous breccia, which rests on basalts and porphyries. The porphyries probably belong to the Basement Complex but the age of the basalt and tuffaceous breccia is unknown. The formation is overlain, probably unconformably, by quartzitic and arkosic sandstones of the Hamamit Formation.

The fauna from the Mukawwar Formation collected between 1637 and 1638·7 m (Carella and Scarpa 1962) is described below in the section on Palaeontology.

On the basis of the ostracod assemblage the Mukawwar is dated as 'Uppermost Cretaceous transitional to Palaeocene', and the depositional environment is described as brackish to shallow marine (see Chapter III, Palaeontology). This conclusion is extremely important from a stratigraphical point of view, for the deposit occurs almost opposite the Série d'Asfan, a group of marine marls, limestones, and sandstones of Maestrichtian to Eocene (?) age that occur north and east of Jeddah, Saudi Arabia and rest unconformably on the Série de Medina, part of the Basement Complex (Karpoff 1957).

The Série d'Asfan is said to be over 1000 m thick and has yielded the following fossils from a limestone bed (identified by Mlle. D. Mongin): *Cardita* (*Venericardia*) *ameliae* Peron, *Cardita* (V.) *ameliae* var. *orfellensis Rossi-Ronchetti*, *Corbula striatuloides* Forbes, *Lucina* cf. *desioli* Ch.—Rispoli, and Gasteropoda indet. Karpoff (1957) stated that this fauna points to a Maestrichtian age and that 80–100 m of unfossiliferous limestone may well be of Palaeocene and Eocene age.

This information, together with the data from Maghersum–1, indicates that the sea was present in the central Red Sea area during late Cretaceous and early Tertiary times, a point hitherto not generally appreciated in Red Sea stratigraphy. Whether the connection was with the Mediterranean or with the Indian Ocean is uncertain. The author favours connection with the Mediterranean, but there is no definite proof of this (Whiteman 1968*a*).

(ii) *Hamamit Formation* (*Eocene or Lower Miocene ?*)

The type locality of this formation is given as the central part of Jebel Hamamit (21° 16′ 50″ N, 37° 01′ 20″ E) by Carella and Scarpa (1962, pp. 14–16). Hamamit sandstones occur at Jebel Hokeb, south of Tehudet; and are present in Maghersum–1 (Figs. 38 and 39).

At the type locality the formation is 226 m thick and consists of sandstones, generally coarse, but

TABLE 7

Correlation chart for Tertiary formations in the Sudan

Nile Valley	Areas east and west of Nile	Red Sea
Integration of Nile drainage		Older terrace gravels Khor Eit
		U
		Pleistocene beds Khor Eit?
Umm Ruwaba Formation and El Atshan Formation	Formation of Jebel Marra, Melit, Meidob, Nakhara, and Bayuda volcanic rocks. Laterite formation age uncertain (Oligocene—Miocene?)	
		U
		Dunganab Formation (post-mid-Miocene unfossiliferous)
		U
Lacustrine and riverain clays and sands, with fossil fish, pigs' teeth etc. Proto-Nile deposits derived from Ethiopian volcanic plateau and East African and Nile Congo divide		Abu Imama formation (mid-Miocene—Tortonian)
		U
Initiation of proto-Blue Nile, Atbara, Rahad, Dinder, Sobat drainage consequent on formation of Ethiopian plateau		Maghersum Formation (A= mid-Miocene—Helvetian)
		U
Hudi Chert with Oligocene molluscan fauna deposits, in an extensive lake in proto-Nile valley	Gedaref erosion surface intrusion of Gedaref crinanites and basalts, 33×10⁶ years K—Ar	Hamamit Formation (?Lower Miocene to? Eocene)
No deposits, Pre-Hudi and Gedaref basalt time predominantly a period of erosion. Marine conditions extended approximately as far south as Aswan and as far west as Aden and Somalia Formation of Late Cretaceous—Early Tertiary Shendi and Togni surfaces		Mukawwar Formation (marine) 190 metres 190 metres uppermost Cretaceous or Palaeocene

...toral		Regional tectonic history		Time-scale
	Abu Shagara Formation? Quaternary fauna	Nile Valley etc.	Red Sea area	Pleistocene
U				U
		Subsidence in the Umm Ruwaba and El Atshan depressions	Subsidence in mid-Pliocene Red Sea connected with Indian Ocean Central trough formed	Pliocene
U				U
...member	Salt, anhydrite interbedded 658 metres		Miocene lagoon extends south from Suez	Miocene U
...member	Gypsum interbedded with clays 54 metres	Tertiary volcanic rocks formed at scattered localities in Sudan		
U				
...member	Reef limestone 62 metres		Marine transgression	
...member	Calcarenite 25 metres			
...member	Conglomerates of igneous rocks and sandstone 50 metres			?
U				
...member ...h fauna)	Marls, sandstone, conglomerates with gypsum veins 111 metres	Downwarping and downfaulting along eastern margin of Ethiopian plateau?	*Globigerina* marl transgression in Suez region and to south; lagoonal conditions poorly developed in Egypt	Miocene Middle
...member ...w fossils)	Gypsum anhydrite with sandstones and shales 859 metres			
...member	Rock salt anhydrite and marls 465 metres			
U		U		
		Climax of Rift Phase 2: separation and formation of Red Sea central trough, separation approx. 80 km		?
Sandstones, quartzitic conglomeratic; sandstones are badly sorted; contains basaltic lava flow at type-locality	226 metres	Extrusion of main Ethiopian plateau lavas from fissures and vents		Miocene L; Oligocene
		U		
		Main uplift of of Arabo-Ethiopian—Sudan swell and formation of Phase 1: separation 130 – 170 km		Eocene U
		U		
				Eocene M
				Eocene L
Maghersum A-member may well be Lower Miocene (Burdigalian) (Whiteman and Bell MS)		Regession of sea. Volcanicity started as early as 69×10^6 years ago, Abbai basalts		Paleocene
				?
				Cretaceous

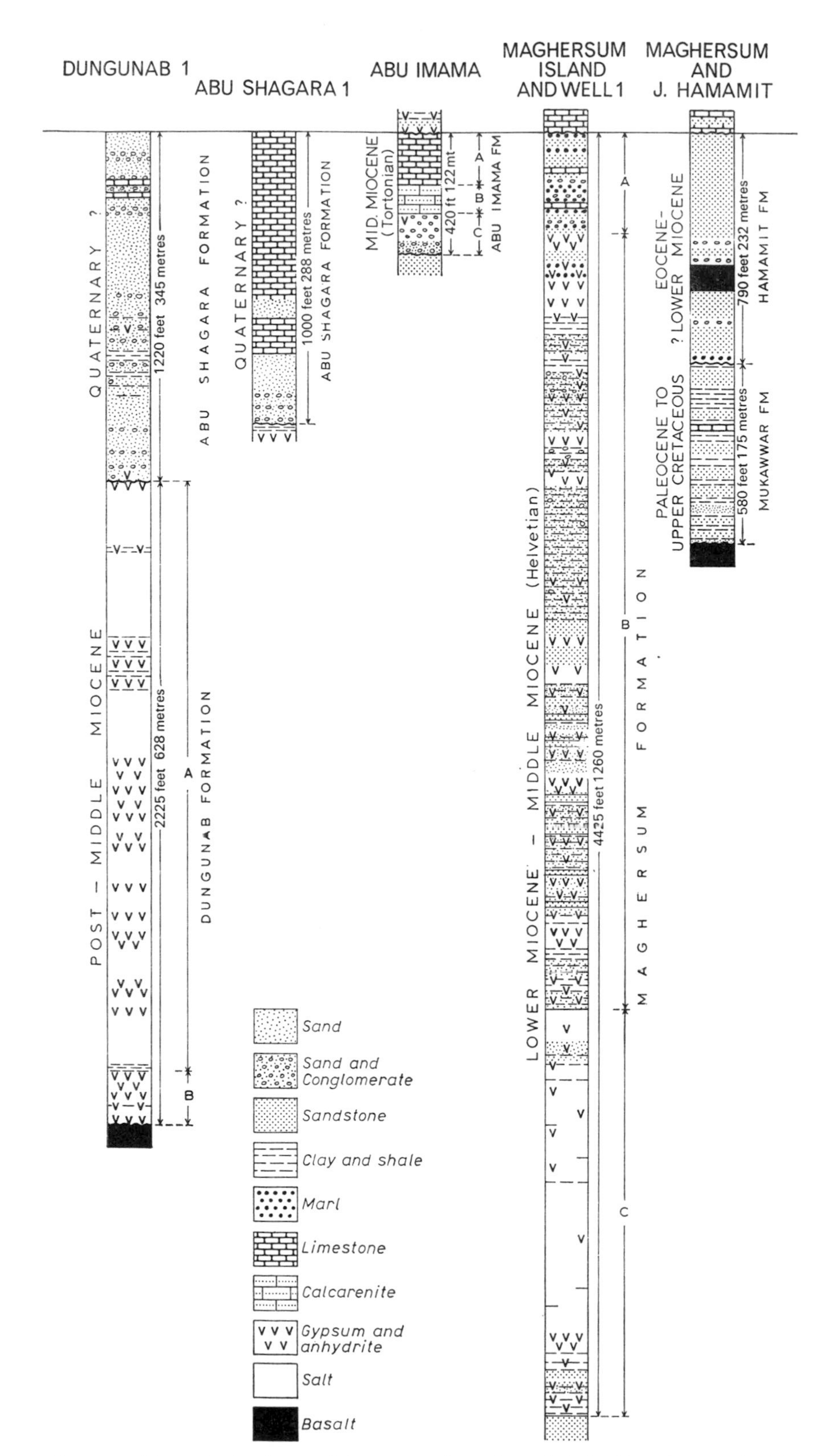

FIG. 39. Mesozoic, Tertiary, and Quaternary formations, Red Sea Littoral, determined in AGIP boreholes, geological sections (See Fig. 38 for location)

sometimes fine, red brown to green, cemented with iron oxides or silica. Occasionally it contains lenses of conglomerate and red shale and is sometimes arkosic. At the type locality a basaltic lava flow, averaging 30 m thick, is intercalated in the sandstone.

At Jebel Hamamit the formation conformably overlies green calcareous shales but in Maghersum–1 these sandstones unconformably overlie Mukawwar shales, marls, and sandstones. The younger Abu Imama Formation truncates the Hamamit Formation at the type locality.

The formation has not yielded fossils and therefore can only be dated on the basis of its stratigraphical position as (?) Eocene to (?) Lower Miocene. In fact no deposits of Middle Eocene to Miocene age have been recognized in the Red Sea Littoral to date, which is indeed unfortunate from a stratigraphical point of view because it prevents us from recognizing the effects of the catastrophic events that took place to the south of Ethiopia and in the Red Sea during this interval of time and which must have affected the Sudan.

(iii) *Maghersum Formation (Middle Miocene Helvetian and ? Lower Miocene*

Fortunately the A Member of this formation has yielded a rich fauna which, according to Carella and Scarpa (1962, p. 13), indicates a Medial Miocene (Helvetian) age.

The type locality of the formation is given as the north-eastern part of Maghersum Island (20°49′ 15″N, 37°16′10″E) and AGIP's Maghersum–1, the coordinates for which are given in the description of the Mukawwar Formation. The formation was drilled in Maghersum–1 between 18 and 1285 m, in Dungunab–1 (1247–1340 m), and in Abu Shagara–1 (below 324 m). Outcrops occur at Jebel Abu Imama, Jebel Dyiba, Jebel Hamamit, Ras Abu Shagara, Jebel Harriam, Jebel Hokeb (although it is not shown as such in Carella and Scarpa sheet 1), Jebel Saghum, Jebel To-Banam, Khor Eit, Gaudiloab, and 'Conical Hills' (Fig. 38).

The formation is 1435 m thick at Maghersum Island, where some 188 m are exposed; the remainder was penetrated in Maghersum–1. The formation is said to vary considerably in thickness and only 66 m were recorded in Dungunab–1.

Carella and Scarpa (1962) divided the formation into the following members (described from top to bottom (Fig. 39).

(1) *A Member*. Marine marls, sandstones, conglomerates, with subordinate thicknesses of limestone constitute the main rock-types. The marls are generally silty, green to red, and fracture conchoidally. The sandstones are mostly fine-grained with quartz elements and a calcareous cement. Conglomerates consist of well-rounded pebbles mostly of igneous origin but pebbles of sandstone and marl also occur.

Gypsum is present in the A Member in veins and nodules. The total thickness of the Member is 111 m. The association of gypsum and marine fossils poses considerable palaeo-ecological problems.

(2) *B Member*. This unit consists of gypsum and anhydrite interbedded with sandstones and sandy shales and contains scattered pebbles. The evaporites are generally finely crystalline and sometimes fibrous, white to grey, and often appear contorted. The sandstones are yellow to red and vary from fine to coarse and contain scattered nodules and lenses of white gypsum. The matrix of the sandstones is either argillaceous or anhydritic.

On Maghersum Island the combined surface and subsurface thickness is about 859 m, a considerable part of which consists of gypsum and anhydrite (Fig. 39). It is unfortunate that information on the proportions of anhydrite to gypsum was not given for these wells because of their differences in specific gravity and their effect on the interpretation of gravity values.

(3) *C Member*. This consists of massive rock-salt with minor bands of anhydrite and silty marls, especially at the bottom of the section. The total thickness is 465 m.

Fossils found in the A Member at Jebel Abu Imama and Khor Eit included a microfauna, which, according to Carella and Scarpa (1963, p. 13), indicates a Middle Miocene (Helvetian) age.

Marine fossils have been recorded in other sections including the type locality but no index fossils have yet been recognized.

The B Member has also yielded poorly preserved but non-diagnostic marine fossils on Maghersum Island and at Ras Abu Shagara.

The fauna of the A and B members is described below (Chapter III, Palaeontology).

According to Carella and Scarpa (1962) the fauna of the A Member has some elements in common with the '*Globigerina* Marls' Formation of the Gulf of Suez, Egypt.

Peters and Bickel (MS 1962) also recorded microfossils of Miocene age from Mukawwar Island. These were identified by Mr. Hughes, who reported: 'The forams in all the Sudan samples from Mukawwar Island represent a shallow water facies. The absence of a planktonic (*Globigerina* and *Globorotalia*) population makes positive dating difficult. However, we believe the occurrence of such forms as *Dendritina* and *Flosculinella* indicate the presence of the Miocene.'

There is some uncertainty about correlating the detailed sections and fossil localities listed by General Exploration Company (Bickel and Peters MS 1962) with the scrambled stratigraphical data presented by Carella and Scarpa (1962) in their stratigraphical logs. However it is probable that all the fossils recorded below by Bickel and Peters are from pre-Abu Imama horizons and occur in the A Member of the Maghersum Formation of Carella and Scarpa (1962).

Bickel and Peters (MS 1962) also recorded fossils from the Abu Shagara Peninsula (Rawaya) where, according to Carella and Scarpa (1962 Sheet No. 1), the Abu Imama and Maghersum Formations crop out at Jebel Abu Shagara. These are included in Sample 2a–2 (see Chapter III). Their age is said to be Miocene.

A microfauna was also recorded by Bickel and Peters (MS 1962) from Jebel Abu Imama from sandstones and sandy limestones outcropping on the eastern limb of the Abu Imama anticline. They include Samples A–31–1, A–31–3, A–31–4, A–31–6 (see Chapter III, Palaeontology).

(iv) *Abu Imama Formation* (*Middle Miocene Tortonian*)

The type locality of this formation is Jebel Abu Imama, north-east flank (21°27′05″N, 36°57′15″E). It crops out also in the Delau Hills, Jebel Dyiba, Jebel Hamamit, Ras Abu Shagara, Mayetib Island, Jebel Harriam, Jebel Saghum, Jebel To Banam, Guadioloab, Khor Eit, and Conical Hills (Fig. 38).

The formation attains a thickness of 137 m at the type locality and the maximum exposed thickness is 143 m at Khor Eit. The formation was also penetrated in Dungunab–1 from 1124 to 1274 m.

Again three members were recognized, described from top to bottom:

(1) *A Member*. The unit consists of reef limestones, grey to yellow, often dolomitized and cavernous. The thickness of the unit is about 62 m.

(2) *B Member*. This unit consists of calcarenites, yellow to pink, with subordinate white marly limestones, and has a thickness of 25 m.

(3) *C Member*. Conglomerates with a marly sandy matrix are the main rock-type. The pebbles are well rounded and consist mainly of igneous rocks, with some sandstones and marl. The conglomerates are lenticular and are present only in the type locality. The C Member is 50 m thick and rests unconformably on the Maghersum Formation.

The fauna of the Abu Imama Formation is listed below (Chapter III, Palaeontology). The presence of *Neoalveolina melo* is taken by Carella and Scarpa (1962) to indicate a Middle Miocene or Tortonian age.

(v) *Dungunab Formation* (*post-Middle Miocene*)

The type locality of this formation is in Dungunab–1 between 396 and 1118 m (21°08′00″N, 37°05′29″ E) and the formation crops out at Khor Eit, Jebel Hamamit, Jebel Dyiba, and Jebel Abu Imama. It has been divided into two members (described from top to bottom):

(1) *A Member*. Massive white rock-salt predominates interbedded with minor grey anhydrite layers (658 m).

(2) *B Member*. This unit consists of gypsum with interbedded clays and sandstones and is some 54 m thick.

The base of the formation is probably unconformable on the Abu Imama Formation and the unconformity can be clearly seen in the Jebel Dyiba structure and at Khor Eit. The Abu Shagara sands probably overlie the Dungunab Formation unconformably.

No fossils have been found but, on the basis of its stratigraphical position, Carella and Scarpa (1962) consider that the formation is post-Middle Miocene.

(vi) *Upper Clastic Group* (Sestini 1965)

'Pliocene Beds' were recorded from Khor Eit by Berry and Sestini (unpublished MS); their exact position with regard to the subsurface formations designated by AGIP is, however, uncertain.

The Abu Imama and Dungunab formations are unconformably overlain by massive weathering

TABLE 8

Tertiary Sequence, Red Sea Littoral (Based on Carella and Scarpa 1962)

Formational name	Predominant lithology		Environment of deposition	Thickness (metres)	Age
Terrace gravels / Elevated reefs Abu Shagara	Boulder beds gravels sands etc.	Reef limestones Lagoonal deposits	Clastic continental marine	+300	? Pleistocene
	Unconformity				
Upper Clastic Group / Elevated reefs Abu Shagara	Boulder beds gravels sands etc.	Reef limestones Lagoonal deposits	Clastic continental marine	+324	? Pliocene
	Unconformity				
Dungunab	A-member	Rock salt with interbedded anhydrite	Evaporites lagoonal	658	? Post-Middle Miocene
	B-member	Gypsum interbedded with clays and sandstones		54	
	Unconformity				
Abu Imama	A-member	Reef limestones cavernous and dolomitized	Marine	62	Middle Miocene (Tortonian)
	B-member	Calcarenites with subordinate marly limestones		25	
	C-member	Conglomerates with marly sandy matrix predominantly igneous pebbles	Clastic continental	50	
	Unconformity				
Maghersum	A-member	Fossiliferous marine marls, sandstones, conglomerates, with subordinate limestone lenses. Pebbles mainly of igneous origin (equivalent to *Globigerina* marls of Gulf of Suez)	Marine	111	Middle Miocene (Helvetian)
	B-member	Predominantly gypsum and anhydrite sandstones and sandy shales	Evaporites lagoonal	859	?Lower Miocene
	C member	Massive rock salt with minor anhydrite and silty marl bands	Evaporites lagoonal	465	
	Unconformity				
Hamamit		Sandstones, red brown coarse to fine, with lenses of conglomerate. Basalt flow at type-locality	Clastic continental	226	? Lower Miocene to ? Eocene
	? Unconformity ?				
Mukawwar		Silty shales interbedded with sandstones, fine to medium-grained. Occasional limestones, depositional environment brackish to shallow marine	Marine brackish	190	'Uppermost Cretaceous or Palaeocene'
	? Unconformity ?				
Unnamed		Basaltic tuffaceous breccia and basalt		Uncertain	Age uncertain
	? Unconformity ?				
Unnamed		Porphyries			? Basement Complex

coarse gravels and sands derived largely from the Basement Complex. Predominantly the boulders and pebbles are composed of metamorphic rocks; the largest boulders are frequently granite. Pebbles of dyke rocks and some M ocene limestone pebbles occur (Sestini 1965).

It is difficult to estimate the thickness of the deposit because bedding is poorly developed. In the Port Sudan area about 1000 ft of sand and gravel has been recorded in a borehole; and the Abu Shagara Formation (Carella and Scarpa 1962), which must be in part the lateral equivalent of Sestini's 'Upper Clastic Group', is about 1267 ft (396 m) thick (Table 8).

Since the 'Upper Clastic Group' has not yielded fossils, its age can only be estimated. The extent of these clastic beds is also uncertain.

(vii) *Durwara–2 and Tokar–1*

Three wells, Durwara–1, Durwara–2, and Tokar–1 were sunk south of Port Sudan between 1961 and 1963 but unfortunately little definite information is so far available.

Durwara–1 well was spudded in June 1961 on a small island south of Port Sudan. The well apparently was dry and was abandoned in 1961 at a depth of 9521 ft.

Tokar–1 was completed in 1963 and was drilled to a depth of 7398 ft. No other published information appears to be available.

Durwara–2 was completed also in 1963 and slightly more general information is available. The well was bottomed at 13 622 ft and high temperatures and some high-pressure, low-volume gas were encountered.

A generalized stratigraphical log with estimated depths has been provided by Dr. Papadia of AGIP for Durwara–2.

DURWARA–2

Lithology	Age	Depth (m)
Reef limestone	Quaternary	0–250
Sand, shale, and gypsum	Quaternary	250–750
Sandstone, shale, and gypsum	Post or ? Upper Miocene	750–2200
Salt	Upper Miocene	2000–3050
Shale, sandstone, and marl with *Globigerina*		
Sand, gypsum, and salt with gas (?) at 3600 m		3050–4050
Basalt with sandstone interbedded. May be top of Basement?		4100–4300?

Despite the general nature of this log two evaporite horizons separated by marine marls with *Globigerina* can be distinguished as in the AGIP wells north of Port Sudan (Whiteman 1965*c*).

(b) Asotriba volcanic rocks, Dungunab sheet 36I

Acidic, post-Nubian volcanic rock crops out as a large mass forming the north–south Asotriba range. Associated volcanic rocks crop out at Jebel Magardi, Jebel Ganaibal, and other smaller vents (Gass 1955, pp. 61–6). Felsite and quartz-porphyrite dykes are associated with volcanicity and can be seen to radiate from the Asotriba centre.

The volcanic rocks and dykes cut the Nubian Formation and therefore are clearly post-Nubian (i.e. post-Jurassic–Lower Cretaceous) in age. Gabert, Ruxton and, Venzlaff (1960), however, class the volcanic rocks as Pre-Cambrian.

The volcanic rocks are mainly rhyolites. They are often flow-banded and epidotized. Jebel Asotriba, Jebel Makim, and Shanwib have been recognized as individual centres intruded through the assimilation granite of the Basement Complex.

The volcanic rocks are all soda-rich, a point of interest when it is considered that similar Tertiary volcanic rocks occur in Saudi Arabia, Yemen, Abyssinia, Eritrea, Somalia (ex-British), Kenya, and in the Equatorial Province of the Sudan, where Wright described albitic rhyolites. Gass and Kennedy suggest that this soda rhyolite is not a derivative from basaltic magma but is due directly to remelting of granitic basement in depth (Gass 1955).

(c) Umm Ruwaba Formation

Two outcrops of 'Umm Ruwaba Series' are shown on the 1:4 000 000 Sudan Geological Survey Map.

The largest outcrop extends along the railway line from Er Rahad, near El Obeid, Kordofan, towards the White Nile at Kosti (Fig. 4). These continental deposits were proved in boreholes for water along the Kosti–El Obeid railway in 1914 and according to Edmonds (1942, p. 29) they were laid down in a depression formed at the same time as the uplift of the Nuba Mountains.

Andrew and Karkanis (1945) and others demonstrated that the deposits extend southward, citing evidence from boreholes sunk between the Nuba Mountains and the White Nile at Kaka, near the Melut bend.

Similar deposits were also proved in boreholes in the Taweisha and Muglad areas west of the Nuba Mountains. The nature of the deposits in the southern extension of the basin southwards from Malakal into the region east of Juba is less well known. As diagnostic fossils have not been found it is not possible to separate the Tertiary and Quaternary deposits. The limits mapped in all these areas must be regarded at tentative.

A second and separate outcrop occurs in the El Atshan Station–Singa–Sennar area. These deposits are described separately below as the El Atshan Formation (Fig. 4).

Andrew and Karkanis (1945, p. 163) proposed the name 'Umm Ruwaba Series' for the thick superficial deposits that occur on the central Sudan plain west of the Kurmuk–Ingessana–Moya ridge and surround the Nuba Mountains on the eastern, southern, and western sides.

According to Andrew (1948 and MS 1943) the Umm Ruwaba sediments consist of unconsolidated sands, sometimes gravelly, clayey sands, and clays. The clays are mainly buff to greyish-white, or greenish-grey. In general the sediments are unsorted and feldspar and biotite often are undecayed. Rapid facies changes are characteristic of the deposit.

Finer-grained sediments are said to occur under the clay plain east of the Sudd and the Bahr el Abiod and Andrew (1948 and MS 1943) mentioned that under the surface clays south of Malakal there are well-sorted sands. Sands are rare, however, in the Jebel Ahmed Aga area where the lithology is predominantly clayey.

Manganese beds occur at Paloich and Wabuit, near Wabuit (Chapter III).

In the Sudd region fine sediments occur consisting of clean washed sands with laminated clays. The sands are mainly without pebbles and are well graded. Iron-stained levels have been recorded. Similar deposits occur in wells in the Yirol area but coarse grits and iron-staining are more common here (Andrew MS 1943). Andrew (personal communication) now considers the sub-Sudd deposits as equivalent to the Yirol Beds.

Over much of the area underlain by the Umm Ruwaba Formation the surface deposits consist of heavy clays with kankar nodules and are dark grey to chocolate in colour. In the Muglad and Umm Ruwaba area 'Qôz' sands rest directly on the clays.

Data on the thickness of the Umm Ruwaba Formation is scanty. At Umm Ruwaba itself the formation is at least 889 ft (278·6 m) thick. Rodis *et al.* (1964, p. 35) stated that the formation attains a thickness of more than 1100 ft (335 m) in eastern Kordofan. The thickness is certainly highly variable as the formation rests unconformably on Basement Complex and other rocks.

Thicknesses of sediments classed as Umm Ruwaba by Andrew (MS 1943) and Andrew and Karkanis (1945, pp. 7–10) are given in Table 9.

The Umm Ruwaba deposits are fluviatile and lacustrine, but probably did not accumulate in one large and continuous lake extending as far north as the rock still at Sabaloka, 45 miles north of Khartoum. Various views have been expressed about the presence of a Lake Sudd covering much of the area now occupied by the Umm Ruwaba Formation (Lawson 1927, Ball 1939, Berry and Whiteman 1968).

In the author's view the Umm Ruwaba deposits were probably laid down in a series of land deltas like the deposits of the Gash Delta, near Kassala. Standing water and lakes existed from time to time dependent on the relationship between evaporation, transportation, and precipitation, date of integration of the river system, river flow, etc. Swamps and lakes such as those of the Sudd may have existed and occasionally these dried up and saline deposits were formed.

Few fossils have been found in the Umm Ruwaba Formation and little can be said about its age except that it is probably Tertiary to Pleistocene.

(d) El Atshan Formation

This formation was proved in eleven bores sunk in the Hawata–Sennar district. The name is taken from El Atshan Station (Ruxton MS 1956).

TABLE 9

Umm Ruwaba Formation, thickness data (Based on Andrew and Karkanis 1945)

Locality	Depth (m)	R.L. of rock head (m)	Lat. N.	Long. E.
Northern area				
Umm Ruwaba	278·6	166	12° 53′	31° 13′
Tendelti	167·6	245	13° 01′	31° 53′
Umm Ushera	102·5	(310)	13° 03′	31° 57′
Selima	85·3	321	13° 01′	31° 17′
Umm Koweika	103·6	288	13° 00′	32° 04′
Khor Murad	128·0	352	12° 34′	31° 24′
East of Nuba Mts.				
Iddl el Dam	84	356	10° 53′	31° 39′
Gereid	65	355	10° 49′	31° 46′
El Yoi	64	336	10° 41′	31° 52′
Jebelein, left bank	29	348	12° 35′	32° 48′
Zuleit, left bank	30	353·5	12° 05′	32° 54′
Qoz Kash Kash	95	(287)	11° 01′	32° 54′
West of Nuba Mts.				
Muglad	116	384	11° 02′	27° 44′
Taweisha W	149	Surface	12° 12′	26° 34′
Taweisha E	122	R.L. not known	12° 21′	26° 33′
Gabeish	139		12° 09′	27° 21′

	Reduced level (m)
High flood level at Malakal	384·41
High flood level at Khartoum (Blue Nile)	375·63
Rock sill at Sabaloka, S. end	359·50
Low-water level at Sabaloka, S. end	362·63

The sediments (sometimes called 'older alluvium') are similar to the Umm Ruwaba Formation as described by Andrew (1948, p. 104). In addition, pebbles of agate and ironstone nodule occur, and kankar is common. Ruxton (MS 1956) recorded a 2-metre band of trachytic pumice at Wad Kabusa (13° 19′ N, 34°25′ E) in a shallow well at 14·4 m, sunk in clays of the central Sudan plains. These are characterized by an abundance of basaltic minerals, such as titanaugite and calcic plagioclase, sometimes with zeolites. Beds of calcrete are commonest at the top of the older alluvium and at the base of the clays. This may indicate an arid phase but no information about precise age is available at present. The clays and kankar beds of Singa and Abu Hugar are most probably Pleistocene deposits and are dealt with later. The relationships of the superficial deposits in the Sennar area are shown in Fig. 40. The age of the El Atshan, like the Umm Ruwaba Formation, is uncertain and best given as Tertiary to Pleistocene.

(e) Hudi Chert Formation

The Hudi Chert outcrops occur in the Nile Valley, the Wadi Muqaddam, and Wadi el Melik areas (Fig. 4). Loose fragments have been recorded west of Dongola and between Wadi el Melik and Wadi Howar. Information about the distribution and origin of this formation is extremely important as it bears directly on the evolution of the Nile drainage (Berry and Whiteman 1968).

The type locality of the Hudi Chert is in a wadi near Hudi Station, some 20 miles east of Atbara. Fossiliferous boulders of Hudi Chert were first found at this locality in 1910 by C. T. Gardner, district engineer of the Sudan Railways. G. W. Graham, the government geologist, submitted two of Gardner's specimens to palaeontologists in England but it was not possible to identify them specifically and to determine their age.

Two fossils preserved in Hudi-type matrix were found by R. C. Skitt in the Zeidab district on the west bank of the Nile some 12 miles south of Atbara. These two specimens are the holotype of the gastero-pod species *Pseudoceratodes rex* n. sp. according to Cox (1932–3, p. 316). A structure resembling a fragment of *Pinna* was also found. Alternatively it was identified as a palm leaf impression. A definite age could not be given to these fossils. G. V. Colchester visited Hudi Station area again and collected a fauna, which was described by Cox (1932–3, pp. 315–48). At Hudi the chert occurs about 50 ft above the high Nile level at Atbara. The country to the east rises about 40 ft in the next 20 miles and there are scattered outcrops of Hudi.

R. C. Wakefield found fossiliferous chert near Ed Debba (Fig. 4) and south of this locality on the east bank of Wadi Muqaddam. Edmonds found chert and chert-like material in the Wadi Muqaddam also. Wakefield also found chert scattered in the area west of Dongola and was of the opinion that not all of it had been transported by man (Andrew and Karkanis 1945, p. 4). He also recorded chert near Jebel Kol ez Zarga (15°39′ N, 36°46′ E) and

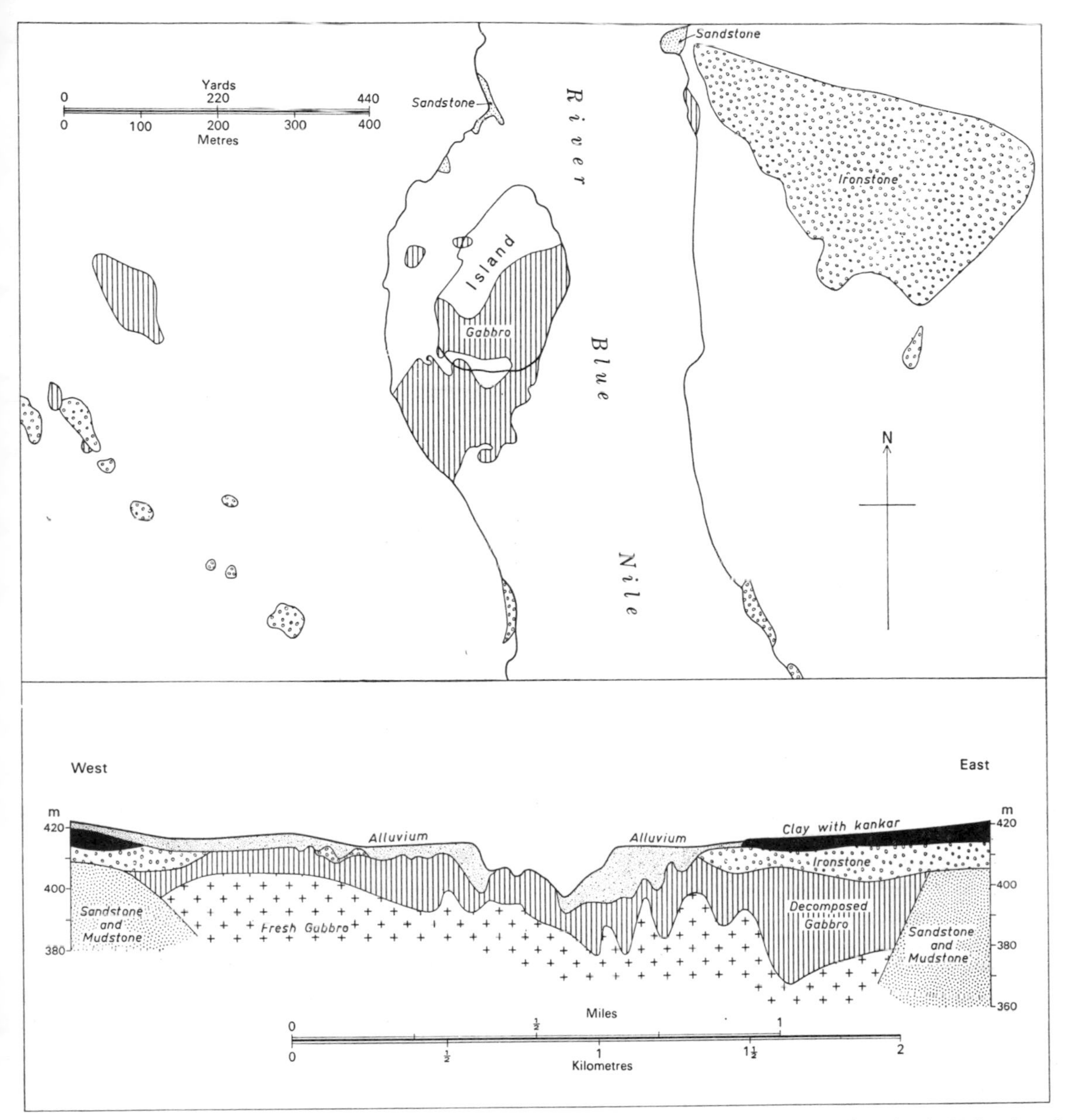

FIG. 40. Superficial deposits, Sennar Blue Nile Province, geological section (Based on unlabelled section in University of Khartoum Geology Library (probably drawn by Ruxton 1956(?) after Grabham)

Dr. Bryant found chert on the west bank of the Wadiel Melik at Nakashush (Andrew MS 1943).

Karkanis (1945) recorded Hudi Chert *in situ* capping hills north of Hudi Station; west of the Nile in the Zeidab area; on Jebel Nakharu north of Berber on the west bank of the Nile; and in the area east of the Sabaloka Jebels, north of Khartoum. Colchester also found chert in this district at Abu Naga (Andrew MS 1943).

At Jebel Nakharu Colchester measured the following succession. A similar section was recorded by the University of Khartoum expedition 1963 (Fig. 33).

JEBEL NAKHARU

	Thickness (ft)
Basalt cap to scarp; overlies thin red pebble horizon at 1180·9 river level	29
Grit, with clay matrix; bed of chert 20 × 40 cm blocks at 1126·4 ft river level	48
Nubian Series mudstone 1107·2 ft river level	19
Nubian Series, sandstone	+25
Flood plain of Nile 1043·5 ft river level	

The Jebel Nakharu Chert is correlated with the Hudi Chert to the south and is overlapped by the thin red pebble bed below the base of the basalts.

Karkanis (1945) also visited Zeidab, west of the Nile and investigated a quartz pebble conglomerate that occurs at the foot of the scarp. It was described by Sandford (1933, p. 303) as 'bound up with fragments of Hudi stone'. These conglomerates are apparently derived from a chert breccia capping the hills. The chert is usually cemented with iron oxide and on Jebel Marafabiya is overlain by sandstone. According to Karkanis (*in* Andrew and Karkanis 1945, p. 5) the breccia has much wider distribution than the chert, and extends south nearly to latitude 16°N, some 20 km south of Hudi.

Other occurrences have been recorded far to the west of the Nile by travellers and officials, but confirmation is needed: the reports were not supported by specimens, and confusion may have risen between the identification of chert and silicified cream mudstone belonging to the Nubian Formation.

The lithological description given by Cox, advised by Campbell-Smith (1932–3; pp. 319–21) is the most complete so far available:

'The Hudi boulders are mostly irregularly ellipsoidal in shape, ranging up to nearly a foot in diameter. In view of Dr. Sandford's observations at Zeidab it is probable that in most cases, at least, they are eroded blocks of rock and not siliceous concretions retaining more or less their original shape. Their surface is usually yellowish-brown in colour; often it is very irregularly pitted, while depressions may communicate with internal cavities. The surface fossils are usually eroded but sometimes, when sheltered in hollows, they project in a very good state of preservation. The rock is variable in texture, but it is always very hard and splintery, frequently possessing a conchoidal fracture. Fractured surfaces seldom have more than a slight lustre and may be quite dull. The colour may be white, grey, yellow, brown, or (occasionally) pink, various shades being irregularly mottled, with rarely any tendency towards parallel banding.

The groundmass of the rock varies considerably in appearance, but three main types may be distinguished. The first type is more or less homogeneous, resembling ordinary flint. The second type also has no definite structure, but is traversed by a network of fine, irregularly distributed veins, frequently stained by ferruginous matter. The third type, which is the most interesting, consist of rounded or subangular grains which are very variable in size and shape and are fused together by a siliceous cement of equal hardness, so that fractures usually pass indiscriminately through the grains. Much larger inclusions are also present, ranging up to about 30 mm in length, although few are longer than about 10 mm; like the smaller grains, these may be either rounded or subangular in shape. Frequently these inclusions, which sometimes differ in colour from the surrounding rock, have an amorphous, cherty appearance. Sometimes they pass imperceptibly into the general groundmass, but at other times they have a well-defined boundary. From the surface of the boulders there frequently protrude what appear to be pebbles cemented into them. When broken across, however, these are seen to be formed of the same type of rock as the inclusions in the interior of the boulders . . .'

A reassessment of the Hudi Chert molluscan fauna using Cox's data (1932–3) is given in Chapter III, Palaeontology.

Although Cox stated with some confidence 'that the shells now described are of Lower Tertiary age, while there is a strong probability that they are referable either to the Upper Eocene or to the Lower Oligoclase,' Bell thinks that the evidence for an Upper Eocene–Lower Oligocene age is unreliable and that the most that can be said is that the fauna has a Tertiary aspect.

Cox (1932–3) also wrote on the origin of the Hudi Chert, pointing out that the fossils, with the exception of a problematical ? *Pinna*-like object (also described as a palm-leaf impression), indicate that the rock was originally a lacustrine deposit. He ascribed the origin of the opaline silica to secondary silicification, comparing the cherts with silicretes that are common in Africa and elsewhere. Concerning the age of silicification, Cox (1932–3) ascribed it to some past period of Tertiary time. He also suggested that the original sediments, which gave rise to the flint-like and veined types described above, were silicified limestone.

The granular type was thought to have been formed by silicification of sandstone and Cox believed that this may have been Nubian Sandstone. There are, however, mineralogical and textural

TABLE 10

Hudi Formation locality details

Locality	Coordinates	Comments	Thickness in feet	Position of base R L in feet
Jebel Laterba	17°18'N 34°09'E	Hudi Chert resting uncomformably on Nubian formation fossilferous (fig 41)	50	1220
Jebel Garatit	17°12'N 34°11'E	Hudi Chert resting unconformably on Nubian formation fans of Hudi boulders form conglomerate with round pebbles of quartz at base of section fossiliferous	17	1258
Jebel Es Sufr	17°05'N 33°52'E	Hudi Chert resting unconformably on the Nubian formation (Fig. 41)	5	1320
Jebel El Kereiba	16°42'N 33°47'E	Hudi Chert resting uncomformably on Nubian formation banked against side of Nile valley and occuring below prominent erosion surface cut across the Nubian Jebels east of Nile valley	5	1378
Jebel Umm Gerab	17°24'N 33°42'E	Hudi Chert conglomerate resting uncomformably on Nubian formation well developed fans of Hudi boulders occur on western side of Jebel	20	Approx. 1250
Two small Jebels south of J. Zaghawi	17°11'N 33°41'E	Hudi Chert boulders in site resting on Nubian formation		
Jebel El Guweir	16°46'N 33°29'E	Hudi Chert conglomerate in site resting on Nubian formation	15	1190
Hudi Station	17°42'N 34°17'E	Hudi Chert in situ bouldery outcrops	30	+50 Nile at Atbara
Jebel Nakhara	18°05'N 32°55'E	Bed of Chert occurs below basalt flow from Jebel Atshan volcano and rests uncomformably on Nubian formation	6	1126
		Above data collected by Bell, Berry, Kheiralla, and Whiteman (1962, 64)		
Hanakat El Keleiwat south of Ed Debba				1040
Jebel Zaghawi	17°17'N	Chert 163·2 ft above flood plain of Nile		1465
South-west of Dongola	17°36'N 29°42'E			960
West of Dongola	19°31'N 28°56'E			960
West of Rahib	17°50'N 26°42'E			1920
West of Laqiya		Worked flakes		1568
West of Khartoum – Kassala track				?1920
		Above data from Andrews (MS 1943, 47)		

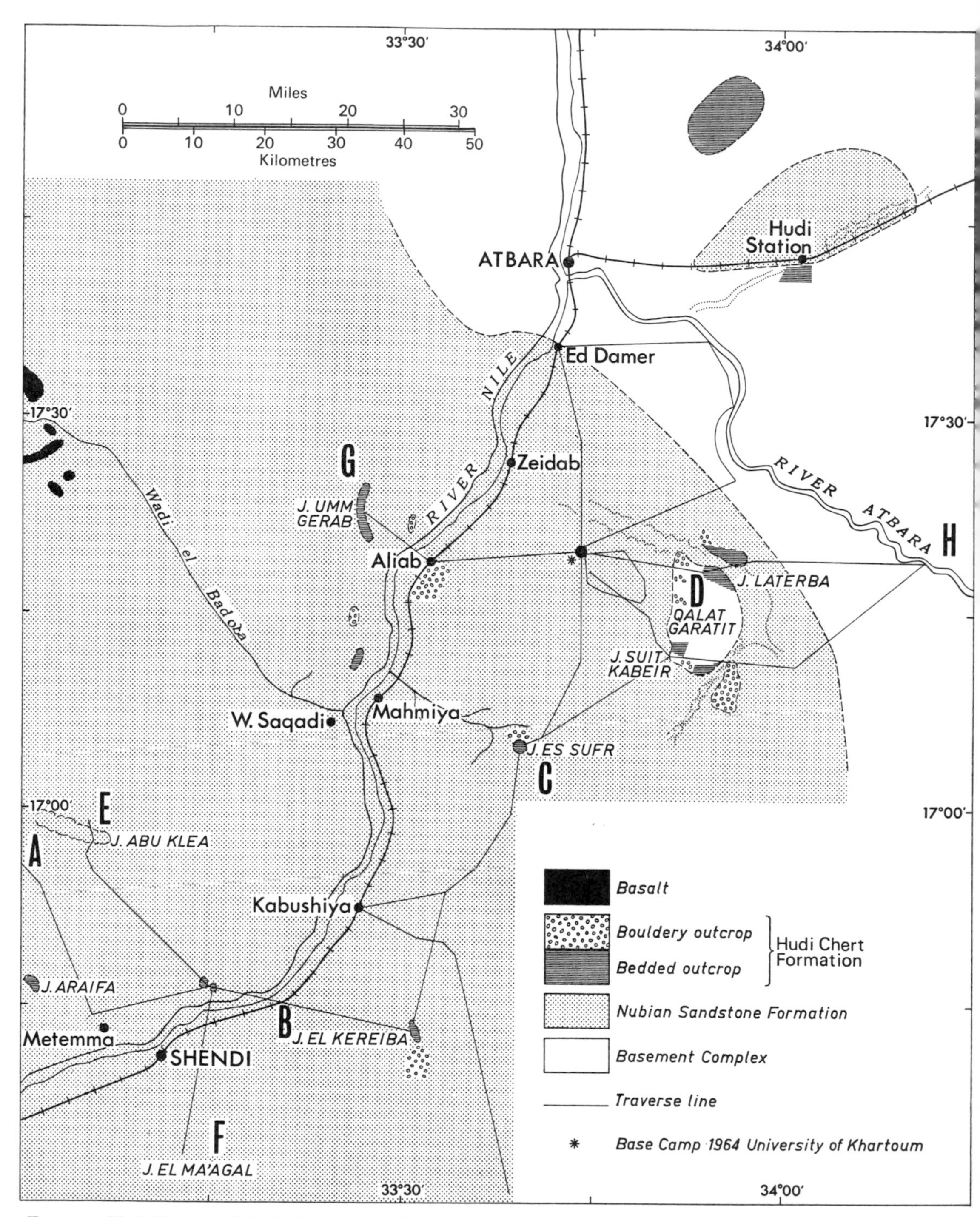

FIG. 41. Hudi Chert and Nubian formations, Shendi–Atbara region. Distribution map (Compiled by Kheiralla, Berry, and Whiteman 1966)

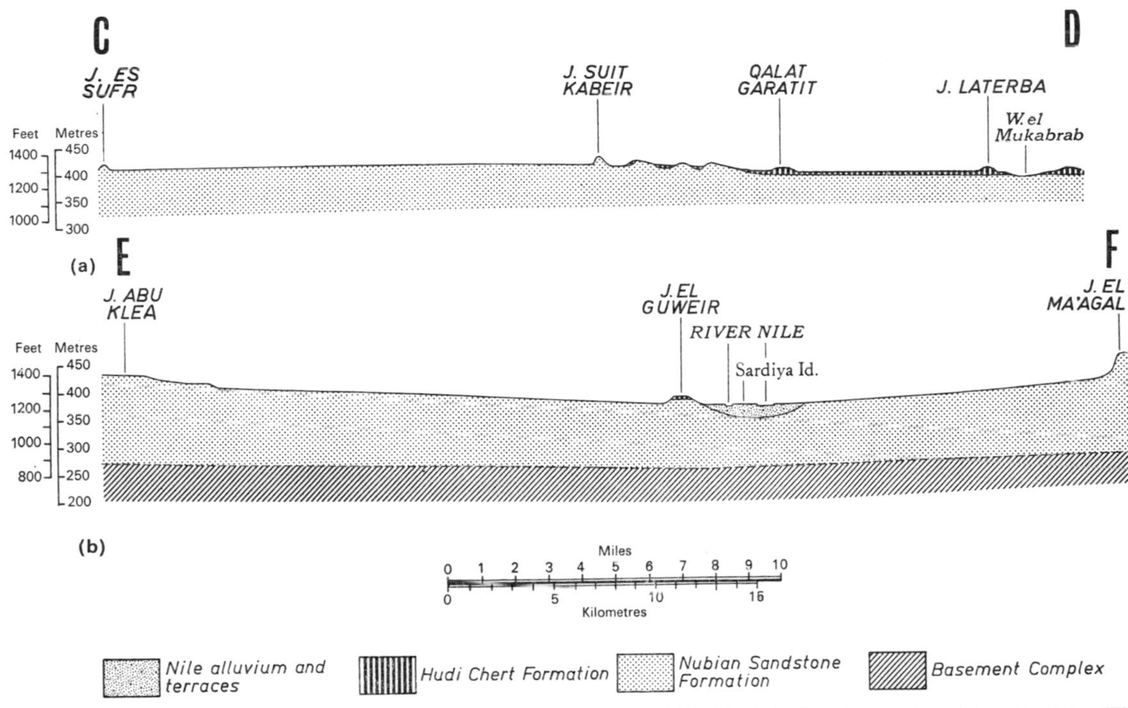

FIG. 42. (a) Hudi Chert Formation, Jebel es Sufr (C)–Jebel Laterba (D). (b) Hudi Chert Formation, Jebel Abu Klea (E)–ebel el Ma'agal (F)

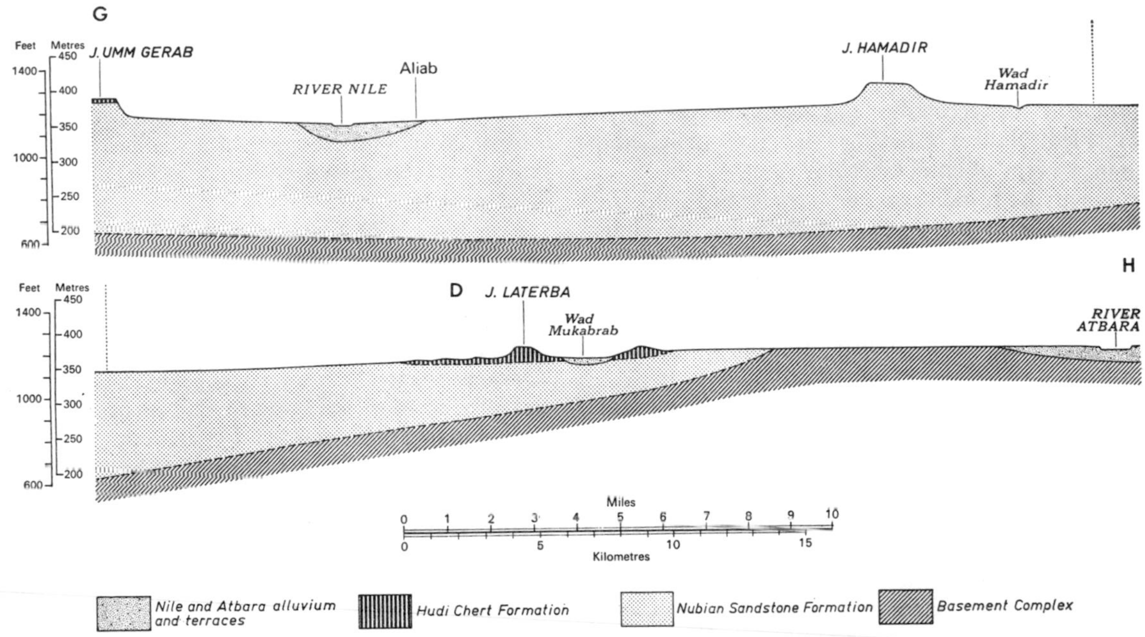

FIG. 43. Hudi Chert Formation, Jebel Umm Gerab (G)–Jebel Laterba (D), River Atbara (H) (Geological section, aneroid levels, and geology by Kheiralla, Berry, and Whiteman 1966)

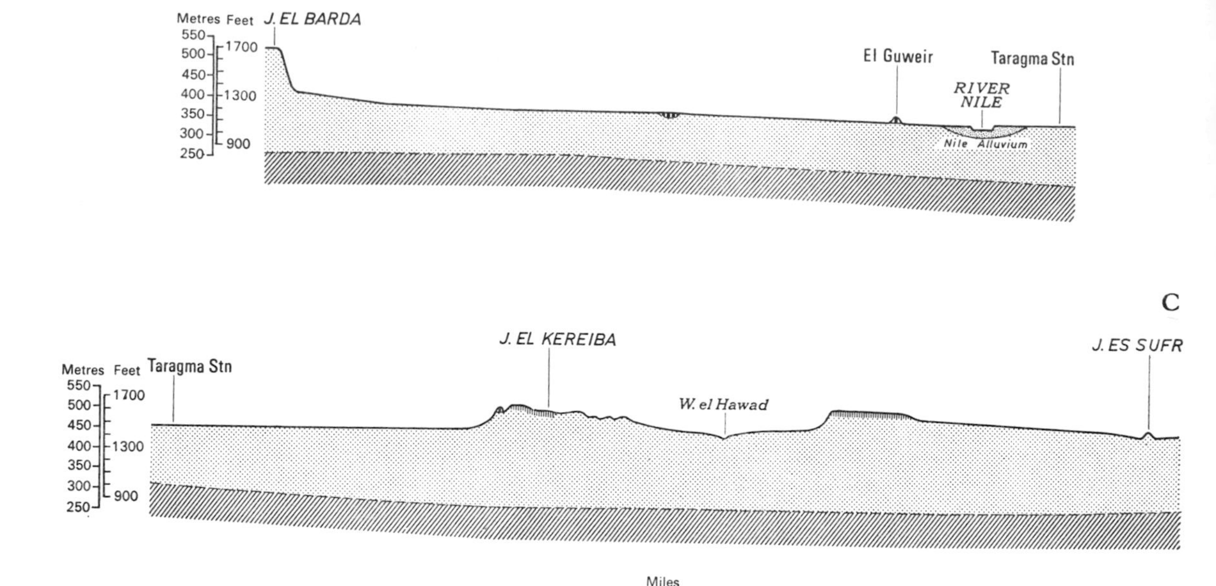

FIG. 44. Hudi Chert Formation, Jebel el Barda (A)–Taragwa–Jebel es Sufr (C). Note the position of the Shendi Surface (Berry and Whiteman 1968) (Geological sections, aneroid levels, and geology by Kheiralla, Berry, and Whiteman 1966)

problems in accepting this last interpretation. Silicified Nubian Sandstone is quite unlike the granular type of Hudi Chert.

Kheiralla, working on the west and east banks of the Nile with the author, investigated the Hudi Chert (Table 10 and Fig. 41). In our view the deposits lumped together as Hudi Chert are polygenetic; some are lake deposits and others may be silicretes produced by similar secondary silicification processes.

The heights at which chert occurs (Table 10) rule out a lacustrine origin for all the chert deposits, unless we are willing to accept that they were deposited in a water body at least 900 ft deep. This seems unlikely.

A shallow lake probably existed in the Atbara–Shendi section of the Nile Valley. In the Jebel Laterba and Jebel Garatit areas (Figs. 41–44) Ustaz Kheiralla Mahgoub Kheiralla and the author mapped a series of embayments in the Hudi Chert outcrops which strongly suggest that deposition took place along the margin of a shallow lake. No shore line deposits were recorded (Berry and Whiteman 1967).

The Hudi Chert at its type locality at Hudi Station was probably deposited in a lake also, as these deposits are most probably *in situ*. Hitherto they were considered to have been transported and thought of as river gravels (Cox 1932–3, Andrew and Karkanis 1945) but reinvestigation by Bell, Berry, and Whiteman in 1962 proved that the deposits are *in situ* and owe their bouldery appearance to weathering in the intensely hot climate of the savana–desert edge.

The author favours a primary origin for the original deposits, the source of silica being local volcanoes and hot springs, perhaps in the Bayuda volcanic field north-east of Atbara. Unfortunately no definite date is available for the age of these volcanoes. They are mapped as Tertiary, but from a reconnaissance study of the flows, craters, etc. much of the volcanicity appears to be recent. Volcanic activity may well, however, have started in the Tertiary.

If the Bayuda volcanoes are accepted as the source of silica we can envisage the following stages in the formation of the Hudi Chert. After deposition in waters of suitable pH the silica, in the form of gel, perhaps including opaline tests of organism such as diatoms, was converted to chalcedonic and micro-crystalline quartz. Eventually the lake dried up. Erosion then took place and with time, weathering, and insolation produced the well-developed bouldery sub-spheroidal outcrops of the Hudi Chert.

In places where the wadis and the Nile and Atbara rivers have cut into Hudi deposits striking gravel spreads have been formed, especially south and east of Atbara, so adding to the difficulty of defining the Hudi outcrops.

(f) Jebel Marra volcanic complex

The Jebel Marra volcanic complex is situated in western and southern Darfur, with the highest part situated some 80 miles, 240° from El Fasher, the capital of Darfur Province. It rises to approximately 3042 m (approx. 10 000 ft) above sea-level (Plates 15 and 16; Fig. 25). According to the Hunting Technical Services report (1958), on which much of this account is based, the volcanic complex occupies some 3740 km². The relief is rugged and youthful, and in places badlands are developed, especially where ash deposits occur. Basaltic and trachytic flows, plugs, beds of pumice, ash, and other tephra form a large, dissected, composite volcanic complex.

The complex appears to have been formed mainly during the Tertiary but explosive and intrusive activity probably extended into the Pleistocene and even into the Recent. The slight amount of dissection of the ash deposits round the central conelet point to this. Hammerton (1966) reported fumarolic activity and hot springs from the inner crater.

Hobbs (1918) described the mountain, and Lynes collected specimens that were subsequently described by Campbell-Smith (Lynes and Campbell-Smith 1921). Campbell-Smith identified quartz trachyte, quartz–riebeckite–trachyte, andesine-bearing kenyte, and mugearite. Sandford (1935, p. 360) recorded trachytes and phonolites from Jebel Marra and pointed out that the volcanic rocks are probably of Tertiary age. Andrew (*in* Tothill 1948 and MS 1943) stated that volcanic activity is generally regarded as having begun in the Miocene when an imposing array of lavas with plugs, cones, and craters was formed in the Jebel Marra region.

The predominant rock is basalt, but trachyte and phonolite are said to be common locally. Pyroclastic rocks occur near the craters, and Jebel Marra itself is described as 'a massive crater 5 km in diameter with two crater lakes, one saline (natron) the other fairly fresh, surrounded by walls of ash and tuff; pumice occurs locally'.

The geologists of Hunting Technical Services (1958, pp. 32–3) provided a detailed account of the Jebel Marra volcanic complex. The detailed description given below is based on this. Lebon and Robertson (1961, pp. 31–2) utilized much of the Hunting Technical Services data and provided a brief geological account of the region. Lebon (1965) also commented on the geology of the region; and Robertson (1965), a director of Hunting Technical Services, described Jebel Marra in a popular geographical article illustrated by excellent photographs.

The following succession was recognized by the geologists of Hunting Technical Services (1958, p. 32):

YOUNGEST

5 *Intrusive phase* (relations with preceding phases uncertain).

4 *Explosive phase* during which extensive deposits of volcanic tuffs were deposited and the Jebel Marra crater formed.

3 *Second eruptive phase* during which basaltic lavas probably in relatively small quantities were erupted.

2 *Erosive interval* during which the newly-formed lava-mountain was subjected to intense sub-aerial erosion.

1 *First eruptive phase* during which lavas occurred more or less continuously.

OLDEST

(i) *First eruptive phase*. Lavas erupted during this phase ranged from olivine–basalt to trachyte and include riebeckite- and aegerine-trachytes. The earliest flows are the most basic and consist of olivine–basalt. They were extremely fluid and flowed long distances down existing wadis and filled in other depressions. Flows were recognized occupying what must have been the pre-existing drainage line of the Wadi Azum and the Wadi Barei, for instance. Some of these early flows are deeply buried beneath earlier alluvial deposits (Hunting Technical Services 1958).

Increasing acidity brought about a corresponding decrease in fluidity and the flows belonging to this phase consolidated nearer their points of eruption. Pre-existing topography, cut in Basement Complex rocks, clearly influenced the direction of flows, for instance between Kalokitting, Tur, and Nyerte (Fig. 25), and the broad outcrop of volcanic rocks extending northwards from the latitude of Guldo may well have occupied a broad regional depression in the topography.

(ii) *Erosion interval.* It is assumed that by the end of the first eruptive phase the volcanic pile had attained approximately its present shape. This must have resulted in higher precipitation and the establishment of new drainage patterns.

Coarse-grained, current-bedded alluvial deposits occur on the west and east of Jebel Marra and in the Hunting Technical Services Report (1958) it is suggested that they were probably laid down in basins on the flanks of the volcano, either before the drainage system had been established or in valley lakes dammed up by lava flows.

Piedmont alluvial deposits were laid down round the foot of the mountain, and some subsidence of the area surrounding the mountain may have taken place at this time, forming the depressions in which the clastics accumulated.

(iii) *Second eruptive phase.* Small quantities of basaltic lava were erupted during this phase compared with the amount erupted during the first eruptive phase. Most of the lava appears to have come from subsidiary vents formed on the flanks of the main volcanic pile. The Hunting Technical Services geologists (1958) reported, however, that locally volumes of lava are considerable, for example, west of Tur basaltic flows overlie piedmont alluvial deposits. Near Kronga lava flows and coarse-bedded alluvial deposits are interbedded (Hunting Technical Services 1958). These younger flows occupy wadi bottoms in general, whereas the older flows frequently cap valley sides.

(iv) *Explosive phase.* The large existing caldera was formed at this time, and on the eastern side of the mountain tuffs, consisting almost entirely of pumice, were laid down. Trachytic tuffs, sometimes interbedded with trachytic lavas, also occur.

Again the accumulation of large quantities of ash must have impeded and interrupted drainage and a new pattern developed (Plate 15). The explosive phase is thought to have taken place in Pleistocene times, but as stated above in view of the small amount of dissection shown by the ash conelet with its crater lake (one of the Dariba lakes), within the caldera, activity may well have extended into the Recent (Plate 16).

(v) *Intrusive phase.* The geologists of Hunting Technical Services (1958) recognized an intrusive phase in the region but its relationship with the eruptive and explosive phases described above is uncertain. A basalt plug intruded into Nubian Sandstone occurs east of Jebel Marra, and many other plugs are known, mainly on the eastern side of the mountain. Basic dykes of uncertain age were also recognized and near the eastern margin of the volcano a fine-grained intrusive rock occurs, composed almost entirely of clear potash feldspar.

Tertiary rocks are represented in the superficial deposits, but without fossils it is impossible to separate them from similar rocks or Pleistocene and Recent Age. Pleistocene alluvial deposits are almost certainly represented here. In certain sections a 'Grey Group' composed almost entirely of volcanic ash and debris formed at the start of the explosive phase is clearly recognizable in these deposits.

According to the geologists of Hunting Technical Services (1948):

'the basic rocks of the volcanic group are extremely hard, fine-grained, and resistant to weathering. Flow-banding is common, but it was rarely observed to indicate smooth flow, turbulent conditions apparently being usual. Vesicular structure varies from small widely separated vesicles to honeycombed structure in which the rock has the appearance of a sintered mass of lava pellets 1 to 2 cm across. Columnar jointing was observed at a few places, but where it was developed megascopic flow structures were absent, the jointing in the basalt with flow structure being irregular. The trachyte lavas are less resistant to weathering than the basalts. Natural exposures of trachyte in upland regions are characteristically rather massive, rounded and bouldery. Vesicular structure is generally absent, while flow structure, though present, is not readily apparent megascopically in contrast to some of the basalts. The tuffs are easily eroded with the formation of a typical "badland topography" which is a useful aid for interpretation from air photographs.'

The culminating feature of the Jebel Marra volcanic pile is probably Uwofuo, one of many lava peaks that form the high plateau north and east of the Dariba lakes.

Andrew (1948 and MS 1943), the geologists of Hunting Technical Services (1958), and Lebon and Robertson (1961) described these lakes as crater lakes. They are situated in a depression some 3 miles in diameter and some 1600 ft (500 m) deep. The dimensions of the depression indicate that it is a caldera, rather than a crater, formed by collapse and explosion during the Explosive Phase. The walls of the caldera appear to be mainly formed of lava with some beds of ash. A small central conelet composed mainly of tephra, ash beds, and some lava flows was formed subsequently. The degree of dissection of this conelet is slight, indicating its youthfulness (Plate 16). The cone is occupied by a crater lake, Lake A, some 357 ft deep according to D. Hammerton of the University of Khartoum Hydrobiological Unit. To the north-east, and occupying much of the north-eastern part of the caldera, is Lake B. This is much shallower, being about 38 ft deep. The depths were recorded by Hammerton in 1964 using echo-sounding apparatus.

Both these lakes are saline. Lake A contains 4200 p.p.m. soluble salts, and Lake B waters contain 14 000 p.p.m. The salts are probably derived from the tuffs (D. Hammerton, personal communication). Hammerton (1966) has reported fumarolic activity and hot springs from Lake A.

The structural setting of the Jebel Marra volcanic complex is not apparent but lies on the prolongation of the Aswa line of Uganda.

(g) Kutum and Melit areas, Berti or Tagabo Hills, Meidob Hills, and isolated volcanic outcrops in Darfur

'Tertiary lavas' are shown north and east of the Jebel Marra volcanic complex on the Geological Map of the Sudan, second edition, 1949, scale 1:4 000 000 (some of the outcrops are omitted on the third edition). North and west of Kutum 'Tertiary lavas' of the Melit area are similarly shown but no description is available for either of these areas. An account by Colchester of the volcanic rocks is said to exist in the files of the Sudan Geological Survey (G. Andrew, personal communication).

The Berti or Tagabo Hills (Fig. 4) consist of trachyte, solvsbergite, and some basalt, and are distinguished by their spine-like forms (Andrew MS 1943). Phonolites are also said to occur in this region (Andrew 1948). Sandford (1935, pp. 353–9) mentioned the occurrence of dissected flows and dykes in the Beri Hills.

Dyke and dissected flows occur in the Meidob Hills, but fresh lavas occur in and around the Malha crater. The lavas are mainly olivine–basalts but some trachytes and phonolites have been reported (Sandford, 1935, pp. 359–60).

Sodium carbonate and bicarbonate-(natron-) rich waters occur at Malha; natron also occurs in the lake at Nakheila (Merga) at Bir Natrun and in a large area around 18° 33′ N, 26° 47′ E, north-east of Bir Natrun. Andrew (1948 and MS 1943) suggested that the natron was of juvenile origin, since it occurs within an area of feeble volcanicity. These saline deposits may be simply evaporites, however (Fig. 61).

Small isolated volcanic outcrops occur in Darfur, and isolated basalts have been recorded to the west in Chad.

(h) Bayuda volcanic field

Grabham (1920) and Gregory (1920) referred in general terms to this volcanic field situated in the bend of the Nile between Merowe, Abu Hamed, and Atbara (Fig. 4). Volcanoes were recognized in this little-known region by the occurrence of pumice that had been transported down the Wadi Abu Dom (Grabham 1920), and in addition an unnamed topographic surveyor had recognized the volcanic nature of the Hosh el Dalam (12° 25′ N, 32° 30′ E) and labelled it as a crater. The range was also seen on *The Times* Africa flight (Gregory 1920).

Andrew and Delany (Delany 1954, p. 16) visited the Bayuda volcanic field in 1957 and provided the following description:

'The main cone, Hosh el Dalam (12° 25′ N; 32° 30′ E) rises some 500 m above the plain level. The initial eruptions appear to have been basalt flows, followed by ejection of ash which built up the cones and was deposited against hills of riebeckite granite or over the gneiss plain. In the Muweilih crater, large boulders of gneiss and granite are enclosed in the lower ash beds and vesicular bombs of basalt are frequent in all the cones in the higher beds.'

The University of Khartoum geological and geophysical Bayuda expedition visited the Bayuda volcanic field for a few days in November 1963. Unfortunately, because of difficulties with transport and water, and inaccessibility, only the southern and western sides of the field were studied. Dr. D. C.

Almond, Professor J. Parry, Ustaz Farouk Ahmed, and Ustaz Badr el Din Khalil visited the area in April 1967. A detailed account is now available (Almond 1969).

More than twenty sizeable volcanic cones and numerous explosion vents and craters occur in the region. There are extensive basaltic lava flows and ash beds.

Unfortunately the topographic maps of the region (1:250 000 sheets 45F and 45G) show little detail, and the positions of some of the cones and craters are wrongly marked. Aerial photographs are available for only a part of the area; Almond has made an interpretation of these (as yet unpublished).

The main volcanic field is centred south of Sani and extends south-eastwards to Hosh es Siddig and Bihadi, and may extend as far south as Jebel Magil. The following cones have been named: Angalafib, Jebel Badi, Umm Qureinat, Jebel Qurein, Hosh ed Dalam (which frequently contains a lake), Washeira, Goan or Harba, Hosh es Siddig, and Bahadi.

In these areas there are extensive lava flows and ash beds resting on a gneissose and granitic Basement Complex. There are well-developed lava flows extending south-west from Sani through the Angalafib area to Muweilih. The fronts of these flows are clearly discernible and rise some 130–150 ft above the pediments cut across the Basement Complex to the west. Spatter cones, lava falls, and caves occur. The surface of the flows is extremely irregular and blocky, and the small amount of dissection of the flows indicates that they are very young.

The Umm Qureinat–Hosh Ed Dalam–Washeira Group form the most impressive part of the volcanic complex; Hosh ed Dalam rises at least 500 m above the pediments to the south and west. This volcano has a conspicuous crater, which often contains water. Numerous other calderas and craters occur in the group.

West of Sani the following isolated cones have been recognized: Jebel Uruf (? volcanic cone), Jebel Umm Khandag, Jebel Mazrub, and Jebel Hebeish. These cones rest on a platform of gneissose and granitic Basement Complex and form isolated masses. Jebel Mazrub rises out of conspicuous series of lava flows on the south and east, and Jebel Habeish is a small explosion crater. The walls of the crater are approximately 70 ft high and consist of tephra and ash beds containing gneiss boulders from the Basement Complex and basaltic lavas. The rim of the crater is little eroded, although on the north-western side a landslip has occurred.

East of Jebel Hebeish there are the remains of a large cone. The western side of this has been largely demolished and the caldera infilled partially by a new ash and lava cone. This jebel is unnamed on the published topographic map.

Lavas occur on the south side of Wadi Abu Siba south-east of Sani, and there are several volcanic cones on either side of the wadi as far south-east as Shimeili or Junkeiwa. These include Sergein, Mersidat, Bararumbo, and the volcanic range with several cones marked on the topographic map, south-west of Junkeiwa.

The volcanic rocks may extend as far south as Jebel Magil, 4 miles north of the Wadi Abu Dom, judging from the morphology of the jebel. The area was not sampled because of its inaccessibility. Several isolated outcrops of basalt were recorded between Umm Merwa and the Atbara Ferry. Most of these appear to be necks intruded into Basement Complex rocks.

Concerning the age of the Bayuda volcanic rocks little can be said except in general terms. They cut the Basement Complex but younger rocks, other than superficial deposits, do not occur in the area.

The unweathered nature of the basalt flows and the freshness of the surface morphological features point to extremely recent volcanic activity. Some of the largest volcanoes, however, especially in the central mass of Jebel Qurein–Hosh el Dalam, may well have erupted in Tertiary times or even later.

It has been suggested earlier that the source of the silica that gave rise to the Hudi Chert (Lower Tertiary age) may have been in the Bayuda volcanic field.

Unfortunately too little time was spent in the area to determine the volcanic history of the region except in a general way. The following sequence is proposed tentatively:

5 *Period of erosion* belonging to present-day cycle.

4 *Second eruptive phase* during which small amounts of lava were erupted around the upper parts of such cones as Goan.

3 *Explosive phase* during which volcanic tuffs were deposited and the large calderas and explosion craters were formed.

2 *First eruptive phase* during which basaltic lavas and some ashes were erupted. Basaltic lava was in a fluid state and flowed considerable distances from the vents. Cones and flows mainly formed during this phase.

1 *Period of erosion* during which an extensive series of pediplains was cut across the Basement Complex.

A detailed history, together with maps and sections, is given by Almond (1969).

Jebel Nakharu. Basaltic lavas cap Jebel Nakharu near Berber east of the Bayuda volcanic field (Fig. 33). Details of the section are given above.

The basalts were recorded by Andrew and Karkanis (1945) and are mentioned by Andrew (1948 and MS 1943). They are more than 54 ft thick below the trigonometrical station and form part of a weathered blocky flow extending from the near-by old volcanic centre of Jebel Atshan.

Again, little definite can be said about the age of the volcanic rocks except to say that they are post-Nubian and probably post-Hudi Chert (? Tertiary) in age. Andrew and Karkanis (1945) correlated the chert-bearing gravels, which occur beneath the basalts at Jebel Nakharu, with the Hudi Chert.

(i) Isolated outcrops south-west and south-east of Laqiya Arbain

Scattered outcrops of 'Tertiary lavas' are shown in this desert area west of the Nile. No details appear to be available, however (Fig. 4).

Farther south in Kordofan, Andrew (MS 1943) mentions that alkaline lavas that occur near Sodiri at Jebel Katul and Abu Ajal may be of Tertiary age also.

(j) 'Tertiary lavas' north of Dongola

Dykes and volcanic necks occur in this region and cut the Basement Complex and the Nubian Formation. On the Geological Map of the Sudan, second edition, 1949, scale 1:4 000 000 they are shown as 'Tertiary Lavas'. They occur in the region between Delgo and Kerma (Fig. 4). Potassium argon determinations made on specimens collected by the author by Geochron Laboratories Incorporated indicate that the necks were intruded in the early part of the late Cretaceous and are of Turonian or earliest Senonian age (for details see section, Mesozoic Formations).

The question arises then, how many more of the isolated outcrops of volcanic west of the Nile are older than Tertiary?

(k) Jebel Kuror

This prominent jebel is situated in the desert between Abri on the Nile and Station No. 6 on the Wadi Halfa–Abu Hamed Railway. It rises over 1600 ft above sea-level and forms a mass surrounded by extensive pediments. Unfortunately the author has not visited it on the ground but, seen from the air, volcanic features such as craters, flows, and an extensive ramifying dyke system appear to be present. Andrew (personal communication) thinks that the dykes are composed of grorudite.

Other large circular and subcircular mound-like outcrops occur in the desert east of Abri and Mursid. These weather black and have jagged outlines, suggestive of volcanic origin. Again they have been seen only from the air.

Little detail is known for the area between Jebel Kuror and the Nile and the Wadi Halfa–Abu Hamed Railway. Basic dykes cut the Basement Complex at a number of localities along the river in the Batn el Hagar. Their age has not been determined, however. The only other published record of volcanic activity in the Batn el Hagar region is the hot spring that occurs at Akasha on the Nile (Grabham 1920, Sandford 1935, Fairbridge 1963). Geological details are not available.

(l) 'The Infra-basaltic sediments', basalts and rhyolites of Khor Langeb

(i) *'Infra-basaltic sediments'*

Mudstones, sandstones, limestones, and dolomites were recorded by Delany (1954, pp. 15–16) from beneath, and interbedded with, basalts of the Khor Langeb–Khor Windi area of the Red Sea Hills, near the Eritrean frontier (Fig. 45).

The sediments crop out along the southern side of Khor Areiwat, in the head waters of K. Durundrun and K. Hamashkoleib, and two isolated outcrops occur south of the Khor Windi between Telorit and Jebel Mandaued (Sudan Geological Survey sheet 46 I). Unfortunately no memoir was issued to accompany this sheet, which was surveyed by Miss Delany in 1951–5. However, the following section for the Durundun area is given by Delany (1954, pp. 15–16):

TOP

	Thickness (m)
Mudstone, finely bedded, massive	0·5
Limestone, slightly baked	0·5
Mudstone, grey	1·0

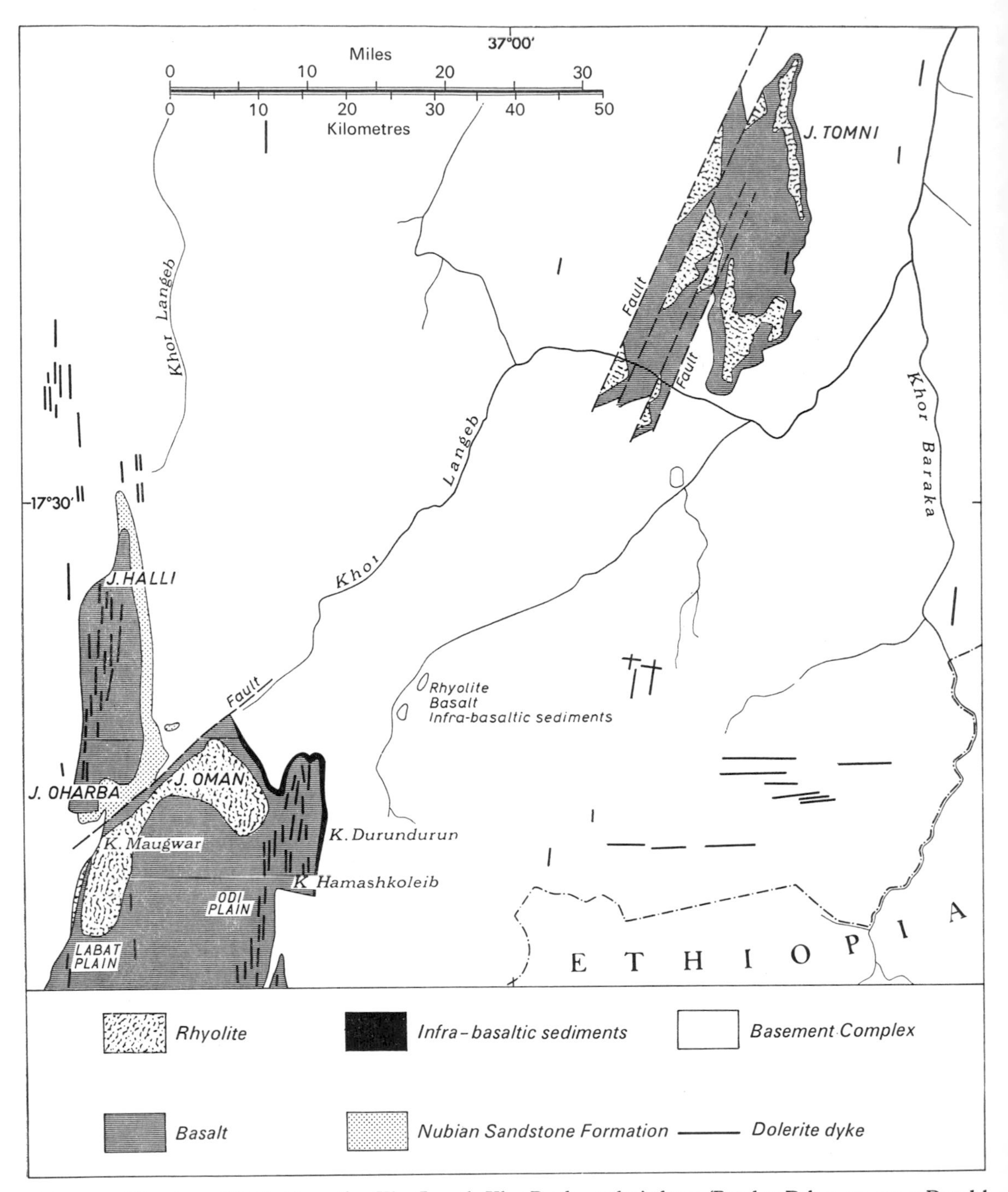

FIG. 45. Volcanics and Nubian Formation, Khor Langeb–Khor Baraka, geological map (Based on Delany, 1:250 000 Derudeb sheet 46 I, Sudan Geological Survey)

Limestone, compact, oolitic grey; few quartz grains which form oolite centres; oolites have generally crystallized around Ostracoda or Foraminifera	1·0
Sandstone, ochreous, yellow argillaceous matrix; highest bed is silicrete resting on yellow ferricrete composed of angular quartz grains with occasional ironstone nodules	
Marram, arkosic sandstone resulting from lateritic weathering of gneiss	4·0
Gneiss belonging to Basement Complex	
BOTTOM	

Unfortunately these sediments have not yielded identifiable fossils and so their age is uncertain.

North-west of Khor Maugwar at Jebel Oharba (17° 13′ N, 36° 34′ E; incorrectly cited as 35° 34′ E), Delany (1954, p. 16) recorded fossiliferous estuarine sediments intercalated with the lowest basalt flows. The section is as follows:

TOP	Thickness (m)
Basalt	
Yellow dolomite	0.2
Limestone, reddish-brown weathering, blue when fresh, fossiliferous	1·0
Marls, blue with gasteropods and plants	0·5
Mudstone, yellow to pink with plants	0·5
Basalt, upper surface of flow altered to coarse marram	
BOTTOM	

Fossils from the limestone bed, identified by experts from the British Museum (Natural History), include teeth, dorsal plates, spicules, and dental plates of crocodile, catfish, and wrasse. The stratigraphical range of the fishes is from Eocene to Recent. The gasteropods from the marls belong to a new species of *Melanoides*.

Barron described rocks of this area (field books in Sudan Geological Survey).

(ii) *The basalts and rhyolites of Khor Langeb*

Grabham (1920) mentioned the occurrence of the basalt in this region and Andrew (1947 and MS 1943) referred to outcrops of volcanic rocks at 17° 15′ N, 36° 15′ E and 17° 30′ N, 37° 15′ E west of the Khor Baraka in the Khor Langeb drainage, Kassala Province. No further details were given, however, except that the volcanic rocks overlie the Nubian Sandstone Formation. Andrew (1947 and MS 1943) apparently considered that these two outcrops are outliers of the main Ethiopian Trap Series, which appears south of Asmara, Eritrea.

Delany (1954, p. 17 and sheet 46I, 1956) described the Khor Langeb–Khor Windi volcanic rocks. There are three main groups of outcrops:

(1) the Labat plain, Odi plain, Jebel Halli–Jebel Oman Group centred on the bend of the Khor Langeb;
(2) three isolated outcrops in the Khor Windi drainage; and
(3) the Jebel Darak–Jabel Anheib–Jebel Adaribab–Jebel Tomni–Jebel Shaba Group situated north and west of the Khor Langeb–Baraka junction.

In the Khor Langeb–Odi Plain–Khor Windi area the following sequence was given by Delany (1954, p. 17 and sheet 46I):

KHOR LANGEB–ODI PLAIN–KHOR WINDI AREA

	Thickness (m)
Faulting on north-west–south-west trend	
Emplacement of Jebel Oharba quartz porphyry	
Extrusion of rhyolites and trachytes	Approx. 150
Intrusion of north–south dolerite dyke swarm	
Extrusion of basalts with three distinct intercalations of banded ash	150
'Infra-basaltic sediments'	14
Nubian Sandstone Formation	
~~ UNCONFORMITY ~~~	
Basement Complex	

Delany (1954) recorded rhyolite on basalt on infra-basaltic sediments among the isolated Khor Windi outcrops.

At Jebel Ohaba (18° 10′ N, 37° 35′ E) riebeckite-granite belonging to the Basement Complex is cut by felsite and dolerite dykes and capped discordantly by basalt.

On Jebel Adaribab (17° 41′ N, 37° 15′ E) the lower slopes are composed of riebeckite-granite of the Basement Complex. The main mass of the jebel is composed of westerly-dipping basalt horizons overlain by trachyte flows and rhyolitic ash, tuff, and lavas. The basalts and trachytes dip to the west at

15° but the rhyolites appear to dip towards a central vent (Delany 1954).

Fossil wood (undetermined) was found between an upper glassy flow and a lower chalcedony-bearing lava.

The faults of the Khor Langeb region trend north-east–south-west in general and control the courses north-eastwards from Jebel Oman (Fig. 45).

(m) Gedaref and Gallabat basalts of the Ethiopian marches

Eight patches of 'Tertiary lavas' are shown on the Sudan Geological Survey Map, second edition, 1949, scale 1:4 000 000 in the Gedaref and Gallabat regions. An isolated outcrop of basalt exposed in the bed of the Atbara above and below the Butana bridge, below Khasm el Girba Dam, is also shown. This is omitted on the third edition, 1964, and only four outcrops are shown in the Gedaref–Gallabat region. The outcrop pattern shown on sheet 55 compiled by Delany (1952) is somewhat different from that shown on the 1:4 000 000 maps. For instance, the three outcrops of the El Magran, Gedaref, and Doka regions are joined up.

On the Geological Map of Africa (1963) the basalts are shown as a lobe continuous with the main outcrops of the Ethiopian Trap Series resting on Adigrat Sandstone, which pinches out into the Sudan, and on Nubian Sandstone (Whiteman 1965).

The most comprehensive account available on the Gedaref and Gallabat basalts was prepared by Ruxton (MS 1956) and was based on mapping done when he was an officer of the Sudan Geological Survey. According to Ruxton (MS 1956)

'most of the basic rocks around Gedaref and Gallabat are fine to very fine-grained crinanites and basalts. The finer-grained types with a doleritic texture are sometimes highly vesicular with zeolite amygdales and are prone to selective decomposition. The zeolite infillings include natrolite, chabazite, stilbite, and phillipsite . . . Most thin sections of these rocks show olivine titanaugite and plagioclase with interstitial analcite. In places phenocrysts of olivine or andesine occur . . . Intrusive trachytic necks and dykes are confined to the outcrop area of the crinanites and basalts south and east of Gedaref. They have appearance of pale coloured silicified mudstones with which they have been confused in the past. The occurrence of sanidine phenocrysts, a peppering of riebeckite, or clear flow banding are diagnostic criteria in the field . . . A neck of trachyte protrudes above the surrounding basalts at Jebel Ukeila (13° 19′ N, 35° 49′ E) and a solvsbergite dyke with horizontal columnar joints normal to its walls runs north-west from here for three kilometres forming Jebel Dahr et Tor. Both the neck and the dyke are intrusive into the basalts.

Far to the south of Gedaref on the open clay plain two large areas are dotted with hills and small domes of solvsbergite and grorudite (Simsim and Jebel Angtub . . . The hills of both areas represent the denuded remnants of large sheets probably intruded between Basement Complex rocks and the Nubian Series (now eroded off them).'

Concerning the mode of origin of the basic rocks of the Gedaref–Gallabat region, it is commonly thought that they are mainly lava flows that were extruded on the Gedaref Sandstone Formation.

According to Ruxton (MS 1956) borehole evidence proves that a substantial proportion of these basic rocks are multiple sheets and irregular intrusions. For instance, Ruxton cites the evidence from the Wadi el Huri bore, which passed through crinanite sills in the Gedarf Formation; and at Tawarit (13° 34′ N, 35° 19′ E) south of Gedaref, some 480 ft (150 m) of Gedaref Sandstone Formation overlies a crinanite sheet exposed at the base of shallow wells.

Very few data are available on thickness of the Gedaref and Gallabat basic rocks. Ruxton (MS 1956) cites a thickness of 650 ft (203 m) penetrated in the Er Rawashda bore without reaching the base of the formation. The crinanites and basalts are said to thin out south-east of Gedaref towards Gallabat to 96–288 ft (30–90 m). The third edition of the Sudan Geological Survey Map, scale 1:4 000 000, still shows two separate outcrops in the Doka and Gedaref areas.

Age of the Gedaref–Gallabat basic igneous rocks. These rocks clearly postdate the Gedaref Sandstone Formation (? Jurassic, Table 5). They predate the present erosion cycle and predate the formation of the Gedaref erosion surfaces cut at about 2076–2336 ft (630–730 m) and believed to be of mid-Tertiary age.

This estimate is confirmed by the potassium–argon date of 33×10^6 years (Oligocene) obtained for Gedaref basalt by Grasty, Miller, and Mohr (1963, p. 101). The structural relationships of the Gedaref–Gallabat basic igneous rocks to the Ethiopian Trap Series is uncertain but Dainelli (1943) suggested, from a consideration of contours on the top of the Basement Complex, that the western scarp of the Ethiopian plateau overlooking the Sudan is of tectonic origin. Mohr (1962, p. 45) confirmed this, citing evidence that at Dul, Beni Shangul, the schists are downthrown 2240 ft (700 m) to the west. Certainly the elevation of the crinanites, teschenites, and basalts of Gedaref is much less than the average for the Trap Series in western Ethiopia.

(n) The Sennar teschenite

During the excavation and borings for the Sennar dam a teschenite mass was discovered. Its shape and dimensions are obscure according to Ruxton (MS 1956) but is certainly intrusive into the Nubian Series. The rock is weathered to a depth of 132 ft (40 m) in places. The relationships as known are shown in Fig. 40. Ruxton (MS 1956) states that the teschenite is allied to the crinanites of the Gedaref area. 'It is texturally a plutonic rock varying from coarse- to fine-grained. The fine-grained types carry abundant olivine and titanaugite, plagioclase, and apatite. The coarse types are poorer in olivine and richer in analcite with much idiomorphic ilmenite.'

Clay with kankar resting on ironstone and alluvium overlie the teschenite-gabbro at Sennar.

(o) The Akobo and Boma basalts of the southern Sudan

The Trap Series of the Ethiopian plateau approaches the Sudanese–Ethiopian frontier in the region south of the Blue Nile and crosses the frontier in the Daga River area. These outcrops are continuous with those of Wollega.

Andrew (MS 1943) states that the Boma plateau, south-east of Pibor Post, Upper Nile Province, is composed of lava and that outliers composed of volcanic rocks occur as far west as Kapoeta and Nagishot in Equatoria Province.

Other outlying volcanic hills on the plain include: Lochiret (5° 57′ N, 34° 10′ E), Kathangor (5° 45′ N, 33° 58′ E), and Ngechela, Kamopir, Karatom (Andrew MS 1943).

The lavas of the Boma region are mainly basaltic but rhyolites, trachytes, and tuffs occur. Sölvsbergite necks form spine-like hills capping the plateau in the Nelichu area (Andrew MS 1943). The Boma plateau lavas are continuous with the Trap Series of Ethiopia in Kaffa Province and extend southwards into Turkana in Kenya.

Baker (1965) includes in the Rift Zone the greater part of Turkana, and faults that must continue into the Sudan are mapped on the Kenya side of the frontier. Further investigation in this region may well throw light on the continuation of the Rift Zone and volcanic rocks along the western margin of the Ethiopian plateau (Chapter IV, Structure).

Willimott (1957), in a description of soils and vegetation of the Boma plateau and Eastern District of Equatoria Province, made reference to the geology of the region using data from reports by Grabham (1937) and Wright (1955). Wright classified the rocks as follows:

3 Volcanic rocks—Upper Series
2 Volcanic rocks—Lower Series
1 Basement gneisses

The Basement Complex gneisses are of the quartz-hornblende type, and the lower volcanic rocks are Tertiary tuffs and lavas. In general, the lavas are greyish-purple, non-vesicular, and fine-grained and overlie a purple and vesicular slag. The lavas are thought to be moderately basic.

The Upper Volcanic Series is much more variable in character and includes vesicular lavas, tuffs, and basalts. They are thought to be trachytic and phonolitic in composition. In general, these lavas are lighter in colour than the Lower Volcanic Series, and frequently contain porphyritic feldspar.

Wright (1955) also recorded bentonites in the Boma plateau area. Most probably they are of the same age as the volcanic rocks of the Ethiopian plateau, i.e. Oligocene.

(p) Jebel et Toriya and Omdurman basalts

The Geological Survey 1:250 000 map of Eastern Khartoum Province (compiled by Delany 1952) shows three small outcrops of basalt within the Nubian Sandstone outcrop west of the Nile at Khartoum. A fourth outcrop has recently been found north of Jebel Et Toriya by Dr. D. C. Almond (personal communication). All four basalts lie along a line 8 miles long trending north-north-east from Jebel Et Toriya in the south to Omdurman in the north (Fig. 46). The best exposures are at Jebel et Toriya. Recently these outcrops have been studied in detail by Qureshi, Almond, and Sadig (1966). Their geological and gravimetric surveys show that the basalt forms a saucer-shaped intrusion about a

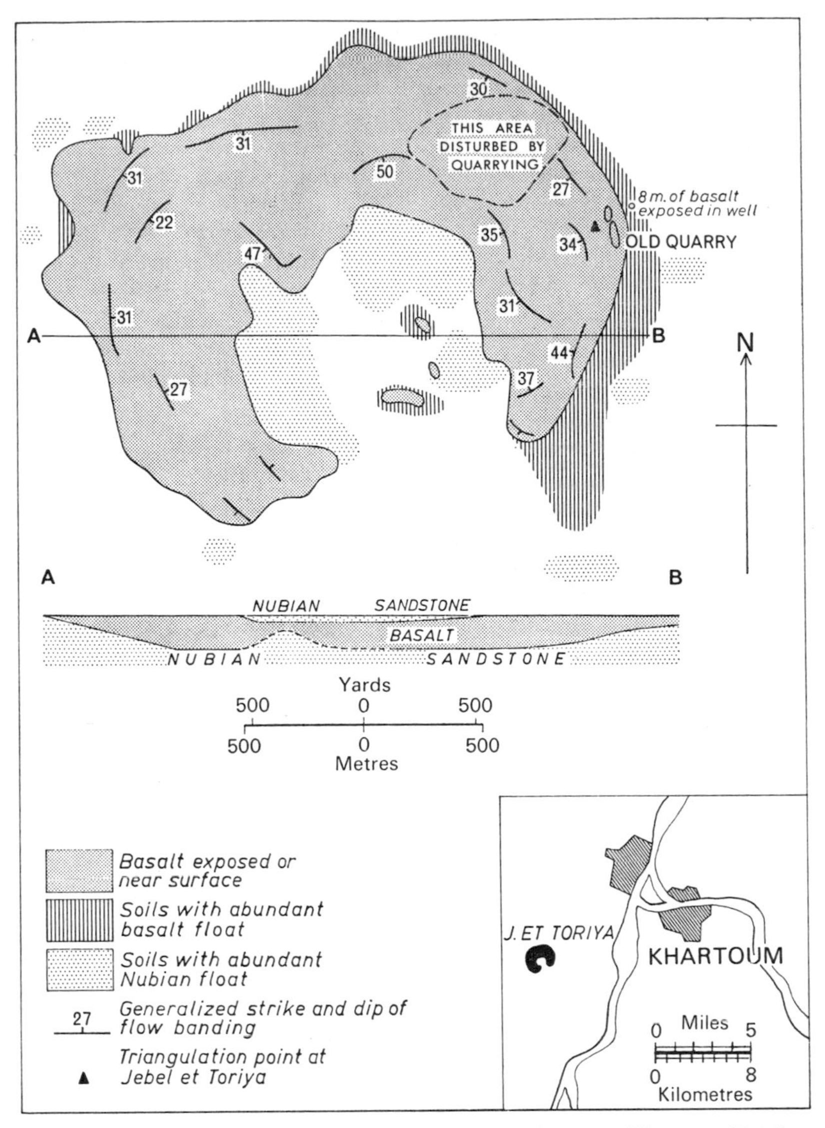

FIG. 46. Jebel el Toriya Basalts, and Nubian Formation, near Khartoum (Based on Qureshi, Almond, and Sadig 1966)

mile across with part of the Nubian Sandstone roof still preserved. The basalt is a flow-banded, olivine-bearing alkaline type with phenocrysts of titanaugite and olivine. Alkaline feldspar and analcite occur interstitially, and there are amygdales of analcite and zeolite. Superficial examination suggests that the other basalts in this area are also intrusions of similar composition.

Jebel et Toriya is now being extensively quarried for road metal. Basalt of this type was used in the construction of the Omdurman road along the River Nile and for dressing on the new Khartoum–Wad Medani road.

A small basalt outcrop occurs immediately west of Omdurman, near the radio mast, it is poorly exposed and has not been dug for stone.

(q) Laterites, lateritic soils, iron-rich deposits, etc. of the southern and northern Sudan

Laterites and lateritic soils are widespread in the southern and central Sudan (Andrew 1948, MS 1943) (Fig. 6). Laterites and similar deposits were also encountered in Northern Province along the route followed by the 1932 Bagnold expedition (Sandford 1935, pp. 269, 366).

(i) *Southern Sudan*

Laterites and lateritic soils form extensive but dissected sheets in the southern provinces. The inselbergs that form such a conspicuous part of the scenery of these areas rise through these deposits as in northern Uganda. In general the iron-enriched surfaces decline northwards, but the sheets are said to rise again towards Nyala and Muglad, where they reach some 1600 ft (500 m) above sea-level. The ironstone occurs mainly on low ground but is rarely found in river valleys, except as ironstone gravels (Dunn 1911; Andrew 1948, MS 1943).

Thick deposits occur east of the Bahr et Jebel near Kerripi, between Nimule and Juba, at 2560 ft (800 m), and along the road from Torit to Juba east of the Lowe Hills. Andrew (1948, MS 1943) thinks that in general the ironstone deposits on the eastern side of the Bahr el Jebel are less extensive than those on the western side and are much more eroded. It is unfortunate that so little is known of the denudation chronology of this interesting area.

West of the Maridi–Yei watershed in western Equatoria a concretionary ironstone sheet caps small hills. It occurs some 16–80 ft above the present clay-covered plateau surface. There is also a much higher erosion surface covered with ironstone in the Yei area. No details were given by Andrew (1948, MS 1943). Near Yambio, at Li Rangu, thin white clay underlies the ironstone (Andrew MS 1943); and in the Yirol area, in Bahr el Ghazal Province, ironstone is said to overlie red 'china clay' and in some places sands. At Tali Post and Yirol ferruginous sandstones and mudstones are overlain by ironstone. Andrew (personal communication) recognized a high-level ironstone cap in western Equatoria and low-level kaolin deposits capped by ferruginous clay sand and pea ironstones. The latter he thinks are late Tertiary to Pleistocene and are restricted to areas south of Bahr el Ghazal.

Further north in Upper Nile Province Andrew (MS 1943) recorded an 'ironstone escarpment' running somewhat irregularly in an east–west direction around the south-eastern flank of the Nuba Mountains. The level of the ironstone in this region is said to be between 1280 and 1344 ft (400 and 420 m). These occurrences are thought to be the northern continuation of the laterite plateau of Equatoria Province.

Little is known of the region geologically. However Willimot (1957, p. 14), in a description of the soil and vegetation, commented on the distribution of laterite. Laterite was observed on the plateau above Top Camp Rest House but lateritization is said to be less on the plateau than on the plain. Near Katchikan, between Bottom Camp and Jebel Kathangor, ironstone occurs in sheets resting on 'pink lavas' and basalt. Pebbles of rhyolite are said to occur in the laterite and Delany (1953) recorded pebbles of trachyte in pea-iron gravel of murram. Wright, according to Willimot (1957), noted that the Basement Complex is also strongly lateritized in this region along the road to Pibor Post.

Other occurrences of laterite and pea-ironstone gravel mentioned by Willimot (1957) are: on the road to Juba; in the Kalimunga area where pea-iron gravel is used for road metal; in the Didinga Hills near the turn-off for Nagishot; and in the Dongotona and Lafit Mountains.

It is difficult to date the lateritization in the southern Sudan. Andrew (MS 1943) regarded the laterites and associated deposits as mid-Tertiary, and cited as support for this the point that ironstone has not been recorded from boreholes drilled into the Umm Ruwaba and similar deposits infilling the depressions. The occurrence of laterite at no great height above the present erosion levels and above the river-terrace systems probably points to Tertiary age also. No bauxite has been recorded in the southern Sudan but it may well occur.

(ii) *Northern Sudan*

North of latitude 11°N the ironstone deposits are said to be discontinuous. Andrew (MS 1943) recorded patches of concretionary ironstone in the region around Sennar. Not all this deposit is *in situ*, however. The age of the deposit is uncertain. He also recorded laterite of uncertain age in the Beni Shangul area along the Sudan–Ethiopian frontier.

Ironstone occurs near the contact of the Gedaref Sandstone Formation and the overlying lavas west of Gedaref but it is not known whether it is a superficial deposit covering the basalt or whether it is the remains of lateritic deposit resting on the Nubian Group and pre-dating the basalt (Andrew MS 1943).

Ironstone gravels occur south of Omdurman (Andrew 1948), and Arkell (1948, p. 6) described a 'non-implementiferous iron cemented conglomerate' some 5 ft thick and a 'fine ironstone gravel' some 4 ft thick at Khor Abu Anga, in Omdurman. The stratigraphical significance of this site is discussed below in the section on Pleistocene stratigraphy. The association of these gravels with Acheulian-type implements implies a Pleistocene age.

North of Omdurman there is a widespread sheet of murram or pea-iron gravel and this, like the Khor Abu Anga ironstone gravels, is clearly derived from iron-rich Nubian Sandstone Formation that crops out on the tops and the upper slopes of the Merkhiyat jebels, north-west of Omdurman, and on the isolated jebels that occur to the north of Omdurman and west of Wadi Seidna.

The ironstone cap of Jebel Surkab, part of the Omdurman battlefield, shows characteristic laterite structures. Iron staining and enrichment extends along the bedding-planes and joints in the Nubian Formation. At the top of the jebel bedding is obliterated in places and masses of botryoidal replacement ore occur. Such a deposit cannot have been formed under present conditions and must have been formed before the evolution of the present drainage.

Ironstone gravel and ironstone occur on the east bank of the Nile on Jebel Direr, Jebel Umm Sa'eifa, Jebel Wad Umm Bashar, Jebel Dura, and Jebel Ibrahim el Kabbeshi. Extensive deposits of ironstone cap some of the Nubian jebels on the eastern side of the Sabaloka dome and all these occurrences once may well have formed part of a continuous sheet. They can have been formed only under climatic conditions different from those prevailing at present and certainly pre-date the present erosion cycle. Exactly how old these ironstones are is uncertain. They are older than the gravels of the Nile terraces, which to the south contain Acheulian-type implements, and it is possible that they were formed during the Shendi–Togni cycle and are late Cretaceous–early Tertiary in age. Iron stains joints at Jebel Aulia; its distribution is controlled by mudstone bands in the sandstones.

The geologists of Hunting Technical Surveys (1964) mapped laterite and ferricrete on the Nubian outcrops on sheet 3, west of Er Rahad, Kordofan at Jebel Howag, Jebel Surfan, etc.

An important piece of evidence bearing on the age of the ironstone occurs north of Shendi at Jebel el Kereiba. A small outcrop of fossiliferous Hudi Chert Formation is banked against Nubian Sandstone at the upper edge of the Nile Valley scarp, a few feet below the edge of the ironstone-capped plateau, the Shendi–Togni surface, which extends eastwards from Kabushiya, Mahmiya, and Aliab towards Wadi Mukabrab. The field relationships (Fig. 44) indicate that there was a depression occupying what is now the Nile Valley even in Hudi Chert times and that the ironstone cap of the Nubian Sandstone is older than early Tertiary (?), the age now given to the Hudi Formation by Cox (1932–3).

In addition to the iron-rich deposits that cap the jebels there are many other similar deposits found on lower slopes in wadis and depressions, or low on divides. The origin of these is often uncertain because it is difficult to distinguish between local contemporary or recent concentrations of iron in the Nubian Formation itself and older horizons. In general, it is safe to say that many of these smaller iron concentrations are younger than the jebel top deposits but where the relief is low there is a distinct possibility that they may represent the lower part of the zone of enrichment.

The iron-ore body that occurs north of the Sudan Portland Cement Company quarry west of the Nile at Atbara, is a replacement body, the iron that replaces marble having been derived from the Nubian Formation that crops out to the south and the west.

Sandford (1935, pp. 366–9) described laterite or 'marram' and other superficial accumulations from the area east of Erdi and Ennedi and between the Wadi Howar and Wadi el Melik in Darfur and Kordofan. He concluded that as marram forms today only on suitable surfaces with a high annual rainfall and annual dry seasons 'we may assume that the old accumulations, now being destroyed by the denudation of an arid region formed under similar climatic conditions about 10°N of this present northern limit of growth in the Sudan.

It follows, therefore, that the old marram is a pre-desert formation, and its presence, actual or inferred, affords information on the pre-desert topography.'

Edmonds thinks, however, that Sandford's interpretation is incorrect and that the red coloration can be explained by evaporation processes under present climatic conditions so long as the sand remains stabilized. 'The whole question of former lateritization in the desert, implying as it does a climate with a high annual rainfall (Sandford 1939, p. 369), seems to me to rest on a misrepresentation of the nature of the red deposits of the desert' (Edmonds 1942, p. 29). Edmonds pointed out that the 'lateritization' process is confined to the sandstones and intercalated shales and nowhere are igneous rocks affected. This is the opposite of what can be expected from what is known of lateritic processes in regions of high rainfall. According to Edmonds (1942) the red iron concentrations imply the continued existence of desert conditions rather than temporary oscillation to a more tropical climate. Another point frequently forgotten in arguments about the origin of laterites, etc., is that they can form in areas of impeded drainage and that they may have been formed when climate was wetter in the Pleistocene.

5. PLEISTOCENE AND RECENT FORMATIONS

Formations of Tertiary, Pleistocene, and Recent age are difficult to separate and correlate in the Sudan because of the scarcity of diagnostic fossils and implements, of palaeoclimatic data, of exposures, and of datable carbonaceous materials.

Depositional history and denudational chronology are best considered under the following headings (Table 9):

(a) Nile Valley†;
(b) areas east and west of the Nile Valley; and
(c) the Red Sea Littoral.

(a) Pleistocene and Recent formations of the Nile Valley

(i) *Umm Ruwaba and El Atshan formations*

These are described in detail above in the section on Tertiary formations. They consist of unconsolidated sands, sometimes gravelly; clayey sands; and clays. A maximum thickness of 1100 ft (335 m) has been recorded.

These formations have yielded few fossils and therefore little can be said about age.

Andrew (1948, MS 1943) pointed out that the Umm Ruwaba deposits are older than the Qôz sands of Kordofan and the dark surface clays of the central Sudan. The Umm Ruwaba and El Atshan formations are best assigned to the Tertiary and Pleistocene.

(ii) *The vertebrate-bearing terrace deposits of the Kosti area, Blue Nile Province*

Very little is known about the Nile deposits between Malakal and Kosti and only a few fossils have been found. Andrew (1948; MS 1943) stated that a bore at Umm Koweika, west of Kosti, yielded fragments of pectoral spines of a large siluroid fish (indeterminate), fragmentary teeth, and reptile and mammal bones. These are said to resemble present Nile forms. No information on depth nor descriptions of the deposit were given.

A molar belonging to the species *Hylochoerus grabhami*, identified by Hopwood (1929, pp. 289–90), was collected by G. W. Grabham from a depth of 45 ft in a well in the District Commissioner's garden at Kosti on the White Nile (Fig. 4). *H. grabhami* is allied to the forest hog of central Africa and is not diagnostic stratigraphically. The lower canine of a small carnivore was also identified. Grabham (*in* Hopwood op. cit.) did not comment on the age of the deposit and, as noted above, it is uncertain whether it belongs to the Nile Pleistocene or Recent deposits or to the Umm Ruwaba Formation.

(iii) *Lake Sudd*

Lake Sudd is supposed to have been a vast lake that covered what is now the Sudd region and the outcrop of the Umm Ruwaba Formation, and to have extended beyond in what are now Upper Nile, Bahr el Ghazal, and Equatoria provinces and into the valley of the Nile in Blue Nile and Khartoum provinces. It was first mentioned by Lombardini (1865), an Italian hydraulic engineer.

The idea was supported by Willcocks (1904) and elaborated by Lawson (1927) and Ball (1939). These writers were impressed by the flatness of the plainlands of the central and southern Sudan and

† The evolutionary history of the Nile drainage with special reference to the Sudan is dealt with separately (Berry and Whiteman 1968). A comprehensive account is not included here.

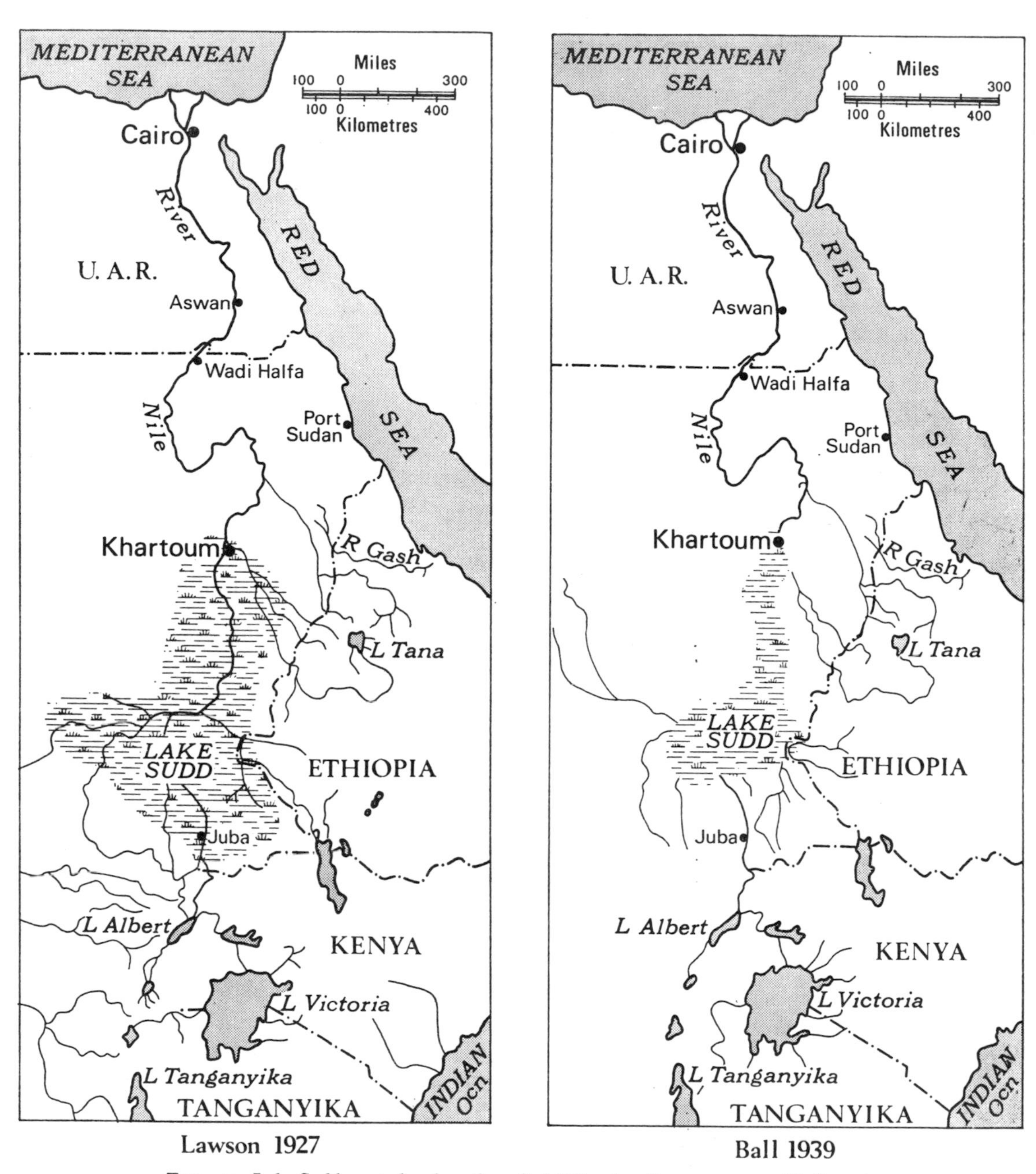

FIG. 47. Lake Sudd, central and southern Sudd (Based on Lawson 1927 and Ball 1939)

recognized that small increases in the level of the Niles would flood extensive areas. They were also searching for a means to explain the idea that the Nile in Egypt is an extremely young river, a concept put forward by Blanckenhorn (1902, etc.), Arldt (1918), etc.

Willcocks (1904) postulated that the ancient lake was 250 miles in length from north to south and that the Blue Nile flowed southwards to join this lake. Arldt (1918) envisaged an even larger lake extending from Darfur to the Ethiopian marches.

Lawson (1927, pp. 253–9) elaborated the Lake Sudd hypothesis and was first to call it by this name (Fig. 47). The essentials of his hypothesis are as follows.

(1) The Sudan above the confluence of the Blue and White Niles as far as Rejaf is a vast plain of aggradation some 700 miles from north to south by some 500 miles from east to west. Lawson (1927) estimated its area at not less than 240 000 square miles or 614 000 km^2.

(2) It is assumed that the deposits covering this plain are all flood plain or deltaic deposits similar to those exposed in the banks of the White Nile.

(3) A consideration of the present climatic and hydrological data leads to the conclusion that the area of a sheet of water which would completely evaporate all the Nile affluents south of Sabaloka at their present discharge is, however, only three-tenths of the area of the alluvial plain. Expressed in another way 'If the swamps of the Sudd were two-and-a-quarter times their present extent, and all the branches of the Nile above and including the Blue Nile were to discharge into a larger Sudd, or, as I have called it, Lake Sudd, the whole of the discharge would be dissipated by evaporation and there would be no flow at Khartoum' (Lawson 1927).

Ball (1939, pp. 74–8) developed the Lake Sudd hypothesis further. He assigned a length of over 655 miles (1050 km) to the lake placing its northern end at Sabaloka near Khartoum and its southern end near Shambe (Fig. 47).

Shambe, on the Bahr el Jebel, is the point where Sudd growth begins to spread widely over the alluvial plains (Fig. 1). Ball (1939) took the 1280-ft (400-m) contour in the Shambe area as marking the probable margin of Lake Sudd. This enabled him to sketch in roughly a position of the margin of the lake using the generalized position of the 1280-ft (400-m) contour shown on the Sudan Topographic Survey and the Egyptian Survey maps. The hypothetical lake is supposed to have had a length of about 655 miles (1050 km) north to south, a maximum width of 331 miles (530 km) east to west, and an area of 88 808 square miles (230 000 km^2). At the confluence of the Blue and White Niles the lake would have had a width of 87 miles (140 km) and a depth of 70 ft (22 m).

Ball (1939) made calculations similar to those of Lawson (1927) and concluded that for a lake of the dimensions given above an average evaporation rate of 3 mm/day over the lake would suffice to dispose of all rain- and river-water entering the lake, since the average annual daily rate of evaporation from open water surfaces at Mongalla, Malakal, and Khartoum are 3·0, 4·5, and 7·5 mm (mean 5 mm) respectively (Hurst and Phillips 1931, p. 60). Shore lines, beach deposits, and terrace features were not recorded by Lawson and Ball, who in fact argued that such features and deposits would have been broken down or redistributed and traces of lake margins obliterated by the tropical rains.

Theoretical calculations of this kind are rarely applicable to geological problems, for many factors cannot be assessed because of inadequate data. The effects of variations in rainfall, humidity, evaporation, etc. during the Pleistocene fluctuations of climate; the volume of the hypothetical basin; percolation factors; the extent of the so-called lake deposits; and the existence of outlets other than Sabaloka are all uncertain. Proof of the existence of lakes in the Nile Valley must be based on the recognition of lake deposits, shore line features, terraces, and fossil shells, and on definite information about sill heights; not on theoretical hydrological calculations as Ball and Lawson advocated.

Since Lawson put forward his ideas in 1927 new information has appeared. The margin of the Umm Ruwaba Formation has been mapped and it is clear that not all the sediments are alluvial-plain deposits as Lawson suggested. The formation is polygenetic, sediments having been swept in by the Nile and its tributaries, and by the khors and wadis draining the Nuba Mountains, the Nile–Congo divide, the Dongatona, Lafit, and Imatong

Ranges of Equatoria, and the Ethiopian marches. The Umm Ruwaba deposits are thought to have been laid down in a series of land deltas similar to the Gash delta of Kassala Province, and to the Sudd at the present day (Berry and Whiteman 1968).

A lake of the size postulated by Lawson (1927) covering some 240 000 square miles is impossible hydrologically and geologically, for this would entail flooding an area considerably greater than that enclosed by the present 1600-ft (500-m) contour and the existence of a water body over 550 miles wide in the latitude of Atbara, Northern Province, dammed back by a hypothetical barrier in the northern Sudan. Even a lake covering the present Umm Ruwaba outcrop, some 94 000 square miles would be hydrologically impossible because at present the Umm Ruwaba outcrop reaches an elevation of over 1600 ft (500 m) in the Abu Gabra and Aweil area of Darfur and Bahr el Ghazal provinces and this exceeds any possible sill height in Blue Nile, Khartoum, and Northern Provinces (Fig. 1).

Previous writers about Lake Sudd have all assumed that the sill height at Sabaloka at the northern end of the Sixth Cataract Gorge was the regulator for the Nile Valley; in fact on the right bank of the Nile it is the lowest point on the broad ridge that extends from the Sabaloka jebels to Jebel Qeili 1963 ft via Jebel Rauwiyan 1950 ft, the rhyolite–ignimbrite Sabaloka plateau 1600 ft, Fururi 1696 ft, Jebel Daruba 1645 ft, Jebel Gunar 1792 ft, Jebel Amriya 1884 ft, and Abu Duleiq 1632 ft which acted as a regulator. The drainage lines, even though they are generalized, show the approximate trend of the ridge (Fig. 48). The southward continuation of the ridge is brought out by the 1600-ft (500-m) contour, which swings in a broad arc approaching nearest to the Blue Nile at Wad en Nau, south of Wad Medani. On the left bank height data are even scarcer than on the east but the ridge can be recognized and swings westwards and southwards from Jebel Rauwiyan 1950 ft (609 m) along the watershed between the Nile and Wadi Muqaddam which reaches over 1450 ft (441 m) near Bir Hukuma. There is little evidence of recent tectonic movement in the region and the physiography of the area probably differed little in Pleistocene times from that of the present.

The maximum height of any lake system in the Nile Valley, if we assume structural stability, is therefore controlled by the minimum height of the Jebel Haraza–Sabaloka–Jebel Qeili ridge. This point lies only a few miles east of the Sabaloka ring complex, where the ridge is 1389 ft (434 m) above sea level (Fig. 48). North of Sabaloka the valley opens out and it is not until Jebel Ali Barsi, north of Kerma, Northern Province, that another barrier ridge crosses the Nile Valley behind which an extensive lake system could have been formed. However this system could only rise a few feet above the 960-ft (300-m) contour that crosses the river near Kerma (Fig. 1).

A Lake Sudd of the dimensions proposed by Arldt (1918), Lawson (1929), and Ball (1939), evidently could not have existed in the Nile Valley, but lakes below 434 m could have been dammed back against the Jebel Haraza–Sabaloka–Jebel Qeili ridge. So far, however, the only lake deposits recognized are at 386 ft and 382 m (Berry 1961, Berry and Whiteman 1968). These are described below and are shown by radiocarbon dating to be of late Pleistocene age.

(iv) *Lake and terrace deposits of the White and main Niles between the Sudd and Sabaloka*

Very little is known about the Pleistocene and Recent deposits of the Nile south of latitude of the Jebel Aulia Dam, and little mapping and levelling has been done, except in the Gezira. Andrews (1912, pp. 110–13), described bones found when caissons were sunk for the Blue Nile bridge at Khartoum. Grabham (1926, pp. 280–2); Sandford (1935, pp. 321–81), and Edmonds (1942, pp. 18–30) mentioned the ? Tertiary 'murram' or ironstone deposits of the area. General accounts of the Pleistocene and Recent history of the Gezira were given in the Soil Conservation Committee's Report of 1944, and by Andrew (1948). Tothill (1946) recorded the results of a study of molluscs from the northern Gezira; he noted that in the White Nile basin the remains of aquatic snails requiring permanent water occur up to the 382-m level.

Arkell (1949, 1953) provided new archaeological evidence bearing on the age of Pleistocene and Recent deposits in the Khartoum area and pointed out in *Early Khartoum* that in Mesolithic times the Blue Nile flood appears to have reached approximately 381·14 m at the Khartoum Civic Hospital site (reference 1946 flood level 377·14 m at Khartoum). Arkell (1953) revised this estimate and postulated a

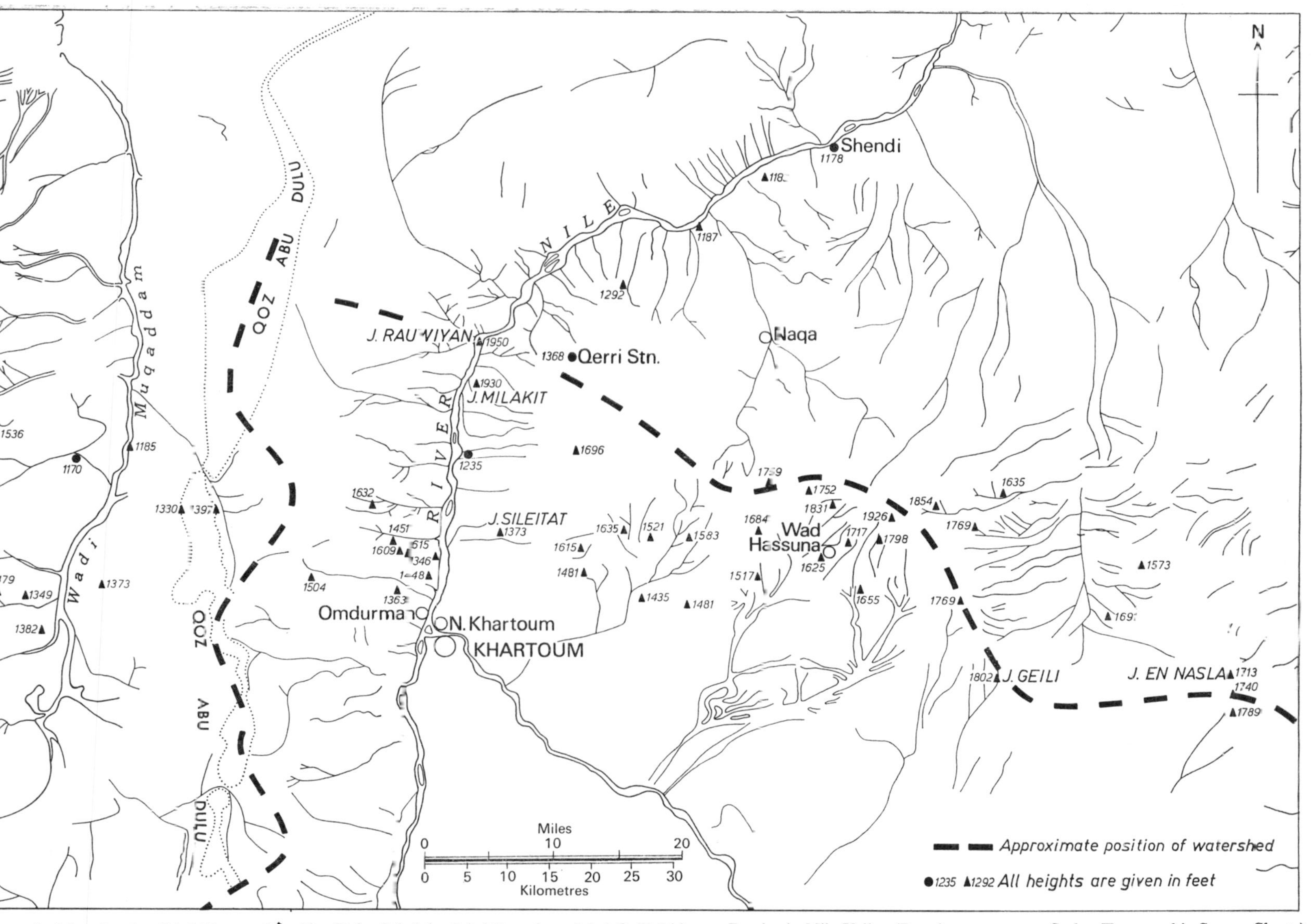

FIG. 48. Map showing Jebel Haraza–Qôz Abu Dulu–Sabaloka–Jebel Rauwiyan–Jebel Qeili Ridge or Barrier in Nile Valley (Based on 1:250 000 Sudan Topographic Survey Sheets)

Nile at Es Shaheinab and Khartoum at least 10 m above the 1946 flood level, i.e. 382 m at Es Shaheinab or 387·14 m at Khartoum in Mesolithic times.

A survey of irrigable areas was made by Sir Alexander Gibb and Partners (1954) and this provided data bearing on the existence of lakes in the White Nile basin. Berry (1959, pp. 4–11; 1962, pp. 7–13) described physical features of the Niles between Sabaloka and Malakal, and the characteristics and mode of formation of the Nile islands.

At least two distinct terraces were recognized in the White Nile basin (Berry 1962, pp. 14–19) between Khartoum, Kosti, and Malakal. Evidence presented in the Sir Alexander Gibb and Partners Report (1954) demonstrates the existence of terrace features south of Kosti. According to Berry (1962, p. 15):

'On the White Nile there are two well-developed terrace levels which do not appear to rise appreciably upstream. One is at a height of approximately 386 m (1235 ft) and the other at a height of 1223 ft (382 m) above sea level at Alexandria. (Based on old Khartoum datum 360 m, under new datum 363 m, all heights are raised 3 m.) The lower terrace extends upstream from the Khartoum area as far as the Melut bend. The higher terrace may be represented in the Khartoum area, and occur intermittently up-stream to a point north of Malakal.'

(1) 1235-ft (386-m) *terrace*. High-level Nile clays have been found in the Khartoum area in excavations near the radio station in Omdurman at about 1235 ft (386 m) and scattered remnants occur at about this height in the Jebel Aulia Basin (Table 11). South of Kosti the surveyed sections of Sir Alexander Gibb and Partners (1954) show that the terrace is a continuous feature. Berry (1962) described this area as follows:

'From Kosti to the Melut bend both terraces are found; the 386 m forming a narrow 1–2 kilometre wide flat in the north, but widening to the south as the lower 482 m terrace becomes less important. In the Melut bend the land near the river is part of the 386 metre terrace; while away from the river there are depressions which form part of the 382 metre terrace. South of the Melut bend the 386 metres terrace appears to be less well defined, though a terrace one or two metres higher (387–388 m) which extends as far as Malakal, may be related to the 386 m levels in the north . . . The 386 m terrace also appears to be flat lying, but to the south of the Melut bend it may either have been tilted by earth movements or covered by additional sediments to raise it one or two metres in height.'

(2) 1223-ft (382-m) *terrace*. The terrace has been recognized in the field in areas south of Khartoum and extends south of Kosti and according to Berry (1962), north of Jebel Aulia dam, the White Nile sediments extend about 9 km west of the river. Beach-like accumulations of gasteropod shells occur at the margin of these clays. The terrace appears to have been degraded somewhat in these areas where the elevation is about 381 m. The edge of the terrace is masked by sandy wash in places and the back is therefore difficult to locate.

South of Kosti the terrace has been recognized in the sections presented by Sir Alexander Gibb and Partners (1954) and is more than 5 miles (8 km) wide on the east bank just south of Kosti. It is 10½ miles (17 km) wide in the El Jebelein area and is well defined as far upstream as the Melut bend. South of Melut and west of the river the terrace has not been recognized.

During 1964 and 1965 Hunting Technical Services, Ltd. conducted a soil survey of irrigable areas along the White Nile for the Ministry of Agriculture, Khartoum. M. A. J. Williams, soil surveyor, kindly collected a series of shell-bearing samples from soil pits in the region. Two samples suitable for radiocarbon dating were selected by the author from soil pits associated with the 382-m lake level (Williams 1966). Whether these represent separate stages in the lake sequence or whether they are bottom deposits formed in the 382-m lake is not clear at present (Fig. 49) (Whiteman 1969*a*).

Some 130 soil pits were put down between Khartoum and Kosti in the area west of the Gezira Manaqil areas. Only five yielded fresh water molluscs and of these only two yielded sufficient quantity for radiocarbon dating. These samples are described below.

SAMPLE NO. 1211

Isotopes Inc. No. I 1486 Age: 11 300 ± 400 years B.P.

Location: 13° 35′N, 32° 41′E, 6 km east of Esh Shawal East

Altitude: 379·45 m

Topography: Level

0–40 cm	Silty clay	Fine granular tending to moderate medium sub-angular blocky structure below 5 cm. Weak horizontal and moderate vertical fissures

TABLE 11

Correlation table for Pleistocene and Recent formations, Sudan Republic

Nile valley				Areas East and West of Nile valley		Red Sea Littoral	Age Whiteman MS 1965 *C date in years B.P. × U – Th date	
Nile silts, Wadi deposits and terraces				Desert deltas, reg, hamada, clay plains, dunes and sand sheets, wadi deposits; Jebel Marra, Melit Meidob, Nakharu, Bayuda volcanics	Qôz belt Post-Meroitic wet phase Dry phase As far S as 13°N — 3000 —	Present-day reefs and lagoonal deposits, Wadi deposits, and terraces	?2100–2600? ?4600–6000?	
Esh Shaneinab Neolithic site (av) main Nile						2m Bench raised reef 3·5-4m Bench raised reef	5253 ± 415*	Recent or Holocene
Charcoal 13m above flood plain (afp) Sudanese Nubia					— ± 7000 — Modification of Qoz during wet phase soils leached	Flooding	7300 ± 250*	
Khor Umar gasteropod site main Nile, Khartoum area							7400 ± 120*	
382m White Nile Lake deposits sample no.1139						of	8730 ± 350*	
Cleopatra bulimoides 12m afp, Sudanese Nubia							9325 ± 250*	
?Esh Shaheinab wavy line pottery, Qeili Station, and Khartoum Hospital sites (mesolithic)	— ? —	— ? —				Red		
Aetheria elliptica 9-10m afp Nubia	Siltation phase (Sebilian) Nubia	Upper clay member	'Gezira clay' Gezira formations		Main phase of Qoz formation (phases based on Warren 1964 *a,b*)	Sea	11 200 ± 285*	Pleistocene: Late
382m White Nile lake deposits sample no 1211 386 White Nile lake?							11 300 +400*	
Corbicula artini 15m afp, Nubia						marsas	11 600 ± 300*	
20m afp; Nubia							14 950 ± 300* ?25 000	
Period of deep scour Shambat B.H.Khartoum	— ? —				— ? —	Cutting of	29 000 ± 300*	
Singa skull and vertibrates, ?Near base of Gezira clay Blue Nile						Red Sea marsas	?38 27 000–	
Khor Abu Anga, Lupemban? Gravels (Cole 1964), Khartoum		Mungata and Gezira member			?Early Qôz phases ?		— ? — Lupemban ?29 27 000– — 30000 —	
Khor Abu Anga, Sangoan Gravels (Cole 1964) Khartoum						(a) 3·5-4m bench (b) 7–8m Mohd. Qol stage (b) 11-12m Shinab stage (c) –16m Mukawwar stage raised reef Mukawwar early Monastirian 18m 60ft stage (Riss–ürm int.)	Sangoan — 50 000 —	
Khor Abu Anga, late Acheulian gravels Khartoum							(a) 75-80 000 (b) 95 000 (c) 91 000 ± 5000×	
50ft terrace Wadi Halfa Acheulian imlements	— ? —						— ? —	
J.Nakharu, Early Acheulian gravels 100ft terrace, Wadi Halfa	Umm Ruwaba and El Atshan formations					Older Terrace Gravels Khor Eit / Abu Shagara formation	Middle	
J.Nakharu, Chellian gravels								
J.Nakharu, Wawa, Nuri Pre-Chelles-Acheul pebble tools							Early	
150ft river terrace and platform 200ft river terrace and platform 300 river terrace and platform Wadi Halfa-S.Egypt					— ? —	— ? — Pliocene beds Khor Eit (Berry and Sestini)	— ? — Plio-Pleistocene	

40–135 cm	Heavy silty clay	Massive 5 per cent $CaCO_3$ specks. Thin coating of silt and very fine sand along cracks. Moderate horizontal and vertical fissures
135–200 cm	Silty loam	Massive 30 per cent $CaCO_3$ aggregates. Occasional vertical and fine subhorizontal cracks

The bulk of the samples were collected from 145 to 170 cm. There are abundant *Cleopatra*, many fragments of cancellate and smooth gasteropod shells, fragments of Nile oyster and ?*Unio*. The base of the shell band is situated at 378·45 m.

SAMPLE NO. 1139

Isotopes Inc. No. 1485 Age: 8730 ± 350 years B.P.

Location: 13° 56′ N, 32° 22′ E, 12 km south of Abu Hibeira

Altitude: 377·6 m

Topography: Level

0–30 cm	Clay	Fine granular to fine subangular blocky structure. Below 10 cm moderate subangular, blocky structure, occasional shell fragments. Much finely divided $CaCO_3$
30–150 cm	Clay	Moderate medium subangular blocky structure. Silt and very fine free sand common in cracks. 8 per cent soft, loosely aggregated $CaCO_3$. Between 130 and 150 cm 2 per cent yellow $CaCO_3$ concretions
150–160 cm	Calcic horizon	Over 60 per cent $CaCO_3$ concretions and friable, aggregated $CaCO_3$ embedded in silt
160–190 cm	Silt	Laminated and slightly warped, highly compacted and very fissible. Planes of lamination very well defined 15 per cent soft $CaCO_3$ aggregates. Trace of iron staining between planes of deposition
190–200 cm	Silt loam	Less strongly laminated than above. Much finely divided $CaCO_3$

The bulk of the shell fragments were collected from 140 to 160 cm and from 180 to 190 cm. Large

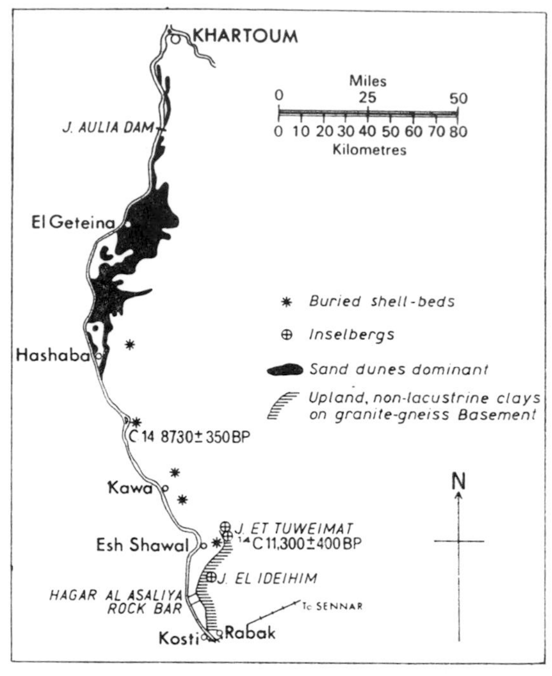

FIG. 49. Radiocarbon dates and shell locality map, Kosti–Shawal region, central Sudan (After Williams 1966. Carbon-14 dates by Isotopes Incorporated for Department of Geology, University of Kharthoum)

Ampullaria fragments were found at both levels. The sample sent for radiocarbon dating was collected from the 180 to 190-cm shell bed with its base at 377·6 m.

(3) *Age and origin of the* 1235-ft (386-m) *and* 1223-ft (382-m) *terraces.* These carbon-14 dates indicate that these terraces are considerably older than deposits at the Neolithic occupation site at Es Shaheinab, some 30 miles north of Khartoum on the main Nile. According to Arkell (1953) the average age for charcoal and shell from this site is 5253 ± 415 years B.P. and he suggested that at this time the Nile flowed some 5 m above the present-day highest flood-level.

Unfortunately no radiocarbon dates are available for the Mesolithic Early Khartoum site at Khartoum Civil Hospital nor for the Wavy Line Pottery site at Esh Shaheinab, but clearly on cultural grounds the sites are considerably older and are higher in both instances than the Esh Shaheinab Neolithic Nile level. Most probably they were associated with a Nile level at least 10 m above the present-day flood level (Arkell 1953, p. 8), i.e. with the 382-m terrace described by Berry (1962).

Arkell assumed that the levels at the Wavy Line Esh Shaheinab site, the Qeili Station, and Khartoum Hospital sites were related to high river levels and presented climatic and hydrological arguments to account for this. In the author's view the 382-m levels are all associated with a lake at 382 m, dammed back by a plug of silt laid down by the Blue Nile north of Khartoum (Berry 1962, p. 17; Berry and Whiteman 1968). Because of the cultural problems involved in making Khartoum Mesolithic older than 11 300 ± 400 years B.P., the date for sample No. 1139 from the 378·45-m level, the author thinks that samples Nos. 1139 and 1211 are best classed as deposits laid down in the 382-m lake.

The age of the University terrace in the Khartoum area is uncertain but it is clearly younger than the Neolithic Esh Shaheinab level. The flattish terrace-like feature that extends southwards and eastwards across Khartoum Airport is not a distinct terrace in the author's view but forms part of the degraded edge of the 382-m terrace. The high-level 'dark clays' and part of the 'high level sands' mapped by

TABLE 12

Pleistocene and Recent sequence with carbon-14 dates, Khartoum region (Based on Berry and Whiteman 1968)

Name	Height (m)	Age (years B.P.)
Esh Shaheinab Neolithic site	377	5253 ± 415
Khor Umar gasteropods	385	7400 ± 120
Sample No. 1139 (382-m lake deposits)	377·6	8730 ± 350
Sample No. 1211 (382-m lake deposits)	379·45	11 300 ± 400
Esh Shaheinab Wavy Line site; Qeili Station site; Khartoum Hospital site Mesolithic		
382-m lake level (Berry 1962)	382	No date
386-m lake level (Berry 1962)	386	No date

Worall (1956), in his soil survey of Khartoum district, probably form part of this degraded terrace (see Table 12).

(v) *Chelles–Acheulian and Tumbian implement-bearing deposits of Khor Abu Anga, Omdurman, and Wadi Afu*

Arkell (1949, pp. 5–52) in *Old Stone Age in the Anglo-Egyptian Sudan* described these deposits. The Khor Abu Anga is a small tributary that joins the main Nile a few hundred yards below the confluence of the Blue and White Niles at El Mogren, Khartoum. The banks of the khor have been quarried extensively for gravel and kankar-bearing clay ('Omdurman lime') and the gravels have yielded many hundreds of hand axes to Arkell and to subsequent investigators.

The implement-bearing gravels occur a few inches above bedrock and slope from 1223·9 ft (382·5 m), outside Omdurman to 1217·09 ft (380·344 m) in the town. These levels are referred to the Old Khartoum Gauge 360 m above mean sea level and they indicate that gravels were laid down by a stream graded to a Nile flowing at 1206–1215 ft (379–380 m). An average high level for the September Nile is 1200 ft (375·4 m) and for the low April Nile it is 1184 ft (370 m) (Fig. 50).

The succession described by Arkell is summarized in Fig. 51. Bedrock exposed in Khor Abu Anga is iron-rich Nubian Sandstone which crops out in places at the surface (layer 1). Outcrops extend westward into the Merkhiyat jebels. This bedrock is followed by decomposed Nubian Sandstone with ironstone boulders and fragments and abundant kankar (layer 2). The age of the kankar formation is uncertain. Arkell (1949) suggested that the kankar was formed during an arid phase. However, the climate need not have been different from that of the present, as kankar and similar deposits form now in the Sudan on the banks of the Nile where water is available and evaporation is high. Fine examples of kankar and cemented sands only a few months or years old have been collected along the edge of the Jebel Aulia reservoir.

'Non-implementiferous iron-cemented conglomerate' succeeds layer 3; it attains a maximum of 6 ft and is clearly part of the superficial deposits.

Layer 4 consists of water-worn coarse gravel and boulders of Nubian Sandstone, some of which are

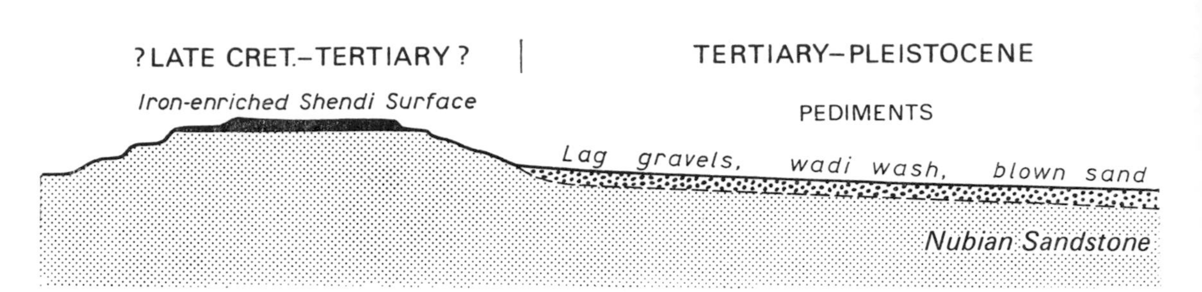

FIG. 50. Tertiary–Pleistocene denudation chronology, Khartoum region. Diagrammatic section (Vertical scale exaggerated to show superficial deposits)

angular. This layer, which contains Chellean, Chelles–Acheulian, and Early and Developed Acheulian-type implements, is widespread and varies from 1 to 3 ft in thickness. The deposit is graded to a Nile that flowed between 379 and 380 m according to Arkell (1949).

Layer 5 is also implement-bearing and has yielded a preponderance of Developed Acheulian and Late Acheulian types. It consists of fine ironstone gravel derived from iron-rich Nubian Sandstone Formation. The gravel is often cemented with clay and kankar. It has been extensively quarried by gravel diggers and varies from 0 to 4 ft thick.

White calcareous soil and calcareous silt, layer 6, overlies the ironstone gravel of layer 5; this is followed by layer 7, composed of red brown sandy clay, which has yielded shells of the land snail *Limicolaria flammata*. Arkell (1949) correlates this deposit with the Early Khartoum Mesolithic deposits described above.

In addition to the true Acheulian implements represented at Khor Abu Anga, a rough ferricrete lance head and other 'Tumbian' or Sangoan implements occur, some of which appear to possess incipient tangs. All these implements are mainly surface finds and the order of superposition is uncertain. Three tools of 'Tumbian' type, according to Arkell, were found *in situ* near bedrock in gravel that had been possibly been relaid.

Cole (1964, p. 160) has expressed the view that 'It is not yet clear that whether a true Acheulian is represented at this site, or whether, as may be more likely, the earlier industry there is Sangoan'. Elsewhere Cole (1964, p. 192) has stated that it includes an Acheulian stage, probably Sangoan, and almost certainly Lumpemban.

At Khor Abu Anga some of the hand axes are similar to those of stage 9 of the Olduvai sequence, and with Arkell the author believes that the 'Tumbian' material has been in some cases reworked. The stratigraphy of the Abu Anga site needs reassessing because in recent years the area has been greatly modified by gravel-digging and excavations for Omdurman lime. In fact, it is extremely difficult now to trace the sequence given by Arkell along the modified khor banks. It is indeed unfortunate that material suitable for radiometric study has not been found at Khor Abu Anga so that the deposit could be placed accurately in the Pleistocene sequence.

The general field relationships of the Khor Abu Anga sequence are shown in Fig. 50. These deposits are clearly younger than the iron-rich late Cretaceous–early Tertiary surface that caps the Merkhiyat jebels and they are younger than the main fans that extend from these jebels towards the Nile. The iron-rich surface is the most important marker in the region and the Nile drainage probably originated on this surface.

If we follow Clark (1965, Figs. 1 and 3), then the Acheulian implement-bearing gravels may be in part Early Gamblian as the implements have Late Acheulian characteristics (Figs. 52 and 53). At Kalambo Falls, Tanganyika–Rhodesia Border, Evolved Acheulian is given a date of 57 600 ± 750

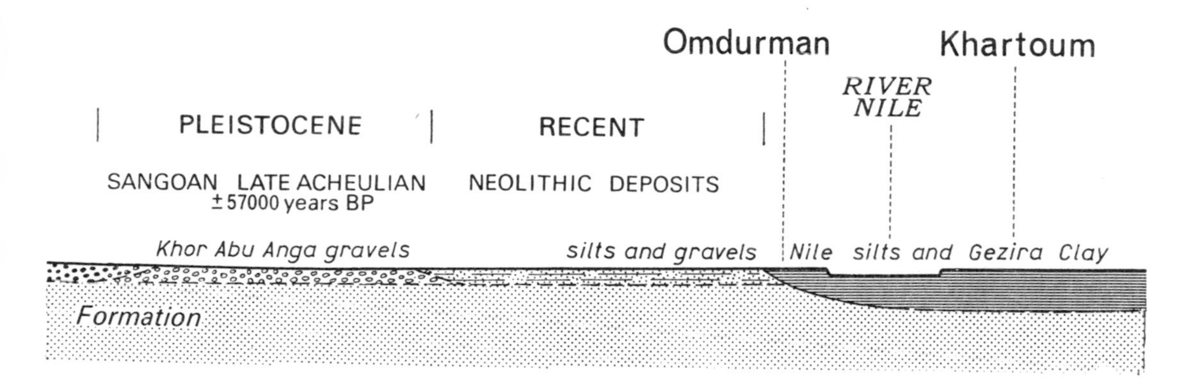

years B.P. and Final Acheulian a date of 52 000 years B.P. (Clark 1965). The Early Gamblian is equated approximately with the Lower Wurm and the Main Gamblian approximately with the Middle Wurm extending from 50 000 to 20 000 years B.P. If the Sangoan of Kalambo Falls and Khor Abu Anga can be taken as contemporaneous, then Sangoan deposits extending from 50 000 to 30 000 years back must be present in the Khartoum area; and if we accept Cole's (1964) statement that Lupemban is present, then there may be deposits present spanning the whole of the Main Gamblian interval.

Silts, clays, and red-brown sands with 'Mesolithic' implements and Neolithic implements and pottery overlie the Acheulian and Sangoan deposits at Khor Abu Anga. Arkell (1940) estimated that they were laid down about 5000 B.C. However, in the author's view they must be older; Cole (1964, p. 170) also places the Khartoum Mesolithic provisionally in the interval 9240–7600 years B.C. *Limicolaria flammata* also occurs in these deposits, and Arkell (1949) thought that this indicates that they were laid down 'in the last wet period'. He did not state, however, whether this is a local expression or whether he was referring to the Makalian wet phases of Kenya Rift-Valley lakes.

If we accept the correlations presented in Table 9 then the Early Khartoum deposits of the Khartoum Civil Hospital site and their equivalents at Khor Abu Anga may have been laid down during Gamblian 3 or Late Gamblian of the east African sequence.

Similar hand axes to those found at Khor Abu Anga were found at Wadi Siru on the west bank of the main Nile about 15 miles north of Omdurman and over a distance of 35 miles between Wadi Siru and the Sixth Cataract at Sabaloka (Arkell 1949, p. 29).

Advanced Acheulian-type implements have also been recorded in the bed and banks of Khor Umar, north of Omdurman on the west bank of the main Nile. The gravels, which are intermittently exposed upstream and downstream of the Wadi Seidna road bridge, are overlain by dark high-level Nile silts and clays, which have yielded *Pila ovata*, *Lanistes carinatus*, and *Limicolaria* sp. These in their turn are overlain by fine red-brown sands and silts with Neolithic implements and pottery sherds. Two isolated finds of Lumpemban-type lance heads were also found loose on the surface.

The snails from high-level Nile silts and clays gave a radiocarbon age of 7400 ± 120 years B.P. They occur above the present-day flood-level of the Nile.

No hand axes have yet been found between Wadi Afu in the Nile Valley, a tributary of the White Nile about 50 miles south of Omdurman, and Uganda. Arkell (1949, p. 30) recorded the following section at Wadi Afu:

7 Ironstone gravel forming a surface spread
6 Soil, red and sandy, to be compared with Bed 7 at Khor Abu Anga
5 Clay, grey, with kankar nodules to be compared with Bed 6 at Khor Abu Anga

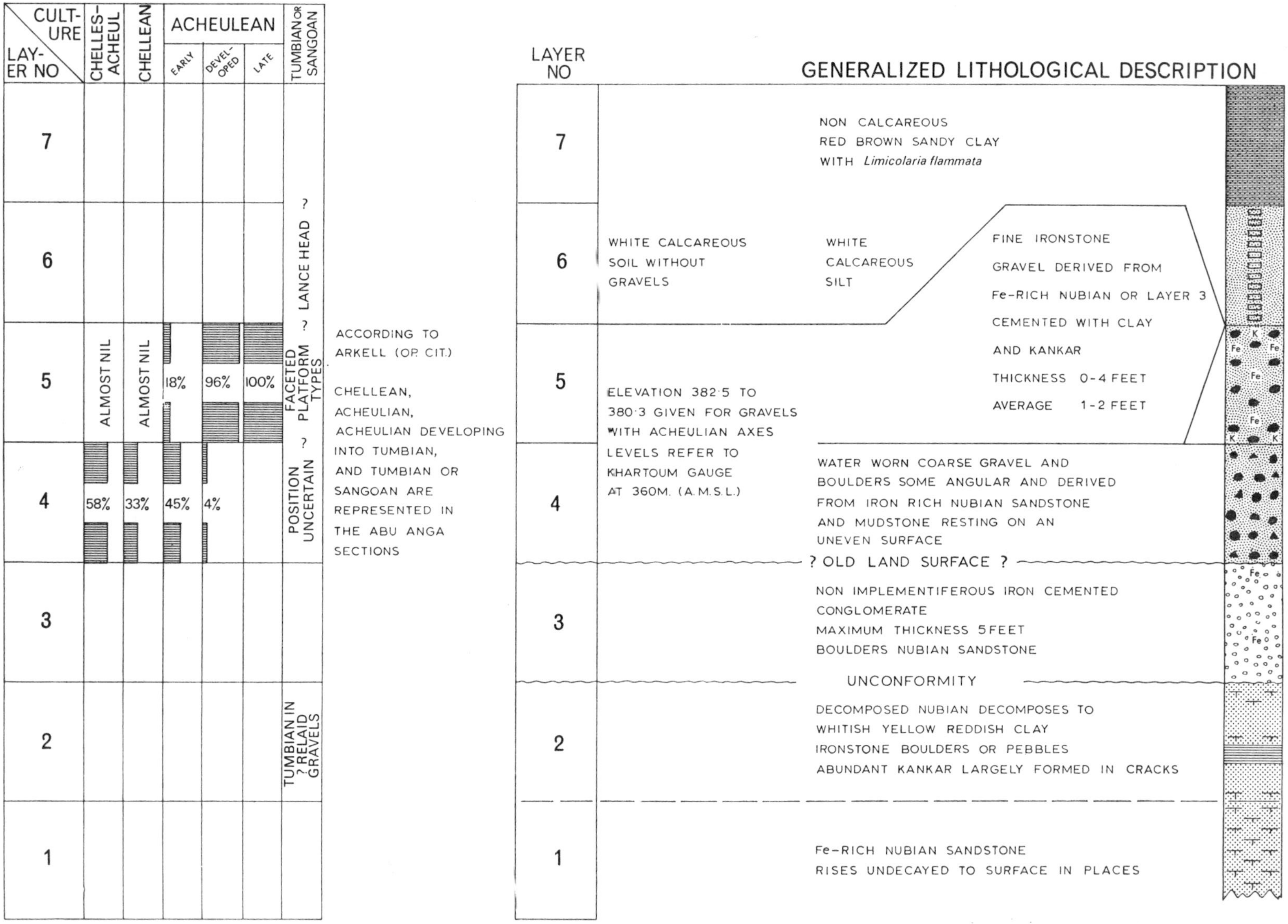

Layer No \ Culture	Chelles-Acheul	Chellean	Acheulean: Early	Acheulean: Devel-oped	Acheulean: Late	Tumbian or Sangoan
7						
6						? Lance head ?
5	Almost nil	Almost nil	18%	96%	100%	Faceted platform types ?
4	58%	33%	45%	4%		Position uncertain
3						
2						Tumbian in ? relaid gravels
1						

According to Arkell (op. cit.)

Chellean, Acheulian, Acheulian developing into Tumbian, and Tumbian or Sangoan are represented in the Abu Anga sections

Layer No	Generalized lithological description
7	Non calcareous red brown sandy clay with *Limicolaria flammata*
6	White calcareous soil without gravels — White calcareous silt — Fine ironstone gravel derived from Fe-rich Nubian or layer 3 cemented with clay and kankar. Thickness 0-4 feet, average 1-2 feet
5	Elevation 382·5 to 380·3 given for gravels with Acheulian axes. Levels refer to Khartoum gauge at 360m. (A.M.S.L.)
4	Water worn coarse gravel and boulders some angular and derived from iron rich Nubian sandstone and mudstone resting on an uneven surface
	? Old land surface ?
3	Non implementiferous iron cemented conglomerate. Maximum thickness 5 feet. Boulders Nubian sandstone
	Unconformity
2	Decomposed Nubian decomposes to whitish yellow reddish clay. Ironstone boulders or pebbles. Abundant kankar largely formed in cracks
1	Fe-rich Nubian sandstone rises undecayed to surface in places

FIG. 51. Khor Abu Anga Sequence (Based on Arkell 1949)

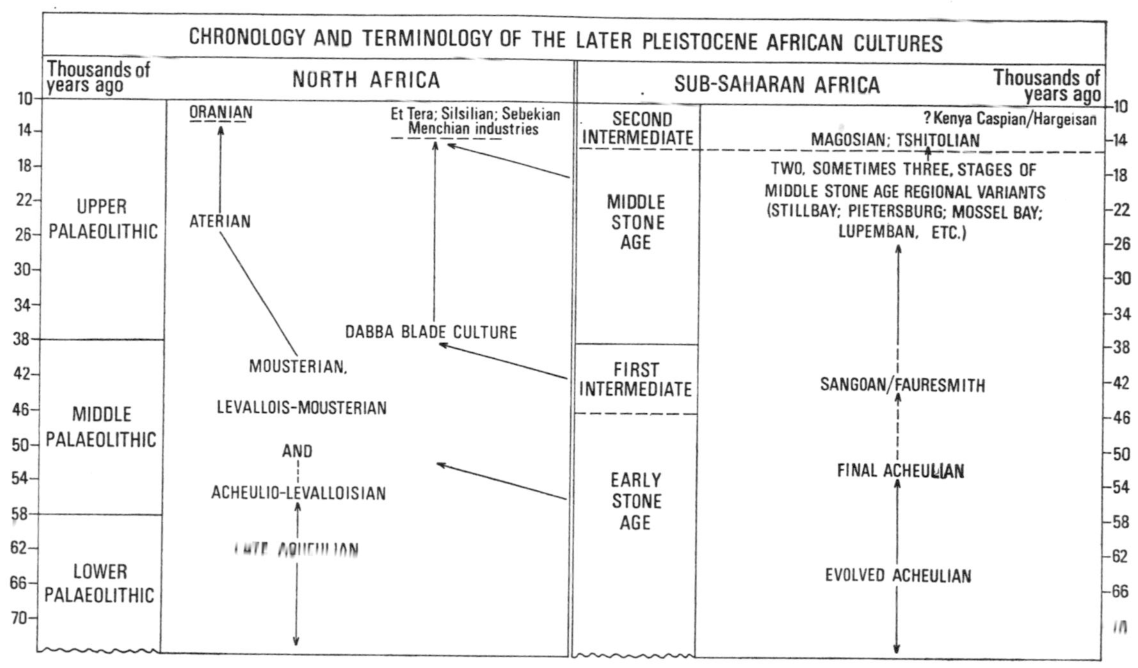

FIG. 52. Chronology and terminology of the later Pleistocene African cultures (Based on Clark 1965)

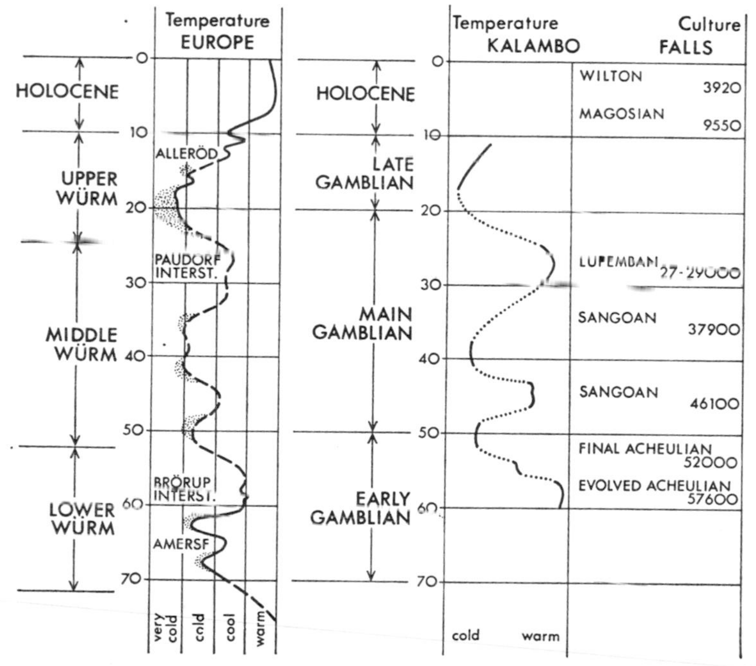

FIG. 53. Comparison of late Pleistocene and Holocene climatic changes, western Europe and central Africa, Kalambo Falls (Based on Clark 1965)

4 Ironstone gravel apparently derived from the break up of layer 1 (? Compare with layer 5, Khor, Abu Anga)
3 Gravel, coarse angular, implement-bearing (compare Khor Abu Anga, layer 4)
2 Clay with kankar, sporadic distribution
1 Nubian Sandstone Formation with ironstone pebbles, (compare Khor Abu Anga, layer 3)

He did not give any thicknesses for these deposits.

In the author's experience there is little evidence for correlating the layers at Wadi Afu with those at Khor Abu Anga. The stratigraphy of the superficial deposits of the Nile Valley is highly complex and correlation must await detailed mapping and the presentation of more evidence. Nor does the author agree with palaeoclimatic conclusions put forward by Arkell (1949). Arkell's view that there may have been a renewal of wet conditions slightly earlier farther north at Khor Abu Anga is not warranted by the field evidence.

Late Acheulian gravels were also identified by Arkell (1953, p. 8) at Esh Shaheinab beneath a black alluvial clay some 6 ft thick and the Mesolithic deposits. Arkell (1953) equated these gravels with these of layers 4 and 5 at Khor Abu Anga.

(vi) *Khartoum Mesolithic deposits at Khartoum and Esh Shaheinab*

There are two well-known Mesolithic sites in the Nile Valley in the Khartoum area; one is situated on a low mound north-east of Khartoum Central railway station and east of the civil hospital and the other is near to the Neolithic site at Es Shaheinab on the west bank of the Nile 50 miles north of Omdurman.

The settlement at Khartoum Mesolithic site, according to Arkell (1949) who excavated and described the site, was small in area and appears to have been situated on top of a sand bank at the edge of the Blue Nile. The soil below the occupation site consists of loose fine buff sand with kankar concretions. The sand is reported to have been wind-blown and may have accumulated as a bank or dune similar to those forming on the banks of the Nile at present. Arkell (1949) thought originally that the settlement was situated at least 4 m above the present flood level of the Nile, but later modified this to over 10 m (Arkell 1953, p. 8). Vertebrates and molluscs collected are listed below (see Chapter III, Palaeontology).

The following palaeoclimatic conclusions were drawn by Arkell (1953):

(1) The river appears to have been bordered by swamps. This is supported by the occurrence of Nile Lechwe, which is purely a swamp dwelling animal. Reed rat, water mongoose, and buffalo were also found and these must live within reach of water.
(2) Most of the other species found are tolerant of conditions other than swamps.
(3) The occurrence of *Limicolaria flammata* indicates an annual rainfall over 400 mm.
(4) The occurrence of *Celtis integrifolia* seeds, the molluscs, and mammals indicates that the annual rainfall was at least 500 mm with a rainy season lasting longer than it does today.

The Early Khartoum site is famous for its human remains, more specifically a human skull and fragments, described in detail by Derry (*in* Arkell 1953). Derry compared the skull with those found at Jebel Moya, southern Gezira, but, as Arkell pointed out, the Jebel Moya material is probably much younger. The Mesolithic site of Khartoum is famous for its pottery, which is among the earliest in Africa.

A Mesolithic occupation site occurs also at Es Shaheinab on the west bank of the Nile about 30 miles north of Omdurman. The site is much eroded and has yielded Dotted Wavy Line sherds and stone artifacts typical of the Early Khartoum culture. The site is near an outcrop of ferruginous mudstone about 1000 yards due west of the Neolithic site described below. According to Arkell (1953, p. 8) the late Mesolithic Nile flowed about 10 m above the modern high-flood level (1946).

(vii) *The Neolithic deposits of Esh Shaheinab*

A number of Neolithic occupation sites have been recognized along the banks of the Nile between Jebel Aulia, south of Khartoum, and the Sixth Cataract. Esh Shaheinab is one of these and was excavated in 1949 by Arkell (1953). The site is situated on the west bank of the Nile at Esh Shaheinab and consists of a gravel ridge about 200 m long between two branches of the Wadi Abu Shush on the north-east edge of the hamlet of Aulad Mesikh. The gravels, which extend for some miles to the north and south, are the remains of a terrace of the Nile. The gravel is composed of small water-worn quartz pebbles derived from the Nubian Sandstone and contains sherds, bones, and artifacts.

As noted above, the Nile is thought to have been about 10 m (382 m) above its present level at Esh Shaheinab (372·0 m) during Mesolithic times and it is thought to have fallen to 5 m (377 m) above the present-day flood level during Neolithic times. Vertebrates and molluscs collected at this site are listed below (Chapter III, Palaeontology).

The absence of swamp-living animals at Neolithic Es Shaheinab, in contrast to Early Khartoum, and the presence of hare, gerbil, ground squirrel, oryx, gazelle, and giraffe, according to Miss Bate (*in* Arkell 1953) suggest somewhat drier climatic conditions than those that prevailed at Early Khartoum. The rainfall, however, was still higher at the time Esh Shaheinab was occupied than it is today. Arkell (1953) recorded *Limicolaria flammata* and *Celtis* seeds, which implies a rainfall of more than 500 mm per year compared with 164 mm, the present annual average rainfall.

Two radiocarbon dates are available for charcoal and shell from Esh Shaheinab. Charcoal from Neolithic hearths at Esh Shaheinab gave an age of 5060 ± 450 years B.P. and shell gave an age of 5446 ± 380 years B.P. At the southern entrance to Sabaloka gorge, 'Levallois'-type implements occur in silts classed as 'Sebilian' by Andrew (personal communication). The silts contain pyroxene, a mineral typical of the Blue Nile deposits and the type Sebilian at Kom Ombo, Egypt.

(viii) *The Singa–Abu Hugar, Blue Nile vertebrate-bearing deposits*

Singa, 200 miles south of Khartoum, is famous for the discovery there of a fossil human 'bush-manoid' skull by Mr. W. G. R. Bond (Smith Woodward 1938, pp. 190–5, and Grabham 1938), and for a fossil buffalo skull and other bones described by Bate (1951).

Arkell (1949, pp. 45–6) described the following section seen in the river bank below the District Commissioner's old house (now the Rest House) where the skull was found.

	Thickness (ft)
Cracking clay with kankar, indistinguishable from the Gezira plain; greeny-brown to reddish at base	29
Unconformity ? marking old land surface	No thickness given
Alluvial sand with kankar	No thickness given
Kankar concretions. ? Water-borne with occasional fossil bones	0·9
Clay	6·4

The clay is situated about 6 ft above the level at which the skull was found. This level is now inaccessible at Singa because of the flooding caused by the Sennar dam.

The following section was measured by L. Berry, S. E. Hollingworth, and the author below the Rest House at Singa:

Locality: Cliff section, left bank of Blue Nile immediately below Government Rest House (Governor's old residence), Singa, sheet 55 K, 13° 10′ N, 33° 57′ E. Examined April 1962.

TOP	Thickness (ft)
Silty, sandy clay, dark brown; grass and tree covered; evidence of land slip; part of extensive terrace	12
Clay, silty sandy, buff brown, structureless, some gasteropods, sample No. 2 at base; some sand and gravel with sharp contact at base	12
Kankar gravel, ¾–1 in; black lime gravel with sandy matrix base of cliff	4
Kankar scree with overlying and underlying patches of 'Nile silt'	5½
Kankar scree	5½
Silt, miceaceous red-brown, ? skin of present day 'Nile silt' resting on kankar	9
Silt, micaceous as above	3
Kankar, gravel and cobbles, 4–5 in, mixed and coated with present day 'Nile silt', bedded	5½
Kankar gravel with present day reetlets; bedded	5½
Datum low-water level Sennar Reservoir April 1962. Skull horizon not exposed	

No fossils other than the terrestial gasteropods noted above were found.

About 25 miles upstream from Singa, is the site of Abu Hugar. Arkell (1949, pp. 46–7) recorded the following section at this site:

TOP	BED NO.	Thickness (ft)
Black cracking clay	5C	6·4
Horizontal banded of kankar nodules	—	—

Green-brown cracking clay containing kankar nodules	5B	8·0
Reddish-brown cracking clay containing kankar	5A	6·4–12·8
Bedded sandy silt containing root-like kankar concretions, in places interbedded with layer 5	4	6·4–12·8
Solid freshwater limestone capping layer 2 (of varying height), antelope teeth and other fossils, quartz flakes	—	0·4
Kankar nodule gravel containing artifacts and lumps of red ochre, hammer stones, and crude flakes	2	4·8–7·4
Greenish clay	1	6·4
Water level		

L. Berry, S. E. Hollingworth, and the author measured the following section at Abu Hugar:

Locality: Cliff, south-west side of meander of Blue Nile, left bank; above 'C' in cultivation on sheet 55–0; 12°52′N, 34°59′E, Abu Hugar, examined April 1962.

Top	Thickness (ft)
Clay, black, silty, soil; evidence of landslip, sample No. 4	6–8
Clay, grey-green with scattered kankar nodules	3–4
Clay, silty, brown with large kankar nodules, many tree roots	3
Clay, silty brown with slip planes; sample No. 3	12
Clay, laminated, grey-brown; vertical joints, black-coated nodules, passes laterally into silty sand about 2 ft thick, sample No. 2 with river derived gasteropods and lamellibranchs, not *in situ*	3
Base of cliff	
Clay, laminated, grey-brown; black-coated nodules; vertical joints	5½
Scree of clay and kankar, small exposures of brown clay vertical joints	5½
Gravels of kankar, well-bedded, forms gently rising bench	5½
Kankar gravels, well-bedded, cross-bedded in places, bone-bearing with hippopotamus teeth washed out from gravels, fossil crocodile skull KG 2316 I-712 *in situ*	5½
Low-water level Sennar reservoir April 1962	
Total	52

Concerning the age of the Singa and Abu Hugar deposits, Lacaille (1953, pp. 43–50) thought that the implements have closest affinity with the Proto-Still Bay industry of Lochard, Southern Rhodesia and classified them as Advanced Levalloisian. If we accept the time-scales proposed by Clark (1965, Fig. 1) and Cole (1963, p. 170), and the designation made by Lacaille (1953), then the Singa implement-bearing beds may have been formed in the interval 38 000–27 000 years B.P. Bate (1953, p. 25) assigned a general age of Upper Pleistocene on the basis of the fauna but pointed out that it may not belong to the latest phase of the period.

To obtain a more precise age for the Abu Hugar and Singa fossiliferous beds the writer submitted a sample of a crocodile tooth (KG 2316/I-712) to Isotopes Incorporated, New Jersey, for radiocarbon dating. The age given was 17 300 ± 2000 years B.P. The analyst commented, however: 'I should not put too much faith in the KG 2316/I-712 sample. The material was far from ideal.'

The crocodile skull was sent to Dr. A. J. Charig B.M. (N.H.) to be identified (see Chapter III, Palaeontology). Dr. K. P. Oakley (personal communication 1963) tested the crocodile bone for collagen and the analyst reported 'nil'. He suggested therefore that the radiocarbon date may have been given by the calcium carbonate (kankar), which impregnates the deposit in which the skull was found. Another complicating factor is that the bone bed is covered with river water during the Blue Nile flood and exposed to extremely high temperatures and high evaporating conditions when the Nile is at low water. There has therefore been plenty of opportunity for impregnation by modern radiocarbon, which could make the deposit appear considerably younger than it really is.

(ix) *The Gezira Formation and the superficial deposits of the Gezira*

The Gezira or land between the rivers is roughly triangular in shape. The apex of the triangle is at Khartoum, the sides are formed by the White Nile on the west, the Blue Nile on the east, and the base of the triangle is drawn arbitrarily at the Sennar–Kosti railway line. The height of the triangle is approximately 200 miles (320 km) and its base 80 miles (128 km). The main formations occurring in the Gezira are shown in Table 11.

The Gezira Formation consists of unconsolidated clays, silts, sands, and gravels. The type locality of the formation is at Ghubshan (14°14′N, 33°02′E) where the formation is 200 ft (70 m) thick. It rests unconformably on the Nubian Sandstone Formation and is overlain by blown sand and other superficial deposits. Gravelly sand, 30 ft (9 m) thick, occurs at the base of the formation at the type locality. This passes upwards into clayey sand, 115 ft (35 m) thick, and the upper part of the formation consists of 55 ft (17 m) of dark clays and silts locally known as 'Gezira clays'. Rapid facies changes are common in the Gezira Formation and three lithological divisions can be recognized:

(iii) Upper Clay Member—'Gezira Clay'
(ii) Lower Sandy Member
(i) Mungata Member

The formation varies considerably in thickness from place to place. It ranges from 15 ft (4·5 m) at Sheikeineiba to more than 365 ft (111 m) at Wad Adam (14°21′N, 32°37′E).

The age of the formation is uncertain. The upper part, the 'Gezira Clay' is probably of Pleistocene age. The fauna of the Gezira Clay was described by Tothill 1946 and is mentioned below (Chapter III, Palaeontology). This clay overlies clays and kankar gravels containing advanced Levalloisian-type implements (Lacaille 1951, p. 49) in the Sennar district, and records of implements near Khartoum led Andrew (1948, p. 107) to suggest that the clays of the central plains were probably post-Lower Palaeolithic but pre-Neolithic in age. Unfortunately Acheulian implement-bearing gravels of the Omdurman area have not so far been found in contact with the Gezira Formation, but the field relationships indicate that the Gezira Formation may well be both late Tertiary and Pleistocene in age. The Khor Abu Anga gravels may be considered then as a facies of part of the Gezira Formation. The two formations were laid down by two separate river systems, the Khor Abu Anga which drains the Merkhiyat jebels, predominantly a Nubian Sandstone catchment, and the Blue Nile which derives its sediment mainly from the Ethiopian highlands.

Few fossils other than those described above by Bate *et al.* (1951) from Singa and Abu Hugar have been recorded from the Gezira Formation. C. W. Andrews (1912) recorded an elephant tooth collected from a depth of 60–68 ft in gravel encountered in the excavations for the Blue Nile bridge, Khartoum. He dated these deposits as Middle Pleistocene. Bate (1951, p. 1), however, wrote that specific identification and geological age are uncertain. Other than this, records are restricted to fish bones and spines of species now living in the Nile.

The lake margin features of the White Nile are cut into the Gezira Formation and these have been dated by the radiocarbon method as late Pleistocene. The Gezira Formation is therefore older than 11 300 ± 400 years B.P.

Concerning the provenance of the formation very little can be said except in general terms. Shukri (1950, pp. 511–34) and Worrall (1956) described the mineralogy of the Gezira Clay and recorded the following minerals: calcic plagioclase, augite, hornblende, biotite, and olivine. This assemblage is similar to that being transported by the Blue Nile today and is obviously a volcanic assemblage derived from the Ethiopian highlands. According to Shukri (1950) the White Nile sediments contain a metamorphic assemblage consisting of sillimanite, kyanite, garnet, epidote, and staurolite. Similar heavy mineral suites have been obtained by the author. Typical Blue Nile minerals, for example pyroxene, especially the purple variety, and zeolites do not occur in the sands and gravels but are common and characteristic of the Gezira Clay (G. Andrew, personal communication).

The provenance of the underlying sands and gravels, etc. is uncertain. It seems unlikely that a volcanic plateau consisting mainly of basaltic rocks could yield so much quartz. Alternative sources for the sand and gravel deposits that make up a considerable part of the Gezira Formation must be looked for. The Nubian Sandstone Formation and the Basement Complex that underlie the Blue Nile basin in the Sudan are the two most obvious ones. The Gezira Formation then, like the Umm Ruwaba Formation, with which it is correlated, is therefore best described as polygenetic, the greatest part of the clay and silt fraction having been swept in by rivers draining the Ethiopian volcanic plateau, and the coarse fraction having been derived from the Nubian Formation and Basement Complex in the Sudan when climate was wetter and erosion more active.

Berry (1964) and Berry and Whiteman (1968) have attributed the formation of the Gezira Clay Member to over-bank flooding by the Blue Nile and its tributaries. The coarser fraction, however, must have been deposited in fans or deltas, and conditions during the late Tertiary and Pleistocene may well have been similar to those prevailing at the present in the Gash delta. We can envisage the Blue Nile and its tributaries forming a series of distributaries feeding fans and small local lakes in the Nile Valley. Sediments accumulated in a subsiding basin so that at Wad Adam, for instance, in the southern Gezira, more than 365 ft of the Gezira Formation accumulated and rock head was depressed more than 262 ft with reference to the sill height at Sabaloka, north of Khartoum.

The fossil gasteropod evidence collected by Tothill (1946) also supports the idea of over-bank flooding. Terrestrial forms are found in general above the 382-m level and species that require permanent water are restricted to the White Nile lakes deposits below this level.

Of the remaining superficial deposits the 'Nile silts' and wadi deposits are the youngest sediments encountered in the area. The 'Nile silt' consists of fine micaceous sands and silts deposited annually after the Nile flood from material carried largely as suspension load. They usually form the banks of the Blue Nile and stand 20–30 ft above low-water level at Khartoum. At the time of the annual flood the waters of the White Nile are dammed back and Blue Nile sediments are deposited a number of miles above the confluence at El Mogren.

Qôz deposits and associated sand sheets of Kordofan extend into the Nile Valley. The qôz are long fixed dunes, which were probably formed during Pleistocene times and are now largely fixed by vegetation. Dunes forming part of the qôz occur in the Nile Valley at Qarrasa and Wad es Zaki (14°28′ N, 32°15′ E), between El Geteina and Hashaba (Fig. 49) and rise some 10–30 ft above the river deposits. The qôz deposits here appear to be younger than the Gezira Formation.

Local sand and silt dunes accumulate behind trees, shrubs, and other obstacles outside the cultivated areas.

(x) *The Chelles–Acheulian and associated Pleistocene deposits of the River Atbara at Khasm el Girba, Sarsareib, and the Butana bridge*

Below Khasm el Girba, now the site of a dam, the river makes a double right-angled bend and then the valley opens out. Extensive superficial deposits occur and these have been dissected to form 'karib' or badlands.

Arkell (1949, p. 35) recorded fossil mammal bones and Chellean-type artifacts from the 33-ft terrace of the Atbara on the left bank upstream from the Butana bridge. He also recorded terraces at 21 ft and 14 ft.

L. Berry and the author in 1966 studied sections exposed on the east bank of the river. No undoubted Palaeolithic implements were found and we were unable to spend sufficient time in the area to obtain a complete picture. The Atbara River is now cutting into a thick series of sands, gravels, and clays exposed from the bridge up on to the plain extending towards Kassala. Terraces are present near the river as Arkell (1949) mentioned, but the terrace-like features that occur on either side of the Kassala track as it climbs out of the valley are erosional features produced by hard and soft bands in the unconsolidated deposits. These stretch eastwards and in the author's view are part of the Pleistocene Gash delta deposits.

The relationship of the Butana sequence to the cracking clays of the Butana to the west was not seen in exposures but appears to be a conformable one. Snails collected from a soil pit dug in the Khashm el Girba area in connection with the irrigation scheme gave a radiocarbon age of 12 150 ± 375 years B.P. The shells were obtained from a depth of 185 cm from a clay layer (15°48′ N, 32°24′30″ E). The Butana bridge deposits, then, excluding the terrace system of the Atbara, must be older than 12150 years B.P., but whether the succession includes Tertiary beds or not cannot be ascertained at the moment.

Downstream beyond Sarsareib, where Wayland (1943, p. 334) reported an early flake culture, for more than 200 miles the geology is poorly known. L. Berry and the author made a rapid reconnaissance trip through part of the area and there are obvious terrace features and wide exposures of thick superficial deposits. At Qôz Regeb the Basement Complex is exposed.

The gravels of the lower Atbara valley contain much chert derived from the Hudi Chert Formation. Upstream the terraces contain agates and the terrace on which the old town of Khashm el Girba was built have yielded many varieties. These may have been derived from the Ethiopian highlands or locally from the Gedaref volcanic rocks (Fig. 4).

If Arkell's identification (1949) of Chellean-type

implements is accepted, then in the Atbara valley there are Middle Pleistocene deposits and this area might well repay a close search for human remains.

(xi) *Terraces between Sabaloka and Atbara round Atbara town and the Jebel Nakharu region*

The Nile and the River Atbara join at the railway town of Atbara, and from Sabaloka on the Nile, and Qôz Regeb on the Atbara, very little is known about the Pleistocene and Recent formations.

Between Sabaloka and Wad Ban Naqa, on the Main Nile, clay and silt terraces occur mainly on the left bank and a profusion of small rocky islands and sand banks are exposed at low water. Again in the stretch from Kabushiya to Atbara, wide clay terraces are developed best on the left bank north of Aliab. Eastwards the valley opens out on the right bank; the jebels become lower and north of Wadi Mukrabab the solid disappears under the silts and gravels of the River Atbara. Extensive gravel spreads occur in the Atbara area. Some of the terraces were described by Sandford (1949).

From Atbara to Berber the Nile runs approximately north and the valley is wide and is flanked by flat to gently-rolling country with occasional low jebels set well back from the river. There are extensive terrace deposits but again these have been studied generally.

At Jebel Nakharu, as noted above (Fig. 33), early Tertiary Hudi Chert resting on Nubian Sandstone Formation and overlain by ? Tertiary basalt is exposed. North of the jebel there are extensive gravel spreads which have yielded artifacts of Chellean type, pre-Chelles–Acheul pebble tools, and Early Acheulian (Arkell 1949, p. 37). Very little information is available concerning the height and edge of these terraces, which must have developed behind the rock sill of the Fifth Cataract.

The Fifth Cataract occurs between Gananita 1050 ft (328 m) and Karaba 1020 ft (320 m) and from thereon northwards and south-westwards the valley is mainly incised in a flat to gently undulating surface cut across mainly metamorphosed Basement Complex rocks. It is not possible to give elevations because there are no spot heights available except along the river. Gravel terraces and silt deposits occur in this section. They are, in general, not extensive and are poorly known.

(xii) *Deposits of the region between the Third and Fourth Cataracts*

The Fourth Cataract, 48 miles upstream from Kareima, is composed of biotite-gneiss. In the Kareima area the valley is steep-sided and the alluvial strip is narrow, varying from 50 to 350 yards on the right bank. The top of the Nubian escarpment on the right bank is covered with thick rounded quartz gravel, which may be a terrace deposit of the Nile.

In this respect it is interesting to note that Arkell (1949, p. 37) recorded 'pre-Chelles–Acheul pebble tools', *in situ*, in gravels on top of Jebel Nuri (1070 ft) on the left bank, some 160 ft above the present high flood-plain of the Nile. 'Typical pre-Chelles–Acheul pebble tools', together with Chellean types, were found on, and probably in, gravel between 100 ft and 50 ft above the high flood-level in the area between Jebel Nuri and the pyramid field. Arkell (1949) also recorded *in situ* a number of large cores, scrapers, and flake tools made from indurated mudstone by the faceted platform technique from a site behind Tangassi market, 12 miles downstream from Nuri. These finds are extremely important from the point of view of denudation chronology because they indicate, if we accept current isotopic dates for the east African sequences, that this section of the valley has been cut since early Pleistocene times and the equivalent of the early Tertiary–late Cretaceous Shendi–Togni surface must be situated well back from the river.

South-west from Jebel Nuri and Tangassi to Ganetti the geology is poorly known except for data collected at 5-mile stations along the University of Khartoum, Bayuda Geophysics Traverse 1963 (Fig. 59).

A limited amount of data is available for the Ganetti–Ed Debba area some 60–75 miles downstream from Merowe. On the descent to the river from Karabat Abu Kuleiwat the track runs across gravels and at least three distinct levels can be recognized. The highest of these terraces certainly lies at least 150 ft above the high flood-level of the Nile, however.

Between Ganetti and Kermakol there is a flat terrace up to 1½ miles wide composed of gravel, sand, and silt; and at El Ghaba, 17 miles downstream on the left bank, at least four terraces are visible above the present flood-plain terrace. Again, unfortunately, no heights are available. At El Baja there are extensive dunes and sand sheets along the left bank of the river and the low terrace is fairly wide as far as El Goled Bahri. The right bank in general appears to be the undercut bank. In the El Khandag area both banks are fairly steep and rise about 40 ft above the cultivation terrace. The Nubian Sandstone crops out on the banks and is capped by rounded quartz gravels with agates. These gravels are greater than 10 ft thick and their base lies about 50 ft above the cultivation terrace. No implements were found, despite a close search north of El Khandaq. The terrace is very wide and certainly extends westwards many miles and northwards into the Dongola area. The gravel cap is exposed on the left bank at many localities as far as Kasr Wad Nimr, near Lebab Island, where the Nubian Formation forms steep bluffs on the river bank.

Northwards towards Dongola there is an extensive gravel and silt terrace marked by a line of bluffs shown on the 1:250 000 sheet 45–A. West of this feature the Nubian Formation is covered with a skin of coarse, rounded, predominantly quartz gravel, at least 5 miles wide, and a dissected terrace back is situated in places 9 miles from the river. Unfortunately no implements were found, but Arkell (1949, p. 37) mentions that 'small late flakes from cores with faceted platforms are to be found sporadically on the gravel above the modern high river.'

North-west of Dongola, G. Y. Karkanis picked up a silicrete sandstone Acheulian hand axe from deposits which, according to Arkell (1949, pp. 37–8), form part of a terrace on the right bank of the Wadi el Ga'ab. This terrace forms part of a much more extensive spread in the depression marked on sheet 45–A Dongola. Arkell (1949) stated that the dotted line on the map bounding this feature roughly corresponds with the 226-m contour, so placing the terrace 2 m above the present high Nile. Wind-worn hand axes belonging to the 'Late Fourth Stage of the Acheulian in Kenya' were found on the gravel that rests on the Nubian Sandstone Formation. The gravel varies from 1 to 6 ft in thickness and its provenance is uncertain according to Arkell (1949). No hand axes were found in the gravel and it is not possible to say whether the hand axes were dropped on the margins of a river (? the Pleistocene Wadi Howar, which now ends in the sand near Jebel Rahib) that ran down the Wadi el Ga'ab, or whether they were dropped on the banks of a backwater of a high Nile. Arkell (1949) postulated that earth movements held back the Wadi el Ga'ab, but in the author's view these gravels and silts may have been laid down in a 'Lake Kerma' that developed behind a natural dam in the Jebel Ali Barsi–Arduan area. Arkell (1949) thought that the Wad el Khawi may have been an alternative post-Acheulian channel situated about 10 miles east of the present Nile. There is little evidence to support this idea. Much more work needs to be done in this region.

Occasional finds have been reported east of the Nile along the Sikkat el Maheila, which cuts across the bend of the Nile from Kareima to Dongola. These include a large number of Levalloisian or Epi-Levalloisian (? Sebilian) artefacts and are taken to indicate that the desert was habitable at that time (Arkell 1949, p. 43).

(xiii) *Deposits of the region between the Third and Second Cataracts*

At Abu Fatima's Tomb, 4 miles north of Kerma, the Kerma basin with its gravel and silt ends, and the valley closes in. The river turns north-west at Simit Island and numerous small islands and banks occur above the Third or Hannek Cataract, which is composed of biotite-gneiss (Plate 9). Terrace deposits and alluvium are narrow and discontinuous, and the river's course is mainly cut across Basement Complex through the Mahas country to Fareig at the Kagbar Cataract. The Third or Hannek Cataract may have acted as a regulator behind which Lake Kerma was dammed back and the Pleistocene sand and gravels, and silts of the Kerma basin accumulated.

The Recent deposits include Nile 'silts' and blown sand. Away from the river, pediments are well developed and these certainly date back to the Pleistocene. The Nile Valley again closes in near Delgo. It is bounded by steep Nubian Sandstone cliffs and volcanic rocks for a few miles north of Delgo and there is little room for superficial deposits.

At Wawa Arkell (1949, p. 43) recorded gravels, east and north-east of the Government Rest House, with 'pebble tools of the pre-Chelles–Acheulian type and rough artefacts of the Chellean type, apparently coming from gravel not more than 15 ft above the present high flood level. One hand axe is either Late Chellean or Early Acheulian in type.'

This evidence is very important because of the later comments of de Heinzelin and Paepe (1964, p. 53) concerning the Wadi Halfa area. They believe that 'During the longest part of the Pleistocene, the Nile did not flow through the Batn el Hagar. Nubia was a large, flat pediplained area wherein the wadis or small rivers were connected to a yet unknown hydrographic system. This agrees with the evidence of an old Plio-Pleistocene Lower Nile which disappears south of Aswan.'

It does not agree, however, with Arkell's evidence from Wawa cited above. If we accept the age-determination data from Olduvai, the occurrence of Chellean implements indicates at least a Middle Pleistocene age, and the occurrence of pebble tools may well indicate Lower Pleistocene deposits. The author feels that de Heinzelin and Paepe are mistaken in their view that the Nile is a very young river in Nubia. Like so many geologists before them, they have allowed the evidence of Egyptian stratigraphy to colour their views concerning the Nile in the Sudan (Berry and Whiteman 1968).

At Sai Island, 15 miles north of Wawa, coarse gravels occur resting on Nubian Sandstone. Again artefacts of 'pre-Chelles–Acheul and Chellean type' occur, particularly south of Jebel Adfu. These are surface finds but Arkell (1949) thought that similar artefacts would be found *in situ* in gravels below the 100-ft terrace of the Nile. Flakes and cores, similar to the Khor Abu Angan 'Tumbian', occur also; so in the Wawa–Sai area most of the Pleistocene may be represented by gravel deposits, etc.

North of Sai Island the river makes a prominent bend eastwards at Abri and enters the Batn el Hagar. The bed and valley are rocky and cut mainly in Basement Complex rocks as far as the Second Cataract. Superficial deposits are narrow and discontinuous.

From a consideration of the heights of gravels in the Nile Valley in the Akasha–Abri–Sai area Arkell (1949, p. 43) postulated that the Nile, down to and including the 100-ft stage, probably flowed straight north in Lower Palaeolithic times, rejoining the present channel just north of Abu Sir at the Second Cataract. In the author's view this hypothesis is based on extremely slender evidence and much more information must be presented before it is accepted. According to Arkell (1949) the occurrence of 'small handaxes of Acheulian type and flakes made with the faceted platform technique from the local Nubian Sandstone suggest that the last men to live along this channel may have produced a Late Acheulian-cum-faceted-platform technique culture.' Again, this idea must be regarded as tentative and more collecting needs to be done, but nevertheless it again points to the incorrectness of de Heizelin and Paepe's conclusion (1964) that the Nile is a very young river which has occupied its valley since the youngest Pleistocene.

To summarize for the Kosha sheet, the denudation chronology shown in Table 13 has been established.

TABLE 13

Quaternary Sequence, Kosha Area, Northern Sudan

Recent:	Deposition of Nile Silts, blown sand, etc.
Pleistocene:	Formation of pediments and wadis, terraces with deposition of gravel and silts (?) with pebble tools, Chellean, and Early Acheulian implements at Wawa and Sai. ? 100-ft channel diversion Akasha to Abu Sir (Arkell 1949, p. 44)
Tertiary:	Formation of pediments and wadi systems; superimposition of Nile drainage
Early Tertiary–late cretaceous:	Formation of Shendi surface (1400 ft) now preserved on jebel tops in the north of sheet

In the Akasha region the valley sides are steep and the river bed is rocky. Rounded pebbles of quartz were noted by the author in the desert along the track to Akasha.

Solecki *et al.* (1963, p. 83) recorded Nile terraces in the Batn el Hagar region, and between Akasha North and Khor Kiding Kong, a distance of 7 miles, terraces have been distinguished at 48 ft (15 m) and 96 ft (30 m) above the present flood plain of the Nile. Both terraces are gravel-covered and have yielded rolled and unrolled assemblages of as yet unidentified industries.

Silts, silver grey, referred to as 'younger silts', were recorded in Khor Kiding Kong, 7½ miles north of Akasha, some 200 yd east of the Nile and 48 ft (15 m) above flood level and 576 ft above sea-level

TABLE 14

Pleistocene and Holocene climatic and carbon-14 data, Sudanese Nubia (Based on Fairbridge 1962, 1963)

Radiocarbon dates B.P.	Comment		Climate	Time scale	
	Stabilization phase	Modern arid phase. Progressive desiccation Middle Kingdom to modern times	Modern arid phase	Late	Holocene
3000	Erosion phase post-Sebilian dissection	Progressive deterioration of living conditions in Nubia and Sahara. Nile bed similar dimensions to present c.5000 years B.P. rainfall averaged 200 mm at 20°N. Neolithic to Middle Kingdom	Mid-Holocene transition phase	Medial	Holocene
7000	Erosion phase post-Sebilian dissection	Rainfall much higher in tropics and subtropics. Nile flow >3 × present. Extensive swamps. Qoz of central Sudan fixed Charcoal 13m above flood plain (AFP)	Interglacial warm phase	Early	Holocene
7300 ± 250 I-530					
9325 ± 250 I-534		*Cleopatra bulimoides* 12m (AFP)			
10 500	Siltation stage (Sebilian)	Marked climatic oscillations. Nile floods to total aridity. Monsoonal rains Sudan *Aetheria elliptica* 9–10m (AFP)	Post-glacial transition phase		Pleistocene
11 200 ± 285 I-531					
11 600 ± 300 I-532		*Corbicula artini* 15m (AFP) silt rises to 195m K. Khiding Kong			
12 000	Siltation stage (Sebilian)	Almost total aridity. Nile almost dry. Kordofan and Eqatoria deserts	Late Würm tropico-equatorial arid phase		Pleistocene
14 950 ± 300 I-533		*Unio wilcocksi* 20m (AFP)			
?20–25 000?					

(about 180 m) (Fairbridge 1963, p. 100). The younger silts are said to rise to a maximum height of 624 ft (195 m) and have yielded *Corbicula artini* in large numbers *in situ*. These shells were dated 11 650 ± years B.P. by the radiocarbon method (Lab. No. I–532).

As Fairbridge (1963) pointed out, the younger silts were formed at the end of the Pleistocene and at the end of the Sebilian Aggradation Phase (the Siltation stage). Fairbridge's conclusions for the Batn el Hagar–Wadi Halfa region are presented in Table 14.

Solecki *et al.* (1963, pp. 80–4) also described the pre-history of the Batn el Hagar and recognized five sorts of localities:

(1) summits and slopes of table mountains;
(2) dykes and outcrops of lava and porphyry;
(3) quartz localities;
(4) Nile terraces (see above);
(5) silt-filled abandoned Nile channels, etc.

Summits and slopes of table mountains. Jebel Brinikol is cited as the best example, and consists of exfoliated gneiss and green schists capped by about 96 ft (30 m) of Nubian Formation. An abundance of Levallois and Mousterian tools, etc., were found. Levallois-like material was also found on Jebel Firka to the south-west.

The dykes and outcrops of lava and porphyry. These yielded abundant Levallois material at two parallel, north–south-trending reddish dyke referred to as site Km 92 on the Wadi Halfa–Akasha road between Khors Murrat and Turkuman (Solecki *et al.* 1963).

Quartz localities. There are a number of weathered quartz veins in the Batn el Hagar, and many types of implements have been recorded in association with them. The age of many of these artefacts is in doubt and it would appear that quartz tools were manufactured and used well into the historic period. They are numerous at both 'A Group' and 'C Group' sites.

Silt-filled abandoned channels occur at several points in the Batn el Hagar including Gemai, Saras, Semna, etc. (Sandford and Arkell 1933, pp. 61–2). In the Saras area there are three silt-filled channels running abreast for about 9 km. Channels meander between high rock outcrops south of Kulb East but are mainly sand-filled. Petroglyphs are common on rocks in many of these channels but are difficult to date.

Numerous other prehistoric sites have been recorded in the Batn el Hagar region but few geological details are given and as yet a synthesis has to be made for the region.

(xiv) *Deposits of the region between the Second Cataract and the Egyptian frontier*

The various international expeditions that have taken part in the 'Save the Monuments Campaign' in Nubia have added greatly to our knowledge of the Pleistocene and Recent deposits of this section of the Nile Valley. A considerable amount of work has yet to be published but already general syntheses have been made (Solecki *et al.* 1963; Fairbridge 1963; and Wendorf *et al.* 1964, 1965). Many new radiocarbon dates have been obtained and a detailed chronology built up. In the author's view, however, the best regional picture is still to be gained from the work of Sandford and Arkell (1933) who, during the field season 1929–30, studied the Nile Valley between Aswan in Egypt and Semna, 40 miles south of Wadi Halfa, Sudan. The main views of Sandford and Arkell (1933) are presented in Table 15. The summary given below is based largely on their work.

Very little can be said about the early Tertiary history of this section of the Nile Valley. The course of the Ancestral Nile and the position of the Tethyan shore line is uncertain but may well have been situated in Eocene times between Aswan and the Sudanese frontier. In Lower Egypt and the Gulf of Suez region uplift and regression marked the end of the Eocene and regional unconformity was developed.

In Oligocene times the southern shore of Tethys was in the Faiyum area and 'Ur Nil' deposits occur as Jebel Ahmar sands and gravels. Farther south the Lower Eocene deposits were trenched by the Ancestral River Nile and the steep-sided deep valley at Luxor was initiated. To the north and east in the Red Sea area major rifting took place, accompanied by extensive volcanic activity and another regional unconformity was developed. No doubt uplift, with consequent incision in the Nile Valley, was related to these movements. During Miocene times the Ancestral Nile became entrenched in the limestones and shales even further and by 'Pontic' times an ancestral Nile Valley was well established. Gigantic landslips occurred along the valley sides, probably as a result of lubrication of underlying shale beds during period of heavy rainfall.

Transgression of the Tethys upon the land marked the ensuing Pliocene epoch and the Ancestral Nile Valley was flooded to a depth of 576 ft (180 m), converting the valley into a ria. The climate at this time must have been wet, judging by the amount of detritus that was deposited in the Nile ria. By the end of Pliocene times it was apparently almost filled to water level. According to Sandford and Arkell (1933), three types of sediment were formed: breccias and conglomerates near the sides of the ria;

TABLE 15

Tertiary Pleistocene and Holocene denudation chronology, climatic data, and cultural sequences in Sudanese and Egyptian Nubia (Based on Sandford and Arkell 1933)

Time scale	Denudation chronology and formations			Climatic data		Cultural sequence	
Recent or Holocene	Prehistory and Historic		Temples and rock pictures Lowering of Semna Cataract 26 ft since 12th dynasty Many channels cut in 2nd Cataract and Semna area Valley cut to present depth in Nubia	Desert conditions	Sand dunes, sand sheets, hamada, etc. Rain rarely occurs today north of latitude 20°N Failure of rain		No details given
Pleistocene	U. Seb. / Middle Sebilian	Degradation phase	Lowest flaking site 33 ft (A.F.P.) at Edfu (Egypt) river 22 ft (A.F.P.) Upper Sebilian at 40 ft Debeira West bars and beaches at 40 ft and 20 ft (A.F.P.) in upper Egypt with Middle Sebilian implements River stood 68–73 ft (A.F.P.) in upper Egypt Kom Ombo River fell, marshes well drained	Desert and semi-desert conditions		Late Palaeolithic	Implements restricted to Nile and near by Evidence of Mousterian derivation
	Lower Sebilian	Silt phase	Silts thicken from north to south 18 ft at Luxor to 100 ft at Wadi Halfa and conceal earlier terraces Southern Nubia — Egypt Mica silts — 10–ft terrace Mica silts — 30–ft terrace River entered 2nd Cataract	Semi-arid conditions	Failure of rain. Wadis chocked Water in upper Egypt wadi floors conform to 50 ft terrace	Middle Palaeolithic	Implements found away from Nile in Eastern and Western deserts Typical 'Egyptian Mousterian' Mousterian
			50-ft terrace widespread no sign of 2nd Cataract 100-ft terrace present between Wadi Halfa and Luxor at same height (AFP)	Abundant rainfall	Water from south east and west mainly from Red Sea Hills	L. Palaeolithic	Acheulian implements Chelles–Acheulian Late Chelles–Acheulian early Chellean
?	'Plio-Pleistocene'		150-ft terrace and river platform present Wadi Halfa to Aswan destroyed by erosion north of Esna 200-ft terrace and river platform not present north of Aswan 300-ft terrace and river platform. No evidence of Basement Complex materials from south ? river course to west of Nile	Abundant water			No implements found
? Pliocene			Nile valley flooded to 576 ft (180 m) forming a ria ? Pliocene river occupied higher levels not preserved		Wet abundant water		
Miocene			Late Miocene uplift. 'Pontic' course established well-developed landslips		Wet abundant water		
Oligocene			Uplift shoreline in Fayum. Jebel Amar sands and gravels; Ur Nil deposits; 1000-ft valley cut at Luxor in Lower Eocene deposits				

finer sediments nearer the centre; and quartz sand from the south derived from the Nubian Sandstone Formation.

The Pliocene ria deposits are easily recognized as far as Esna and extend as far as Aswan. No Pliocene deposits have been recognized south of there however, and in the Sudan the Pliocene river either flowed at higher levels, which have now been eroded away, or, followed a separate course, of which all trace has now disappeared. It is unlikely that such a course remains undiscovered. The evidence of the Nubian-type quartz sand is proof of the existence of a river flowing from the south and, as mentioned earlier, a case can be made for the existence of Oligocene, Miocene, and Pliocene Niles in the Sudan and adjoining countries.

The 300-ft terrace or rock platform is the earliest river feature of which traces survive, and patches of quartz gravels with some flint and chert occur. In the Abu Simbl area this feature is 6 miles wide according to Sandford and Arkell (1933, p. 19). Another important point is that no materials derived from the Basement Complex, which crops out south of Abu Sir Rock and the Second Cataract, have been recognized in the 300-ft terrace deposits. This applies to the 200-ft and 150-ft terraces also, and Sandford and Arkell (1933) suggested that the Nile may have made its way from Dongola to Wadi Halfa on a course a few miles west of the present outcrop of ancient rock. In the author's view, however, there may well have been a thin but extensive Nubian Sandstone cover extending much further south than at present.

River platforms and terraces may be followed northward from Wadi Halfa at 300 ft, 200 ft, and 150 ft to Aswan, and their width and development changes according to the nature of the solid rocks and the poorly consolidated deposits across which they were cut (Table 5). No implements have been found in any of these river terraces nor on the platforms, and so ages are uncertain; Sandford and Arkell (1933) suggested a Plio-Pleistocene age.

The earliest definite Pleistocene deposits recognized below the Second Cataract are the 100-ft terrace deposits which have yielded Early and Late Chellean and Chelles–Acheulian implements (Lower Palaeolithic or Middle Pleistocene).

The total absence of implements, however, from the 150-ft terrace deposits has led Sandford and Arkell (1933) to suggest that a very long interval must have elapsed between these two river stages. The 100-ft stage may be recognized from Wadi Halfa to Luxor, more or less at the same height above the flood plain. Below this at one locality, at Dhimit in Egypt, they recognized an intermediate terrace at 75 ft (Sandford and Arkell 1933).

The 50-ft terrace is widespread and has yielded Chelles–Acheulian and Acheulian implements. After 50-ft terrace times, according to Sandford and Arkell (1933), the river entered the Second Cataract for the first time and as a consequence of increasing aridity the wadis became choked and the Nile bagan to aggrade, depositing large quantities of micaceous silts with occasional beds of fine gravel. These silts are referred to 'Sebilian Silts'. They are thickest in the Wadi Halfa region where they are more than 100 ft thick; they thin towards Luxor, where they are about 18 ft thick.

In Upper Egypt two terraces at 30 ft and 10 ft were formed and have yielded respectively implements classed as Mousterian and 'typical' Mousterian of Egypt. The relationships of these terrace gravels and the Sebilian Silts are uncertain. These gravels may never have been formed in the Sudan and Upper Egypt, or they may have been destroyed, or are buried beneath younger deposits.

Discussing the climate of Middle Palaeolithic and Lower Sebilian times, Sandford and Arkell (1933, p. 85) expressed the view that at the close of early Palaeolithic times there was a marked deterioration of climate, and the wet conditions that characterized Miocene, Pliocene, and early Palaeolithic times gave way to semi-arid conditions in Nubia, although wetter conditions may have prevailed north of Wadi el Alaqi.

After the 10-ft terrace stage in Egypt, which appears to have been deposited by sporadic currents, rain failed completely in Nubia in Lower Sebilian times. However, full desert conditions do not appear to have set in until much later, for dunes of blown sand do not appear to be associated with the Sebilian Silts, although blown sand covers much the silts on the left bank. Large areas of Sebilian Silts remain intact in Upper Egypt, for example the Kom Ombo plain, and wadis have been cut across them; because of this, Sandford and Arkell (1933) have postulated that rainfall had virtually ceased there by Upper Sebilian times. Supporting evidence here is that Sebilian implements have not been found on the desert surfaces away from the Nile, whereas Lower and Middle Palaeolithic implements are, especially between the Nile and the Red Sea.

During Recent times intense aridity developed and the river, largely fed by the Ethiopian monsoon, cut its valley to its present depth, eroding some 26 ft at the Semna Cataract since the Twelfth Dynasty (Sandford and Arkell 1933). Sand dunes and sheets and hamada developed away from the Nile. (See Fairbridge (1963) for alternative explanation of the Semna levels.)

Fairbridge (1962, pp. 108–10 and 1963, pp. 96–107) described the pattern sedimentation in the Nile Valley in the Wadi Halfa area, elaborating with the aid of radiocarbon dates the climatic data presented by Sandford and Arkell (1933). This information is summarized in Tables 14 and 15. Fairbridge also discussed the origin of the Sebilian Silts. He rejected Ball's catastrophic hypothesis (1939) that the breaching of the rock regulator at Sabaloka, the Sixth Cataract, caused a vast efflux of waters from Lake Sudd (a hypothetical lake), and that these waters caused the silts to be swept downstream and to be deposited below the Second Cataract. As an alternative to this hypothesis he suggested that the Sebilian Silts are the products of numerous flash floods and that the sediments were formed in a semi-arid or arid environment in the central Sudan.

It is estimated that the Siltation Phase was initiated some 20 000–25 000 years ago, and that from this time to about 10 500 years B.P. was a time of almost total aridity in Nubian times with semi-arid conditions extending into Ethiopia and Equatoria. Fairbridge (1962) postulated that the Nile at times almost ceased to flow and that a transitional (oscillitory) phase occurred between 12 000–10 000 years B.P. The evidence on which the last point is based is not clear, however.

The Recent or Holocene age opened with a great increase in rainfall, and the period 10 000–7000 years B.P., roughly equivalent to the Mesolithic (Fairbridge 1962), was a time of high rainfall in the tropics and sub-tropics. Fairbridge estimated that the flow of the Nile was at least three times that of today's normal flow. This period was followed by a progressive deterioration of living conditions in Nubia and the Sahara in general (7000–3000 years B.P., Middle Holocene). Dissection took place and the Nile gradually shrank to its present proportions. During the late Holocene (3000–0 years B.P. Middle Kingdom–Modern times) totally arid conditions developed in Nubia, and the Nile levels were similar to those of the present day.

Many of Fairbridge's conclusions have been challenged recently by de Heinzelin and Paepe (1964, pp. 48–50), who pointed out that although they are based to some extent on radiocarbon dates, the pattern of sedimentation and climatic fluctuation presented is based mainly on a theoretical curve constructed for the Nile flood, plotted against a temperature curve deduced for temperate regions and a curve showing eustatic variations of sea level (Fairbridge 1962, pp. 108–10).

'In the absence of any relevant geomorphic levelling, profile description, or geological observations we cannot agree with such interpretations. Fairbridge completely underestimated the complexity of the fluviatile sedimentation which built the terraces. The concept of one single silt formation stage cannot be accepted, and derived theoretical considerations are questionable.

The concept of one siltation stage, or one silt deposit, is not borne out by the facts. The Nile terraces are, as they normally must be, the result of complex sedimentation processes, as is now the case in the bed of the river itself' (de Heinzelin and Paepe 1964, p. 1048–50).

De Heinzelin (*in* Solecki *et al.* 1963, p. 91) and de Heinzelin and Paepe (1964) have also challenged the conclusions of Sandford and Arkell (1933) concerning the recognition of the high terrace and river platform (300 ft, 200 ft, 150 ft) in Nubia. They have stated that these terraces 'are no more than structural platforms, pedimentation surfaces or wadi gravels'. L. Berry and the author, however, agree with Sandford and Arkell's (1933) interpretation of the field data; because the rocks are horizontal it does not mean that river erosion surfaces do not exist (Berry and Whiteman 1968). The striking flatness and extent of these high-level surfaces, together with the presence of quartz gravels, some flint and chert pebbles (Sandford and Arkell (1939, p. 12), and observations by the author), all indicate a riverain origin for these surfaces.

Their equivalents south of the Batn el Hagar are uncertain because of the difficulty of correlating and the lack of height data. They are, however, clearly younger than the Shendi–Togni surface (late Cretaceous–early Tertiary) and must have been formed at the time when the extensive Tertiary pediments were being formed in the valley further south. Most probably the reason why the 300-ft,

200-ft, and 150-ft terraces and river platforms are so well developed below the Second Cataract is because of the softness and easily-eroded nature of the Nubian Sandstone over which the valley passes with one break to near Aswan.

De Heinzelin and Paepe (1964) have recognized four stages in the development of the Nile terraces at 96 ft (30 m), 64 ft (20 m), and 38 ft (12 m). In addition, pediments and gravels have been found at higher levels. A contrast in composition between the gravels of the pediments above 506 ft (152 m) elevation and the terraces below this is noted. Below the 506-ft level the gravels contain agate, jasper, and flint, which is clearly of southern provenance.

Wendorf *et al.* (1965, p. 28) gave the following geological sequence for the superficial deposits of Sudanese Nubia:

8 A fourth, and final interval of aggradation of Nile silts and sands to a maximum of 5 m above the flood plain, or an elevation of 126 m (Qadras unit) with a subsequent and significant interval of erosion, followed by aggradation of the modern flood plain

7 A period of desiccation with erosion of the channel 11 m to an elevation of 123 m

6 A third interval of Nile aggradation of sands and silts to a maximum elevation of 134 m, or 13 m above the flood plain Arkin unit)

5 A period of desiccation, again with the Nile channel incised at least 20 m, to an elevation of 120 m

4 A second interval of Nile sedimentation recorded by a series of silts, soils, sands, and gravels to a maximum level of 141 m, or 20 m, above the modern flood plain (Sahaba unit)

3 A period of desiccation when the Nile channel was incised 20 m or more, to an elevation of 130 m; in some localities a non-calcareous brown soils formed over this eroded surface

2 The first series of true Nile silts, soils, fluvial sands, interbedded dune deposits, and gravels to a maximum elevation of 157 m, some 36 m above the modern flood plain (Debeira unit)

1 A pre-Nile sequence of oxisols, pediments, gravels, dune sands, and vertisols, divisible into several units

A composite profile of Nile sediments in Nubia produced by Wendorf *et al.* (1965) is presented in Fig. 54. Archaeological sites and radiocarbon dated samples are located on this section. Unfortunately Wendorf *et al.* did not discuss the conditions controlling accumulation of the Nile sediments but concentrated mainly on archaeological data. Clearly, however, from the sequence presented above, the interpretation differs fundamentally from that given by Fairbridge (1962 and 1963) and to some extent from that given by Sandford and Arkell (1933). These three geologists associate periods of silt-accumulation with periods of desiccation. In the author's view this is correct. A good example of this type of aggradation can be studied at present in the Gash delta near Kassala in the Sudan and, of course, there is the working model of the Nile itself. Very little silt is deposited by the Nile in the Sudan except as occasional over-bank floods; in fact the flood tends to scour rather than deposit in the Sudan and much of the suspended load is swept down to the delta or trapped in marginal basins. The scouring of the Sennar reservoir during the flood may be cited here also. The Sebilian Silts were not formed under flood conditions, and like Fairbridge (1963) the author feels that they must have been formed during arid periods when the river was sluggish and incapable of transporting its suspended and traction loads in Sudan and Upper Egypt. It is indeed unfortunate that such problems remain unresolved, for so much of the area in question has already disappeared under Lake Nasser.

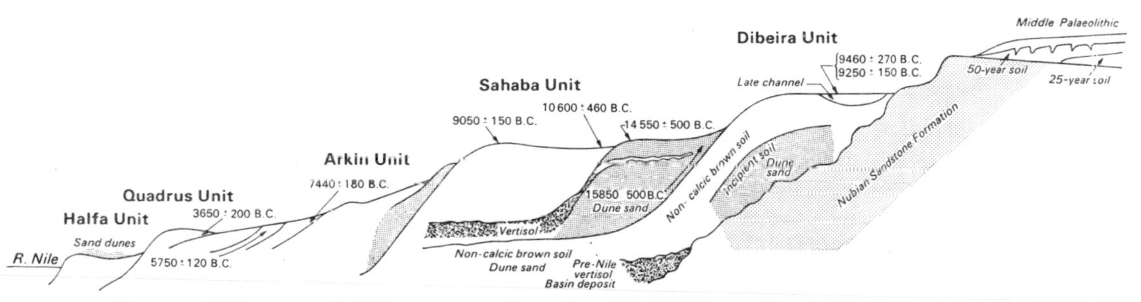

FIG. 54. Pleistocene and Holocene composite sections, Wadi Halfa area (Based on Wendorf *et al.* 1965)

(b) Pleistocene and Recent formations of areas east and west of the Nile

(i) *Qôz deposits of Kordofan and central Sudan*

Probably the most extensive of the Pleistocene and Recent deposits in the area west of the Nile are the Qôz or Kordofan sands. Originally the word 'qôz' was applied to the dune-like accumulations of sand in Kordofan (Edmonds 1942) but latterly this term has been used to include the sand sheets as well, and in places is used almost as a '*pays* name'.

Qôz is widespread in the central Sudan, especially in Darfur, Kordofan, and Northern Province, where it forms extensive gently-rolling sheets, and fixed dunes (Fig. 55). These are now stabilized partly by a special flora, partly by a thin surface crust of iron oxide, and partly by fine silt and clay.

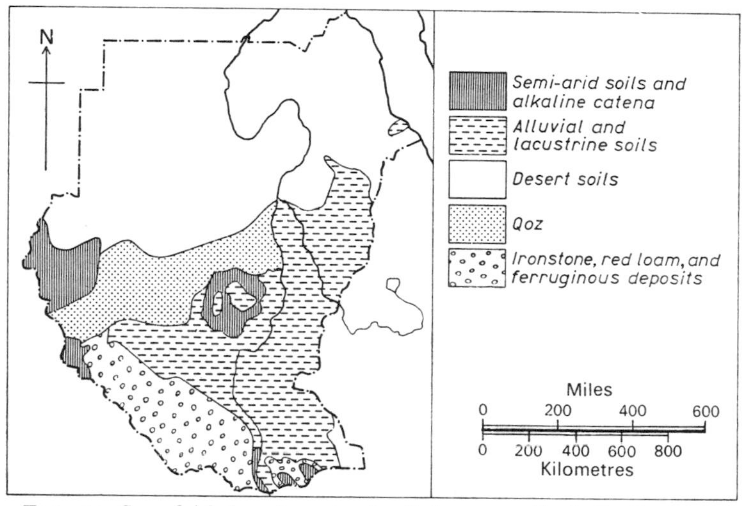

FIG. 55. Superficial deposits and soils, Sudan, generalized distribution map (Based on Sudan Topographic Survey data. *Note*: The extent of alluvial and lacustrine soils is too large and generalized)

The sands consist of well-rounded quartz of grains of Nubian origin and vary in colour from pale buff to deep red. According to Edmonds (1942) the colour is due to iron staining of the grains and extends only to a shallow depth, sometimes as little as 4–5 ft. Below this the sand is often colourless. This is due to surface concentration of iron in the upper part of the deposit.

The qôz occurs in a belt where rainfall varies from 200 mm in the north or less to 700 mm in the south (Fig. 3) and obviously could not have formed under these climatic conditions. Nowhere at the present time could such large quantities of sand accumulate at these latitudes. A climatic shift of about 6° or more of latitude must be postulated to account for these 'fossil' desert deposits.

A range of forms similar to those found in the Libyan and Sudanese deserts to the north occurs, but everywhere the sharpness of the topography has been modified and rounded. Longitudinal and transverse dunes are still recognizable, however. The longitudinal dunes are often up to 140 ft high and the transverse dunes up to 50 ft high. The dunes were formed by northerly winds, as is indicated by sand blasting and polishing on the northern sides of jebels. Sand has piled up on the northern sides of jebels as face dunes and extends round the sides as flanking and tail dunes, which in places coalesce. Barchans have not been recognized but transverse dunes frequently interfere and form an interlocking pattern such as that of the 'Kheiran' north-west of Bara (Edmonds 1942). Water occurs at a shallow depth in the Kheiran and obviously has controlled the development. Warren (1964, p. 7) recognized crescentic dunes said to be open to the wind.

According to Andrew (MS 1943) the northern limit of the qôz is indefinite because in the desert the iron crust has been removed and vegetation is largely absent. Lebon (1965, Figs. 2 and 5) attempted

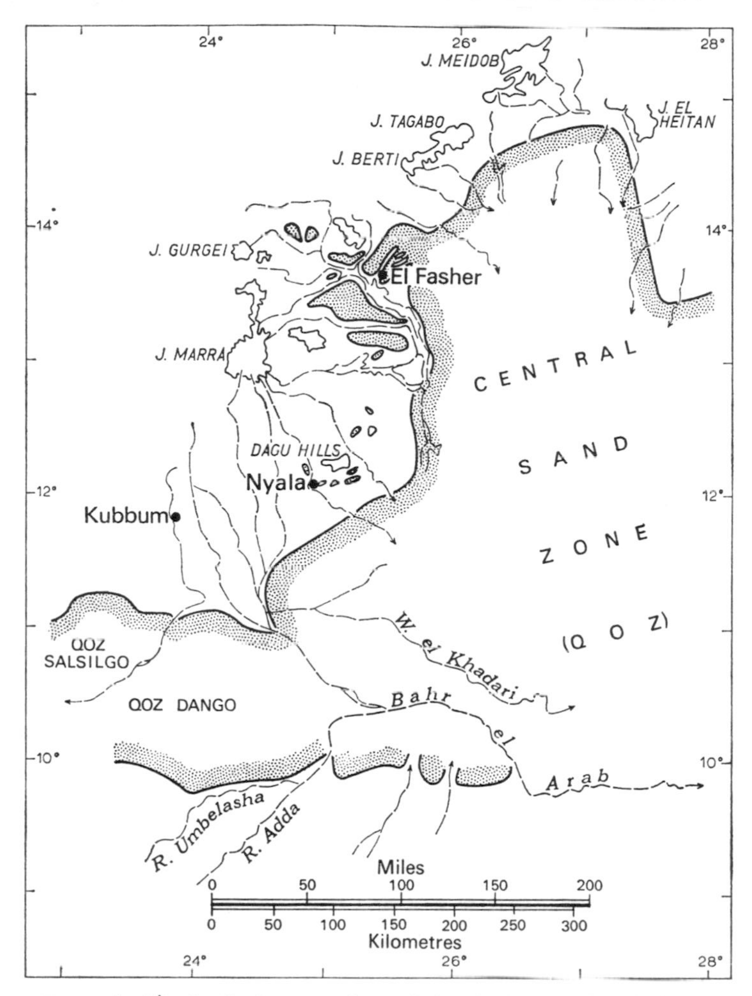

FIG. 56. Qôz distribution according to Lebon (1965) (Based on Sudan topographic and geological maps (various scales) and Sudan Topographic Survey Trimetrogon photograph U.S.A.A.F.)

to delimit the qôz belt, but it should be borne in mind that the two figures of Lebon's mentioned above are of a reconnaissance nature based largely on photo-interpretation of U.S.A.A.F. Trimetrogon photographs (vertical scale 1:40 000) taken during 1943–4. The standard of definition is variable and in the author's view the separation of qôz and blown sand in this way becomes highly subjective, especially along the northern margin of the qôz belt. Andrew (MS 1943), for instance, gives Curoguro (19° N, 24° E) on the Sudan–Chad frontier as the northern limit of qôz but Lebon (1965, Fig. 2) places it south of Jebel Marra about latitude 11° N. Lebon's limits also differ from those shown in Andrew (1948) and much of what Andrew includes as qôz in the Wadi Howar region is designated wind-blown sand by Lebon.

In the east the northern limit of qôz on the left bank of the Nile is about the latitude of Atbara, where the well-developed Qôz Abu Dulu ends. This large qôz is a more or less continuous seif-like feature extending for almost 200 miles. Qôz deposits occur on both banks of the Nile north of Khartoum, but are less extensive and less common on the east bank. This is quite clearly related to the distribution of the Nubian Sandstone Formation, the outcrop area of which is much more restricted on the right bank of the Nile in these latitudes.

In the south and west both Andrew and Lebon agree on the limits of the qôz just north of the Umbelasha tributary of the Bahr el Arab (Fig. 56) and this belt extends across Darfur and Kordofan with its southern boundary passing north of the Nuba Mountains. The eastern boundary is mainly the White Nile, although between Hashana and El Geteina qôz-like deposits occupy considerable areas on the east bank. Williams (1966) pointed out, however, that in this area the sand may have been derived from the White Nile and 'fossil' distributary system of the Blue Nile (Fig. 49). Williams also postulated inversion of relief, channel deposits having been raised into dunes by the wind, levees removed by deflation, and former channels filled in and levelled off. These sand spreads are said to cover 300 000–400 000 acres.

Concerning the origin of the qôz, Grabham (1935, p. 275) suggested that the sand had been blown south from the desert and that the black clays locally known as 'cotton soils' of the central Sudan were formed in the same way; in fact, one was supposed to represent the coarse and the other the fine fraction of the deflation process. Grabham (1935) derived these deposits from the weathering of the Basement granites. Edmonds (1942), however, pointed out that the sand is of Nubian type and obviously represents the weathering product of this group. The dune sands overlie calcareous clays of unknown derivation, and Mesolithic and Neolithic artefacts are common on some of the dunes. Andrew (personal communication) believes that the qôz are of local origin; whereas the dunes are derived from the south-east, east of the White Nile and south to south-west, west of the White Nile. Warren (1964*a*, *b*) recognized two main dune-forming phases in Kordofan and proposed the following sequence (Fig. 57):

5 *Present phase of intermediate rainfall.* Lakes and rivers dried out or shrunk

4 *Short wet phase.* Lakes between the dunes and hollows

3 *Short dry phase.* Sand reworked as far south as 13° N and fresh sands added from north. Most of the high dunes formed during this period

2 *Wet phase*, long or *intense*. Dunes of preceding phase lowered by wash; soils leached and consolidated, silty terrace deposits formed; streams flowing in qôz belt

1 *Dry phase.* Extensive dune formation with sands blowing southward to beyond 11° N

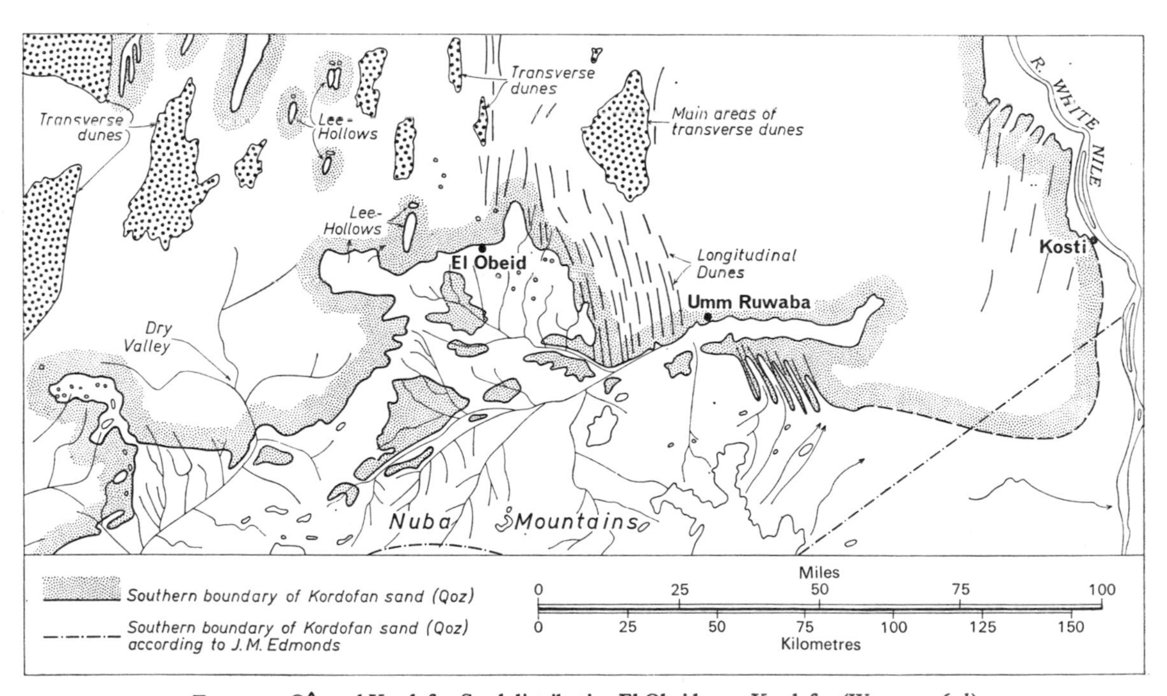

FIG. 57. Qôz and Kordofan Sand distribution El Obeid area, Kordofan (Warren 1964*b*)

In the author's view, qôz deposits were no doubt formed during earlier dry phases in the Pleistocene, but the main period of formation was during the Sebilian (Middle Palaeolithic) Dry phase recognized in Nubia (Tables 15 and 16). According to Fairbridge (1962, 1963) this was initiated some 20 000–25 000 years ago. The ensuing long or intense wet phase (2), when the dune forms were modified, probably took place some 10 000–7000 years B.P., roughly in Mesolithic times, when, according to Fairbridge (1962, 1963), Nile flow in Nubia was at least three times the present flow. During the late Holocene (3000–0 years B.P., Middle Kingdom–Modern times) total aridity developed in Nubia and the second phase of qôz formation in Kordofan is probably referable to this period. Our knowledge of climate during Meroitic times in the Sudan is limited, but Warren's wet phase in Kordofan may have taken place in late Meroitic times when the climate was probably wetter than at present.

Surface drainage is limited in the qôz as much of the rainfall is absorbed. Earlier drainage lines are frequently blocked by qôz and, in silty and clayey hollows, shallow seasonal lakes (rahads and fulas) often occur. According to Andrew (personal communication) an early drainage line was dammed by sand at El Fasher. Andrew also noted that in the early stages of qôz formation lakes were much more common than at present, and that there are deposits of shelly freshwater limestone and diatom beds interbedded with qôz sands in parts of Darfur and northern Kordofan, as there are in Chad and Tibesti.

If we follow Sandford and Arkell (1933) and Fairbridge (1963), then we must place the onset of intense aridity in Nubia at the end of the Lower Palaeolithic (Tables 15 and 16). In early Sebilian (Middle Palaeolithic) times rain is supposed to have failed in Nubia and as a consequence the wadis and the main Nile became choked with silts and sand below the Second Cataract and in basins such as the Kerma basin above. Fairbridge (1963) postulated that this Siltation Stage began some 25 000–20 000 years ago and that much of Darfur, Kordofan, and even Bahr el Ghazal and Equatoria experienced desert and semi-desert conditions. Much of the qôz must have been formed at this time. We can envisage a southward swing of the trade wind belt and sand derived from the Nubian Sandstone outcrops was blown southwards and south-eastwards across Northern Province, Darfur, and Kordofan.

The great effectiveness of the present-day trade winds is clearly visible, especially from the air, at the present time; great sand shadows and sheets lie in the lee of many jebels and much of the Bayuda area in the bend of the Nile has been swept clean of sand (Plate 12).

Sandford and Arkell (1933) have postulated that before the Sebilian Siltation Phase, as far back as Pliocene and Miocene times, rainfall was high in Nubia and Upper Egypt and it therefore would seem unlikely that the qôz deposits of the central Sudan are older than Wurm (late Pleistocene).

(ii) *Sand dunes and sheets*

Such deposits are not extensive in the northern Sudan. Lebon (1965, Fig. 2) has attempted to show their distribution, but again this could be done only in an arbitrary way, for the U.S.A.A.F. Trimetrogon photographs were used as the basis for interpretation. For example, on Lebon's map the extremely striking patches of blown sand that spread over the jebels on the right bank of the Nile between Ed Debba and Kareima are not shown, nor are some of the well-known sand-spreads in the Red Sea Hills.

Extensive sand dunes and sheets occur in the Wadi Howar region of northern Darfur and between Selima and Jebel Uweinat, where long seif dunes have been mapped (1:250 000 Topographic Series). These form part of the famous Great Sand Sea of the Libyan Desert.

(iii) *Reg and Hamada*

Much of the desert of the northern Sudan is floored by reg and hamada, especially in the mountainous areas, and great areas are composed of stripped surfaces or are covered with ventrifacts and angular lag gravels (Plate 18). The remarkable force and constancy of the winter trade winds has produced some striking rock-carved features, but extreme forms, such as yardangs, are not common. This may be because the desert is comparatively young in the northern Sudan and southern Egypt.

(iv) *The clay plains*

The great clay plains of the Sudan are perhaps the most striking feature of the geomorphology of the country.

The clays of the Gezira have been described above. These, like the clays of the Rahad, Dinder valleys, etc., are clearly alluvial deposits laid down as Tothill (1948), Berry (1962), and (Berry and Whiteman 1968) have suggested by over-bank floods from the Blue Nile and its tributaries. Grabham (1935) postulated an aeolian origin, but this is clearly incorrect.

These 'clays' are remarkably uniform; their clay content is from 50 to 60 per cent and there is very little coarse material, except for small calcium carbonate concretions, which weather to a dark colour at the surface and become whiter with depth. Gypsum crystals also occur, and these may well have been formed at the time the sediment was laid down. Towards the White Nile, the clay fraction increases, no doubt reflecting the lower rate of flow of the distributory systems that deposited the clays.

On either side of the Blue and White Niles, there are clay and silt terraces that are many miles wide locally. Westwards along the White Nile, north of the latitude of the Nuba Mountains (Fig. 55), these terraces are bordered by the qôz belt but east and north of the Blue Nile they adjoin the Butana clay plain.

The Butana clay plain is developed east of the Nubian Sandstone outcrop mainly on the Basement Complex rocks. A friable clay loam, sometimes with dark small calcareous concretions, is developed in the Butana. The clays are many feet thick in places (more than 40 ft was excavated in canals on the eastern margin of the Khashm el Girba irrigation scheme) but they thin out and become much coarser, eventually passing into clayey sand and gravels near the edges of pediments. Many of the Butana Basement Complex jebels have distinct light-coloured weathering aureoles around them. These are particularly conspicuous from the air and clearly demonstrate the process by which the clays were formed (Plate 10).

Sheet wash and wadis have redistributed these clay soils but they are not primarily alluvial soils as Lebon (1965, p. 48) has stated. The clays are largely impervious except near jebel masses, and in the, rainy season, especially after storms, large areas are inundated to a depth of a few inches. The clays are wetted down to a depth of several feet by water seeping and infilling the cracks that develop in these clays during the dry season. The result is that during the 'kharif' large areas are turned into seas of mud which are almost impassable. Northwards, Butana-type clays pass into desert and semi-arid soils in the foothills of Jubal el Bahr el Ahmer.

Southwards and south-westwards from the Gezira the clay plain is developed extensively on the Umm Ruwaba outcrop. Much of it near the great rivers of the Bahr el Abiod system is flooded annually and local soil types have developed. During the rains great areas are inundated, as in the Butana. Because of the polygenetic nature of the Umm Ruwaba Formation there is much more lithological variation in these areas and in places the soils are sandy.

Throughout the clay plain the soils are mainly alkaline, although soils rich in humus develop in the Sudd region. South-westwards the clay-plain soils give place to latisols that extend across the Nile–Congo divide. These iron-rich soils and laterites of the southern Sudan have been described earlier. Clay plains of a more limited extent occur in the Nuba Mountains.

A considerable amount has been written about the origin of these heavy dark alkaline clays—'cracking clays' or 'cotton soils'. Grabham (1909) pointed to their similarity with Indian regur and in a number of papers and reports discussed their distribution (Grabham 1934, 1942). He proposed an aeolian origin for the clay and thought that it was more or less contemporaneous with the formation qôz of Kordofan.

Anyone who has experienced a 'haboob' or 'dust storm' in the central clay plains cannot fail to be impressed and overawed by the vast quantities of fine clay silt and fine sand that are transported; and quite clearly haboobs play a large part in redistributing the clays and silts. Sheet floods play a much more important part in the formation of the clay plains, however. In the Butana the deposits of the plains are part of a series of pediment deposits developed on Basement Complex rocks. The clays and silts pass laterally into sands and gravels around the residual jebels (bornhardts) and in the areas in between the weathering products are distributed by sheet floods and to some extent by haboobs and by wadis. The Butana grass patterns described by Worrall (1959) and Berry and Ruxton (1960) indicate that mass movements are taking place and have a noticeable effect in the development of the clay plain. The clay plains of the Nuba Mountains were similarly formed, but water played a more important part here because of the higher rainfall.

As Berry (1964) has pointed out, the reworking of the clay in the central Sudan may have been more active in the Pleistocene, and the current cycle of soil development was probably initiated first in the Holocene wet phase (Tables 15 and 16).

Elsewhere in the areas underlain by Umm Ruwaba and Gezira formations the clay plains are obviously of alluvial origin and sheet wash and river action have caused redistribution of sediments.

(v) *Desert deltas, Wadi deposits, etc.*

Other Pleistocene and Recent superficial deposits include the Gash delta deposits; the tuff and ashes and alluvial of Jebel Marra region; the alluvium and terraces of large wadis such as Wadi el Arab, Wadi Azum, and the Khor Abu Habl, etc.

The Gash delta deposits have been described in detail by Swan (1956). The Gash rises as the Mareb near Asmara, Eritrea, and, like the Blue Nile, descends to the Sudan plain carrying large amounts of clay, silt sand, gravel, etc. The volume of water is insufficient to reach the Atbara, although it may well have joined this river during wet periods in the Pleistocene. At present the river forms an extensive desert delta near the town of Kassala (Plate 10).

Pleistocene and Recent alluvial deposits most certainly occur in the Jebel Marra region but without fossils it is not possible to separate them from the Tertiary deposits. The geologists of Hunting Technical Services (1958) have recorded a grey group of soils that occur near the surface and are composed almost entirely of volcanic ash and debris, and these and other tephra must be of Holocene age. Extensive terrace systems are associated with wadis such as the Wadi Azum, draining the Jebel Marra region (Plate 17). The Wadi Howar, which extends across northern Darfur, also has extensive alluvial deposits of Pleistocene and Recent age (Sandford 1935).

(c) Pleistocene and Recent deposits of the Red Sea Littoral

Overlying the Tertiary deposits of the Red Sea Littoral are a series of interdigitating marine and continental deposits. These have been described by Darwin (1842), Buist (1854), and Crossland (1907, 1911, 1913); in Sudan Geological Survey reports by Grabham and others; by Carella and Scarpa (1962), Berry (1964), Berry, Whiteman, and Bell (1966), etc.

(i) *Continental facies*

Extensive wadi systems drain El Jubal el Bahr Ahmer and debouch on to the Littoral forming a series of fans and deltas, composed of gravels, sand, silt, and clay. In some of the major wadis Andrew (MS 1943) has recorded a boulder conglomerate overlain by a thin series of silts and clay as in Khor Arbaat, and in places there are freshwater limestones, which cement the clastics to form calcrete conglomerates (Plate 21).

Large quantities of sediments now reach the sea only in the Baraka (Tokar) delta in the south and in the Khor Arbaat delta near Port Sudan. Considerable quantities of sediment are occasionally swept down the Wadi Oku, which drains the northern part El Jubal el Bahr el Ahmer.

The Baraka has a large catchment, with part of its head waters in the Asmara area, and this, together with the higher rainfall of the southern jebels, has enabled the khor to build up an extensive silt delta. The khor is a seasonal stream and flows between 40 and 70 days a year. The annual discharge varies from 205 to 968 million m³. During heavy floods large quantities of silt and sand are carried by the Baraka and as much as 10·6 per cent of the weight of water is suspended sediment.

The Baraka or Tokar delta is 45 miles long and is gradually being extended seaward. South of the Baraka delta small streams deposit sand and gravel. The sediment-rich waters prevent active coral growth near the shore and the sediments have been built up by the waves into bars and spits. Behind these, lagoons have developed in places.

North of the Baraka or Tokar delta several large wadis such as Khor Garab, Khor Arbent, and Khor Arbaat form fans and deltas of gravels, sands, and silts as they debouch on to the plain. The quantity of sediment brought down by these intermittent streams is insufficient to prevent coral growth, except for Khor Arbaat, which has a maximum annual flow of up to 49 million m³. Large quantities of sediment are swept down to the sea by this khor.

Between Khor Ashar and Port Sudan the coastal area is commonly a flat sandy or clayey plain, which rises gradually inland towards low mounds of eroded raised-Reef Complex that stand some 6 to

9 ft above high-water mark. Wind-blown sand and dunes occur along this part of the coast.

Few wells or boreholes penetrate the coastal plain sediments in the Port Sudan area but at Khor Mog a thick series of clayey grits and pebble beds underlies silts and sandy gravels, and shell sand also occurs. The following succession was noted by Andrew (MS 1943).

	Thickness (ft)
Silts, sandy gravels	9
Sand, shelly	13
Gravels, sands, clays, generally uncemented	745

In the absence of diagnostic fossils it is not possible to say how much of this succession belongs to the Tertiary or the Quaternary.

North of Port Sudan aridity increases rapidly and north of Ras Abu Shagara rain falls only occasionally. Few streams ever reach the coast, except after very heavy storms. Spreads of sand and gravel and small dunes occur along this section of the coast, for example at the mouth of Khor Eit, where some fine examples of barchans were noted in 1962.

Near the sea, a salt crust consisting of common salt and gypsum forms as a result of intense evaporation and capillary action. This results from the very high rock and soil surface temperatures that prevail in the area. Berry and Cloudsley-Thompson (1960) recorded 182·5° F (83° C) in September, for instance.

(ii) *Marine facies*

Pleistocene and Recent marine deposits along the Littoral of the Sudan are predominantly reef complex deposits, as defined by Newell and others (1951, 1953). Lagoonal reef rock and fore-reef facies have been recognized. These deposits form the major features of the coastline north of the Baraka delta; they rise up to 51 ft (16 m) above high-water mark and extend inland for more than 4 miles. They constitute the Emerged Reef Complex of the Red Sea Littoral (Berry, Whiteman, and Bell 1966). Sestini (1965, Fig. 5) shows a belt of emergent reefs along the coast but this is incorrect. Much of the area mapped by Sestini as reef consist of lagoonal clastics, some of which are reef-derived.

The thickness and seaward extent of the raised reef deposits are uncertain. It is possible that reef growth may have continued seaward as Pleistocene sea levels dropped but any deposits of this type are now either eroded or obscured by modern reef or water. The exposed thickness of the emerged reef on the mainland is approximately 12 m while on Mukawwar Island some 5½ miles off the coast (Fig. 38) emerged reef, banked against Tertiary formations, occurs 51 ft (16 m) above high-water mark. Abu-Shagara-1, A.G.I.P.'s borehole, appears to have passed through 608 ft (190 m) of reef complex limestones, all of which are assigned to the Pleistocene.

In general, emerged reef deposits are found at lower elevations south of Port Sudan and only small patches of reef facies have been noted in this southern area. North of Port Sudan, the height of the emerged reef is fairly constant rising to a maximum of 32–38 ft (10–12 m). The Emerged Reef Complex rests unconformably on both Red Sea Hills clastic facies of Mesozoic age and calcareous marine facies of Tertiary age. They rest unconformably on the ? late Jurassic basaltic flows, sills, and coarse clastics in the Khor Shinab area. The emerged reef deposits are not folded, nor are they faulted, and the coastal area appears to have acted as a stable but drifting block since at least the last part of the Pleistocene and during Recent times.

Well-defined untilted benches occur on the reefs, especially north of Port Sudan, and these are best explained as having been formed during eustatic changes in sea level.

A series of *Tridacna gigas* Linné samples were collected for radiocarbon analysis (Berry, Whiteman, and Bell 1966) and this data, together with height data, and a ^{230}Th–^{234}U date, suggest that we are dealing with an elevated reef of ? Riss–Würm age and that the benches were cut into the reef, rather than with a series of overlapping reefs formed at different levels.

(iii) *Summary of radiocarbon data and a tentative Pliocene, Pleistocene, and Recent succession* (Berry, Whiteman, and Bell 1966)

A summary of the radiocarbon evidence and a tentative succession for the Sudan Red Sea coast is presented in Table 16. It has been suggested that the emerged reefs of the Sudan coast are relatively modern features that owe their elevation to fault movements (Darwin 1942; Crossland 1907, 1911, 1913). However, the field evidence and the radiocarbon dates suggest otherwise, and the origin of reefs and associated deposits can be explained better in terms of eustatic changes in sea-level. It is clear that tectonic movements have played a large part in shaping off-shore topography but these movements appear to have taken place before the formation of the Pleistocene emerged reefs and benches.

1. *Pliocene Beds*. After the deposition of the marine Pliocene Beds, which themselves contain fringing and patch coral reefs, earth movements associated with the formation of the Rift System occurred. These produced small folds and faults. Dips of 10°–20°, predominantly seaward, are common in these rocks.

2. *Older terrace gravels*. Following these movements, and perhaps contemporaneously with them, the older terrace gravels, such as those that occur at Khor Eit, were deposited. Some of these occur high on the coastal plain and appear to have been tilted as at Khor Eit, but this is not clearly discernible everywhere because of the poorly bedded and massive nature of these gravels. They are in part the 'Younger Clastic Formation' of Sestini (1965). Some of the older terrace gravels are obviously younger than the marine Pliocene deposits but there is no clear evidence of their age.

The size of the boulders and gravels that make up these deposits indicates that they were formed during a much wetter climate and they may have been formed during interglacial or pluvial stages during the Pleistocene. They may indicate also continued uplift in the Jubal el Bahr el Ahmer of the shoulders of the Red Sea Depression. In the present state of knowledge it is not possible to be more precise, especially now that Flint (1959) has rejected so much of the East African sequence established by Wayland, Leakey, and others on climatic evidence.

3. *Mukawwar stage 16-m emerged reef*. At this stage sea levels must have been over 16 m higher than those of the present. The radiocarbon dates indicate that the reef is more than 37 000 years old. If we follow the general synthesis presented by Fairbridge (1961), then the elevation of the reef suggests that it was formed during the early Monastirian. The main Monastir terrace of Europe and Talbot terrace of America are both a product of this high sea-level and correlate well in height (15–20 m). If this correlation is valid the terrace must be over 100 000 years old. This date was confirmed subsequently by a ^{230}Th–^{234}U date, RSH 36, which indicated an age of 91 000 ± 5000 years B.P. (for details see Berry, Bell, and Whiteman 1966).

4. *Shinab stage 11 to 12-m emerged reef or bench*. At many points on the coast the emerged reef rises to about 11 m above high-water mark, and on Mukawwar Island a well-marked bench occurs at 12 m. The emerged reef is older than the limit for the method of radiocarbon analysis used, so we are unable to tell whether new reef growth occurred at this time or not. The height of the reef, however, and its general state of preservation indicate that it also may be a Monastirian feature. Terraces and benches at this height have been reported elsewhere in the Red Sea area. The bench at Marsa Fijja is covered with washed gravels from the Red Sea Hills.

5. *Mohammed Qôl stage 7 to 8-m bench*. South of Mohammed Qôl large areas of emerged reef are cut by a well-developed bench at 7 m above high-water mark, the surface of which is weathered and somewhat degraded. On Mukawwar Island a well-marked bench at about 1½ to 8 m occurs on the east coast and is a prominent feature on much of the island. A similar bench also occurs at Khor Shinab and a number of other localities. No radiocarbon dates have been obtained for these benches. Their height suggests a correlation with the late Monastirian terrace of Europe (6–8 m) and the Pamlico of north-east North America (90 000–95 000 years B.P.).

6. *Dissection of the reef*. The large and rapid drop of sea-level of the Würm–Wisconsin emergence caused large areas of the reefs to be exposed. Field evidence shows that the 3·5–4 m benches are found in positions which indicate that they post-date the dissection of the reefs, although this feature may in places coincide with the 3·5–4 m epi-Monastirian terrace. With the lowering of sea-level, new stream valleys were cut through the reefs, and streams became graded to levels that may have been over 100 m below present sea-level. During this period wadis, which were to become marsas, were cut through the emerged reef, and erosion deepened and extended the boat channel. There is some evidence that a wetter climate prevailed for part of this period and weathering of the upper surface of the exposed reef was accelerated.

7. *3·5-m bench*. As the ice sheets melted in post-glacial times sea-level rose, the boat channel of the Sudan coast was flooded, and the incised wadis became drowned valleys or marsas. Coral growth appears to have kept pace with the rise in sea-level

and the modern reefs probably date from this time. The location of the benches at 3·5 m on many small inlets and inside the marsas indicates that they are post-glacial in age. This bench was formed probably about 5200 years ago. Its surface is slightly dissected and weathered, perhaps suggesting a period of rather more intense rainfall after its formation.

8. *2-m bench.* The next definite stage recognized on the Sudan coast is the 2-m bench. This bench is comparatively fresh and is found on almost all parts of the coastline. Suakin is built on it and it occurs on the shores of many bays and inlets. Its form and position suggest a recent origin and it can confidently be assigned to the Abrolhos terrace stage of Fairbridge (1958, 1961), dated at 2600–2100 years B.P. This bench is common in most Red Sea areas and has been noted in many parts of the world.

9. *Present-day reef and deposition.* Some dead coral remnants and notches suggest that a sea-level of about 0·75 m higher than the present one may have occurred since the 2-m bench was formed. This is a minor stage. The main recent events are the infilling of the heads of marsas with sediment, the growth of the present fringing reef, and the notching of the 2-m reef.

Fossils from the Emerged Reef Complex have been described by S. V. Bell and are listed in Table 16.

TABLE 16

Pleistocene and Holocene events; radiocarbon and $^{230}Th/^{234}U$ dates, Red Sea Littoral, Sudan (Based on Berry, Whiteman, and Bell 1966)

Local sequence Red Sea coast, Sudan	Radiocarbon age of Red Sea specimens (years B.P.)	Suggested correlation based on major eustatic events	Estimated age (after Fairbridge 1961) (years B.P.)
	RECENT		
Present-day reefs and lagoonal deposits; wadi alluvium and terraces			
2-m bench. Cut into raised reef; fresh surface, unweathered and slightly dissected; well developed along Red Sea coast, e.g. at Suakin		Abrolhos submergence (Recent Dunkirk) (1·5–2 m). Well developed in Indian Ocean, Far East	2100–2600
3·5–4-m bench. Cut into raised reef; surface weathered, covered with loose shell rubble a few inches deep; gullied on seaward margin by rainwater; best developed Marsa Arus and Marsa Shinab	22 600 ± 600 Marsa Shinab, RSH. 138. *N.B.* Specimen probably contaminated and older than ^{14}C age. Dates raised reef and not cutting of bench	Older Peron bench stage (3–5 m)	4600–6000
	Würm or Wisconsin glaciation		
Formation of the Red Sea marsas and erosion of structurally-controlled boat channel		Würm–Wisconsin emergences	10 800–+75 000

TABLE 16—*continued*

Local sequence Red Sea coast, Sudan	Radiocarbon age of Red Sea specimens (years B.P.)	Suggested correlation based on major eustatic events	Estimated age (after Fairbridge 1961) (years B.P.)
	Interglacial		
? 3·5–4-m bench. Classified as Older Peron above but the two terraces might be coincident on open coast		Epi-Monastirian terrace. Princess Anne, or Oulijian terrace (3·5 m)	75 000–80 000
Mohammed Qôl stage 7–8-m bench. Cut into raised reef on Mukawwar Island. Well developed at Mohammed Qôl and Khor Shinab		Late Monastirian or Pamlico terrace (6–8 m)	90 000–95 000
	PLEISTOCENE *Riss/Wurm or Sangamon*		
11–12-m Shinab stage. Emerged reef and bench well developed at Marsa Shinab—and represented by abandoned cliff line at approx. 12 m on Mukawwar. Capped at Marsa Fijja by igneous and metamorphic gravels derived from Red Sea Hills Basement Complex	40 000 Khor Shinab. RSH. 142. Dates reef rock and not the feature	Late Monastirian	95 000 ?
16-m Mukawwar stage Raised reef set in embayment on Tertiary escarpment; dissected, weathered surface, thick rubble top. 13-m to 18-m bench and cliff, with beach rock west side of Mukawwar	37 000 *N.B.* Dates reef rock 91 000 ± 5000 ($^{230}Th/^{234}U$)	? Early Monastirian 60-ft stage (18-m) (Fairbridge 1961, p. 133)	100 000
Older terrace gravels, Khor Eit			
			FOLDING AND FAULTING
	PLIOCENE		
Pliocene beds, Khor Eit			

IV

Stratigraphical Palaeontology

THE following account is based largely on published information. The fossil fauna and flora of the Sudan Republic is poorly documented, little systematic collecting has been done, and the formations have yielded few fossils. Stratigraphical conclusions therefore can only be limited.

1. PALAEOZOIC

(a) Lower Palaeozoic

Lower Palaeozoic fossils have not been described from the Sudan; although intrusions of this age and younger volcanic rocks have been dated by the potassium–argon method from the Basement Complex formations of Sabaloka, near Khartoum. Metasediments of (?) early Cambrian age crop out in this area.

(b) Upper Palaeozoic

Sandford (1935) recorded *Archaeosigillaria vanuxemi* and a lepidodendron flora from sandstones in the Jebel Uweinat area, Northern Province. Menchikoff recorded *Archaeosigillaria* aff. *vanuxemi* Kidston from these sandstones, and on the basis of this flora these rocks have been assigned to the early Carboniferous (Burollet 1963). It should not be forgotten, however, that the lepidodendroid flora was identified by W. N. Edwards (Sandford 1935), not from specimens, but from field sketches by Sandford; and that *Lepidodendron* is characteristic of coal measures (late Carboniferous). Related forms range into the Upper Devonian (Neaverson 1955). Also *Archaeosigillaria* ranges from Devonian to Lower Carboniferous in North Wales and the English Lake District (Neaverson 1955); and Andrews (1961, p. 251) noted that both sigillarian and lepidodendroid structures occur together on *Archaeosigillaria primaerva* White.

The Jebel Uweinat sandstones are thus best thought of simply as late Palaeozoic, possibly late Devonian to Carboniferous in age.

2. MESOZOIC

The Mesozoic formations of Sudan include the Nubian Sandstone, the Gedaref, and the Mukawwar formations.

(a) Nubian and Gedaref formations

Silicified wood is common in the Nubian Formation but only at Jebel Dirra, 47 miles east of El Fasher, has a flora been collected and described. The following species were recorded by Edwards (1926): *Weichselia reticulata* (Stokes and Webb), *Frenelopsis hoheneggeri* (Ettinghausen), and *Dadoxylon aegyptiacum* Unger.

Of the three species, *Dadoxylon aegyptiacum* is a form species and therefore has little stratigraphical value, and the other two species indicate simply a Cretaceous age for the Jebel Dirra Beds.

Weichselia is characteristic of the Wealden in Europe but also occurs in the Aptian of the Isle of Wight, England, and the Albian of Germany. Seward (1933) assigned an early Cretaceous age to *Weichselia* but it has been reported from the Upper Cretaceous (Cenomanian) of Baharia Oasis, Egypt, by M. Hirmer.

Frenelopsis occurs in the Lower Cretaceous of Europe, the United States, and Greenland but *F. hoheneggeri* (Ettinghausen) has been recorded from the Upper Cretaceous (Cenomanian) of Portugal and Bohemia, and from the Turonian of France.

According to Edwards (1926) the flora of the Jebel Dirra sandstones is at first sight early Cretaceous (Neocomian or Barremian), but clearly a late Cretaceous age cannot be ruled out.

Since Edwards (1926) described the Jebel Dirra flora no other plant fossils have been described from the Nubian Formation in the Sudan. However, potassium–argon dates for basalts cutting the Nubian Formation at Fareig, Northern Province, indicate that the Nubian in this area is probably pre-late Cretaceous in age.

The Gedaref Formation has not yielded fossils in the Sudan but eastwards it passes under the Antalo Limestone of Upper Jurassic (Oxfordian) age in Ethiopia. The formation is classed therefore as (?) Jurassic (pre-Oxfordian).

(b) Mukawwar Formation

Carella and Scarpa (1962) have recorded the existence of marine 'uppermost Cretaceous or Palaeocene' deposits in Maghersum-1 borehole (20°49′03″ N, 37°17′00″ E) at drilled depths of 1560–1761 m. They assign this group of deposits to the uppermost Cretaceous transitional to Palaeocene on the basis of the ostracod assemblage recorded from a cored sample between 1637 and 1638·70 m. The recorded fauna includes mollusc casts, echinoid spines, fish teeth, small pyritized foraminifera, ostracods, *Brachycythere* sp., *B.* cf. *ledaforma* (Israelsky), *Isohabrocythere* sp., *Buntonia* sp., and *Bairdia* sp.

The depositional environment is described as 'brackish to shallow marine' by Carella and Scarpa (1962). This is questioned, however, because the presence of echinoid spines, if they are *in situ* rules out the possibility of a brackish environment. Echinoids are exclusively marine animals, as are the ostracods *Brachycythere*, *Bairdia*, and *Buntonia*.

3. TERTIARY

The fossiliferous Tertiary formations of the Sudan include the Hudi Chert Formation, the sediments of the Red Sea Littoral, and the infra-basaltic sediments of Jebel Oharba, Khor Langeb (17°13′ N, 35°34′ E).

(a) Hudi Chert Formation

Cox (1932–3) described eleven species from the Hudi Chert Formation. Ten of these are non-marine gasteropods and one is a marine lamellibranch doubtfully identified.

Of the ten species of gasteropods, seven species were described for the first time; one could not be determined beyond generic level; and only two are known from elsewhere. These are *Pseudoceratodes mammuth* (Blanckenhorn) and *P. irregularis*.

The marine lamellibranch is a very doubtful record of *Pinna* alternatively described as (?) a fossil palm leaf.

Both *P. mammuth* and *P. irregularis* were recorded from pebbles found in the Wadi Samur, 90 miles south of Cairo; *P. mammuth* was found in association with Eocene oysters in the pebbles.

Pseudoceratodes mammuth was also recorded from pebbles found in the Abbasia district, north-east of Cairo, supposedly derived from the Jebel Amar Quartzite. This formation has been assigned to the Oligocene by some authors and to the Pliocene by others. Cox (1939) also pointed out that none of these Egyptian fossils was found *in situ*.

As the age assigned to the Hudi Chert Formation depends on the stratigraphical range of *P. mammuth* and *P. irregularis*, the most that can be said about the age of the formation is that it has a Tertiary aspect.

Cox (1932–3) commented that he was able to state 'with some confidence that the shells now described are of Lower Tertiary age, while there is a strong probability that they are referable either to the Upper Eocene or to the Lower Oligocene'. This is no longer acceptable.

The following fauna was recorded by Cox (1932–3) from the Hudi Chert Formation. Lamellibranchia: *Pinna* ? sp. or ? palm leaf. Gasteropoda: *Pila colchesteri* Cox, *Pila* sp., *Pseudoceratodes mammuth* (Blankenhorn), *P. irregularis* (Blankenhorn), *P. rex* Cox, *Lanistes grabhami* Cox, *Hydrobia ? sudanensis* Cox, *Achatina* (*Burtoa*) *hudiensis* Cox, Gen. et sp. indet. Cox, *Planorbis siliceus* Cox, and *P. nubianus* Cox.

(b) Infra-basaltic sediments, Khor Langeb

Delany (1954) recorded a fauna from a thin limestone and marl at Jebel Oharba (17°13′ N, 35°34′ E) supposed to occupy the same stratigraphical position as the Hudi Chert. Teeth, dorsal plates, spicules, and dental plates of a crocodile, catfish, and wrasse were recorded from the limestone. The underlying marl is said to contain gasteropods, including an undescribed species of *Melanoides*. Commenting on this fauna, Delany stated that the species of *Melanoides* is similar to specimens collected from unconsolidated deposits in the Bara depression and that catfish and wrasse range from Eocene to Recent. The presence of catfish and wrasse is used to support a Tertiary age but as no specific determinations were made, this is unwarranted in the author's opinion.

(c) Tertiary formations: Red Sea Littoral, islands and Suakin Archipelago

Carella and Scarpa (1962) recognized six formations on the Red Sea Littoral of the Sudan. These formations (described above) are:

Abu Shagara Formation	Pleistocene
Dungunab Formation	Post-Middle Miocene
Abu Imama Formation (A, B, and C members)	Middle Miocene (Tortonian)
Maghersum Formation (A, B, and C members)	Middle Miocene Helvetian and Lower Miocene
Hamamit Formation	Lower Miocene (?) Eocene
Mukawwar Formation	Uppermost Cretaceous or Palaeocene (discussed above)

(i) *Hamamit Formation*

The Hamamit Formation has not yielded fossils but is assigned to the Eocene–Miocene interval because of its stratigraphical position (Carella and Scarpa 1962, p. 15).

(ii) *Maghersum Formation*

The overlying Maghersum Formation is divided into A, B, and C members. The A and B members are fossiliferous but no fossils have been recorded from the C member. The following fossils have been identified.

Maghersum Formation (A Member): Lamellibranchs, gasteropods, echinoids, foraminifera, *Neoalveolina* sp., *Bolivina arta* Macfadyen, *B. scalprata* Schwager. miocenica Macfadyen, *Hopkinsina bononensis* (Fornasini), *Uvigerina rutila* Cushman and Todd, *U. schwageri* Brady, *U. tenuistriata* Reuss sp. *gaudryinoides* Lipparini, *Globigerinella aequilateralis* Brady, *G. subcretacea* (Lomnicki), *Globigerinoides helicinus* (d'Orbigny), *G. sacculifer* (Brady), *Globoquadrina dehiscens* (Chapman, Parr, and Collins), *Miogypsina* sp., and *B. scalprata* Schwager sp. *miocenica* Macfadyen.

Maghersum Formation (B Member): Gasteropods, Bryozoa, fish teeth, Foraminifera, *Rotalia calcar* (d'Orbigny), *Nonion boeanum* (d'Orbigny), Peneroplidae, and Miliolidae.

The fauna of the A Member is said to indicate a Middle Miocene (Helvetian) age (Carella and Scarpa 1962, p. 13). However, it contains only two species in common with those of the Clysmic area, Egypt, and they are not diagnostic of any particular part of the Miocene sequence in that area (Fig. 58).

The A Member of the Maghersum Formation contains *Miogypsina*, *Globigerina*, *Globoquadrina*, and *G. dehiscens*. This combination indicates an early Miocene (Burdigalian) age if we use the range data

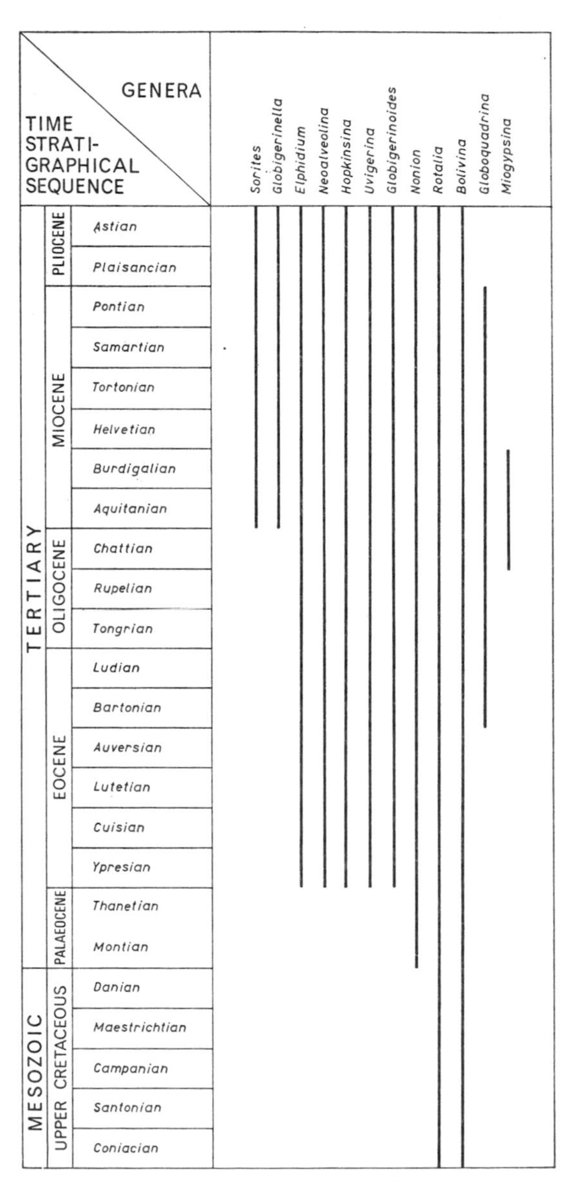

FIG. 58. Mesozoic and Tertiary distribution chart, Foraminifera, Red Sea Littoral (Compiled by S. V. Bell)

presented in Grimsdale and the *Treatise on invertebrate paleontology* (Loeblich and Tappan 1964).

The fauna of the B Member is not diagnostic and the C Member is unfossiliferous. Carella and Scarpa point out that the B and C members are probably early Miocene in age. If the interpretation mentioned above is accepted they may well be Aquitanian (early Miocene).

(iii) *Abu Imama Formation*

The Abu Imama Formation has yielded the following fauna and flora: corals, Mollusca, echinoids, Melobesiae (calcareous Algae), ostracods, fish remains, Foraminifera, *Neoalveolina melo* (Fichtell and Moll), Heterosteginae, Amphisteginae, and Globigerinidae.

The assemblage indicates that the Abu Imama Formation was laid down under marine conditions, and the presence of *Neoalveolina melo* (Fichtell and Moll) indicates a medial Miocene age (Carella and Scarpa 1962, p. 11).

(iv) *Dungunab Formation*

The Dungunab Formation has not yielded fossils. The uppermost beds are assigned to the Tertiary by Carella and Scarpa (1962).

(v) *Age-determinations*

Age-determinations were made by Sestini (1965, p. 1459), who described the coastal sediments in the area of Khor Eit, Khor El Kuk, Jebel To-Banam, and Jebel Sahgum. He recognized the Maghersum Formation of Carella and Scarpa (1962) but included the A Member in the Khor Eit Formation, which he considered to be transitional between the underlying Maghersum Formation and the overlying Abu Imama Formation.

Sestini (1965, p. 1459) stated that the presence of *Heliastraea* cf. *reussiana* (D'Arch, Edw. B H.) is indicative of a Middle Miocene age. These corals occur in reefs below the A Member of Carella and Scarpa, which is included in the Khor Eit Formation by Sestini. It is not clear, however, whether these reefs are equivalent to the B or C members of Carella and Scarpa. Since a reconsideration of the fauna reported by Carella and Scarpa indicates a Lower Miocene (Burdigalian) age for the A Member of the Maghersum Formation, the Middle Miocene age for the B and C members indicated by Sestini is doubtful.

Sestini (1965, p. 1461) reported that the upper third of the Khor Eit Formation, broadly equivalent to the lower part of the Abu Imama Formation of Carella and Scarpa (Sestini 1962, p. 1457, Fig. 3), contains 'small coral colonies, oyster banks *Ostrea latimarginata* Vredenburg; *Ostrea gryphoides* (Scholtheim) and abundant *Cardita* sp.; less common are *Venus* sp.; *Chlamys senatoria* Gmelin, *Barycypraea* sp., *Strombus* sp., *Conus* cf. *brevis*, and other unidentified mollusks'. This fauna is said to indicate an early to Middle Miocene age.

At a higher stratigraphical level, a patch reef 2 m thick contains *Heliastraea ellisiana* Defr. and *Eliastraea defrancei* Edw. & H. These corals are said to be common in the Miocene deposits of the Eritrean coast north of Massawa.

Sestini (1965, p. 1461) reports that the identified corals, *Barycypraea*, *Chlamys senatoria*, and *Ostrea gryphoides* suggest that the Khor Eit Formation is of Middle Miocene age.

As the A Member of the Maghersum Formation, here regarded as being of Lower Miocene (Burdigalian) age, the age of Sestini's Khor Eit Formation is in doubt. The sequence described by Carella and Scarpa therefore seems to be correct with the A Member of the Maghersum Formation being of Lower Miocene (Burdigalian) age; and the Abu Imama Formation, on the basis of Sestini's fauna, being Middle Miocene.

The Abu Imama Formation has been divided into seven units by Sestini (1965, p. 1463). At the base, Unit 1 is reported to contain *Quinqueloculina*, *Triloculina*, pancroplids (peneroplids?), rotalids, *Neoalveolina melo* Fitchtel and Moll, and *Lithothamnion* fragments. No fauna is reported from Units 2–5 but Unit 6 contains *Operculina complanata* Defrance, *Operculina complanata* var. *heterostegina* Silvestri, lamellibranchs, and echinoderms (*Clypeaster*). The following fauna is recorded from Unit 7: *Lithothamnion*, miliolids, rotalids, paneroplids (peneroplids?), *Neoalveolina*, molluscs, bryozoans, algal filaments, echinoid spines, and coral fragments.

Sestini (1965, p. 1464) stated that 'The macro- and micro-fossils in the Abu Imama limestones of the area studied are indicative of early to middle Miocene age, with a preference for the middle Miocene'. No fauna is reported from the Dungunab Formation, which overlies the Abu Imama unconformably.

(vi) *Fauna reported by the General Exploration Company of California*

The General Exploration Company of California recorded Miocene fossils from sediments exposed on Mukkawwar Island, the Rawaya (Abu Shagara) Peninsula, and the mainland (Bickel and Peters MS. 1962). Twenty-one rock samples were collected from Mukawwar Island, of which eleven contained a microfauna of definite Miocene age and six contained *Flosculinella* of Burdigalian (Middle Miocene) age (Bickel and Peters 1962, p. 2). Bickel and Peters (1962, p. 7) also stated that 'The foraminifera in all the Sudan samples from Mukawwar Island represent a shallow water facies. The absence of a planktonic population makes positive dating difficult. However, we believe the occurrence of such forms as *Dendritina* and *Flosculinella* indicate the presence of Miocene.'

The faunas reported from the samples collected from Mukawwar Island are as follows.

MUKAWWAR ISLAND

Sample	Age	Fauna
M/3	Miocene	*Quinqueloculina* *Elphidium* *Rotalia papillosa* *Dendritina* *Flosculinella* *Clavulina* ostracods
M/4	Indet.	*Elphidium*
M/5	Pliocene?	ostracods foraminiferal casts
M/6	Pliocene?	*Quinqueloculina* foraminiferal casts gasteropods echinoids
M/7	Miocene	*Dendritina* *Flosculinella* *Clavulina* *Elphidium* *Flintina floridana* *Quinqueloculina* ostracods gasteropods lamellibranchs
M/8	Indet.	ostracods foraminiferal casts
M/9	Miocene	*Flosculinella* charophites ostracods gasteropods corals
M/10	Indet.	echinoid spines corals
M/11	Miocene	ostracods *Quinqueloculina* *Flosculinella* foraminiferal casts
M/12	Indet.	ostracods
M/13	Miocene	*Elphidium chapmani* *Rotalia papillosa*
M/14	Indet.	pyritized stems
M/15	Miocene	*Elphidium chapmani* *Eponides* cf. *ornatus* *Rotalia papillosa* ostracods
M/17	Miocene	*Amphistegina lessoni* *Bolivina brevior* *Quinqueloculina*
M/18	Miocene	*Amphistegina lessoni* *Elphidium* *Gaudryina* echinoid spines
M/19	Miocene	*Flosculinella* *Quinqueloculina* *Pyrgo* foraminiferal casts ostracods echinoid spines
M/21	Miocene	*Rotalia papillosa* *Elphidium chapmani* *Quinqueloculina* *Spiroloculina planulata* *Elphidium* sp. ostracods
M/23	Miocene	*Flosculinella* *Elphidium* *Quinqueloculina* echinoid spines

Of the samples collected from the Rawaya (Abu Shagara) Peninsula, two contained microfossils and one macrofossil fragments. Only one of these samples yielded a fauna of definite Miocene age.

RAWAYA PENINSULA

Sample	Age	Fauna
2a/1	Indet.	gasteropods cup corals
2a/2	Miocene	*Elphidium chapmani* *Amphistegina lessoni* echinoid spines ostracods lamellibranchs gasteropods
2b/1.	Indet.	macrofossil fragments

Twenty samples were collected from the mainland and only one of these yielded identifiable fossils. This came from Jebel Abu Imama. The following identifications were made (Bickel and Peters 1962).

MAINLAND

Sample	Age	Fauna
A 30/1	Indet.	fossiliferous reef limestone
A 30/2	Indet.	fossiliferous limestone
A 30/3	Indet.	fossiliferous limestone
A 30/5	Indet.	megafossil noteo
A 31/1	Indet.	foraminiferal casts
A 31/2	Indet.	fossiliferous limestone
A 31/3	Miocene	*Amphistegina lessons* *Globigerina bulloides* gasteropods corals
A 31/4	Indet.	gasteropods lamellibranchs
A 31/6	Indet.	*Elphidium*
C 21/1	Indet.	echinoid spine microfossil fragments
C 21/3	Indet.	fossiliferous limestone

Amphistegina lessoni indicates a Miocene age.

4. PLEISTOCENE AND RECENT FORMATIONS

Records of Pleistocene fossils from the Sudan are not numerous and few of them are of diagnostic stratigraphical value. Marine fossils occur in the Emerged Reef Complex (Berry, Whiteman and Bell 1965) and the Abu Shagara Formation of Carella and Scarpa (1962). Carella and Scarpa (1962, p. 8) reported the following fauna: Fungi sp., *Meandrina* sp., *Orbicella* sp., *Cardium* sp., *Arca* sp., *Pecten* sp., *Ostrea* sp., *Conus* sp., *Turritella* sp., echinoids, ostracods, Peneroplidae, Miliolidae, *Elphidium crispum* (Linné), and *Sorites* sp.

Carella and Scarpa stated that the fauna has not been studied in detail and tentatively assigned the formation to the Quaternary.

Berry, Whiteman, and Bell (1965, p. 142) recorded the following fauna from the 'Emerged Reef Complex' (in part equivalent to the Abu Shagara Formation of Carella and Scarpa): Gasteropoda: *Canarium gibberulum* Linné, *Chicoreus anguligerous* Lamarck, *Clanculus pharoanius* Linné, *Conus princieps*, *C.* aff. *sumatrensis*, *Cypraea arabica* Linné, *C. turdus* Lamarck, *Cypraea* sp., *Pterocera bryonia* Gmelin, *Strombus fasciatus* Bornemann, *S.* cf. *tricornis* Lamarck, *Strombus* sp., and *Vasum cornigerum* Lamarck. Lamellibranchia: *Anadara antiquata* Linné, *A. radiata* Reeve, *Barbatia fusca* (Brugiére), *Barbatia* sp., *Cardium leucostomum* (Bornemann), *Glycimeris pectunculus* Linné, *Lima lima* Linné, *Spondylus aculeatus* (Chemnitz), *Tridacna gigas* Linné, and '*Venus*' *reticulatus* Linné. Coelenterata: *Coeloria lamellina* (Ehrenberg), *Fungia fungites* (Linné), *Galaxes fascicularis* Linné, and *Stylophora* cf. *erythraea* Marenz. Miscellaneous: cuttlefish bones, echinoids (sand dollar type), and echinoid spines.

Radiocarbon dating of *Tridacna gigas* Linné and ^{230}Th–^{234}U dating of corals from these emerged reefs indicates that these beds are Pleistocene (Monastirian) in age.

Records of fossils of continental origin are not abundant from the Sudan. The recorded fossils have little or no diagnostic stratigraphical value, even though they may be of great interest in other fields of palaeontology.

Vertebrate remains were reported from the excavations made during the construction of the Blue Nile bridge at Khartoum by Andrews (1912). The fauna included elephant, hippopotamus, a small giraffe, antelope (? *Tragdaphus*), and siluroid fish.

Hopwood (1929) reported on a small fauna collected by Grabham at Kosti. This included a small carnivore and a new species of fossil pig, *Hylochoerus grabhami* Hopwood. Hopwood noted that the state of mineralization of these fossil remains was similar to that reported by Andrews at the Blue Nile bridge locality.

Bate (1951) described a fauna collected from two localities, Singa and Abu Hugar. The fauna recorded is as follows.

	freshwater snail
Crocodilus niloticus Laurenti	Nile crocodile
Hystrix astasobae Bate	extinct porcupine
Equus sp.	equine
Rhinoceros sp. (large)	rhinoceros
Hippopotamus cf. *amphibius* Linné	hippopotamus
? Sivatherine	? short-legged giraffoid
Oryx sp.	oryx
? Antilopine or ? Caprine	extinct antelopine
Gazella sp.	gazelle
Antilope sp. (? Hippotragine)	antelope (large)
Antilope sp. (about the size of Grant's gazelle)	antelope
Homoioceras singae Bate	extinct long-horned buffalo

This fauna was probably collected from the same horizon as the Singa skull, first described by Woodward (1938) and redescribed by Wells (1951); both the fauna and the skull are associated with Palaeolithic stone implements.

A crocodile skull collected by Berry and Whiteman in 1962 from these deposits at Abu Hugar was sent to the British Museum (Natural History). A. J. Charig (personal communication) commented that the specimen is considerably larger than *Crocodilus niloticus* Laurenti, and 'is rather different in form. Indeed, it is of the same order of size as a skull of *C. porosus* Schneider (estuarine or salt-water crocodile, range from India to Fiji) which may well represent the largest Recent crocodilian recorded, supposedly about 33 ft long. However, there are again differences in form, especially in the width–depth ratio'.

Perkins (1965) described three faunal assemblages from Sudanese Nubia discovered by the Columbia University and New Mexico Museum expeditions to Sudanese Nubia. The oldest faunal assemblage is from the Upper Pleistocene silts, (C^{14}-dating 14 950 ± 300 years B.P., Fairbridge 1962). It is as follows:

Clarias lazera	catfish
Clarias sp.	
? *Synodontis* sp.	
? *Clarotes* sp.	
Lates niloticus	Nile perch
Crocodilus niloticus	crocodile
Gazella ? *dorcas* spp.	gazelle
Bos sp. (*primigenius* type)	wild cattle
Alcelaphus buselaphus	hartebeeste
Phacochoerus aethiopicus	wart-hog
Hippopotamus amphibius	hippopotamus
Equus asinus africanus	ass
Unio wilcocksi	freshwater mussel

The other two faunas are of 'A- and C-group' age and are dated at 3000 B.C. and 2240–2150 B.C. respectively. The 'A-group' fauna was collected from opposite Wadi Halfa by UNESCO scientists and made available to Perkins for identification and study. The fauna included:

Clarias sp.	catfish
? *Synodontis* sp.	
? *Clarotes* sp.	
Lates niloticus	Nile perch
Struthis camelus	ostrich
Alopocher aegyptiacus	Egyptian goose
Lepus aegyptus	hare
Gazella ? *dorcas* spp.	gazelle
Boris sp.	
Capra hircus	domestic goat
Equus asinus africanus	ass

This assemblage has been reported in historic time. The lack of domesticated animals is notable.

Perkins (1965) also examined the fauna collected by the Scandinavian joint expedition in Nubia and recorded

Oris aries platyura-aegyptiaca	sheep
	goats
Bos sp. (*primigenius-brachyoceros* type)	cattle

This fauna indicates that the domestication of animals had taken place by 2240–2150 B.C. in this region.

The Khartoum Hospital site (Mesolithic in age) yielded a large fauna (Arkell 1949). Derry (in Arkell 1949, p. 15) listed the following vertebrates: crocodile, porcupine, hippopotamus, wart-hog, and buffalo.

The fish remains identified by Trewaras (in Arkell 1949, p. 17) include the following: *Polypterus* sp., *Labes* sp., *Clarias* sp. (including *C. lazara*), *Synodontis* sp., *Clarotes* sp., *Lates* cf. *niloticus* (Linnaeus), *Tilapia* sp., and *Hydrocyon forskalii* Cuv. Trewaras (*in* Arkell 1949, p. 17) also stated that 'Besides the above Mr. Arkell has distinguished spines of *Bagrus* sp.'

The reptilian remains were studied by Swinton (*in* Arkell 1949, pp. 17–18), who reported a fauna including *Crocodilus* sp., *Python* sp., *Varasnus* sp., *Trionyx* sp., and *Testudo* sp.

Bird remains were rare at the Khartoum Hospital site, and only one of the three bones found could be identified. This was referred to *Plectropterus gambensis* (Linn.), spur-winged goose (Arkell 1949, p. 18).

Connolly (in Arkell 1949) identified the following molluscan fauna. Land snails: *Zootecus insularis* Ehrn., *Limicolaria flammata* Caill., *L. flammata* Caill. var. *candidissima* Shutt., *Ampullaria wernei* Phil., *Lanistes carinatus* Oliv., *Trochonanina* sp., *Pupoides sennaarensis* Pfr., *Bulinus truncatus* Andouin, and *Bithynia sennaarensis* Kust. Freshwater mollusca: *Cleopatra bulimoides* Oliv., *Viviparus unicolor* Oliv., *Aetheria elliptica* Lam., *Aspatharia rubens* Lam., *A. rubens* Lam. var. *caillaudi* Mts., ? *A. wahlbergi hartmanni* Mts., ? *Chambardi locardi* Bgr., *Mutela angustata* Sow., and *Corbicula africana* Krs.

The mammalian fauna described by Bate (*in* Arkell 1949, p. 18) consists of twenty-two species. No domesticated species were recorded. It was concluded from this evidence that the people occupying the Khartoum Hospital site were hunters and fishermen.

The fauna recorded is:

Hyaena cf. *hyaena*	hyena
Canis? lupaster	? Egyptian wolf-jackal
Afilax cf. *paludinosus*	? water mongoose
Mungos sp.	mongoose
Panthera cf. *pardus*	? leopard
Felis sp. (cf. *ocreata*)	wild cat (large)
Hystrix sp.	porcupine
Thryonomys arkelli Bate	Arkell's rat
Rattus (*Mastomys*) cf. *coucha*	? white-nosed rat
Arvicanthis sp.	spiny field rat
Hippopotamus cf. *amphibius*	hippopotamus
Phacochaerus sp.	wart-hog
Onotragus cf. *megaceros*	? Nile lechwe (Mrs. Gray's cob)
? *Adenota leucotis*	? white-eared cob
Antilope sp. (large)	antelope
Antilope sp.	antelope
? *Qurebia* sp.	? oribi
Antilope sp. (another small sp.)	small antelope
Syncerus cf. *aequinoctialis*	? north-eastern buffalo
Equus sp.	equine
Diceros cf. *bicornis*	? black rhinoceros
Loxodonta cf. *africanus*	? African elephant

Arkell (1953) published the results of his investigations at Es Shaheinab, a Neolithic site 30 miles north of Omdurman. The faunal lists reported from this site are similar to those recorded from the Khartoum Hospital site with the significant difference that the remains of three domesticated species were found. The fauna is as follows.

Freshwater Mollusca. Univalves frequenting shallow water: *Ampullaria wernei* Phil., *Lanistes carinatus* Oliv., and *Melanoides tuberculata* (Muller). Univalves frequenting deeper water: *Viviparus unicolor* Oliv., and *Cleopatra bulimoides* Oliv. Bivalves: *Aetheria elliptica* Lam, *Aspatharia marnoi* Jickeli, *Mutela nilotica* Sow., *Unio* (*Nitia*) *teretiusculus* Phil., *U.* (*Horusia*) *parreyssi* Phil., and *Cyrena* (*Corbicula*) cf. *artini* Pallary. Land snails: *Zootecus insularis* Ehrn., *Limicolaria flammata* Caill., and *L. kambeul* Brugière.

Arkell (1953, p. 10) stated that most of these species were identified by C. P. Castell of the British Museum (Natural History).

Fish reported from Shaheinab include *Clarias* sp., *Synodontis* sp., and *Lates* cf. *niloticus* (Linnaeus).

The reptiles include *Crocodilus* sp., *Python* sp., *Varanus* sp., *Trionyx* sp., and *Testudo hermanni* (identified J. C. Battersby).

A total of 111 bird bones were collected but had not been identified at the time of publication (1953).

Bate recorded thirty-two species of mammals (*in* Arkell 1953) of which twenty-nine were wild species and three domesticated species. These include:

Wild species:

Cercopithecus cf. *aethiops* (Linnaeus)	grivet monkey
Canis spp.	jackal
Canis ? cf. *aureus soudanicus*	jackal
Hyaena cf. *hyaena* (Linnaeus)	striped hyena
Panthera cf. *leo* (Linnaeus)	lion
Panthera cf. *pardus* (Linnaeus)	leopard
Felis cf. *lybica* Forster	African wild cat
Civetticus sp.	African civet
Mellivora sp.	honey badger
Lutrine	otter
Genetta cf. *tigrina* (Schreber)	genet
Herpesteo (*Myonax*) *sanguineus* (Ruppell?)	black-tipped mongoose
Hystrix sp.	porcupine
Lepus sp.	hare
Tatera cf. *robusta* (Cretzschmar)	gerbil

Euxerus cf. *erythropus* Geoff.	ground squirrel
Hippopotamus cf. *amphibuius* (Linnaeus)	hippopotamus
Phacochoerus sp.	wart hog
Oryz sp.	oryx
? *Hippotragus* sp.	? roan antelope
	antelope (not determined)
Strepsiceros cf. *Strepsiceros* (Pallas)	greater kudu
Gazella sp. (cf. *rubifrons* (Gray) red fronted)	gazelle
Gazella sp.	gazelle
Sylvicapra cf. *grimmia* (Linnaeus)	bush duiker
Syncerus or *Homoiceras*	buffalo
Giraffa sp.	giraffe
Rhinoceros sp. *Diceros bicornis* group	black rhinoceros
Loxodonta africana (Linnaeus)	African elephant
Domestic species:	
Capra sp.	dwarf domestic goat
Capra or Ovis sp.	twisted-horned goat or sheep
? *Ovis* sp.	? sheep

While engaged in a study of the Palaeolithic site at Khor Umar, near Jebel Surkab, north of Omdurman, Whiteman, Berry, and Bell discovered the following non-marine molluscan fauna in silty clays overlying implementiferous gravels (? Late Acheulian) on the south side of Khor Umar: *Limicolaria* sp., *Pila ovata* Olivier, and *Lanistes carinatus* Olivier.

In a study of the origin of the 'Sudan Gezira Clay Plain' Tothill (1946*a*) compared the habitats of living terrestrial and freshwater molluscs with the 'semifossil' occurrences in the uppermost 6 feet of the Gezira Clay. This comparative analysis was used to draw climatic conclusions in support of his theory of the origin of the Gezira Clay Plain. The fauna was identified by comparison with a named collection of living species sent to Tothill by the British Museum (Natural History); the 'difficult specimens' were sent to Connolly at the British Museum (Natural History) for identification. The reported semifossil fauna is *Cleopatra bulimoides* Olivier, *Ampullaria wernei* Phillipi, *Eanistes carinatus* Olivier, *Limicolaria flammata* Cailloud, *Viviparus unicolor* Olivier, *Melanoides tuberculata* Muller, *Corbicula artini* Parr, *C. consobrina* Caillaud, *C. radiata* Parr, and Unionidae.

In a discussion of the age of the alluvial clays in the western Gezira, Williams (1966) recorded an age of 11 300 ± 400 years B.P. (carbon-14 date by A. J. Whiteman) for a sandy loam containing *Lanistes carinatus* Olivier and *Cleopatra bulimoides*. Another carbon-14 date of 8370 ± 350 years B.P. was obtained for specimens of *Ampullaria wernei* Phil. from a similar horizon in Khor Umar, near Omdurman by the author.

Tothill (1947) described the nature and habitat of some of the species recorded in his study of the Gezira Clay. The fauna described is as follows: *Lanistes ovum* Trosch, *L. carinatus* Oliv., *Ampullaria wernei* Phil., *Burtoa nilotica* Pff., *Limicolaria flammata* Caill., *L. festiva* Mts., and *L. kambeul* Bruguiere.

In an attempt to rationalize the great number of species of *Ampullaria* reported from the Sudan, Bell (1966) worked out the synonomy for species recorded as *Pila ovata* (Olivier).

A clay occurring in the broader part of the valley at Erkowit in the Red Sea Hills has yielded fossil gasteropods (Tothill 1946*b*). By comparison with the geographic distribution of living forms Tothill (1946*b*) concluded that at the time of deposition of the clay the climate was comparable to that of Malta and Palestine today. The fauna listed is as follows: Land snails: *Xerophila* sp. and *Carastus abyssinica* Pfr. Pond snails: *Planorbis herbini* Bgt., *Bulinus truncatus* Aud., and *Limnaea caillaudi* Bourg. Aquatic snail: *Melanoides tuberculata* Mull.

V

Structure

THE tectonic structures of the Sudan may be divided into three groups (Whiteman 1968*b*):

(1) Basement Complex structures;
(2) Post-Basement Complex structures (excluding Rift Valley structures); and
(3) Rift Valley structures.

1. BASEMENT COMPLEX STRUCTURES

Little is known about the structure of the Basement Complex formations in the Sudan. Few observations are available concerning direction and dip of foliation, etc., and little is known about regional trends. General trends and photolinear trends are shown in Fig. 4.

(a) Equatoria, Bahr el Ghazal, and southern Darfur

Trends are said to be constant over wide areas of Equatoria Province and to vary from 160° to 180°.

The Assua Mylonite Zone enters the Sudan near Nimule but has not been mapped in detail in the field in the Sudan. The Maadi Quartzite folds are parallel to the Assua trend and these probably continue as far as Kidi (Fig. 4). Andrew, Auden, and Wright (1955) mapped quartzites (probably of the Maadi Formation) along ridges bearing 320° as far as Kajo Kaji, then on Jebels Kaiti, Dalili, Rimi, and Kura, which trend in a more northerly direction. A line of jebels continues as far as Loago and Perisak on the Juba–Yei Road, where the trend changes to 340°. The ridges, and therefore probably the structure, can be traced as far as Kidi (Fig. 4).

Details of the Assua Mylonite Zone in the Sudan are not available but it has been described in the Kitgum district near the Sudan–Uganda frontier by Almond (1962). He noted that the mylonite zone is complex and the degree of mylonitization variable. Markedly different rock types and structures occur on either side of the mylonite zone; but not enough is known of the geology to estimate the amount of lateral displacement. The direction of movement is also uncertain but appears to be sinistral (Almond, personal communication). Originally it was cited as dextral (Almond 1962).

The Assua Mylonite Zone was probably formed in late Pre-Cambrian times, and it is against this structure that the Miocene-to-Recent Rift faults end near Nimule (see below).

The structure of the Nile–Congo–Chad divide is largely unknown between Juba and Hofrat en Nahas. This region is shown as a swell separating the Chad, Congo, and Sudd basins by Holmes (1965, Fig. 763), and on the Provisional Structural Map of Africa (Furon *et al.* 1958) northerly structural trends have been sketched in for the 'P I' Basement Complex. These appear to be theoretical in some places or are projection of trends in Congo. Andrew, Auden, and Wright (1955) did not show such trends, and in the Hofrat en Nahas area the trends shown by Furon *et al.* (1958) differ in places from published trends by at least 15°.

(b) Central Sudan

The predominant trend of the Basal schists and phyllites in the Gedaref–Sennar region is 002°, and this is repeated in the Ingessana Hills (Fig. 10).

In Kordofan the strike of the metamorphic rocks is predominantly north to north-north-east but east–west trends have been recorded (Mansour and Samuel 1957; Hunting Technical Surveys 1964).

The quartzo-feldspathic gneisses of the Sabaloka area are composed mainly of biotite, sometimes garnetiferous gneiss; granitic gneiss, hornblende-gneiss, and amphibolites that form the Basement Complex generally trend between 292° and 365° and foliation dips at 60°.

Outcrops of Basement Complex rocks are few and scattered in the Gezira. At Jebel Moya and Jebel Saqadi, near Sennar, Iskander (1957) recorded metasediments dipping east at 70° and striking east-north-east–west-south-west. On the west bank of the Nile, south of Jebel Aulia Dam, Jebel Bereima consists of highly brecciated felsite striking north–south and dipping steeply to the east.

(c) Jubal el Bahr el Ahmer

The Odi schists of Derudeb sheet 46I trend east-north-east and north-east; in the Kassala district to the south the grey gneiss trends roughly north–south and east–west joints are well developed.

A conspicuous east–west-trending dyke system occurs east of the Tehilla ring complex in the Derudeb area. Towards Haiya Junction north of the Derudeb sheet this system is extensively developed and trends east-north-east.

In the Mohammed Qôl area northern Jubal el Bahr el Ahmer foliation and lineation are well developed; the regional strike direction in the 'Primitive System' ranges from east–west in the centre of the sheet to north-east–south-west to north–south in the northern part of the sheet (Fig. 18). These rocks are folded into broad anticlines and synclines and the fold axes plunge in a northerly direction. An east–west dyke swarm occurs in the 'Primitive System' (Ruxton 1956).

The Nafirdeib Group, which unconformably overlies the 'Primitive System', is folded in the north of the map, near Gebeit Gold Mine, in open anticlines and synclines with fold axes that plunge at low angles and bear 030°. At To-Olak (36°24′ N, 20°30′ E) well-developed fracture cleavage is imposed parallel to the axial planes of the folds. Small thrust faults are common (Ruxton 1956).

At the south of the Mohammed Qôl sheet Ruxton (1956) noted that the Nafirdeib Formation is again folded into broad antiforms and synforms plunging at low angles along axes trending east-north-east.

Acid, intermediate, and basic dykes cut the Basement Complex formations of the Jubal el Bahr el Ahmer. They form a striking reticulate system of late Pre-Cambrian age. Directions vary but are predominantly 070° and 340° (Fig. 18). At Jebel Hamashaweib a basic dyke trending east–west gave an age of $740 \pm 48 \times 10^6$ years. As this system is situated on the shoulders of the Red Sea Depression it is possible that the dykes were formed during an early tension phase in the development of the Red Sea Rift System (Whiteman 1965).

Ruxton (1956, p. 328) stated that all the Basement Complex formations are faulted and that the region south-west of Jebel Erba is 'a mosaic of tilted fault blocks with prominent escarpments and shallow dip slopes'. The easterly trending fault shown south of Jebel Erba might be a very large fault according to Ruxton (1956). Large khors often follow the major faults, and fault breccias as much as 200 ft wide are known. Ruxton thinks that these faults are tensional but unfortunately the movements are undated. Sestini (1956) has mapped a number of similar photolinear trends in this area. At localities to the north fault blocks with Nubian Sandstone occur; these faults are post-early Cretaceous and may well be Tertiary in age. Probably they are associated with the formation of the Red Sea Rift Depression. General trends on the Deraheib and Dungunab sheets are shown in Fig. 18.

Gass (1955) recorded that the dominant strike of the Oyo Series of the Dungunab sheet is north-east although, in the north-east part of the area, the strike swings due north.

In the south-eastern part of the area a number of east–west faults have been mapped. As the younger formations are not affected by these faults they are thought to be Basement structures. In nearly all cases these structures are tear faults, which show limited lateral movement (Gass 1955).

(d) Northern Province

Information on the general tectonic trends is mainly restricted to outcrops along the Nile and to Jebel Uweinat.

Burollet (1963) described the Basement Complex formations of Jebel Uweinat. The jebel itself forms part of a large ring complex of post-Carboniferous age intruded into folded metamorphics (Fig. 20). The predominant trend is north-easterly. No structural data are available for Basement rocks between Jebel Uweinat and the Nile.

Marbles, interbedded with highly foliated gneisses, occur in the Sudan Portland Cement Company's quarry on the west bank of the Nile near Atbara. The beds strike at 070° and dip at an average of 35° to the east-south-east. Folding is complicated, the marbles having acted as incompetent beds.

Kabesh (1960) described in some detail the geology of the mica-bearing area of Rubatab, downstream from Atbara. Large areas of Kabesh's map are structurally incorrect, however, and the area needs reinvestigating, using modern structural techniques. The predominant trend is slightly west of north.

North of Rubatab and the Fifth Cataract the schists and slates with interbedded marbles trend in a meridional direction, but in the Umm Nabari (Omm Nabardi) gold area, near Station No. 6, the trend of the schists, calcareous schists, and rhyolites varies between 045° and 090°. The predominant trend is east–west.

There is little information available concerning the Basement rocks of the section from Abu Humed to the Fourth Cataract. The Fourth Cataract is composed of biotite-gneiss seamed with quartz veins striking between 000° and 030°.

From a point just above the confluence of the Wadi Muqaddum and the Nile to near Abu Fatima's Tomb near Dongola the Nile flows over Nubian Formation and nothing is known of the Basement formations. Near Abu Fatima's Tomb the trend is meridional but at the Third or Hannek Cataract the gneisses strike east–west across the river.

At the Kagbar Cataract between Arduan Island and Delgo the strike is about 337°. Gneisses crop out beneath the Nubian Formation that forms the banks. In the Delgo area the regional trend is again meridional as it is at the Amara Cataract where both banks of the Nile are composed of calcareous schists. Hornblende-schists crop out in the Akasha region and the strike is predominantly north–south (Fig. 22).

The Tangur Cataract, between Akasha and Ambigol, is composed of southward-dipping foliated gneiss. In the Sarras area there are a series of broad antiforms and at Murshid the meridional trend of the schists controls the river direction. Structural data for the diorite masses of the Second Cataract area are not available.

(e) Western Sudan

A small amount of structural data is available for this region. Afia and Widatalla (1961) published information on Hofrat en Nahas collected by the Nile–Congo Divide Syndicate, and by T. D. Guernsey and P. E. Fairbairn, who investigated the copper deposits for Anglo-American Corporation. The foliation of the 'Older Plains Group' strikes at 083° and dips generally northwards at 30–80°. Local variations occur in the Grey schist subdivision of this group and foliation strikes from 015° to 054° and dips 30°–65° north-west. In the 'Younger Group' the quartzites strike from 355° to 040°. Dips are generally steep and to the west (Fig. 23).

Hunting Technical Services (1958) recorded the strike of rocks along the Zalingei–Nyerte Road in the Jebel Marra region as varying from 090° (northerly dip) to 045° (north-westerly dips). The schists are said to be highly folded isoclinally (Fig. 25).

Lotfi and Kabesh (1962) attempted to differentiate a pattern of geosynclinal development for the Jubal el Bahr el Ahmer, but in the author's view this is premature. Clearly these rocks are geosynclinal deposits that have had a long history of folding, intrusion, and metamorphism but until more isotopic dates are available, and the regional correlations secured, the evolution of the geosynclinal belt and its relationship to the craton cannot be worked out.

In conclusion: much more information is needed before structural patterns can be distinguished for the Sudan.

(f) Ring complexes and Newer Granites

Ring complexes and 'Newer Granites' are common in the Basement Complex of the Sudan. The distribution of the main bodies is shown in Figs. 4 and 18. According to Ustaz Badr el Din Khalil, Department of Geology, University of Khartoum who studied the tin and wolfram deposits associated with these rocks, there are 9 rings and 109 Newer Granite masses known so far (Tables 17 and 18).

'Newer Granites' in the Sudan were first recorded from the Qala en Nahl area by Wilcockson and Tyler (1933). Andrews (1948) mentioned the occurrence of feldspathoidal soda syenites and riebeckite-granites, south of 6° N near Kadogli and Talodi in the Moro Hills of the central Sudan; Delany (1955), Gass (1955), and Kabesh (1962) recorded many more.

The Newer Granites and ring complexes are a group of mainly granitic and syenitic rocks intruded into the Basement Complex. They are high-level magmatic intrusions with sharply defined transgressive contacts and are commonly the youngest Basement Complex rocks in an area. The Sabaloka ring complex was dated as early Cambrian ($540 \pm 25 \times 10^6$ years) and many of the other rings and Newer Granites may well be of this age. The Jebel Uweinat Ring Complex is, however, said to be ? post-Lower Carboniferous in age (Burollet 1963).

TABLE 17

List of ring complexes known up to 1968 in Sudan

No.	Name and/or coordinate	Size	Remarks	Recorded by
18	Umm Shibrik ring complex 21° 35′ N, 36° E	1½ miles²	Formed of two concentric annular members, the central block is formed of hornblende–biotite-granite and the outer ring is composed of syenite	Gass (1955)
27	Salala ring complex 21° 19′ N, 36° 13′ E	11 miles²	Composed of two definite circular masses with two different centres. Aegerine–riebeckite–syenite	I. G. Gass (1955)
63	20° 50′ N, 36° 10′ E	—	Unmapped ring. Located by Delany on aerial photographs. Near a Nubian Sandstone outlier shown on Mohammed Qôl sheet. (Kabesh 1962)	Delany (1955)
64	20° 35′ N, 36° 05′ E	—	Unmapped ring complex. Observed on aerial photographs by Delany (1955). Corresponds in location to Jebel Deirurba To Hadal mapped by Kabesh later (1962) as pink-red granitic mass, i.e. Younger Granite?	Delany (1955)
65	20° 25′ N, 36° 03′ E	—	Unmapped ring complex. Identified by Delany (1955) on aerial photographs. Same location as Jebel Lakeb mapped by Kabesh (1962) as pink-red granitic mass, i.e. Younger Granite	Delany (1955)
66	20° 18′ N, 36° E	—	Unmapped ring complex. Location on aerial photographs by Delany (1955). Mapped as pink-red granitic mass by Kabesh (1962)	Delany (1955)
58	Tehilla ring complex 17° 45′ N, 36° 05′ E	7·0 miles by 5·5 miles = 37 miles²	Quartz–orthoclase granitic ring-dyke 1–2 km in width. The granitic intrusion of the ring-dyke was preceded by a gabbroic intrusion. The ring-dyke is vertical having an oval shape trending NE–SW	Delany (1955)
67	Sabaloka ring complex 16° 14′–16° 27′ N 32° 47′–32° 35′ E	300 miles²	The ring-dyke consists of porphyritic microgranite. Width varies from 50 m to 2 km. The emplacement of the ring-dyke was preceded by extrusion of basic and acidic lavas and pyroclastics and accompanied by cauldron subsidence. The shape of the ring complex is oval with main axis trending NE–SW	Delany (1955)

TABLE 17—*continued*

No.	Name and/or coordinate	Size	Remarks	Recorded by
69	Queili ring complex 15° 31′ N, 33° 47′ E	Eliptical in shape with axes 2 × 1 miles	Two or more rings and central mass of riebeckite-aegerine-syenites. Intrusions of the two ring-dykes and the central mass were preceded by extrusion of rhyolites. According to Delany (1955) the sequence is as follows: 1. Rhyolites (earliest) 2. First syenite ring 3. Essexite 4. Second syenite ring 5. Central quartz-syenite plug 6. Sodic dykes (latest) The ring complex is elliptical. The major axis runs NE–SW. (See Chap. III for further details)	Delany (1955)

TABLE 18

List of Newer Granite masses known up to 1968

No.	Name and/or coordinates	Remarks	Recorded by
1	J. Uweinat 22° N, 25° E	Granites, syenites, monzonites ? Post-Carboniferous	Sandford (1935) and Burollet (1963)
2	J. Kissu 21° 30′ N, 25° 10′ E	Granites, syenites, monzonites ? Post-Carboniferous	Sandford (1935) and Burollet (1963)
3	J. Shendib 22° N, 36° 17′ E	Granophyric granite 40 miles²	Gass (1955)
4	Wadi Oyo 21° 55′ N, 36° 03′ E	Pyritiferous biotite granite 1 mile²	Gass (1955)
5	Macruff Mine 21° 54′ N, 36° 06′ E	Pyritiferous biotite 1 mile²	Gass (1955)
6	J. Rebaidit 21° 43′ N, 36° 20′ E	Hornblende-syenite 5 miles²	Gass (1955)
7	J. Olak 21° 43′ N, 36° 24′ E	Fine-grained biotite-granite 4 miles²	Gass (1955)
8	J. Mog 21° 51′ N, 36° 24′ E	Biotite-granite 6 miles²	Gass (1955)
9	J. Eiwiet 21° 46′ N, 36° 17′ E	Biotite-granite ½ mile²	Gass (1955)
10	J. Harginab	Biotite-hornblende-granite 2 miles²	Gass (1955)
11	J. Romaeit 21° 44′ N, 36° 03′–21°	Biotite-granite 3 miles²	Gass (1955)

TABLE 18—*continued*

No.	Name and/or coordinates	Remarks	Recorded by
12	J. Haitem 21° 44′ N, 36° 31′ E	Biotite-granite 1¼ miles²	Gass (1955)
13	J. Shi-it	Biotite-granite	Gass (1955)
14	K. Fordikwan	Fine-grained biotite-granite 6 miles²	Gass (1955)
15	K. Delatit 21° 41′ N, 36° 39′ E	Muscovite-granite 1½ miles²	Gass (1955)
16	J. Arit 21° 37′ N, 36° 24′ E	Perthitic biotite-granite	Gass (1955)
17	J. Karam Arit 21° 35′ N, 36° 11′ E	Granophyric riebeckite-granite 4½ miles²	Gass (1955)
18	Umm Shibrik 21° 35′ N, 36° E	Ring complex. Hornblende-biotite-granite and syenite 1½ miles²	Gass (1955)
19	K. Kimoikwan 21° 39′ N, 36° 13′ E	Biotite-granite ¾ mile²	Gass (1955)
20	Bir Saritri 21° 35′ N, 36° 13′ E	Biotite-granite ¾ mile²	Gass (1955)
21	K. Saritri 21° 33′ N, 36° 12′ E	Muscovite-biotite-granite 7 miles² in area	Gass (1955)
22	J. Karaiaweb 21° 32′ N, 36° 03′ E	Biotite-granite ½ mile²	Gass (1955)
23	J. Misieu 21° 29′ N, 36° 01′ E	Red granite—no coloured minerals 1 mile²	Gass (1955)
24	J. Ankur 21° 22′ N, 36° 01′ E	Syenite 3½ miles²	Gass (1955)
25	J. Odarikul 21° 20′ N, 36° 02′ E	Biotite-microgranite	Gass (1955)
26	K. Mafdieb 21° 16′ N, 36° 03′ E 21° 21′ N, 36° 05′ E	—	Gass (1955)
27	Salala ring complex 21° 19′ N, 36° 13′ E	Two definite circular masses. Aegerine–riebeckite–syenite	Gass (1955)
28	K. Odar Aweb 21° 18′ N, 36° 13′ E 21° 19′ N, 36° 19′ E	Biotite-microgranite 3 miles²	Gass (1955)
29	J. Odar Aweb 21° 17′ N	Biotite-granite 1½ miles²	Gass (1955)
30	J. Erbairi 21° 17′ N, 36° 19′ E	Biotite-quartz-syenite 6 miles²	Gass (1955)
31	K. Haio 21° 14′ N, 36° 18′ E 21° 17′ N, 36° 29′ E	Muscovite-biotite-granite 8 miles²	Gass (1955)
32	K. Shinab	Biotite-granite 1½ miles²	Gass (1955)
33	J. Umarad 21° 14′ N, 36° 15′ E	Granophyric biotite-granite 5½ miles²	Gass (1955)
34	Bir Sawa 21° 18′ N, 36° 04′ E	Granite—no coloured minerals 5 miles²	Gass (1955)

TABLE 18—*continued*

No.	Name and/or coordinates	Remarks	Recorded by
35	J. Todar 21° 10′ N, 36° 02′ E	Porphyritic muscovite-granite. 20 miles² in area	Gass (1955)
36	J. Eimo 21° 06′ N, 36° 44′ E	Muscovite-granite 12 miles²	Gass (1955)
37	J. Tagoti 21° 06′ N, 36° 10′ E	Microcline-granite 8 miles²	Gass (1955)
38	K. Mishdad	Biotite-granite 12 miles²	Gass (1955)
39	K. Yahamib	Biotite-granite ¼ mile²	Gass (1955)
40	K. Lolignet	Muscovite-biotite-granite 4 miles²	Gass (1955)
41	K. Kamoikwan	Biotite-granite ½ mile²	Gass (1955)
42	K. Surt	Biotite-microgranite ½ mile² in area	Gass (1955)
43	J. Hamashaweib 20° 55′ N, 36° 58′ E	Microcline-granite	Kabesh (1962)
44	The area around K. Asir	Microcline-granite; continuation of 43	Kabesh (1962)
45	The area around Tosherer	Continuation of the outcrop of microcline-granite above	Kabesh (1962)
46	The area around K. Haieit	White granite—supposed to be younger than microcline-granite, smaller in area	Kabesh (1962)
47	J. Erba area	Continuous outcrop of pink-red granite, forms the characteristic escarpment of the Red Sea Hills. The pink-red granite is supposed to be younger than Kabesh's white granite and microcline-granite (47–53)	Kabesh (1962)
48	J. Eirawe		
49	J. Elangweib		
50	J. Oda		
51	J. Hadarachati		
52	J. Nakwat		
53	J. Gumadariba		
54	J. Lakeb area	Pink-red granite, in the western part of Mohammed Qôl sheet (54–56)	Kabesh (1962)
55	J. Mishallieh area		
56	J. Deirurba To- Hadal area		
57	18° 45′ N, 36° 30′ E	Younger Granite?	Geological Survey (unpublished)
58	Tehilla ring complex 17° 45′ N, 36° 05′ E	Quartz-orthoclase-granite ring-dyke 1–2 km in width. The granitic ring dyke intrusion was preceded by a gabbroic intrusion. The ring-dyke is vertical in dip, having an oval shape, axis 11 km and 9 km in length trending NE–SW	Delany (1955)
59	J. Tatab 17 17° 40′ N, 36° 30′ E	Recorded as an unfoliated granite. Could be a Newer Granite	Delany (1955)
60	J. Oduarn 17° 55′ N, 37° 20′ E	Soda granite	Delany (1955)
61	61° 40′ N, 37° 25′ E	Soda granite	Delany (1955)
62	Stony Hills 17° 15′ N, 37° 25′ E	Unfoliated granite; could be Younger Granite	Delany (1955)
63	20° 50′ N, 36° 10′ E	Unmapped ring complexes. Observed only from aerial photographs (63–66)	Delany (1955)
64	20° 35′ N, 36° 05′ E		
65	20° 25′ N, 36° 03′ E		
66	36° E–20° 18′ N		

TABLE 18—*continued*

No.	Name and/or coordinates	Remarks	Recorded by
67	Sabaloka ring complex 16° 14′ N, 32° 47′ E 16° 27′ N, 32° 35′ E	See Table 17	Delany (1955)
68	J. Seleitat 17° 50′ N, 32° 30′ E	Soda granite	Delany (1955)
69	Qeili ring complex 15° 31′ N, 33° 47′ E	See Table 17	Delany (1955)
70	J. El Uheirish 15° 15′ N, 34° 48′ E	Unfoliated granites. Could be Younger Granite?	Delany (1951)
71	J. Mundara 15° N, 34° 30′ E	Soda granite	Delany (1951)
72	J. Ghar el Watawit 14° 50′ N, 35° 15′ E	Unfoliated granite similar to No. 70	Delany (1951)
73	J. Megamis 14° 35′ N, 35° 15′ E	Unfoliated granite similar to No. 70	
74	14° 18′ N, 34° 47′ E	Soda granite and unfoliated granite	Delany (1951)
75	J. Abu Khudud 14° 10′ N, 34° 15′ E	Soda granite	Delany (1951)
76	J. Ban Balos 13° 30′ N, 35′ E	Unfoliated granite similar to No. 70	Delany (1951)
77	J. Marfa 16° 25′ N, 30° 10′ E	Syenite	Delany (1951)
78	J. Haraza 15° N, 30° 30′ E	Soda granite with acidic (rhyolite) lavas	Delany (1951)
79	J. Kershangel 14° 45′ N, 30° 10′ E	Soda granite	Delany (1951)
80	J. Aku Hadid 14° 37′ N, 30° 10′ E	Unfoliated granite similar to No. 70 and others above	Delany (1951)
81	J. Maxmum and J. Werkat 12° 05′ N, 33° 30′ E 12° 15′ N, 33° 52′ E	Unfoliated granite similar to No. 70 and others above	Delany (1951)
82	J. Geifrat 12° 55′ N, 33° 18′ E	Unfoliated granite discordant. Could be Younger Granite	Delany (1951)
83	J. Moya (station) 13° 25′ N, 33° 20′ E	Unfoliated granite discordant. Could be Younger Granite	Delany (1951)
84	J. Saqadi 13° 33′ N, 33° 08′ E	Unfoliated granite discordant. Could be Younger Granite	Delany (1951)
85	14° N, 32° 42′ E	Unfoliated granite discordant. Could be Younger Granite	Delany (1951)
86	El Jebelein 12° 35′ N, 32° 48′ E	Unfoliated granite discordant. Could be Younger Granite	Delany (1951)
87	J. Khon 13° 13′ N, 31° 40′ E	Unfoliated granite discordant. Could be Younger Granite	Delany (1951)
88	J. Ed Dair 12° 30′ N, 30° 42′ E	Soda granite	Hunting Technical Services (1964)

TABLE 18—*continued*

No.	Name and/or coordinates	Remarks	Recorded by
89	J. Dumbeir 12° 35′ N, 30° 48′ E	Syenite	
90	12° 47′ N, 30° 50′ E	Soda granite	Hunting Technical Services (1964)
91	11° 43′ N, 29° 30′ E South of Salara		
92	11° 30′ N, 29° 20′ E Near El Logowa		
93	11° 32′ N, 31° E Near Abu Gubeiha		
94	11° 38′ N, 33° 30′ E Near Sheikh Idris		
95	11° 10′ N, 29° E		
96	11° 03′ N, 29° 30′ E Near El Berdab	Shown in the legend of the geological map of southern Sudan as syenites and riebeckite-granites separated from the Pre-Cambrian Basement Complex formations. They are probably Younger Granites and syenites	Andrew, Auden, and Wright (1955)
97	10° 58′ N, 29° 50′ E Near Kadugli		
98	The Area around Katcha 10° 41′–10° 52′ N 29° 24′–29° 58′ E		
99	Moro Hills 10° 45′–10° 52′ N 30°–30° 28′ E		
100	10° 33′ N, 30° 20′ E Near Talodi		
101	10° 30′–10° 38′ N 30°–30° 16′ E		
102	5° 40′ N, 31° E Near Tingalu		
103	5° 22′ N, 31° 30′ E Near Rego		
104	5° N, 31° 33′ E Near Lado		
105	5° 04′ N, 33° 40′ E N. of Kapoeta	Similar to Nos. 91–101 above	Andrew, Auden and Wright (1955)
106	4° 48′ N, 31° E West of Juba		
107	4° 18′ N, 33° 55′ E Around Nagishot		
108	4° 20′ N, 33° 53′ E		
109	4° 09′ N, 33° 49′ E On Sudan–Uganda Borders		

The plutons and ring complexes are of varying size (Tables 17 and 18) and are usually oval in shape. The small masses usually show only one intrusive phase, whereas the large ones and some of the ring complexes are multi-phase intrusions.

The emplacement of the ring complexes at Sabaloka and Jebel Qeili (coordinates are given in Table 17) was preceded by the extrusion of great quantities of predominantly acidic lavas and tuffs. These are still preserved within the ring structure at Sabaloka. The volcanic rocks here appear to have been extruded from vents aligned along the ring fracture, which extended to the surface, and from simple central volcanoes. The more extensive later phases of volcanicity at Sabaloka are related to cauldron subsidence when much more magma was available. At Jebel Qeili the different phases that make up the ring-dykes were intruded along fractures zones of different radii.

High-level intrusive rocks occur as dykes, irregular sheets, and diatremes; and in places shallow surface cauldrons have been recognized.

The Younger Granites are rarely foliated and show few internal structures. Occasionally they show primary banding in the central zones. All are well jointed. Great sheet joints are common and weathering has produced characteristic 'boiler-plate' and 'onion-skin' types (Plate 4B).

2. POST-BASEMENT COMPLEX STRUCTURES

Pediplanation, which followed the development of the Basement Complex structures in the Sudan and lasted until early Carboniferous, or even late Jurassic time, was probably the result of more than one erosive phase, and there may have been intervening periods of deformation. As a result, unmetamorphosed fossiliferous early Palaeozoic rocks and much of the late Palaeozoic and early Mesozoic are unknown in the Sudan. The early Cretaceous Nubian Formation rest unconformably on the Basement formations over much of the country, except in the west and extreme north-west, where Palaeozoic rocks intervene between the Basement and Nubian formations.

Early Cambrian rocks occur at Sabaloka (K–Ar age (? cooling date): $540 \pm 25 \times 10^6$ years) and may occur elsewhere in other ring complexes. In Egypt in the Atbai Range even younger Palaeozoic rocks have been found (Table 3), and in Sudan isotopic analysis will probably reveal rocks of similar ages. In the author's view early Palaeozoic rocks may form not an inconsiderable part of the Basement Complex, especially in the northern Sudan. The extent of the 'Pan-African Orogeny' (Kennedy 1965) in Sudan is not known. Cahen and Snelling (1966) show the 'Katangan Belt' running into the southern Sudan in the Albertine Rift zone and the Mozambiquian probably runs into Ethiopia and Sudan from Kenya. Its limits are not known however (Whiteman 1969*a*). The distribution shown on the International Tectonic Map of Africa (1968) has no factual basis.

Concerning the form and amount of uplift of the Basement Complex, unfortunately the lack of topographic data, erosion, and the strongly unconformable nature of the Nubian Formation–Basement contact prevents isohypsals being drawn for much of the country. Dainelli (1943), however, produced such a map for Ethiopia and parts of this cover the eastern Sudan. According to Mohr (1962, p. 152), the maximum observed value of uplift (not taking into consideration erosion of the pre-Trap Mesozoic rocks, Mesozoic uplift of the rocks, and Rift Valley movements) is about 8960 ft (2800 m) at Mt. Soira in southern Eritrea. These values decrease westwards via Senafé, where they are about 7040 ft (2200 m), to about 4800 ft (1500 m) in the Lower Mareb or Gash valley. In Tigré values again decrease in the west from 800 ft (2500 m) at Amba Teru in Western Geralta to 2560 ft (800 m) in the Tekeze or Setit valley, to less than 1920 ft (600 m) north of Lake Tana.

Faults are supposed to limit the Ethiopian uplift to the west (Mohr 1962, Fig. 13; Dainelli 1943), but as much of the area is largely unexplored geologically these faults must be regarded as mainly theoretical. The Basement over much of the frontier zone could just as well have been down-warped as down-faulted. Faults bounding the plateau are not shown on the Sudan Geological Survey 1:4 000 000 maps (1948 and 1963 editions); nor are they shown on the Geological Map of Africa (1963 edition); nor, except for a small section north of Lake Tana, are they shown on the Structural Sketch Map of part of the Eastern Rift zone (Scale 1:5 000 000, published 1965) of which Mohr was part author. The trends of the isohypsals shown on Mohr's map in the Sudan–Ethiopian frontier zone are mainly theoretical in the author's view. More geological and topographic data are needed before such lines can be drawn with assurance for Sudan.

Elevation of the Basement Complex decreases in general north of Asmara, although there are large areas almost as high as the plateau surface near Asmara. North and west of the Baraka the land is much lower, but geology and structure are largely unknown, and as no post-Basement time-stratigraphical units are present it is impossible to deduce from the little data available the shape of the sub-Mesozoic–Basement Complex surface.

The area appears to be situated in a structural low with respect to the structurally high Asmara area, for normal faults downthrow the Nubian Formation, infra-basaltic sediments, Tertiary rhyolites, and basalts in Khor Langeb, east of Derudeb. The faults are discontinuous, trend at 020°, and are possibly Tertiary structures associated with the Rift System (Fig. 45).

The structure of the rocks that form the Sudan plainlands is known only in the broadest way. As noted above, structural details are available for a small part of the Basement Complex; dips in the overlying sedimentary formations are often obscure and low, so that it is very difficult to determine structural patterns except on regional maps and then only in a general way. Dip observations are particularly difficult to make in the Nubian Formation because of cross bedding, which is common; because marker bands are rarely traceable over long distances; and because of the lack of accurate topographic maps. Likewise dips are difficult to determine in the Umm Ruwaba, El Atashan, and Gezira formations, which consist mainly of unconsolidated or semiconsolidated deposits.

Several broad structural elements may be defined from the outcrop patterns on the generalized geological map of the Sudan (Fig. 4). In the south-west the Nile–Congo–Chad divide is anticlinal, being situated between the Congo cuvette, the Umm Ruwaba, and the Chad basins. Its structural relationships with the N'Dele plateau of the Central African Republic are uncertain, as are the Basement Complex structures of the Sudan and the Congo.

The Umm Ruwaba Formation was laid down in two synclinal basins, which merge north-east of Juba. The western structure occupies much of the Bahr el Arab basin and runs via Abu Gabra, north of Meshra er Req, via Duk Faiwil, where it merges into the eastern syncline north of Torit or Kapoeta. The centre of the eastern basin appears to run via the Pibor Post area, Akoba, east of Malakal, where the depression is occupied by the Nile Valley between Gelhak and Renk. It then curves around the Nuba Mountains towards Bara and El Obeid (Fig. 4). Dips in the Umm Ruwaba depression must be exceedingly flat and the stratigraphy is not well enough known to enable marker beds to be traced from borehole to borehole. However, the regional map shows that the Umm Ruwaba depression is synclinal with respect to the Ethiopian plateau and to the Nuba Mountains.

The Nuba Mountains are an area of Basement Complex rocks almost surrounded by sedimentary formations (Fig. 4). Considerable areas exceed 1600 ft (500 m) and small areas around Rashad exceed 3200 ft (1000 m). Again, little is known about structure but clearly the Nuba uplift took place after the deposition of the Umm Ruwaba Formation. The area is still active tectonically and forms part of an earthquake zone. The earthquake of 9 October 1966 had its epicentre at Dumbeir village near Er Rahad in the northern Nuba Mountains (Qureshi *et al.* 1966).

Well-developed erosion surfaces occur in the Nuba Mountains at about 3000 ft—the Nuba surface—and at Dilling, north of the Salara Hills, at 1600 ft (500 m). The Khor Abu Habl system is cut into this surface. Unfortunately the ages of the Nuba surfaces are unknown, but they must be considerable; at an estimate, based mainly on their position compared with the Erkowit surface on the east side of the Nile Valley, which predates the Shendi–Togni late Cretaceous–early Tertiary surface, they may well be Cretaceous or even Jurassic in age. The Dilling surface is much younger and is developed as an extensive pediment around the north end of the Nuba Mountains. Again, its age is uncertain but tentatively the author has correlated it with the Shendi–Togni surface of late Cretaceous–early Tertiary age of the Nile Valley further north.

If this correlation is correct the Nuba uplift must have taken place first in pre-early Tertiary (? pre-Oligocene) times. Later movements occurred, so enabling the Umm Ruwaba and younger deposits to accumulate. The area, as noted above, is active tectonically.

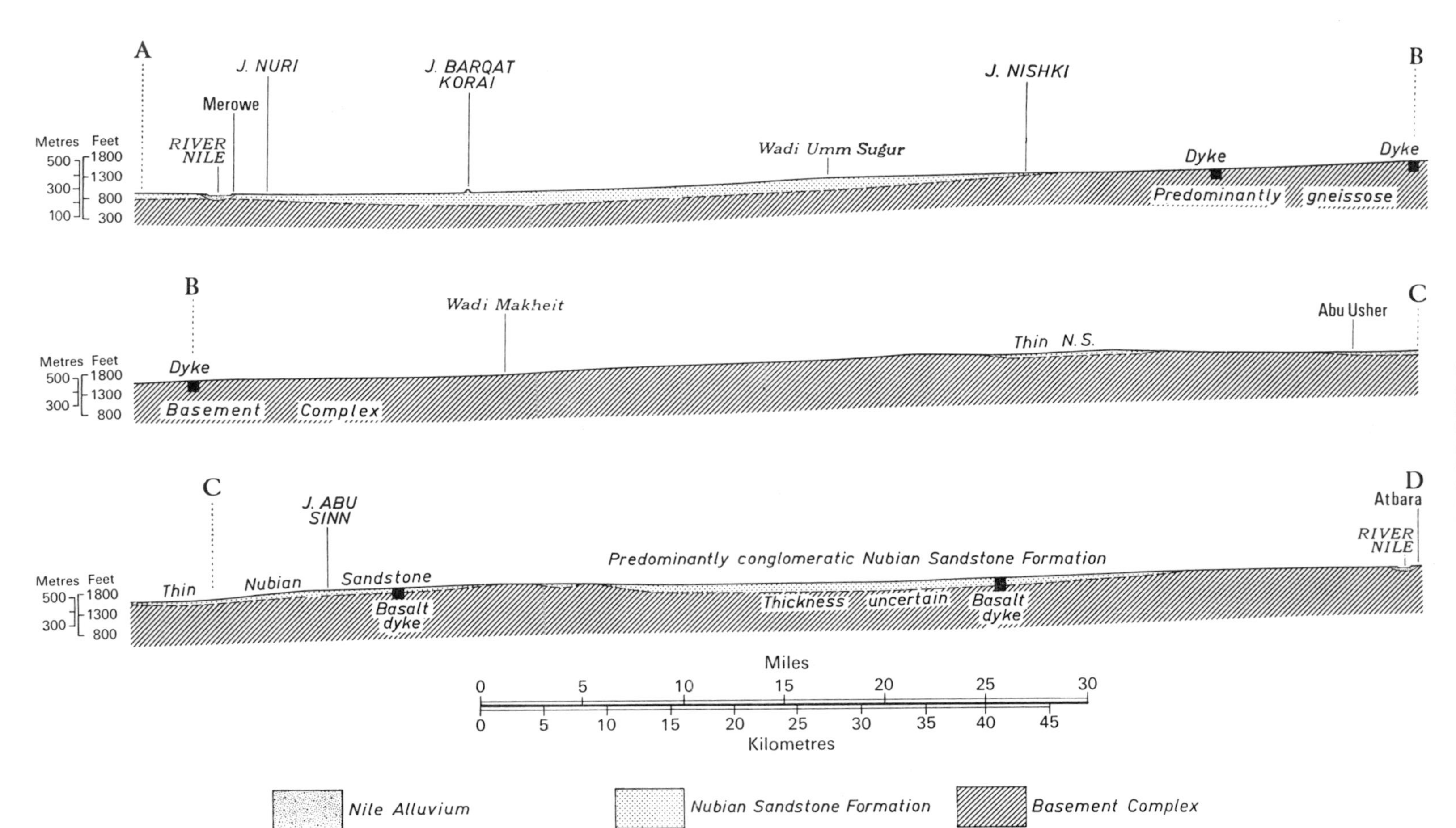

FIG. 59. Bayuda geological section, Jebel Nuri–Atbara (Data by University of Khartoum geophysics group 1964)

North-west of Dilling lies the Nahud outlier composed of the Nubian Formation. This outcrop is probably synclinal and occupies a saddle between the Nuba uplift and the Wadi el Melik-Sodiri uplift, which brings up Basement Complex rocks (Fig. 4).

To the west the Nubian outcrops, which extend from Suq el Gamal, west of Hamrat es Sheik to the Wadi Howar, are synclinal. The Meidob Hills volcanic rocks occupy the central part of this zone (Fig. 4).

The Nubian outcrops east of the Wadi el Melik–Sodiri axis must be synclinal also. These form the Ed Debba–Khartoum–Sennar lobe. The gently synclinal area, in which Gezira and El Atshan formations were laid down, forms part of this lobe, and the course of the Blue Nile in this section may have been determined by the structure. Intervening between the southern section of the lobe and the Bara–Kosti synclinal zone lies an indeterminate area of Basement rocks, which is probably anticlinal. Another such area must lie east of the Khartoum–Sennar Nubian lobe and west of the Gedaref area in the Butana (Fig. 4).

The sinuous nature of the Nubian–Basement contact in these areas is due to both structural and depositional dips. It is well known that the Nubian Formation buried a surface of considerable relief, in places as much as 1500–2000 ft. This topography is developed on a regional scale (Fig. 32) and accounts for some of the sinuosities and irregularities of outcrops.

North-east of the Ed Debba–Omdurman axis lies another poorly defined gentle structure. This runs from Sennar via Jebel Kuror through the Bayuda volcanic field, which may be part of a separate uplift. Details are again uncertain but the outcrop pattern indicates a fairly well defined but gently dipping south-western limb to the structure and a less clearly defined north-eastern limb. The lobe of Nubian Sandstone extending from the Station No. 6 area on the Wadi Halfa–Khartoum Railway may be fault controlled (Figs. 4 and 22). The gentle anticlinal nature of the Bayuda area is illustrated in Fig. 4. The trend of the axis of this structure is uncertain also because at present only one line of levels has been run across this part of the bend of the Nile (Fig. 59).

According to Gregory (1921) the long straight section of the bend of the Nile from Abu Hamed towards Ed Debba is fault-controlled. No faults have been located in the field, however, and in the author's view the explanation offered by Gregory is incorrect.

Furon (1963, Fig. 3) also postulated fault-control for the south-eastern part of the S-bend of the Nile. He stated that 'the Nile through the Sudan clearly follows NE–SW and NW–SE faults which caused the river's rectilinear route'. This is incorrect. No faults have been found in the field, and in the section Kosti–Khartoum–Atbara the course of the Nile is not fault-controlled; in fact the river transgresses the Ed Debba–Khartoum–Sennar structure and is superimposed (Fig. 4 and Plate 12).

Faults controlling the course of the river have not been recognized in the S-bend of the Nile between Atbara and Ed Debba, and from Dongola to Wadi Halfa the river is superimposed. Superimposition took place from a Nubian or younger cover.

Faults have been described by de Heinzelin and Paepe (1965) in the Nile Valley between the Second Cataract and the Egyptian frontier. Block fault-control of the course of the Nile below the Second Cataract is advocated (de Heinzelin and Paepe 1965, Map 4). However, fault-control of many of the physiographic features as shown on this map cannot be proved in the field. In the author's view no major faults affect the Nubian Formation. The formation is well jointed and minor faults with only a few feet displacement occur in the Wadi Halfa region. Certainly the contact between the Nubian and Basement Complex formations in the Abu Sir region of the Second Cataract is unconformable (Plate 5); it is not a faulted one. De Heinzelin and Paepe (1965) cited as evidence of faulting the fact that the Nubian Formation crops out at a lower elevation than the Pre-Cambrian rocks of the Second Cataract region. A better explanation of this is that the pre-Cambrian rocks form part of an undulating topography and that the relationship of the Nubian and pre-Cambrian formations are strongly nonconformable.

The sub-parallel courses of the Atbara below Khashm el Girba, the Rahad, the Dinder in the Sudan, and the Blue Nile all point, in the author's view, to some overall structural control. The Blue Nile follows a course along the Khartoum–Sennar synclinal axis for many miles, and the courses of the Rahad and Dinder are sub-parallel. As the Atbara River flows across largely clay-covered Basement

Complex formations below Khashm el Girba it is not possible to decide whether its course is structurally controlled or not, but the rough parallelism of the courses of these rivers points perhaps to a general structural control.

3. RIFT STRUCTURES

The African Rift System affects three regions in the Sudan (Whiteman 1968*a*, *b*):

(a) the region around Nimule and Juba on the Bahr el Jebel;
(b) the Sudan–Kenya frontier zone immediately west of Lake Rudolf; and
(c) the Jubal el Bahr el Ahmer, the Red Sea Littoral, and off-shore belt of the Red Sea.

(a) Nimule–Juba and the Sudan–Uganda–Congo frontier region

No rift system faults have been mapped in the Sudan (Andrew, Auden, and Wright 1955) but immediately south of the frontier such structures have been mapped by officers of the Uganda Survey (Almond 1962; Hepworth 1964; and Uganda Geological Map 1961, scale 1:1 250 000).

The fault pattern between the north end of Lake Albert and the Sudan frontier is irregular. Faults, instead of paralleling the sedimentary trough as they do in the Lake Albert area, cut across the contact between the Nile sediments and the Basement Complex until at Rhino camp they are again parallel to the sedimentary trough and trend at 025°. East of Yumbe the boundary faults change direction to 070° and continue towards Nimule where they end abruptly against the Assua Mylonite zone. The sediments filling the trough thin down and pinch out at Nimule also. The southern boundary is unconformable and apparently unfaulted.

The gravity pattern in northern Uganda is irregular and areas of closed anomalies straddle the frontier zone. The long linear anomaly of the Lake Albert depression dies out northwards and a smaller closed anomaly occurs around Rhino camp. These anomalies are clearly related to the sediments filling in the down-faulted and down-warped zone. The Rhino camp anomaly dies out towards Nimule. A closed anomaly of similar size (−130 mgal) to that at Rhino camp occurs in Congo, south of Yei, and straddles the frontier. No gravity work has been done in the southern Sudan.

Hepworth (1964) described the following stages in the evolution of the Albert depression in West Nile Province.

STAGE 1 'P III Peneplane' extended across northern Uganda into Congo towards Atlantic. The watershed was situated in Kenya and rivers flowed east to the Indian Ocean and west to the Atlantic. P III was a mature surface with residuals of P II and lateritic mantle.

STAGE 2 The Rift Depression was initiated by down-warping in the west and faulting took place in the east along north-east-trending faults. The Rhino camp fault and Pakwach depression formed.

The 'Rise to the Rift' developed 20 miles east of the Lake Albert depression with the reversal of the Kafu River (and other westward flowing streams) towards Lake Victoria and Lake Kioga, leaving several short streams to flow into the new depression.

STAGES 3 and 4 The western boundary fault began to operate and the primitive plateau was raised above the plain in a series of abrupt movements interrupted by quiet periods during which erosion surfaces developed related to the base level of the Rift Floor and presumably the early Albert Nile which captured the east–west flowing rivers.

A combination of tilt and north-easterly strike caused the river to flow northwards.

During Stages 1–4 the Primitive Rift Depression was formed. This was some 20 miles wide with its western side some 1000–2000 ft higher than the eastern side, which remained at the original height of the P III Peneplain.

STAGE 5 Long static phase during which the Red Plain was able to mature at a constant elevation of 3400–3200 ft; about 10 miles by 50 miles in size.

To the east lay a fairly shallow lake about 10 miles wide which covered the rest of the floor of the rift.

STAGE 6 Renewal of sinking along the eastern faults with deposition of gravels on the third erosion surface below the Red Plain forming beach deposits along the shores of a lake in which clays and diatomites were laid down.

STAGE 7 PLIOCENE, LOWER, AND MIDDLE PLEISTOCENE
Second period of stability during which the Rift Lake filled with sediments and became dry land. Infraformational laterites were formed at least three times. The lake then filled with sediments to the 2350-ft level and shore lines, stacks, and bluffs were cut.

Eventually the lake swamps dried up and the deposits lateritized. Kaiso and Kisegi Beds of Lower and Middle Pleistocene age and Epi-Kaiso Beds.

STAGE 8 Lake rift faults formed in the south-eastern part of the area running in north-easterly direction; the western faults generally downthrow to east.

Reduction in size of lake with levels at 2350, 2250, and 2150 ft. Development of lakes related to breaking down of rock barrier at Nimule.

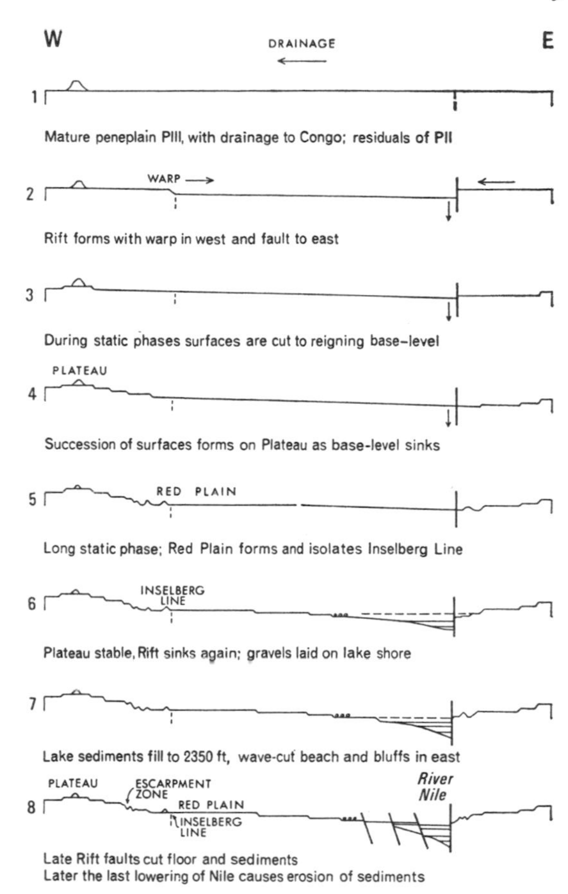

FIG. 60. Albert depression, West Nile Province, Uganda, Sections showing evolution of rift structures according to Hepworth (1964)

This sequence of events is shown in Fig. 60. The youthfulness of the depression and the late date at which the Albert Nile and Bahr el Jebel head waters were connected are the most important points arising from this analysis (Berry and Whiteman 1968).

Earthquakes are common in the southern Sudan and since 1960 at least four tremors have been experienced in the following places (*Sudan Morning News*, 10 October 1966; information from meteorological service): Juba, 20 May 1960; Juba, 4 January 1961; Kaya–Yei area, ? August 1962; Yambio, 3 March 1966. Rejaf Mountain, south of Juba, is known locally as the 'Trembling Mountain' because of the frequent earth tremors felt in the vicinity.

Little can be said about Basement control of the Western Rift fault pattern. Unfortunately there is no information available concerning regional folding and lineation along the Nile–Congo divide except in the area mapped by Hepworth (1964). He reported Kibalian cataclastics and mylonite zones west of the Twol fault and Lake Albert depression and a rough parallelism between these zones and the Rift structures. In places the Rift faults follow the grain of the Basement Complex formations and in others they break across it, like a rough break made across the grain of a piece of wood.

The Rift structures end abruptly at Nimule, and apparently there is no connection between the Eastern Rift zone of Kenya and the Albertine Rift structures. Cahen and Snelling (1966) relate rift and Katangan Basement structures (see above).

On the east bank of the Bahr el Jebel parallelism of the Imatong, Dongotona, Lafit, and Didinga ranges is particularly noticeable. These ranges strike out into the plains of southern Sudan in a way that suggests fault control. Unfortunately little is known of the geology of these high ranges (Mt. Kinyeti attains ±10 200 ft), but on the Uganda side of the frontier Almond (personal communication

1965, unpublished structural map of north-western Uganda) has not shown any major faults. The Basement is divided into acid and pyroxene-gneisses, charnockites, and dioritic masses (Geological Map of Uganda, 1961). The lineation trends in general in a west-north-westerly–north-westerly direction, roughly paralleling the Assua Mylonite zone; it swings to a north-westerly direction north of Kitgum Matidi, and to a more northerly direction in Karamoja adjacent to the Kenya–Ethiopian frontier.

Unfortunately information about the denudation chronology is scanty for this area and so it cannot be used to help us date uplift, etc. L. C. King (1962, Fig. 162) plotted Gondwanaland (Jurassic), post-Gondwanaland (Cretaceous), African (early Cenezoic) and post-African (late Cenozoic) erosion surfaces along the Sudan–Uganda–Kenya–Ethiopian frontier zone, but in the author's opinion these designations cannot be accepted in the absence of a geological map, of contours, and in many areas of even rudimentary spot heights. It is not clear whether we are dealing with unwarped or upwarped surfaces (Berry and Whiteman 1968).

(b) Sudan–Kenya frontier zone

Immediately west of Lake Rudolf, Baker (1965) noted as a result of mapping and photo-interpretation that the greater part of Turkana, from the Uganda escarpment and the Turkana monocline in the west to the line of Thomson's Falls (south-east end of Lake Rudolf)–Lake Stephanie in the east, must be included in the Rift zone. The Rift zone then in the latitude of Lake Stephanie is about 180 miles wide compared with 30 miles to the south. This widening is part of a general divergence and has given rise to 'a series of broad stepped platforms with zones of close horst and trough faulting' (Baker 1965). The faults obviously continue into the Sudan and no doubt they affect the lavas of the Boma and Akobo plateaux. No details are available, however, and the only geological map of the region available does not show faults (Andrew *et al.* 1955). Whether the western edge of the Ethiopian plateau is fault-bounded or not is unproved at present.

(c) Jubal el Bahr el Ahmer, Red Sea Littoral, and off-shore belt of the Red Sea

An account of the stratigraphy of these areas is given in Chapters III and IV (Stratigraphy and Palaeontology). It is summarized in Tables 7, 8, and 16. The areas are situated athwart the western portion of the gigantic Afro-Arabian swell or anticlise, the axis of which runs through the Mohammed Qôl area on the Sudan side of the Red Sea and through Jeddah on the Arabian side. The central part of the swell is trough-faulted and forms part of the Red Sea Depression. The evolution of this structure is discussed elsewhere (Whiteman 1968).

(i) *Jubal el Bahr el Ahmer*

The Basement Complex structures have been described above and little more can be said concerning the younger structures. Delany (1954), Gass (1955), Ruxton (1956), Kabesh (1962), etc. recorded outcrops of Nubian Formation in various parts of the region and most of these are fault-bounded (Fig. 34). Few of the outcrops have been mapped and the most that can be said at the moment is that the faults are post-Nubian (i.e. post-early Cretaceous).

Ruxton (1956) mentioned that the region south-west of Jebel Erba is a mosaic of tilted fault blocks and that in places breccias as much as 200 ft wide are known. The age of the faults is uncertain and many of them can only be dated as post-Basement Complex. Sestini (1965, Fig. 10) has plotted many photolinear trends in the Basement formations but which of these are faults and which are major joints was not determined in the field.

Erosion surfaces are well developed in the Jubal el Bahr Ahmer but very little is known about their age and they cannot be used to help in elucidating structure, except in a general way.

Gass (1955, p. 82) stated that the faults that downthrow the Nubian Formation on the Dungunab sheet are generally aligned north–south and in most cases downthrow to the west, which is the opposite of the majority of faults in the Red Sea Depression. Gass does not think that these faults are connected with the Red Sea faults. In the author's view, however, they were probably formed during the uplift of the Afro-Arabian swell. According to Gass (1955) vulcanicity occurred in the area in the Tertiary. Soda-rich volcanic rocks, predominantly rhyolites, of the Asotriba Formation are of this age according to Gass, but Gabert, Ruxton, and Venzlaff (1960) classed these rocks as Pre-Cambrian. Gass, however, pointed out that soda-rich rhyolites of comparable age are common in neighbouring Saudi Arabia,

Table 19

Summary chart of stratigraphy and tectonic history, Pre-Cambrian to Palaeocene, of Gulf of Suez and northern Red Sea (Based on Said (1962) and various Egyptian Geological Survey publications)

Formation	Description	Conditions of deposition	Tectonic history	Time scale	
Esna Shale 15-40m	Shales and thin chalky limestone. Axis of deposition along Sinai littoral. Age: Palaeocene to ? Maestrichtian	Marine →	Active block faulting.	Palaeocene	
Chalk ±250m	Chalky limestones and chalk absent from some blocks. Senonian deposits occur as far south as 26°N	Marine →	Active block faulting Variable pattern of sedimentation basins with axes bending NE–SW developed at north and of gulf	Dan Maes Sen	Cretaceous
Santonian –Turonian formations 0–436m	Sandstones, shales with limestone and marls predominant in lower part. Axis of deposition of trough trends N–W along Sinai littoral. Not present in some blocks	Marine →	Deposited in submarine trough which over-floods E and W shoulders	San Tur	
Cenomanian formations 0–277m	Shales, gypseous limestones with abundant fossils. Predominantly calcareous sedimentation	Regional transgression →	Gulf is recognizable block faulting isopachs show trough	Ceno	
Nubian formations ±500m	Sandstones and shales of Nubian facies derived from Arabo–Nubian shield	Regional regression ←	Isopachs show trough. Shallow-water conditions, marine in the north	Lower	
Jurassic formations	No details for Gulf of Suez Marine sandstones and limestones at Khashm El Galala Sandstones of Khashm El Galala. Fluvio-marine conditions at entrance to gulf	Regression ← Marine → Marine →	Subsidence localized in shallow arm of sea, maximum extent in Middle Jurassic	U M L	Jurassic
Triassic formations	Marine transgression in Sinai	Marine transgression →		Triassic	
Permian formations	Sandstones of limited distribution			Permian	
Ataqa formations	Sandstones with *Lepidodendron sigillaria* etc.	Regression ←	Subsidence and formation of sedimentary trough parallel to Suez trend	Carboniferous	
Umm Bogma formation	Sea extended as far south as Hurgada; abundant marine fossils	Transgression →			
	Sandstones of uncertain age		Early history	Dev–Cam.	
Basement complex			Unknown	Pre–Camb.	

Table 20

Summary chart of stratigraphy and tectonic history, Palaeocene to Pleistocene, of Gulf of Suez and northern Red Sea (Sources as for Table 19)

Formation	Description	Conditions of deposition	Tectonic history	Time-scale
Pleistocene formations 1500m	Gravels, fan deposits, emerged reefs, raised beaches E. side Plio-Pleistocene gravels 15 000m in Belayim wells W. side Plio-Pleistocene gravels thinner c.240m		Subsidence and Pleistocene eustatic fluctuations caused temporary swamping of isthmus with faunal mixing	Pleistocene
Tilting, faulting, and regional unconformity				
Pliocene formations (no thickness data)	Marine facies S. of Jebel Zeit Oolite with *Clypeaster* and *Schizaster* — Gravels and grits with anhydrite and *Pecten vasseli, P. fischeri, Ostrea gryphoides* Marls and limestones with Red Sea shells — Indian Ocean shells proved as far north as 28° 30′N	Terrestrial facies N. of Jebel Zeit Gravels, grits, with anhydrite bands; occasional bands of *O. gryphoides* indicating transgression from south, active subsidence in fault basins, and formation of Isthmus of Suez		Pliocene
Subsidence with influx of Indian Ocean				
Lithothamnion Limestone ±130m	Calcareous marls rich in algae, sometimes reefal; thins towards north. Some fresh-water deposits with *Planorbis, Bulimus* etc.	← Lagoonal to fresh-water	Area has a tendency to stand high	Miocene: Upper
Evaporite Formation ±595m	Jebel Hamman Faraun 595 m Gypsum with calcareous marls 70 Shells with *Brissopsis* and *Lucina* 55 Anhydite, massive 170 marls with thin gypsum and *Echinocyanus, A. lessoni, Aequipecten seniensis* 120 Massive Gypsum 180 (Tawila no 2 > 3360m)	← Lagoonal Marine → L ⇄ M L ⇄ M ← Lagoonal	Subsidence in lagoon and arm of sea behind a well-developed sill at Ayun Musa; block faulting	Miocene: Middle
Globigerina Marls > 490m	Upper Marls crowded with *Globigerina* interbedded with Anhydrite contans *Carcharodon* and Anhydrite Modules *Atruria aturi* Bed	Marine → L ⇄ M	Sill at Ayun Musa. Deposition in actively subsiding fault basins	Miocene: Middle
	Lower Calcareous sandstone and gypsum shales (*G. Maris* attain 2920m on Sinai side)		Deposition in actively subsiding fault basins	Miocene: Middle
Basal Sands and Conglomerates ± 220m	Boulder beds including granitic and metamorphic types from Red Sea hills, flints and basalts from Oligocene flows; sill formed at Ayun Musa	Regional transgression →	Deposition in actively subsiding fault basins	Miocene: Lower
Regional unconformity, major rifting, and volcanicity				
Oligocene formations (no thickness data)	Extrusion and intrusions of lava flows, sills, sheets, dykes, etc. Mainly olivine-basalts and diabase Fluviatile sands and gravels; abundant silicified wood. Faulting mainly N-W but some N-S and N-E	← Regional regression Fluviatile conditions	Rift movement at end of Oligocene produced modern shape of gulf. N W trend predominated period of volcanicity	Oligocene
Gypseous Marls ± 125m	Sandstones and marls some gypsums and with bands of salt	Shallow water and lagoonal	Uplift and regression	Eocene: U
Upper Lutetian ± 400m	Grey-white chalky limestone	Marine →	Subsidence of E. and W. blocks	Eocene: M
Lower Lutetian ± 600m	Limestone, chalky crystalline, limited to western side of gulf, which acted as submerged half-graben	Regional transgression →	Active subsidence on west side of gulf — Faulting and uplift on east side in Tih and Egma blocks	Eocene: M
Chalky Limestone 39-212m	Chalky limestone with flints unit missing from some blocks. Axis of deposition along Sinai littoral	Marine →	Active block faulting; non-deposition in some areas at N. end of gulf	Eocene: L
Esna Shale 15-40m	Shales and thin chalky limestones. Axis of deposition along Sinai Littoral. Age: Paleocene to ?Maestrichtian	Marine →	Active block faulting	Palaeocene

Yemen, Ethiopia, etc. No information is given concerning the connection between volcanicity and faulting.

(ii) *Red Sea Littoral*

The Mesozoic, Tertiary, and Quaternary rocks of this region have been described above and stratigraphical information is summarized in Tables 3, 7, 8, 16, 19, and 20.

Both faults and folds occur on the Red Sea Littoral and details of these are given below. The broader aspects of Rift structure such as the origin of the great escarpment of the Jubal el Bahr el Ahmer and the Red Sea Depression are discussed separately.

Khor el Mar'ob structure. Khor el Mar'ob is situated at 21°49′N, 36°48′E and is the most northerly structure for which we have information on the Red Sea Littoral of the Sudan. Willis *et al.* (1961) recorded dip reversals in the coastal sequence, indicating the presence of a pseudo-anticline similar to the Shinab structure described below. Details were not given however.

Mersa Shinab structure. This faulted anticline runs north-west–south-east closely paralleling the coast for 14 miles, from 3 miles south-east of Mersa Shinab (Fig. 38) to a point just south of Mersa Abu Imama. The structure is interrupted by a cross fault north-west of Mersa Halaka, and south of this point there is no evidence of westerly dip for 2 miles (Willis *et al.* 1961). The structure continues southwards and Carella and Scarpa (1962) have divided it into two: the Jebel Abu Imama and Jebel Dyiba anticlines. The axis of the structure is sinuous and forks at its northern end. An easterly-trending nose extends from near its centre to just south of Mersa Halaka. The flanks of the structure are broken by normal faults and the strike is erratic. Relatively steep dips (15–30°) occur on the eastern flank but gentler dips (10–15°) occur on the western limb (Willis *et al.* 1961, Carella and Scarpa 1962, and personal observations).

The oldest formation that crops out in the Jebel Abu Imama area is the Maghersum Formation, consisting mainly of gypsum. Basalts overlain by the Maghersum Formation are exposed south of Mersa Shinab (Carella and Scarpa 1962). The basalts are Tertiary according to these authors but, as noted above, a potassium–argon date of late Jurassic–early Cretaceous has been obtained for them by the author. The rocks in which the basalts occur should not therefore be classed as Maghersum Formation.

South of the Jebel Dyiba structure at Jebel Hamamit there is an eroded faulted anticline some 2–6 km wide (Carella and Scarpa 1962). Red sandstones of the Hamamit Formation are the oldest rocks exposed. North–south and north-east trending faults occur. Willis *et al.* (1962) ascribed the origin of the Shinab structure largely to slumping resulting from the solution of certain members of the evaporite sequence (Fig. 39). Clearly the heads of Marsas Abu Imama, Halaka, and Shinab are solution features, and it is suggested that solution took place, probably along a tilted salt bed striking roughly parallel to the coast. Solution progressed down-dip below the insoluble cover, causing it to subside into a pseudo-syncline and producing a pseudo-anticline along the zone between dissolved and undissolved portions of the soluble bed. The author agrees with Carella and Scarpa (1962), who think that the Shinab structure is primarily of tectonic origin. The superficial structures developed subsequently.

Carella and Scarpa (1962), basing their conclusions mainly on geophysical subsurface and submarine data, limited the Shinab structure to the east by a normal fault which they have drawn along the steep submarine escarpment. The position of the western fault is shown in Fig. 38 and is indicated by the steepening of the Bouguer contours near the heads of the marsas and by the seismic refraction data. The intervening anticlinal horst is thought to run from the Abu Imama area to Abu Shagara (Fig. 61). The westernmost fault trifurcates north of Jebel Hamamit and the outermost and central faults form the eastern and western boundary fault of Dungunab Bay, a synclinal zone situated between the Abu Shagara anticlinal horst zone coincident with a rise in the Basement Complex and the edge of the Tertiary and Mesozoic rocks. These faults trend mainly at 325°. North-east–south-west- and north–south-trending structures occur but these are less common.

Faults between Ras Abu Shagara and Marsa Salak. Ras Abu Shagara and Maghersum Island are said to be separated by a fault zone which trends from the southern tip of Ras Abu Shagara to a point just north of Mohammed Qôl, where it probably continues along the khor, which drains down to Mohammed Qôl (Willis *et al.* 1961, Carella and

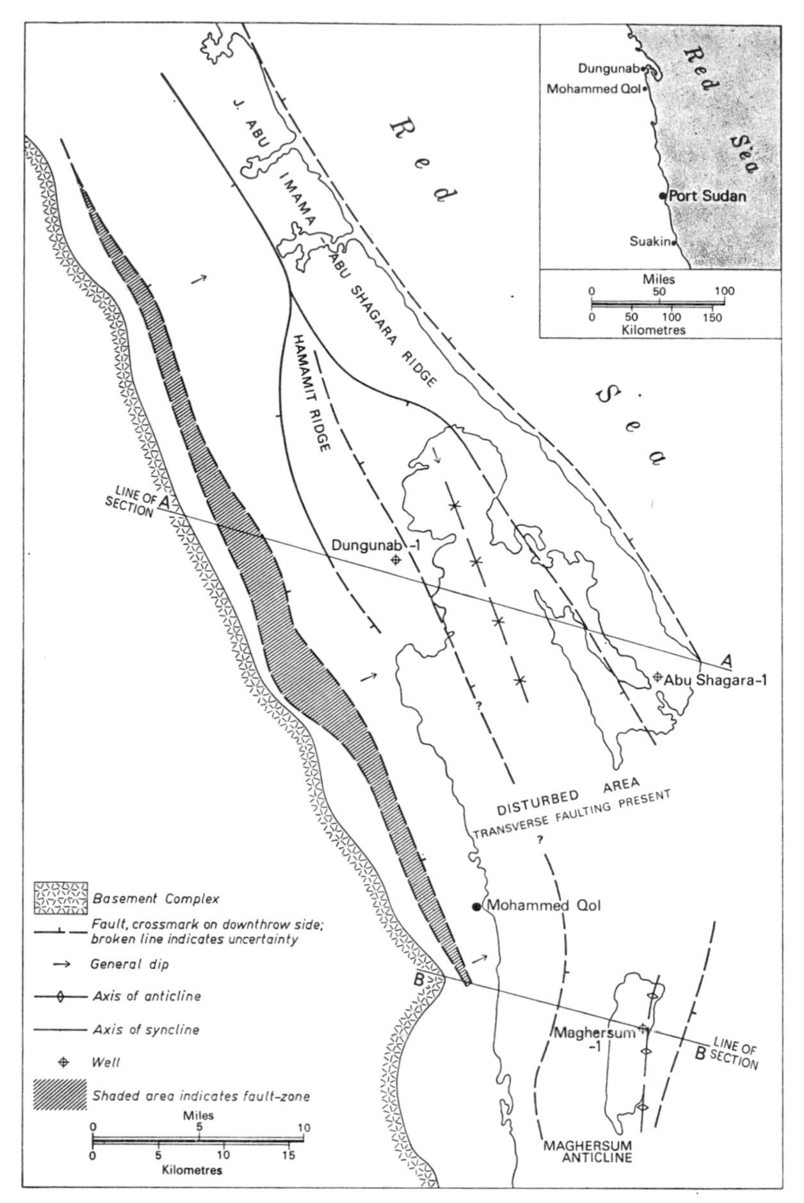

FIG. 61. Dungunab–Maghersum area, Red Sea Littoral, structural sketch map (Based on Carella and Scarpa 1962)

Scarpa 1962, personal observations). Conspicuous magnetic and gravity anomalies occur along the line of this fault. Carella and Scarpa (1962) think that it is a transverse fault, but their map does not show any displacement of the western boundary fault of Dungunab Bay. Willis *et al.* (1961) have shown a right-handed shift of the synclinal axis of Dungunab Bay and the synclinal axis west of Mukawwar Island. They do not show a shift of the Tertiary-basement contact, however.

Sestini (1965) has shown a fault a little to the south of the above-mentioned structure which displaces the Tertiary–Basement contact by about 6 miles.

Willis *et al.* (1961) have also plotted three faults in this zone trending at 065°, and the southernmost fault shifts the Tertiary–Basement contact by at least 5 miles. The movement appears to be dextral. These faults have not been mapped on the ground, outcrop patterns having been determined from aerial photographs. Clearly they are worthy of careful investigation because if they are truly tear faults they alter the currently accepted structural style for the central Red Sea area.

An east–west trending fault is thought to run through Marsa Salak (Willis *et al.* 1961), and Sestini (1965) has marked a 'morphological alignment' in this area. It is not possible because of lack of photographic coverage to be certain whether the Tertiary–Basement contact has been displaced, but the occurrence of Basement Complex due north of the outcrop Abu Imama and Maghersum outcrops north of Khor Ballobab may well indicate a small displacement.

No folds have been mapped on the mainland from Mohammed Qôl to Marsa Salak, but off shore Willis *et al.* (1961) and Carella and Scarpa (1962) have mapped the Mukawwar Island anticline and the Dungunab Bay syncline.

Mukawwar Island anticline. South of Ras Abu Shagara the regional trend of the Jebel Abu Imama–Abu Shagara structure changes from north-west–south-east to the north–south Mukawwar Island trend. This probably takes place at the Ras Abu Shagara–Mohammed Qôl fault zone.

Mukawwar Island lies about 5 miles off the coast, east of Khor Inkeifal. It is 7 miles long and 1½ miles wide and consists of the west flank and the south and north plunges of a true closed anticline. The axis runs from north to south and the beds dip mainly from 10 to 15°. The oldest beds at outcrop are mudstones and evaporites of the Maghersum Formation. The eastern flank of the anticline is hidden by the sea but seismic data indicate that a normal fault about 2 km from the axis with downthrow to the east bounds the structure on the east. At the northern end of the anticline there are a series of parallel faults striking north-west around the plunge (Figs. 61 and 62). Willis *et al.* (1961) believe that this is a tectonic structure, not a superficial structure like the Jebel Dyiba and Abu Imama anticlines. They have postulated that a right-handed lateral displacement along a major lineament, which is supposed to run off shore parallel to the coast from off Ras Abu Shagara to Suakin and thence up the Baraka Valley, produced the Mukawwar anticline. In the author's view this explanation is unacceptable. First the stress pattern is incorrect, and secondly AGIP subsurface and seismic data indicate clearly that the Basement Complex is domed and that the folding is related to the faulting. Most probably the Mukawwar fold is a drape structure produced as a result of block faulting and uplift in the Basement Complex.

The structural interpretation for Dungunab Bay and Mukawwar presented by Berry, Whiteman, and Bell (1966, Fig. 2 (*a*)) should be considered out-dated. It was based on topographic data from the British Admiralty charts and a limited amount of field data.

Jebel Eit–To Banam anticline. Sestini (1965) described this structure as a symmetrical anticline with a north-north-east–south-south-west axis extending from Eit Well to Jebel To Banam. On the map, however (Sestini 1965, Fig. 4), the axis north of Eit Well is shown as almost meridional, swinging to 350° south-west of Jebel To Banam and then to approximately 220° on Jebel To Banam. Dips vary from 5 to 30° along the structure.

In the Eit jebels a system of normal faults trending at 010° (given as north-west–south-east by Sestini, 1965, p. 1468) repeat the Miocene Maghersum and Abu Imama formations. These faults have throws of about 160 ft. The gravels (Sestini's Upper Clastic Group) that occur east of Jebel Eit are tilted and appear to be faulted, so indicating that faulting continued into Pleistocene times (Table 16). However, the regional continuity of marine benches notched in the Pleistocene Monastirian Elevated Reefs of the Littoral indicates stability in the latter part of the Pleistocene (Berry, Whiteman, and Bell 1966).

At Jebel Saghum (Fig. 38), north of To Banam, the Miocene Maghersum and Abu Imama are broken into small blocks, and small homoclines dip towards the east (Sestini 1965). Collapse structures such as those described by Willis *et al.* (1961) occur at Jebel Saghum.

There are few outcrops of Tertiary rocks south of Khor Eit, and as far as the Eritrean frontier much of the area is obscured by Pleistocene reef complex deposits near the coast and Pleistocene and late Tertiary gravels, etc. inland.

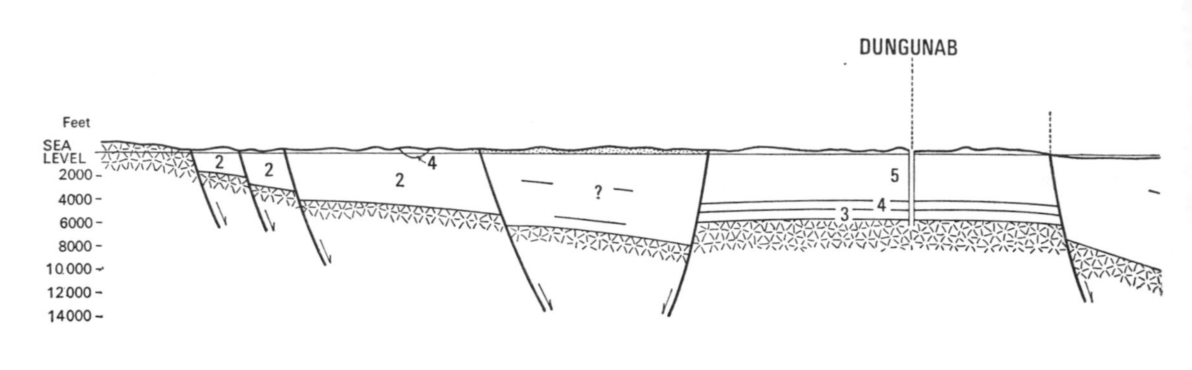

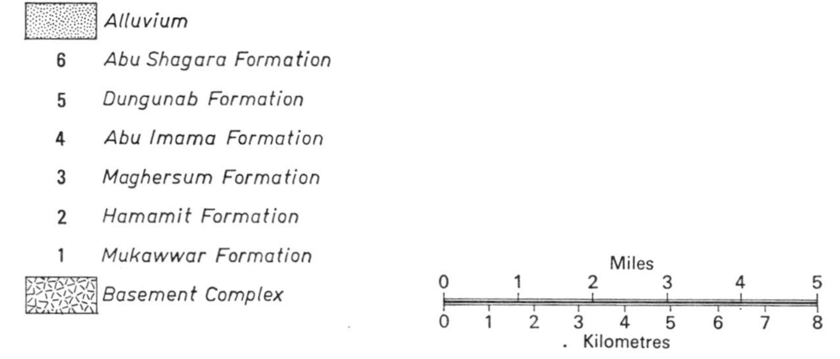

FIG. 62. Dungunab–Maghersum area, geological sections (Based on Carella and Scarpa 1962)

(iii) *Off-shore belt: structural interpretation of submarine near-shore topography*

The near-shore platforms are cut by the submarine extensions of the marsas or havens in a number of places. Crossland (1913) believed that the marsas had been caused by faulting: 'the harbours and other fissures in the coral limestone, etc. of both coastal and barrier reefs are due to some secondary faulting' (Crossland 1913, p. 115). This idea is unacceptable; the marsas and their seaward extensions are clearly erosional forms, the different shapes having developed on the different rock-types involved. The incised valley courses were cut in response to the Würm (and earlier?) low sea-levels and the valleys flooded by the subsequent post-Glacial rise in sea-level (Berry, Whiteman, and Bell 1966). At Khor Shinab the following sequence of events is postulated:

1. Superimposition of Khor Shinab from gravels and wadi deposits on to the Tertiary beds and Pleistocene Monastirian emerged reef.
2. Excavation of the deep valley in response to the lowering of base level during the Würm–Wisconsin emergence.
3. Excavation along the strike by solution and erosion of the bedded gypsum deposits which dip seaward at 15–20° at this locality.
4. Submergence and recolonization by fringing reefs and partial infilling of the head with lagoonal deposits.

Several platforms, bounded by steep slopes in some cases, occur in the off-shore belt. Paucity of data on the charts of the British Admiralty Hydrographic Department and vigorous Recent coral

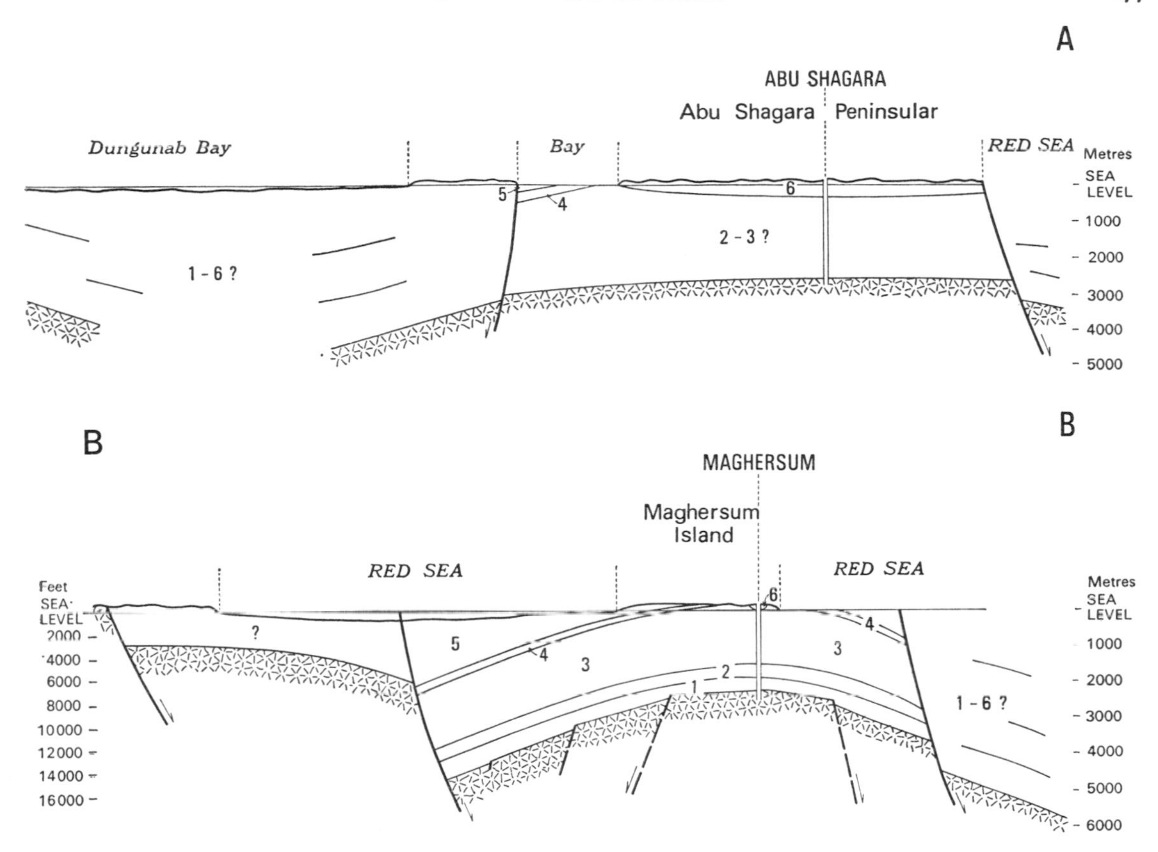

growth make it difficult to recognize minor details such as benches. Nevertheless, some generalizations and useful comments on structure may be made.

The largest area enclosed by the 50 fathom line along the Red Sea coast is the Suakin Archipelago. This is an area of shallow water with many coral inlets, reefs, and banks (Fig. 81). Outside this area to the east and north-east the reef and banks rise from water 200–400 fathoms deep. The eastern side of the archipelago is fairly steep and straight, running at approximately 330°. North of Tamarshiga Island, the margin of the archipelago swings north-west and west towards the mainland in a series of scalloped edges so that just off Suakin there is over 200 fathoms of water.

The form of the Suakin Archipelago suggests block faulting, with major faults following the eastern boundary at 330° and minor faults controlling the northern and north-eastern edges. There is no evidence of the age of these faults except that AGIP's well, Durwara-2, is reported to have passed through 300 m of limestones on Durwara Island off Trinkitat. This suggests a considerable period of calcareous deposition. Reef growth has spread over the area, producing the present topography, which has resulted from a balance between sedimentation and subsidence.

A major feature of the Sudan coast-line is the boat channel, a depression that is often over 40 fathoms deep and separates the barrier reef from the shore. Large sections of this boat channel are structurally controlled. The channel south of Suakin is over 40 fathoms deep but it becomes shallower and less definite towards the Baraka delta.

North of the Suakin archipelago the Towartit reefs form an extensive bank, rising from 300 to 400 fathoms on the seaward side and with a well-developed boat channel 40–44 fathoms deep on the landward side (Fig. 81). These reefs are again probably fault-bounded. North-west of the Towartit reefs deep water occurs close inshore, and the shape of this embayment again suggests fault control.

North-east of Port Sudan the 50- and 100-fathom lines form a large embayment extending northwards west of Wingate reef. The depression continues northwards off the coast as far as Marsa Fiji, being delimited on the east by barrier reefs. At Marsas Fiji and Arus deep water again occurs close inshore and topography is complex. Barrier reefs, separated by a boat channel with a depth of more than 30–35 fathoms and some soundings greater than 208 fathoms, occupy the section from Marsa Fija to Marsa Salak; again this section is probably structurally controlled. Solution of Miocene evaporites may also have taken place.

Northwards from Marsa Salak the 50- to 100-fathom lines swing off shore to enclose Mukawwar Island and Ras Abu Shagara. Deeper water of 300–400 fathoms lies east of this line and the marked change in direction of the 100-fathom line and the bottom topography indicate fault control. Details of the structure of this region are shown in Fig. 62.

From Ras Abu Shagara to Haleib the coast swings northward in a gentle curve. Fringing and barrier reefs are developed close inshore. Admiralty charts provide little data in this area but once more there is an indication of general structural control of the coast and reef pattern.

To summarize for the Red Sea Littoral and off-shore belt: folds are of minor importance and faults with north-west–south-east trends are dominant, if we are to judge from the exposures between Khor Eit and Khor Mar'ob. Some east–west and north-east–south-west trending faults occur and some of these might be right-handed tear faults. Confirmation in the field is needed of this last point because if this interpretation is correct then the accepted structural style for the central Red Sea must be changed. No major individual faults are known, however, but combined throws in place total 3200–4800 ft (1000–1500 m).

Anticlines roughly parallel the dominant fault-trend and are mainly drape folds, developed in response to block movements in the Basement Complex. Superficial structures associated with solution of evaporites occur as at Jebel Dyiba–Khor Shinab.

How much the major fold structures are due to tensional or epeirogenic forces is not clear, for regional arching of the whole region occurred along the Afro-Arabian Red Sea axis before and during faulting.

The faults, irrespective of direction, cut the anticlines, and beds as young as the Upper Clastic Formation of Sestini of Plio–Pleistocene age are involved. The littoral seems to have been stable during the late Pleistocene (Berry, Whiteman, and Bell 1966).

Little is known of the off-shore structure except in the Dungunab–Ras Abu Shagara area (Fig. 61) but faulting appears to be dominant.

(iv) *The contact of Tertiary and Mesozoic sediments with Basement Complex*

On many maps and diagrams this contact is shown as a faulted one, for example Drake and Girdler (1964), Holmes (1965), and Girdler (1965), to quote recent examples, and the faults are considered to form part of the boundary fault system of the Red Sea Depression. However, for the greater part of its course in the Sudan, wherever visible, the contact is nonconformable. Auden (1958 *in* Quennell) emphasized that as far as was then known the contact between the Tertiary coastal plain sediments with the Basement is not primarily a faulted one: 'but is rather of the nature of a marginal overlap across an irregular indented topography'. In all the places where the contact has been seen by the writer it is clearly an unconformable one, and as early as 1935 Grabham recognized this.

Sestini (1965, Fig. 10) shows both faulted and unconformable contacts for the Littoral between Dungunab Bay and Suakin. The greater part of this contact is, unconformable. A series of discontinuous faults with north-easterly and north-north-westerly trends are shown in the section west of Port Sudan to a point west of Marsa Salak. Most of these faults were plotted from aerial photographs, etc. and were not checked on the ground and, as Sestini has pointed out, should be considered as tentative. There are a number of inconsistencies in Sestini's interpretation. For example, the Tertiary–Basement contact south-west and west of Jebel

Saghum is shown as faulted on the general map (Sestini 1965, Fig. 10) but unfaulted on the detailed map (Sestini 1965, Fig. 4), and from the outcrop pattern shown on the latter if faults are present they must be drawn east of the positions shown on the general map. Immediately south of this area the Dungunab Formation lies unconformably on the Basement Complex according to Sestini (1965).

Faulted contacts between Tertiary and Basement are shown west of Dungunab Bay by Sestini (1965, Fig. 10). This is based on Carella and Scarpa (1962, Fig. 6, Section 1) who there show a faulted stepped contact (Figs. 38 and 61) based on an interpretation of the Bouguer gravity data. These two authors place the position of the western margin of the fault belt 2 km east of the Tertiary–Basement contact along the line of section 1 (Figs. 38 and 61). As is shown in these diagrams the fault zone does not intersect the Basement rocks. In the author's view, until more evidence is available the contact here should be regarded as unfaulted also.

The most comprehensive information so far available concerning the Tertiary–Basement contact was presented by Willis *et al.* (1961) in three detailed photogeological maps, which were in part checked in the field. The sheets extend, with occasional breaks owing to lack of photo-cover, from Marsa Marob in the north to Marsa Akakiyai in the south. Faulted contacts between Tertiary and Basement rocks are shown in the area south-east of Marsa el Marob, where the formation may be affected by a series of *en echelon* faults, trending approximately 330°. No other faulted contacts are shown, except for the possible tear faults mentioned above, as far south as Marsa Arakiya (Fig. 38).

The overall structural picture that emerges from the foregoing details is that the Red Sea Depression in the Sudan section is not fault-bounded (Whiteman 1965, 1968). A consideration of the stratigraphical, structural, and geophysical evidence leads to the conclusion that the great escarpment of Jubal el Bahr el Ahmer is not a fault scarp, nor a fault-line scarp, as so many geologists have shown on generalized maps and diagrams.

The author's views concerning the stratigraphical and structural history of the area are summarized below. Details are presented in a separate paper on the origin of the Red Sea Depression (Whiteman 1968).

1. The Afro-Arabian swell or anticline is an ancient structure that has been rising probably since late Pre-Cambrian time. The Jebel Hamashaweib, Sudan dykes indicate that tension prevailed in the area in late Pre-Cambrian time. Its axis runs through the Mohammed Qôl area and Abu Hamed on the Sudan side of the Red Sea and Jeddah and Er Riyad on the Arabian side.

2. In the Gulf of Suez area a zone of subsidence developed in Carboniferous times and by ? Jurassic–Cretaceous times this had extended to the central zone, the Red Sea Depression having developed between two marginal monoclinal flexures situated near the present Sudanese and Arabian shores.

3. The swell continued to rise reaching a culmination in late Eocene times. The main Ethiopian plateau lavas were extruded in the Oligocene and the climax of Rift System faulting took place in ? Oligocene and early Miocene times. Many of the faults that cut the Sudan littoral sediments must have been initiated at this time. The separation of the African and Arabian plates started.

4. Subsidence, no doubt accompanied by faulting, continued on the down-side of the Sudan monoclinal zone resulting in the accumulation of more than 14 000 ft of sediments in the Suakin Archipelago. The sediment probably filled the trough from shore to shore (Fig. 82).

5. The escarpment of Jubal el Bahr el Ahmer was initiated by a process of pediplanation operating on the marginal monoclinal flexure, in similar manner to the Lebombo monocline of southern Africa described by L. C. King and others.

6. In response to changing base level, consequent on subsidence, erosion surfaces developed in the Jubal el Bahr el Ahmer.

7. In Miocene times lagoonal conditions apparently prevailed throughout the Red Sea with evaporites being laid down from shore to shore.

8. Intense faulting took place in Pliocene and early Pleistocene time culminating in the formation of the central trough and the intrusion of ultrabasic and basic high-velocity dykes, etc. derived from Upper Mantle described by Drake and Girdler (1964). By this time the Red Sea Depression had attained much of its present shape.

9. In Pliocene times also, the connection with the Mediterranean over the Ayun Musa sill was closed and the Bab el Mandeb opened, making the connection with Indian Ocean.

10. During late Pleistocene time the area was apparently reasonably stable and undeformed marine benches were cut into Monastirian elevated reefs.

The total amount of separation in the Central Red Sea which appears to have taken place since separation of the African and Arabian plates in Late Eocene time is between 130 and 170 km (Tramontini and Davies 1969), and unpublished gravity data compiled by University of Khartoum geophysics group and worked up by Dr. I. R. Qureshi (MS 1970) indicates that plates have separated by a similar amount off Suakin, Sudan.

VI

Ground-water

1. INTRODUCTION

GROUND-WATER is the most precious of Sudan's minerals. Its availability makes it possible for pasturalists and agriculturalists to occupy thousands of square miles of semi-arid and arid land that would be otherwise uninhabitable.

North of a zone running east–west through Abu Hamed, rain rarely falls and the country is mainly desert (Figs. 2 and 3). Away from the river in this zone, wells are frequently the only source of water. Southwards a tropical continental type of climate prevails, except in the extreme south where the climate is best described as continental equatorial. Both these climatic zones, however, have a distinct dry season, which increases in intensity and duration northward. From October to June humidity decreases over much of the country and is as low as 10 per cent at some stations in the northern Sudan. High insolation, and consequently high ground temperatures, combine to give very high evaporation rates. Grabham calculated that the evaporation rate is about 1 cm/day in the dry season in the central Sudan.

To conserve water, haffirs (open earth-dammed reservoirs) have been constructed in many parts of the country, but their effectiveness is greatly reduced by evaporation and erratic distribution of rains. Drinking water is often carried on camels or donkeys, 5, 10, or even 15 miles, and is sent out from town or village boreholes in tanker lorries and sold *en route* at a price that is sometimes above 2 piastres (about 5 pence or 5 U.S. cents) for a 4-gallon tin. In the Red Sea Littoral it has been calculated that at times the shell-fishermen of Dungunab spend something like 30 per cent of their income on buying fresh water, which they mix with brackish well-water. Great numbers of people, therefore, are dependent on ground-water for their livelihood.

It is difficult to estimate the annual amount of water abstracted from the various aquifers in the Sudan but Rodis *et al.* (1964) have estimated that about 600 million gallons are withdrawn annually from the Nubian and Umm Ruwaba formations in Kordofan alone. Abdl Salaam (1966) estimated that 1464 million gallons are abstracted annually in the Gezira from the Nubian and Gezira formations. To place a real value on this water is not easy but if we value it at Kordofan peak dry-season prices (say 2 piastres for 4 gallons) then Gezira and Kordofan well-water is worth about £S 10 000 000 per annum (£S1 = 23*s.* 11*d.*). Another way of estimating the value of water is to charge it at the Central Electricity and Water Board's town rate of 25 millemes per m^3. Some 15 per cent of Khartoum's supplies are obtained from underground resources, and at the CEWA rate this is worth about £S 8 000 000 per annum. Even if we value water at only one-tenth of the town and country values mentioned above, the value of water extracted for the country as a whole must still run into many millions of pounds per annum.

2. HISTORY OF RESEARCH

Among the earliest published references to ground-water geology in the Sudan was the discovery of water at Station No. 4 at a depth of 100 ft some 77 miles from Wadi Halfa and at Station No. 6 some 55 miles further on. According to Churchill (1899) in *The River War* these discoveries accelerated by nearly one month the completion of the Desert Railway. The water here occurs in the Nubian Sandstone at a depth of about 100 ft (Dunn *in* Grabham 1934).

Grabham (1909, 1934, etc.) described the wells of the north-eastern Sudan and general features concerning ground-water. Chemical analyses of a number of desert well-waters were given. Sandford (1935) discussed sources of water in the north-western part of the Sudan. Many of these desert wells are very old and some, as at Naqa, Northern Province, date from Meroitic times.

Ground-water is the subject of a number of Sudan Geological Survey unpublished reports written both before and after independence, by Andrew (1950), Delany (1950, 1951), Karkanis (1950, 1952), Auden (1954), Mansour and Samuel (1957), and Kleinsorge and Zscheked (1958). Other data and well lists are given in annual reports of the Sudan Geological Survey (see Bibliography). Widatalla (1960) gave a general account of the ground-water resources of the Sudan, and Mansour (1963) described in general terms the water-resource problems of Kordofan Province.

Comprehensive articles dealing with the availability of ground-water in Kordofan Province were presented by Rodis, Hassan, and Wahadan (1963, 1964), Rodis and Iskander (1963); and Doxiadis and Partners (1964) gave an account of land and water use in Kordofan. Doxiadis and Partners (1964, 1965, 1966) also repared for the United Nations Special Fund and Food and Agriculture Organization a series of articles on well fields, hafir hydrology, hydrometric records, and the Rahad Lake, etc. in Kordofan.

Abdl Salaam (1966) completed his M.Sc. thesis, University of Khartoum on the ground-water geology of the Gezira, and Kheiralla (1966) completed an M.Sc. thesis on the Nubian Sandstone Formation of the Nile Valley between latitudes 14° N and 17°42′ N (Ed Dueim to Atbara) with special reference to the ground-water geology. Similarly, Farouk Ahmed in his M.Sc. thesis (1968) dealing with the geology of Jebel Qeili igneous complex, Butana, has dealt with the ground-water problems of this region.

In 1966 changes in legislation were introduced by the Sudan Government and a new water authority was proposed for the Sudan. This body, known as the Water Resources and Rural Development Corporation, is to combine the duties and activities of the Land Use and Rural Water Development Department, their drilling division, and the hydrogeological section of the Sudan Geological Survey, and is to concentrate its activities in a 'Campaign against Thirst' between latitudes 10 and 16° N. Boushi and Whiteman (1968) discussed the ground-water geology of the Nubian Formation.

3. GENERAL GROUND-WATER GEOLOGY

The main water-bearing formations of the Sudan in order of importance are:

(a) Nubian Sandstone Formation
(b) Umm Ruwaba Formation
(c) Gezira Formation
(d) Pleistocene and Recent unconsolidated deposits
(e) Sedimentary formations of the Red Sea Littoral
(f) Basement Complex Group

(a) Nubian Sandstone Formation

The lithology of the Nubian Formation has been described in detail above. Briefly it consists of bedded, flat-lying or gently dipping conglomerates, grits, sandstones, sandy mudstones, and mudstones. The following lithological types have been recognized in the Khartoum–Shendi area (Kheiralla 1966) and no doubt similar types occur in other areas.

1. Pebble conglomerates
2. Intraformational conglomerates
3. Merkhiyat sandstones
4. Quartzose sandstones
5. Mudstones

Ground-water is commonly found in the pebble conglomerates and in the Merkhiyat and quartzose sandstones. It is rarely found in mudstones (unless they are fractured) or in intraformational conglomerates.

The thickness of the Nubian Formation is variable but apparently rarely exceeds 1500 ft.

The Nubian Formation is by far the best aquifer in the Sudan and on average, away from the permanent rivers, the water-table lies between 300 and 400 ft below the surface (Abdl Salaam 1966, Kheiralla 1966). Recharge takes place from the Nile and other water-courses and from rainfall, especially in the southern part of the outcrop, where the rainfall is heavier and more constant (Fig. 3). Unfortunately very few data are available on pumping-water levels and yields, and little can be said quantitatively about the effect that water abstracted so far has had on lowering the water-table.

(b) Umm Ruwaba Formation

The Umm Ruwaba Formation consists of unconsolidated sands, sometimes gravelly, clayey sands, and clays. The sediments are poorly sorted and are more than 1000 ft thick in places. Thickness varies from place to place, since the formation rests unconformably on Basement Complex and other rocks. The deposits are thought to be of fluviatile and lacustrine origin and were laid down in a series of interlocking land deltas and fans. 'Yields' are highly variable because of marked lithological variations and the arbitrary way in which they are obtained. In general, however, they are moderate to low, as the sediments are commonly fine-grained. In addition, water is sometimes saline. Water is frequently found between 300 and 400 ft below the surface. Like the Nubian Formation, the Umm Ruwaba aquifer is recharged by rain-water and by water seeping in from wadis and the White Nile.

(c) Gezira Formation

The lithology of the Gezira Formation is described above. Beneath the Gezira Clay the formation is, in general, highly porous and permeable. Ground-water occurs under open aquifer conditions in lens-shaped, largely interconnected sand and gravel bodies. The sands are sometimes clayey and pebbly but generally consist of fine- to coarse-grained clean sand. The best producing horizons are clean gravelly sands. The thickness of saturated strata in the Gezira Formation is between 60 and 80 ft and the depth to the top of the saturated zone varies from 20 ft (7 m) near the banks of the Blue Nile to 148 ft (52 m) in the centre of the Gezira.

Yield data are inadequate for the Gezira as a whole. Apparently only one standard pumping-test has been made and the figures listed as yields have been obtained by 'bailing' the well and are in many cases misleading. In the absence of standard pumping-test data, for all but one well, the best estimate of the producing capacity of wells in the Gezira Formation are obtained from the time taken to fill storage tanks. Using this type of information the Gezira Formation can be said to produce between 500 and 10 000 gal/h. The pumping test on S.G.S. B.H. 1751 Hasaheisa demonstrated, however, that the formation can produce 15 000 gal/h with relatively small draw-down and rapid recovery. In general, the Gezira Formation is a good aquifer, especially near the rivers, but it is of limited extent.

(d) Pleistocene and Recent unconsolidated deposits

The Pleistocene and Recent unconsolidated deposits consist of gravels, sands, sandy clays, silts, and clays laid down in a variety of environments. Here are included wadi deposits, qôz and sands sheets, lake or rahad deposits, river terraces, alluvium and valley fill, desert deltas such as the Gash delta at Kassala, etc. (Plate 10). Their water-bearing capacity is highly variable and chances of sinking successful wells can be assessed only if detailed local knowledge is available. Some large valley-fill deposits, where the deposits may be 150–300 ft thick, may yield on average 5000 gal/h. Sands like the qôz of Kordofan, which vary in thickness from a few feet to over 100 ft, are in general poor producers. They do not usually bear water except locally where interbedded silts have produced perched water-tables.

(e) Sedimentary formations of the Red Sea Littoral

Generalized lithological descriptions of the Cretaceous–Recent deposits of this region are given in Tables 7, 8, and 16. Very little is known about the water potential of these strata. The presence of salt and anhydrite complicates exploration, for the beds are folded and faulted and in many places covered by superficial deposits, especially in the Port Sudan area, where the greatest demand for water lies.

Near the coastline freshwater–sea water interface problems exist and Widatalla (1960) has stated that salinity frequently increases with time, presumably because of overpumping and drawing in of sea-water.

(f) Basement Complex

In general, the Basement Complex Group produces little water except where the rocks are deeply weathered, or are well jointed or fractured. Few boreholes produce from the Basement Complex, but hand-dug wells excavated in weathered zones and in weathering aureoles surrounding some jebels, for example Jebel Moya, produce limited quantities of water of variable quality, some of which is, however, saline.

Run-off collects in places in enlarged joints at the surface in the Basement rocks. Some of these store large quantities of water, which last well into the dry season.

4. GROUND-WATER GEOLOGY, NORTH-WESTERN SUDAN

Sandford (1935) is still the main source of information on this region, and part of the account of this little-known section of the Sahara which follows is based on his work.

Over the greater part of this region rain is extremely rare (Fig. 3), but evidently there is enough for surface run-off to accumulate in fans and wadi deposits, even as far north as the Jebel Uweinat area (22° N). Rainfall increases southwards from the Wadi Howar, which according to Sandford (1935) is the southern border of the true desert in the north-western Sudan.

(a) Jebel Uweinat

Surface water occurs at Ein Dua at the level of the plain that surrounds the jebel. The water, which was reported by Hassanein Bey in the later 1920s, lies among spheroidally weathered granite boulders, and according to Sandford (1935) is derived from rain that falls occasionally and seeps downward through cracks and joints. A similar supply occurs at Ein Zuwaia. At near-by Karkur Murr water is thrown out as a spring, probably along an unconformity in sandstones. It accumulates in gravel and sand in the wadi floor.

Water also occurs in sands and gravels among masses of fallen rock at Karkur Talh, where there is sufficient for *Acacia* (talh) to maintain itself. Water-holes were also recorded by Hassenien Bey at Jebel Archenu, Erdi, and Ennedi.

(b) Wadi Howar

From its head-waters to 15°45′ N, 24°10′ E (Fig. 4) the wadi sides are composed of granite,

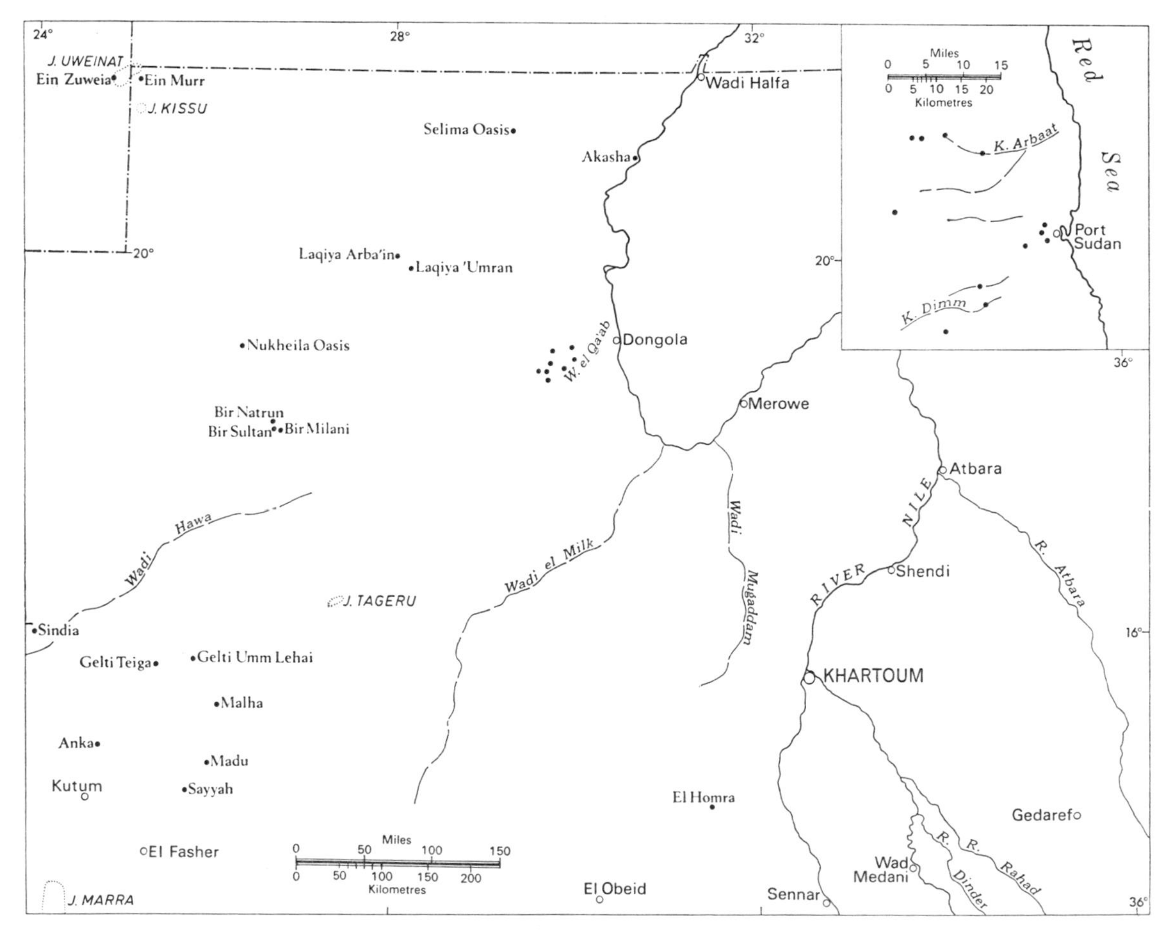

FIG. 63. Wells and oases of the northern Sudan (Based on Sandford 1935)

schist, and gneiss, with occasional patches of Nubian Formation; thence downstream it flows across Nubian Sandstone (Sandford 1935). During the rains Sandford recorded pools of water in the upper part of the Wadi Howar, Lake Tundur being the largest of these.

In spate, large quantities of sand and gravel are transported and there is a well-developed valley fill. Sandford calculated that 1·5 km^3 per annum of water pass into the superficial deposits that floor the wadi in most years. This water is extremely important because the wadi fill carries water down from 2880 ft (900 m) to 1690 ft (530 m) above sea-level and recharges the Nubian Formation. Sandford (1935) has suggested that the water in the southern part of the Libyan desert is derived from this and similar sources. The disposition of the Basement Complex and younger rocks, however, makes this improbable (Fig. 26). In the author's view much of the water in the Libyan and Western Deserts must be 'fossil' water of Plio–Pleistocene and earlier age.

North of the Wadi Howar, in the sand-covered areas, occasional patches of *Acacia* occur in small mud pans, indicating the near-surface presence of water. According to Sandford (1935) the mud serves to reduce evaporation and conserve moisture in the porous rocks below.

Between Wadi Howar and the Nile a few oases and wells occur. These include Selima Oasis, Laqiya el Arba'in, Laqiya 'Mran, Nukheila Oasis, Bir Natrun, Bir Sultan, etc. (Fig. 63).

Selima Oasis owes its origin to structure, and erosion has cut down close to the water-tables. The Laqiya water-holes are associated with anticlinal flexures in the Nubian Formation. These appear between Nukheila (Merga) and Laqiya. At Laqiya Umran the water-hole is in the middle of a broad depression formed by the erosion of the core of an anticline. The structure appears to pitch slightly south-eastwards and water is found at the foot of the associated scarps.

The Bir Natrun water-holes lie at the foot of a north–south scarp which extends towards Nukheila (Merga) Oasis. The beds in this area dip east and south-east and the rocks are deeply wind-eroded. There is no evidence of folding or faulting.

(c) Jebel Meidob and Berti Hills

The movements of ground-water of the lava plateaux of Meidob and Berti are governed mainly by the interbedding of porous and less porous flows. The water that occurs in calderas and craters appears to be derived partly from rainfall, partly from the zone of saturation, and partly from deep-seated volcanic sources.

The Malha crater on the west side of Meidob contains a salt lake. On the crater sides springs occur at the contact of the Nubian Formation and the Basement Complex. Sandford (1935) mentions that the level of the lake is reputed to change suddenly and this is occasionally accompanied by strange noises from below.

5. GROUND-WATER GEOLOGY, KORDOFAN AND ADJACENT PARTS OF DARFUR

Kordofan Province occupies some 146 890 square miles of the central Sudan in the savana belt (Fig. 1). It is underlain by Basement Complex (mainly Pre-Cambrian), Nawa Formation (? Palaeozoic–Mesozoic undifferentiated), the Nubian Formation (? Lower Cretaceous), the Umm Ruwaba Formation (? Tertiary to Pleistocene) and Pleistocene superficial deposits such as cracking clays and laterites, the latter perhaps being as old as late Cretaceous or early Tertiary.

Some 175 drilled wells located in 75 water-yards have been drilled in Kordofan (data up to 1963). In 1962 approximately 600 million gallons were abstracted largely from the Nubian and Umm Ruwaba formations according to Abdullah (*in* Rodis *et al.* 1964). In addition there are many thousands of hand-dug wells, but these often run dry after the rainy season. Very little hydrogeological data is available concerning hand-dug wells. More than 90 per cent of the productive drilled wells in Kordofan are situated in water-yards where there are often 2–5 companion wells commonly less than 295 ft (90 m) apart. The wells range in diameter from 4 to 16 in and in depth from 50 ft (15 m) to 1200 ft (366 m). An average 'yield' per well is given as 1000 gal/h (Rodis *et al.* 1964). Very few wells have been test pumped in the normal sense, however, and yields quoted must not be regarded as absolute yields because again most are simply bailing estimates.

According to Rodis *et al.* (1964) the Nubian and Umm Ruwaba formations contain five discrete ground-water bodies (Fig. 64). Recharge takes place from adjoining Darfur; and by infiltration through the superficial deposits and wadi fill. A recharge rate of as little as 0·002 in/year of rain would probably more than balance the current abstraction rate.

Regional movement of ground-water through Basement Complex formations is very small and limited to weathered zones, but in the Umm Ruwaba and Nubian formations this is considerable and in south-western Kordofan it is towards the Bahr el Arab. In eastern Kordofan movement is towards the Nile (Fig. 64). Discharge is by evapo-transpiration, quantitatively very small; by lateral outflow from the province; and by withdrawals from wells.

For 1961 it is estimated that about 550 million gallons were abstracted. On the basis of the simple measurements that have been made there appears to have been no appreciable changes in water levels.

(a) Basement Complex

The Basement formations and metamorphic rocks consisting mainly of augen and acid gneisses, schists, metasediments, greywackes, acid and basic volcanic rocks, foliated granites and granodiorites,

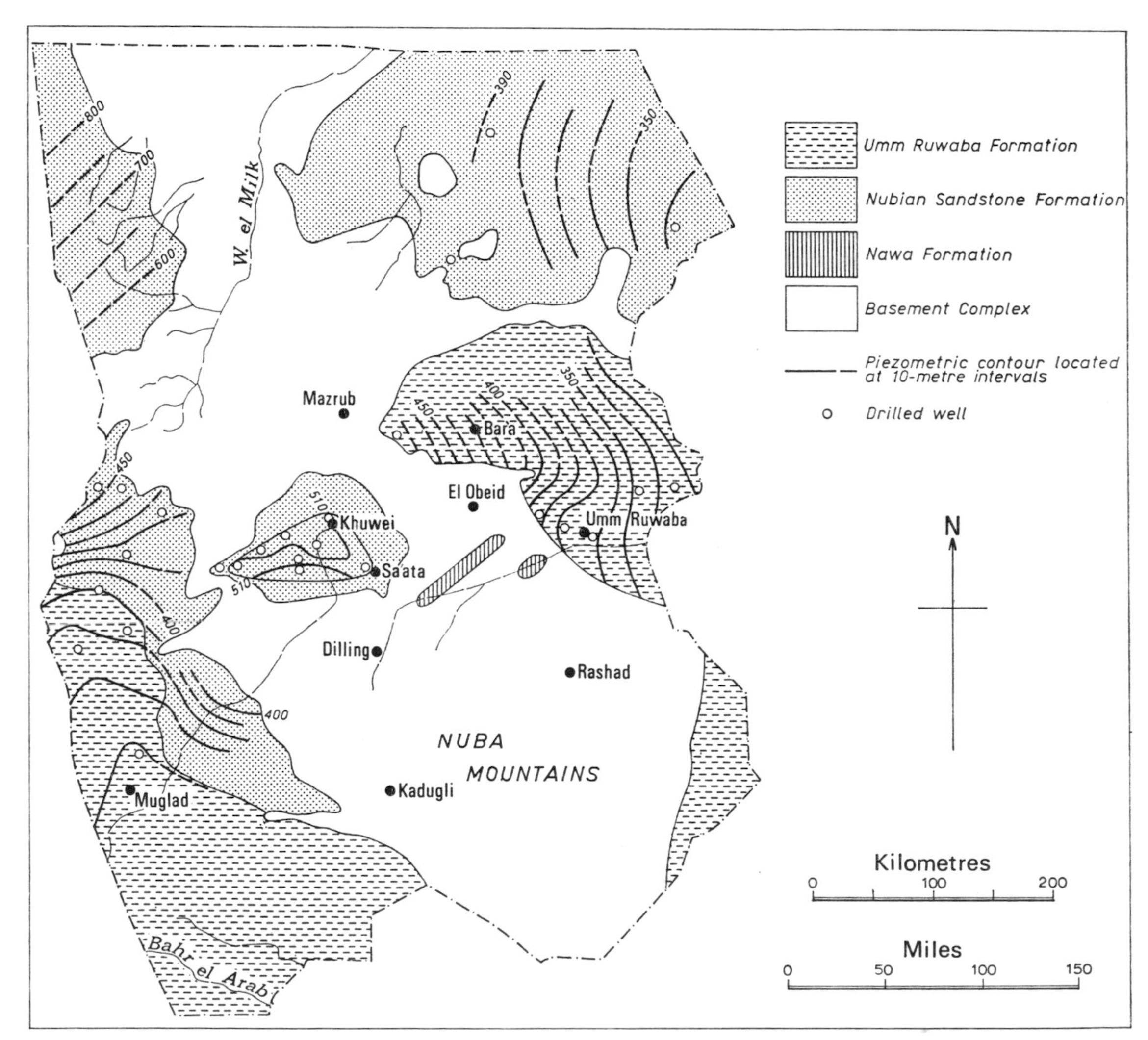

Fig. 64. Standing water-level contour map and formations, Kordofan (Based on Rodis *et al.* 1964)

younger granites, and syenites. These rocks are largely impermeable but water accumulates in weathered zones, joints, and other lines of weakness. Quantities of water extracted from wells sunk in the Basement Complex are small and many wells run dry for a part of the year, especially in low rainfall areas. In the Nuba Mountains, however, where rainfall is higher, the Basement formations have yielded moderate quantities of water, as at Dilling and Kadugli (800–1200 gal/h). Details of water-supply problems in the Nuba Mountains on the Rahad sheet 66A and Talodi sheet 66E were given by Mansour and Samuel (1957).

Little data is available concerning the chemical quality of Basement Complex waters. Some are as hard as the Umm Ruwaba waters and in the north-western two-thirds of Kordofan Basement water is frequently salty or brackish (Rodis *et al.* 1964). In general there is relatively little circulation and water is of poor quality, except in areas where the rainfall is high.

(b) Nawa Formation

Few wells penetrate this formation, which has limited distribution, and its water potential is largely unknown.

(c) Nubian Formation

The lithology of the Nubian Formation is described above. The strata are in general flat-lying or dip gently northwards. There are abrupt lithological changes, and extensive beds of mudstone occur (Fig. 37). Predominantly the Nubian consists of sandstone. It is silty or clayey in places and is occasionally conglomeratic.

Water occurs mainly in the sandstones and conglomerates but jointed and fractured mudstones may contain water. According to Rodis *et al.* (1964) water is frequently under low artesian pressure and the water bodies are interconnected in the individual lenses of Nubian Formation and extend for as much as 31 miles (Rodis *et al.* 1964). Near outcrops of Basement formations where the Nubian Formation is less than 200 ft thick water is usually absent. The depth to the zone of saturation in Kordofan varies from about 197 ft (60 m) to 525 ft (160 m) below well-head. Again information about yields is inadequate because testing has been done mainly by the 'bailing' method. Rodis *et al.* (1964) cite 'yields' of between 700 and 1200 gal/h, with some wells producing 3800 gal/h.

Nubian water is characterized by low dissolved solids (100–340 p.p.m.) and low total hardness (60–264 p.p.m.) (Fig. 65). Some waters are fluoride-rich according to Rodis *et al.* (1964).

Ground-water of the Nahud outlier of the Nubian Formation was studied by Rodis and Iskander (1963). The main zone of saturation underlies an area of 2000 square miles and has a maximum thickness of about 1000 ft. The water is under slight artesian pressure but individual aquifers are of limited extent because of variations of permeability.

Depth to water ranges from 200 to 400 ft and has been measured in some wells once a year since 1932. No significant decline in the water-table has taken place, and Rodis and Iskander (1963) have pointed out that even if no recharge were to occur withdrawals at the 1961 rate of approximately 125 million gallons per annum would only lower the water-table 1 ft in 167 years. Because of the large surface area of the Nahud outlier, recharge at the rate of 0·004 in/year would probably suffice to sustain the quantities abstracted at present.

The chemical quality of the Nahud outlier water is good and low in dissolved solids. The temperatures of the water sampled ranged from 91° to 93°F (Rodis and Iskander 1963).

(d) Umm Ruwaba Formation

This is a series of fluviatile and lacustrine deposits. Its general lithology has been described above. In places in eastern Kordofan it has been proved to be more than 1100 ft (335 m) thick.

Aquifers are often of limited extent in the Umm Ruwaba Formation because of the numerous facies changes that take place. In places the Formation thins down over the Basement Complex and as a consequence in a number of areas, especially in eastern Kordofan, little or no water is found.

The chemical qualities of the water from the Umm Ruwaba Formation are shown in Fig. 65. Total dissolved solids range from 420 to 3000 p.p.m. and averages 1050 p.p.m. The water is harder also than that of the Nubian Formation, but differs regionally. Low hardness is found locally in the Umm Ruwaba in the extreme eastern and western parts of the province.

According to Rodis and Iskander (1964) 'yields' in the Umm Ruwaba vary from 600 to 1500 gal/h.

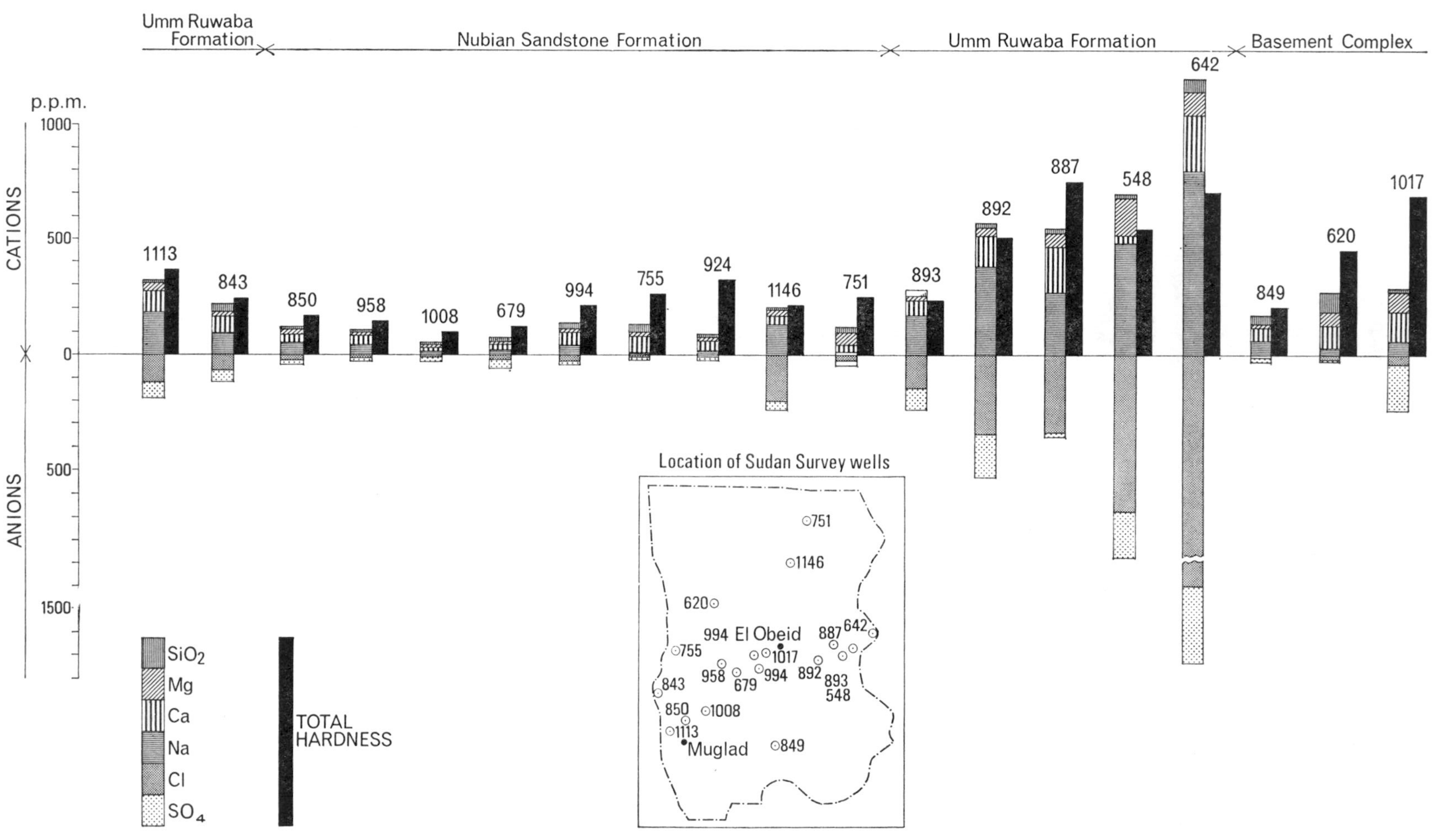

FIG. 65. Chemical properties of ground-water, Kordofan (Based on Rodis *et al.* 1964)

Where the Umm Ruwaba Formation and the Nubian Formation are associated, Rodis and Iskander (1963) believe that they are essentially continuous hydraulically. The chemical differences between Umm Ruwaba water and Nubian water do not support this view.

(e) Pleistocene and superficial deposits

Superficial deposits are thick and extensive in many parts of Kordofan. Rodis *et al.* (1964) estimated that the northern two-fifths of the province are covered with coarse residual desert soils and sand dunes, many of which are active. South of this area are the qôz sands, fixed dunes, and these interfinger southwards with residual clays.

The thickness of these deposits is highly variable. The qôz sands are said to be more than 148 ft thick in places, and where sands rest on clay such deposits are water-bearing. The ground-water bodies near Muglad, El Fula, El Odaiya, and Abu Zabad, which yield as much as 700 gal/h, are said to be of this type (Rodis *et al.* 1964).

Water commonly occurs in wadi fill. Again thicknesses and yields are highly variable; the water is generally unconfined and wells are often dry soon after the rains. In places, slope-wash deposits yield-water and the unlined wells at Jebel Kagmar, which yield about 500 gal/h, are said to be of this type.

Laterite and latisols are common in Kordofan. They range in thickness from a few feet to as much as 50 ft but are unimportant as sources of water.

Karkanis (1965) described the geohydrology of adjacent parts of Kordofan. He concluded that the ground-water body in the Umm Ruwaba and Nubian formations is to a large extent homogeneous and continuous, and that there is a general movement of water from north to south. The ground-water is said to be of good quality, with total dissolved solids ranging from 100 p.p.m. in the north to 500 p.p.m. in the south.

The Basement Complex forms small discontinous aquifers with low ground-water potentialities, but few data are available (Karkanis 1965).

6. GROUND-WATER GEOLOGY OF THE GEZIRA

The Gezira is situated between the Blue and White Niles, and is arbitrarily bounded to the south by the Sennar–Kosti railway line. Despite the fact that it is the main irrigated area of the country it depends for much of its drinking water on underground resources. This is because many of the canals are infected with *Bilharzia*-carrying snails and because of sanitation problems.

(a) Basement Complex

The oldest rocks exposed in the Gezira are the Basement Complex. These include metasediments such as quartzites, marbles, graphitic slates, pelitic schists, and gneisses; intrusive rocks such as the charnockitic quartz-diorite of Jebel Moya, porphyritic granites, felsite dykes, etc. The outcrop area of the Basement Complex is small, however, and the formations produce little water unless the rocks are deeply weathered or are well jointed. The impervious Gezira clays prevent much water from seeping into Basement formations. No boreholes produce from the Basement Complex but there are a few hand-dug wells excavated in weathered zones and in weathering aureoles that surround some of the jebels as in the Jebel Moya area. Commonly this water is saline.

(b) Nubian Sandstone Formation

The Nubian Formation unconformably overlies the Basement Complex and consists of conglomerates, grits, sandstones, sandy mudstones, and mudstones. The maximum proved thickness in the Gezira is about 800 ft at Soba, near Khartoum. The total thickness of the Nubian Formation in the Khartoum area is probably about 1500 ft. There are large variations in thickness, however (Fig. 32). These have resulted from the deposition of Nubian on a well-dissected Basement Complex land surface. This last point is important from a geohydrological point of view because the aquifers are sometimes discontinuous and in some cases form isolated cells.

The Nubian Formation is at present the second most important aquifer in the Gezira but clearly it has much greater potentialities. According to Abdl Salaam (1966) approximately 10 per cent (some twenty-eight boreholes) produce from the Nubian Group and some 30 per cent produce from both the Nubian Formation and the Gezira Formation.

The Gezira Formation and the Nubian Formation are hydrogeologically interconnected, especially via the weathered top of the Nubian Formation. The

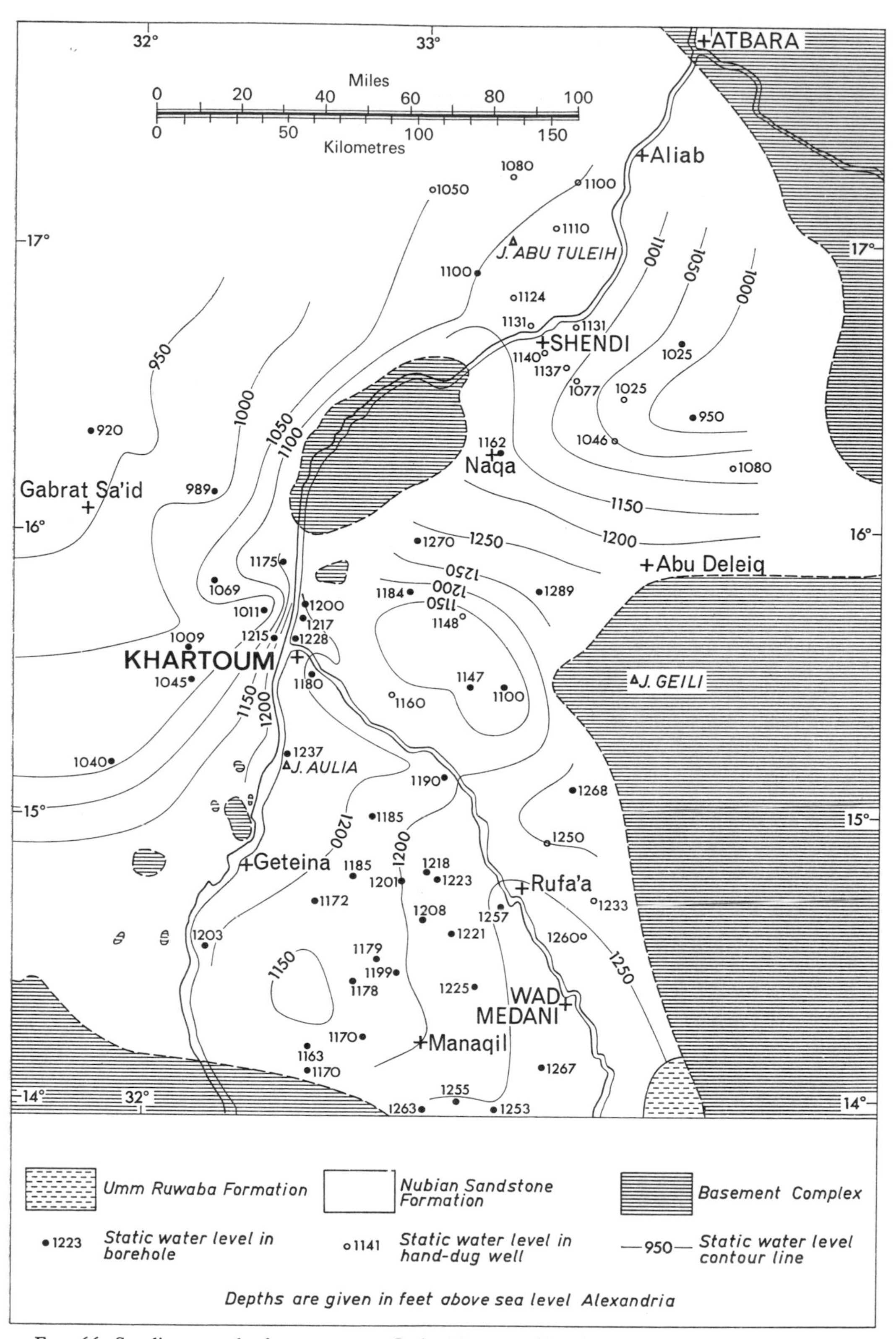

FIG. 66. Standing water-level contour map, Gezira–Khartoum–Shendi region (Compiled by Kheiralla, Abdl Salaam, and Whiteman 1966)

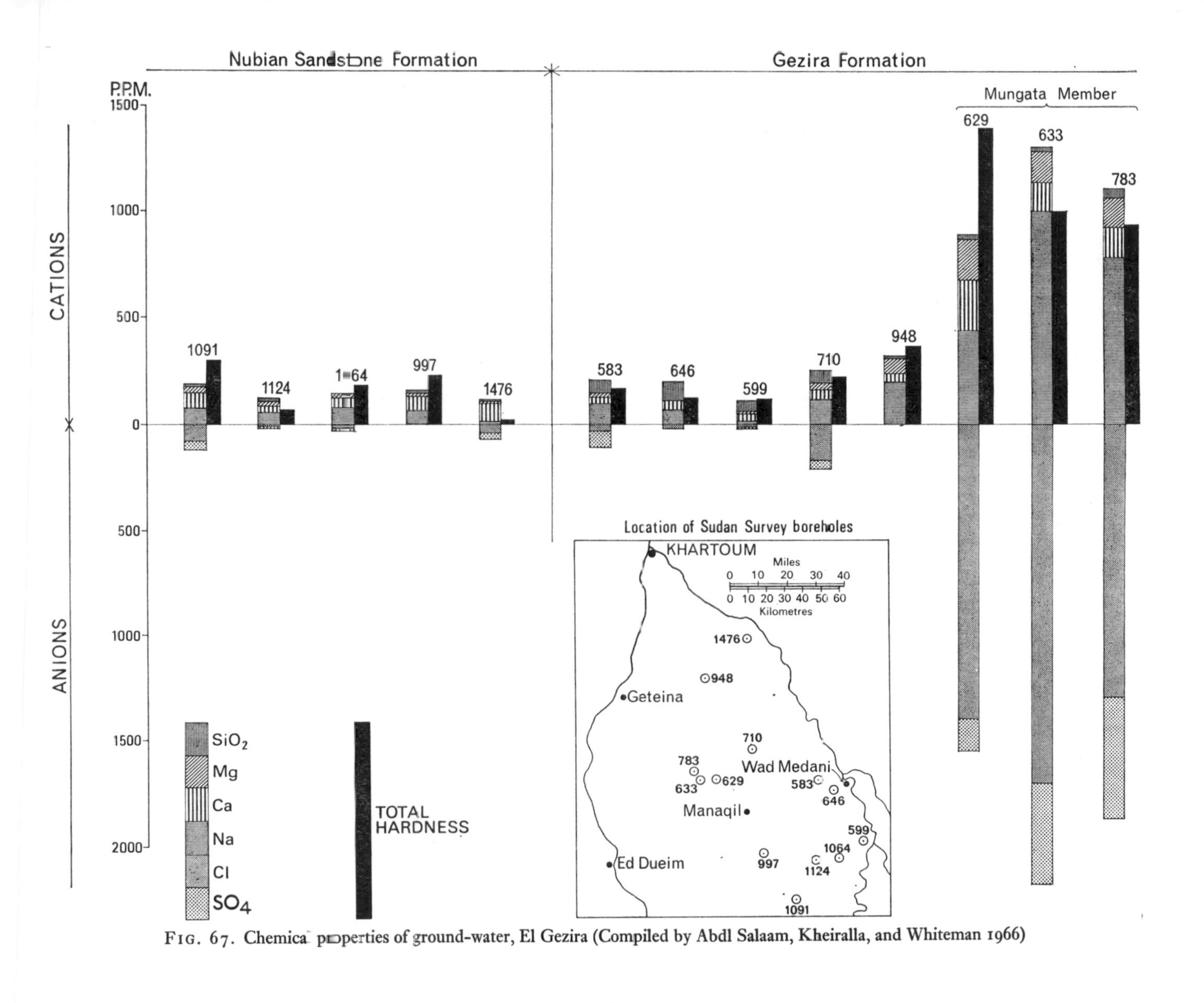

FIG. 67. Chemical properties of ground-water, El Gezira (Compiled by Abdl Salaam, Kheiralla, and Whiteman 1966)

porosity, permeability, and thickness of the Gezira Formation are therefore important factors governing recharge of the formation. A list of boreholes producing from the Nubian Formation is given in Abdl Salaam (1966, Table 8).

The water-bearing beds are mainly sandstones and conglomeratic sandstones. These are of limited extent because of frequent facies changes into mudstones. Water sometimes occurs in fissures in mudstones but in general these beds yield little water. In places ground-water in the Nubian Formation occurs under slight artesian pressure, especially in sandstones that are partially enclosed by mudstones. The average saturated thickness of the Nubian Formation penetrated in the Gezira is 140 ft (49 m) according to Abdl Salaam (1966). Depth to water in the Nubian Formation in the Gezira is variable but the standing water-levels are higher nearer the rivers, indicating that they are the main source of recharge (Fig. 66). Depth to the top of the zone of permanent saturation varies between 80 ft (24 m) at Soba and 190 ft (58 m) at Amrat el Sheikh Hagu.

Again very little can be said about total yields, for the figures given have nearly all been obtained by bailing the wells. The average reported 'yield' is 1000 gal/h, which is very low for a formation with the lithological characteristics of the Nubian. This is clearly an underestimate, as a pumping test conducted in Omdurman has shown. There, with a constant draw-down of 55 ft after 80 min, the formation produced 25 000 gal/h. The water-level recovered in 20 min.

The chemical quality of Nubian water is shown in Fig. 67. It is generally of good quality and the water has a low mineral content. Total dissolved solids vary from 200 to 560 p.p.m., average 390 p.p.m. Similarly, hardness is low and in general it is temporary (14–294 p.p.m., average 146 p.p.m.). Ground-water from the Nubian Formation is frequently clear, odourless, and tasteless. In the southern Gezira the Nubian Formation is abnormally saline in boreholes at Sheikh Daw and Umm Degheina, but the high salinity may be derived from the overlying Gezira Formation (Abdl Salaam 1966).

(c) Sennar Teschenite

At Sennar, intruded into the Nubian Formation is a teschenite mass of limited extent. Its chief interest in connection with ground-water is that it occurs in the foundations of the Sennar dam.

(d) Gezira Formation

The Gezira Formation is the main aquifer of the Gezira: 167 boreholes sunk in the Gezira up to 1964 produce from this formation. The lithology of the Gezira Formation has been described above. Excluding the Gezira Clay the formation is generally highly porous and permeable. Ground-water occurs under open aquifer conditions over much of the region, and even though the water-bearing strata are commonly lens-shaped they appear to be interconnected. The main water-bearing strata consist of fine-, medium-, and coarse-grained sands. The beds are sometimes clayey and gravelly and the best producing horizons are clean, gravelly sands.

The average saturated thickness of an aquifer in the Gezira Formation is 60–80 ft (18–25 m). Depth to water-table varies from 20 ft (7 m) near the banks of the Blue Nile to 148 ft (45 m) in the centre of the Gezira. In the Gezira the most striking feature of the zone of saturation is the 'water-curtain' effect associated with the Nile (Fig. 68). The standing water-level near the rivers is higher than in the surrounding areas. This is discussed below and can be explained simply because of the reduction in effective head by frictional and capillary effects in the pore-spaces of the aquifers, and the existence of buried Basement hills.

Again yield data are inadequate for the Gezira Formation because only one standard pumping test has been made in the whole area (Abdl Salaam 1966). To try to establish the yield of an aquifer by bailing the well is rarely possible. Abdl Salaam (1966) has estimated from times taken to fill storage tanks that the formation can produce between 500 and 10 000 gal/h.

Ground-water from the Gezira Formation varies from good water with a low mineral content to extremely poor, saline water, which is unfit for human consumption. In general, however, the Gezira Formation water is of fair chemical quality with total dissolved solids varying from 200 to 870 p.p.m. (average 420 p.p.m.). Hardness ranges from 24 to 270 p.p.m. Water from the Mungata member of the Gezira Formation is commonly highly mineralized and total dissolved solids vary from 1000 to 20 000 p.p.m. (average 5000 p.p.m.). Hardness is also high, and is due to the presence of sulphate radicals. It ranges from 256 to 2660 p.p.m. (average 900 p.p.m.) (Fig. 69).

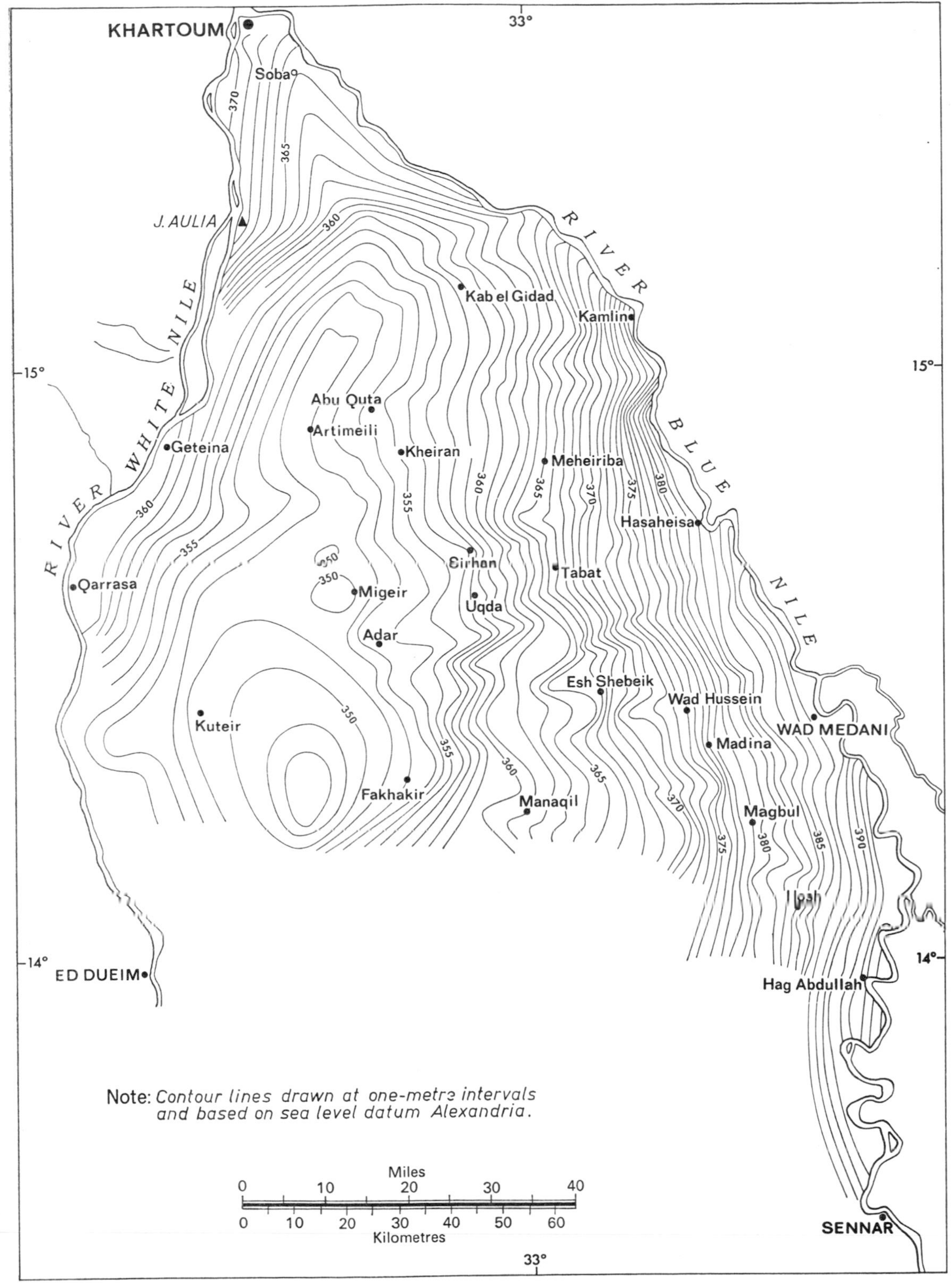

FIG. 68. Standing water-level contour map, El Gezira (Compiled by Abdl Salaam and A. J. Whiteman 1966)

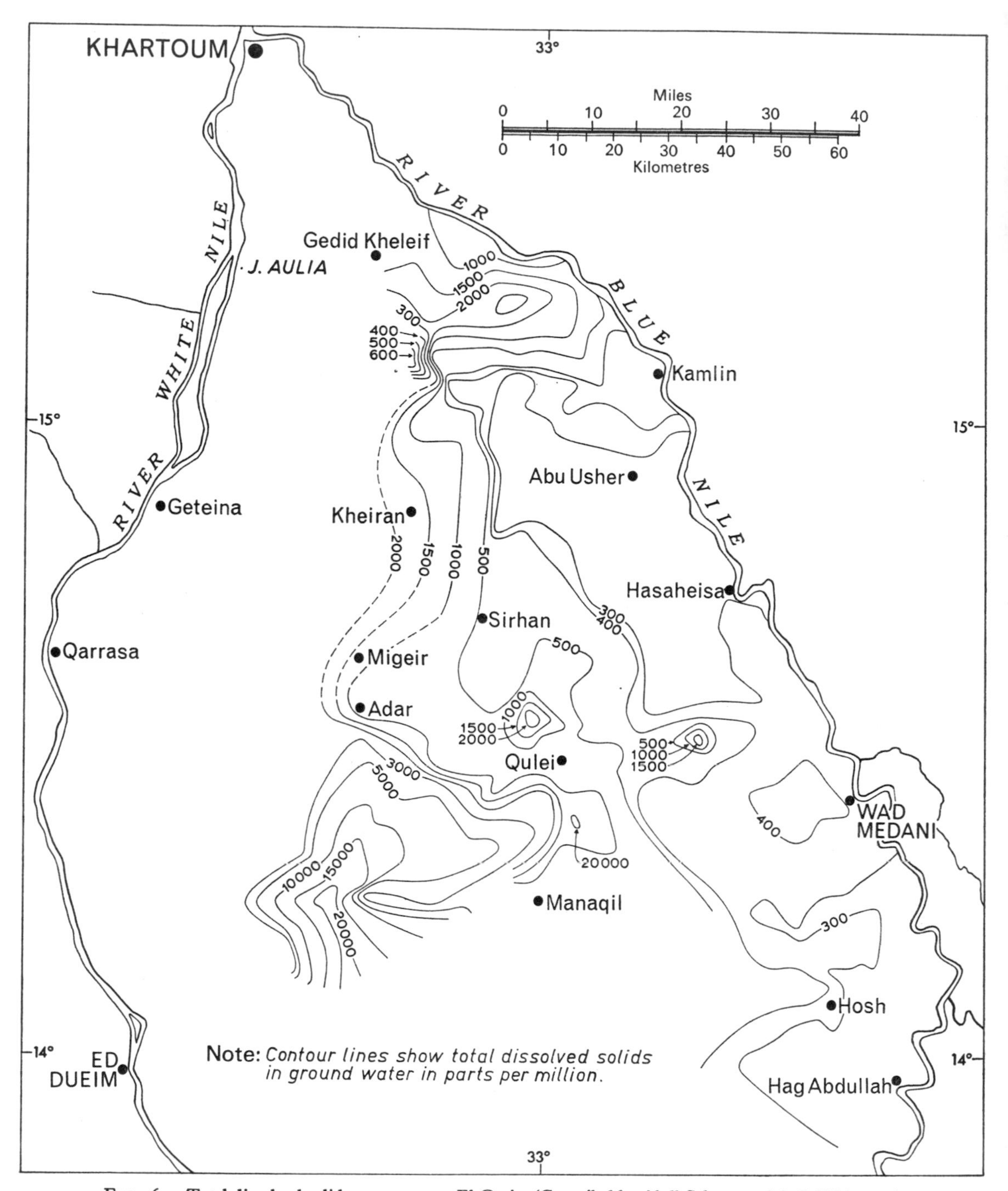

FIG. 69. Total dissolved solids, p.p.m. map, El Gezira (Compiled by Abdl Salaam and A. J. Whiteman)

In the northern part of the Gezira ground-water from the Gezira Formation is also saline, but here the cause is not at present known (Abdl Salaam 1966).

Abdl Salaam (1966) estimated that some 665 million gallons per annum are produced from 500 hand-dug wells and 167 boreholes that penetrate the Gezira Formation.

(e) Superficial deposits

The superficial deposits of the Gezira are listed in Table 11. Their water resources are limited because the deposits lie above the zone of permanent saturation. Perched water-tables occur in khor and wadi deposits. Locally these provide small supplies but frequently during years of poor rainfall they dry up.

(f) Movement of ground-water, recharge and discharge

The general movements of ground-water can be deduced from the contour map of the water-table shown in Fig. 68. This map is generalized and shows the lowest water-table because some 79 per cent of the measurements were made in the dry season between 1950 and 1964. Andrew (1950) pointed out that 'the static water level in Blue Nile rises upstream and falls towards the west' and that 'there is ample evidence to show that north of 14° N the static water level is falling from the Blue and White Niles in regular contours until a trough is formed a distance of 20 miles east of the White Nile.'

Abdl Salaam (1966) noted that water moves approximately at right angles to the contour lines from a 'high' of 1275 ft (390 m) at Wad el Haddad upstream on the Blue Nile to a 'low' of 1194 ft (365 m) downstream at Soba (Fig. 68). It also moves from a high of 1294 ft (382 m) at Hasaheisa on the Blue Nile westwards to a low of 1145 ft (350 m) at Migeir in the central Gezira. The ground-water contours fall west of the Blue Nile at regular intervals with a gradient 1 in 2000. Ground-water moves from a high of 1177 ft (360 m) near the White Nile eastwards to a low of 1145 ft (350 m) at the centre of the Gezira with a hydraulic gradient of 1 in 4000. The 1144-ft (350-m) contour forms a trough situated approximately 20 miles (32 km) east of the White Nile. Quite clearly the Blue Nile is influent and is the main reason for the formation of the pronounced ground-water mound or curtain (Fig. 70, etc.).

The smoothly-sloping water-table probably indicates that there is no appreciable recharge from rainfall and the configuration of the water-table is largely controlled by differences in head between the intake zones of the rivers and the discharge points.

The trough formed east of the White Nile, according to Abdl Salaam (1966), is due to the lower permeability of formations towards the centre of the Gezira (the distribution of the Mungata Member, which is clayey, roughly coincides with the trough); and to the fact that recharge comes mainly from two opposite directions (Fig. 70).

Recharge takes place mainly by underflow from the Blue Nile and to a lesser extent from the White Nile. The Blue Nile is largely floored by sandy deposits and water seeps into the Gezira sands and deeply weathered Nubian Group. Seepage is thought to be less on the White Nile because the beds are apparently more clayey and silty. Aquifers are also replenished from rain falling on near-by outcrops of Nubian Sandstone, but there are few quantitative data available.

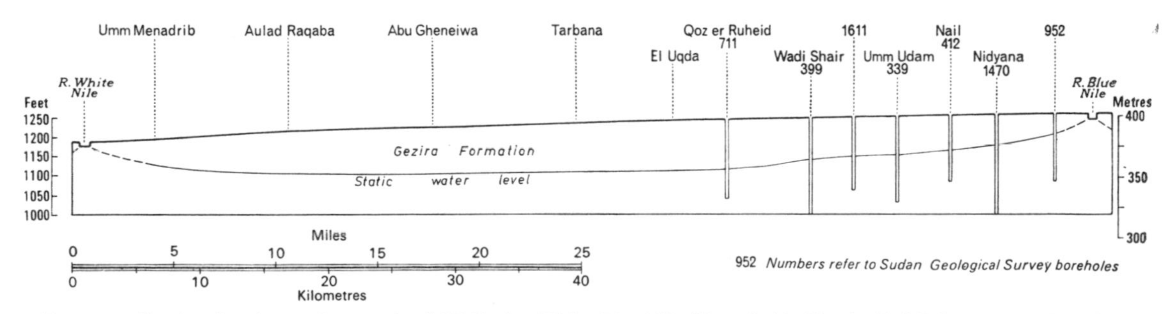

FIG. 70. Section showing static water-level, El Gezira, White–Blue Nile (Compiled by Yassin Abdl Salaam and A. J. Whiteman 1966)

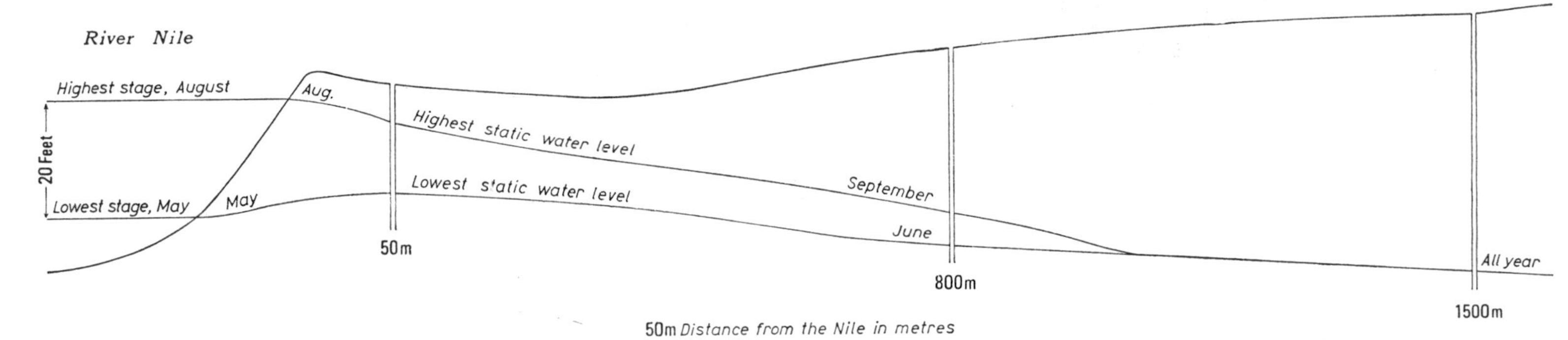

FIG. 71. Fluctuations of static water-level near Nile bank during annual Nile Flood Based on Grabham 1934. Compiled by Abdl Salaam 1966)

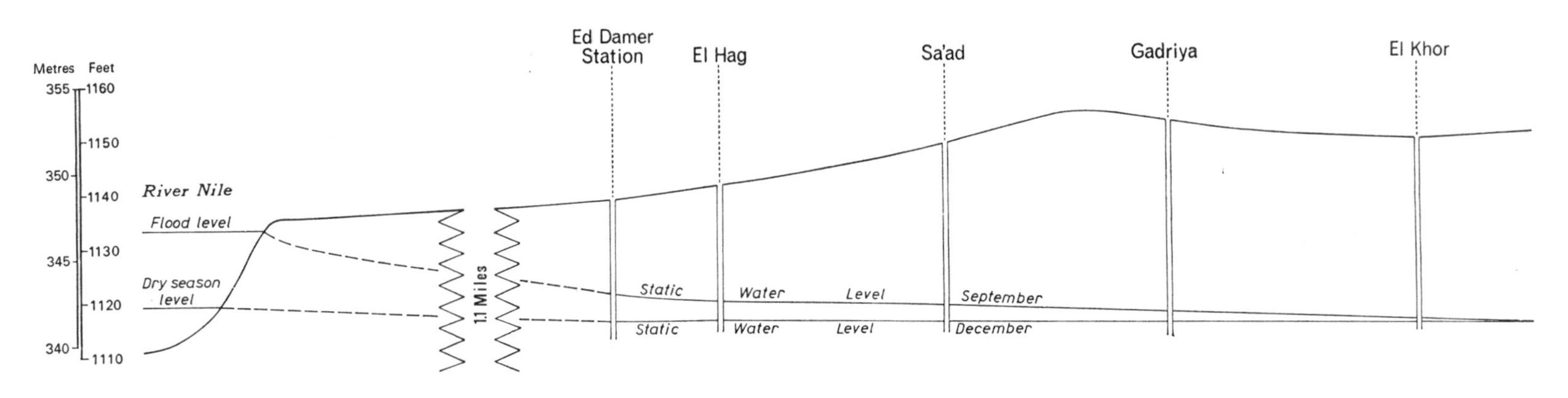

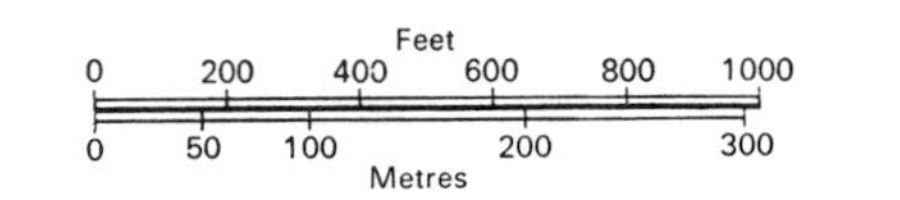

FIG. 72. Seasonal fluctuations, static water-level, Ed Damer region, Northern Province (Compiled by Kheiralla 1966)

Discharge takes place from the zone of saturation by evapotranspiration and evaporation. Near the rivers this effect may be strong, and in many places the banks are encrusted with caliche or kankar deposits as a result. However, compared with artificial discharge by pumping from wells, this is small. Abdl Salaam (1966) has estimated that roughly 464 000 000 gal are withdrawn annually from the Gezira wells and boreholes.

7. GROUND-WATER GEOLOGY OF THE NILE VALLEY BETWEEN ED DUEIM AND ATBARA

(a) General

This region was studied in detail by Kheiralla (1966) who mainly worked on the Nubian Formation and its geohydrology. Over much of the area studied good quality water is available from the saturated zone of the Nubian Formation. Some water is also obtained from the Basement formations and from the superficial deposits. There are numerous hand-dug wells that produce from the Basement formations but these have very low yields. No boreholes produce from the Basement formations.

Perched water bodies occur in the fill in some wadis. For example Abu Duleiq, located on the left bank of the Wadi El Hawad, obtains its supply from a perched water body in the khor bed. The khor drains an area of about 135 miles^2 and receives an average annual rainfall of about 213 mm. The khor bed is about 600 ft wide and the fill consists of very coarse, gravelly clean sand. It rests on gritty, pebbly, and clayey Nubian Formation. Water is available in shallow wells at depths of 10–15 ft. The yield is good for this type of well and lasts throughout the year. In places, however, water from these shallow sources is saline.

Shallow hand-dug wells produce from the Nile superficial deposits and provide a major source of water for many riverain villages. North of Khartoum, along the American road to Khogalab, tube wells produce from the Nile superficial deposits and weathered Nubian Formation; numerous farms and gardens are irrigated from these sources.

In the Nubian Formation ground-water is commonly found in pebble conglomerates, and beds of the Merkhiyat and quartzose sandstone types. It rarely occurs in mudstones and intraformational conglomerates. However, subartesian conditions frequently prevail in sandstones overlain and underlain by thick mudstones.

The depth to the saturation zone in the area studied (Fig. 66) varies from 40 to 735 ft, and is closely related to lithology as well as distance from the Nile. The saturated zone is largely continuous. The Manaqil and Khartoum basins are linked (Fig. 66) and there is also continuity between the Naqa and Khartoum basins west of the Nile. Continuity is broken east of the Nile by the Sabaloka–Wad Hassuna–Jebel Qeili ridge, which forms a subsurface barrier extending broadly east–west across the area. Its crest is higher than the standing water levels in the Naqa–Khartoum basins east of the Nile, and as a consequence boreholes drilled along it are dry.

(b) Nile water curtain

The Nile water curtain is the major hydrogeological feature of the Nile Valley. It is characterized by gentle but unequal hydraulic gradients which flatten out away from the river (Figs. 66 and 70). The levels near the river fluctuate with the seasonal rise and fall of the Nile (Figs. 71, 72, 77B).

(c) Recharge

The Nubian Formation of the Nile Valley is recharged from three main sources: direct percolation of rain-water falling on Nubian outcrops; seepage through the beds of the khors and wadis; and seepage from the bed of the Nile.

The beds of the Main Nile and the Blue Nile are composed of coarse sands, and the Main Nile flows over rocks of the Nubian Formation from Khartoum to Ed Damer (Fig. 73) excluding the Basement rocks of the Sabaloka inlier. The White Nile has a more gentle gradient and the bed is said to be composed of finer material. According to Kheiralla (1966) this accounts for the fact that the hydraulic gradients are less along the White Nile than along the Blue and Main Niles (see above). The saturated zone in the Nile Valley is also recharged by underground flow from the Nubian Formation in north-eastern Kordofan (Rodis *et al.* 1964).

(d) Discharge

Again the amount attributable to evapotranspiration is small and the greatest discharge results from pumping. Unfortunately the records of the amounts of water abstracted from hand-dug wells or boreholes are not kept, and only a very rough estimate can therefore be made. Kheiralla (1966) pointed out that the amount of water abstracted is broadly related to the density of the animal population in the area. Using average figures for the quantity of water that each animal drinks at one time, he estimated that more than 2368 million

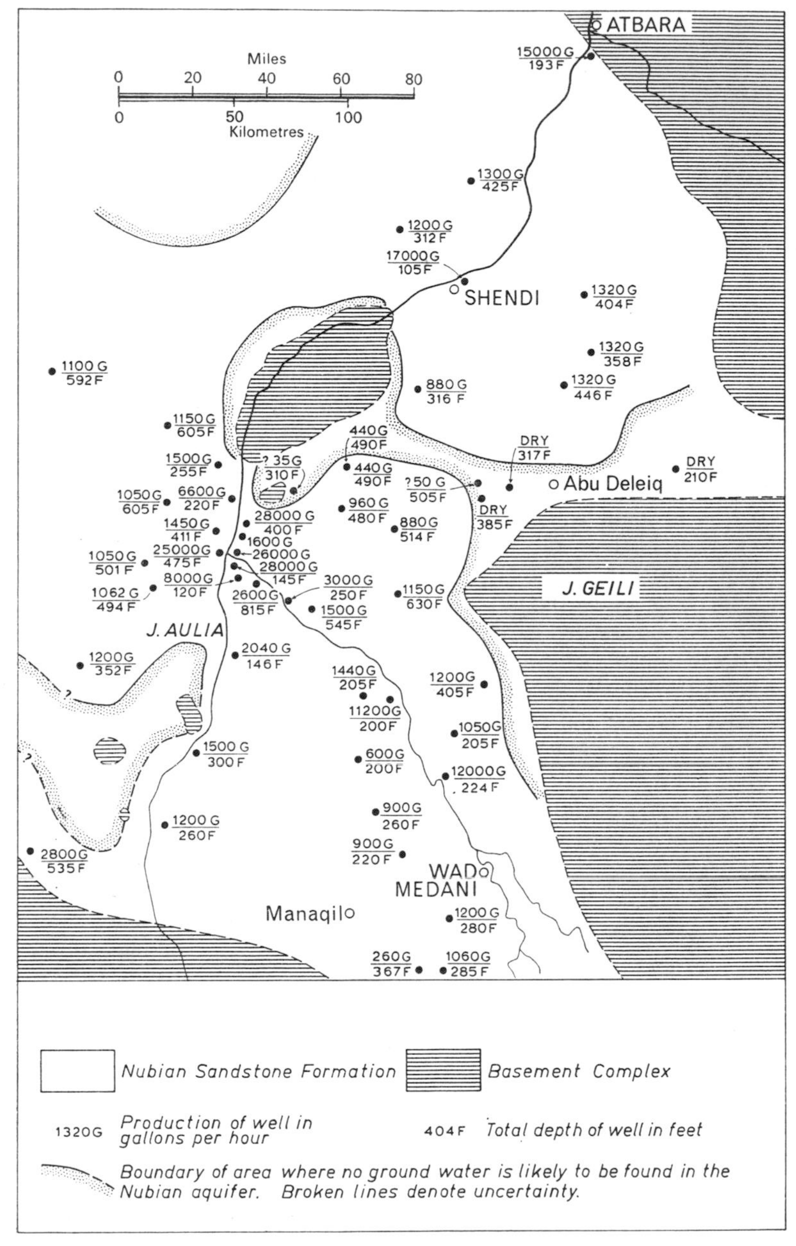

FIG. 73. Approximate yields and total depth, Nubian Sandstone Formation wells, Gezira–Khartoum–Shendi areas. Nile Valley (Compiled by Kheiralla 1966 and A. J. Whiteman)

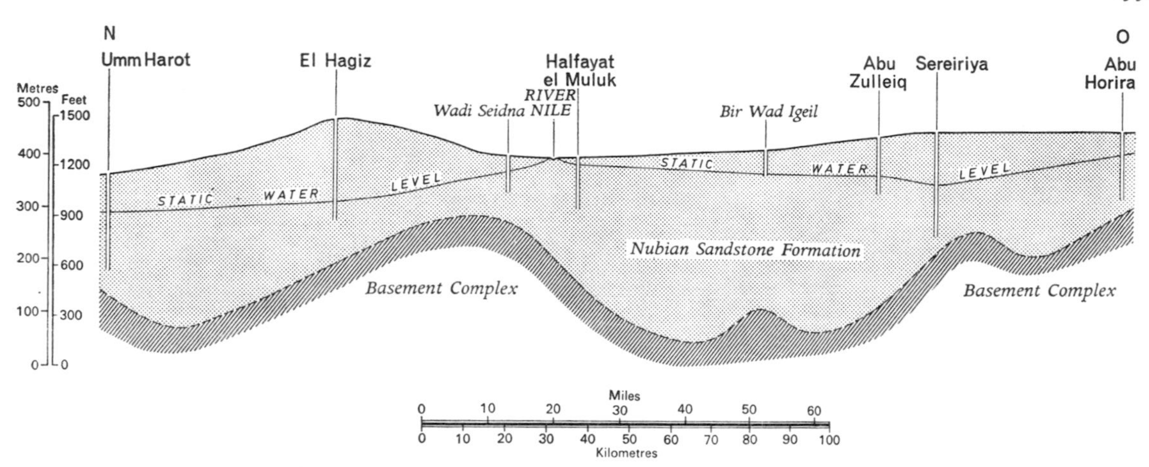

FIG. 74. Geological section showing Nile water curtain, Umm Harot–Abu Horira, Khartoum region (Compiled by Kheiralla 1966 and A. J. Whiteman. See Fig. 78 for location)

gallons per annum must be abstracted annually from the Nubian aquifer in the area studied.

(e) Yields

Pumping tests were carried out for only 4 per cent of the boreholes studied by Kheiralla (1966) and as a consequence little can be said about yields and reserves except in a general way (Fig. 73). The pumping tests at Ed Damer, Rufaa, and Umm Dubban show that the yield may vary considerably. At Ed Damer the yield was 15 000 gal/h with a draw-down of only 7 ft. The well recovered in 10 min. At Rufaa the well was pumped at a rate of 10 000 gal/h, which resulted in a draw-down of 10 ft, and the initial standing water-level was attained within 60 min. However, at Umm Dubban the well yielded only 1500 gal/h and the draw-down was 16 ft. The coefficients of transmissibility for Ed Damer and Rufaaa are 76 000 and 44 000 gal/day per ft. The coefficient for Umm Dubban cannot be calculated properly, but according to Kheiralla (1966) is very low. Variation in transmissibility in

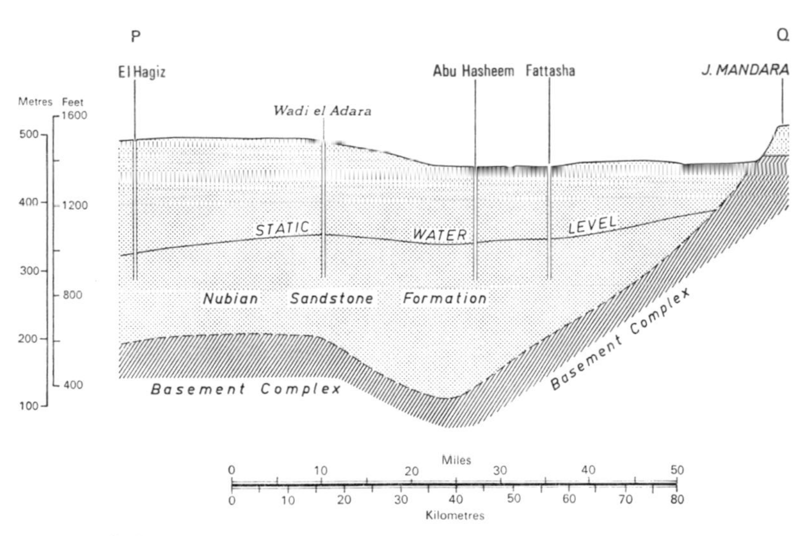

FIG. 75. Geological section showing static water-level, west of Khartoum (Compiled by Kheiralla 1966. See Fig. 78 for location)

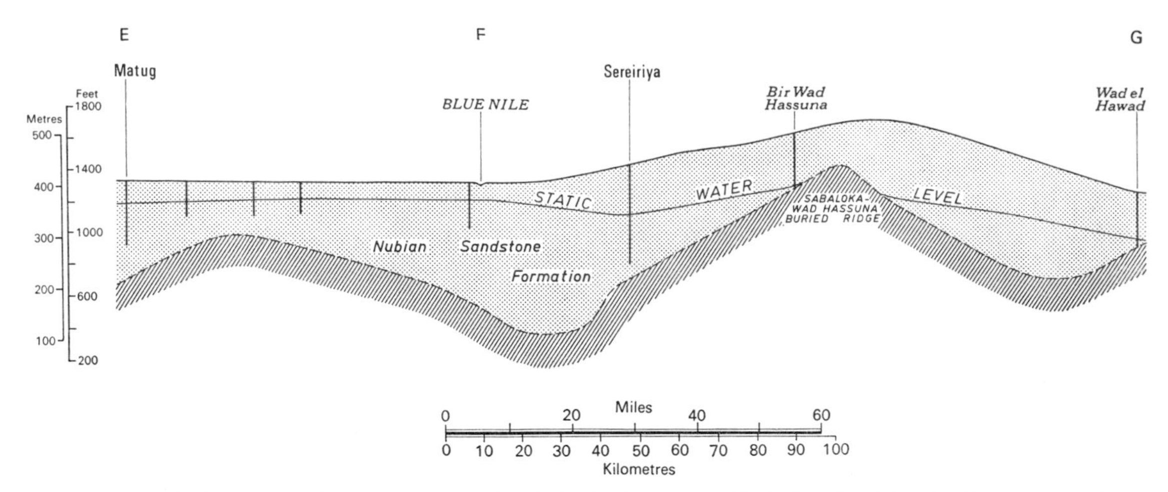

FIG. 76. Geological section showing static water-level, Nubian Formation, El Gezira–Blue Nile–Bir Wad Hassuna–Wad el Hawad (Compiled by Kheiralla 1966. See Fig. 78 for location)

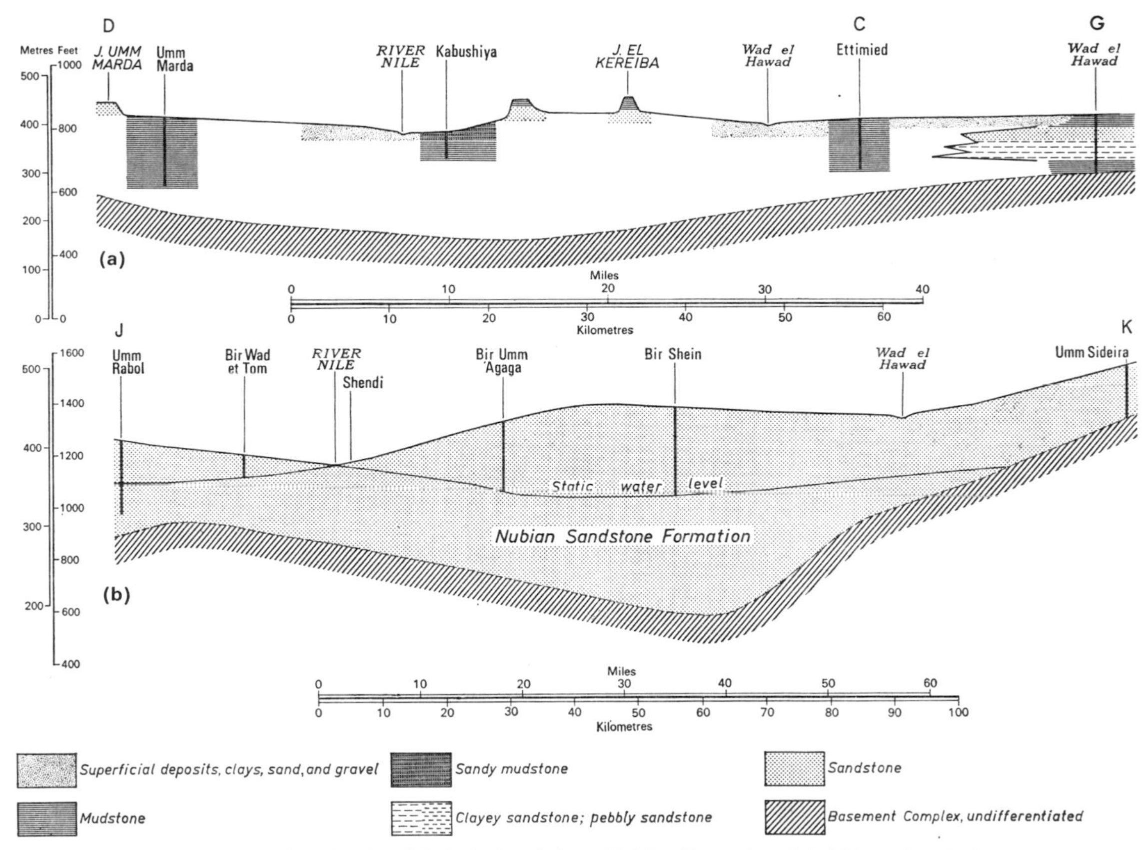

FIG. 77. (a) Geological section showing lithological variations, Nubian Formation, Jebel Umm Marda–Kabushiya–Wad el Hawad (See Fig. 78 for location). (b) Geological section showing Nile water curtain, Umm–Rabol–Shendi–Umm Sideira (Compiled by Kheiralla 1966. See Fig. 78 for location)

the Nubian Formation is clearly related to variation in lithology.

(f) Chemical quality

The total dissolved solids in the Nubian Formation for the greater part of the area studied by Kheiralla (1960) ranges from 150 to 736 p.p.m. There are, however, places where the values are higher, as at Et Timeid (SGS 956) where the value exceeds 1500 p.p.m. Apparently this is due to the fact that water is confined between impervious mudstone and Basement Complex and that circulation is low. High values elsewhere in the Nubian Formation can be explained by contamination from the Basement Complex or the superficial deposits.

For ground-water from the Nubian Formation in the area studied total hardness is generally low; it averages 257 p.p.m. and ranges from 100 to 760 p.p.m.

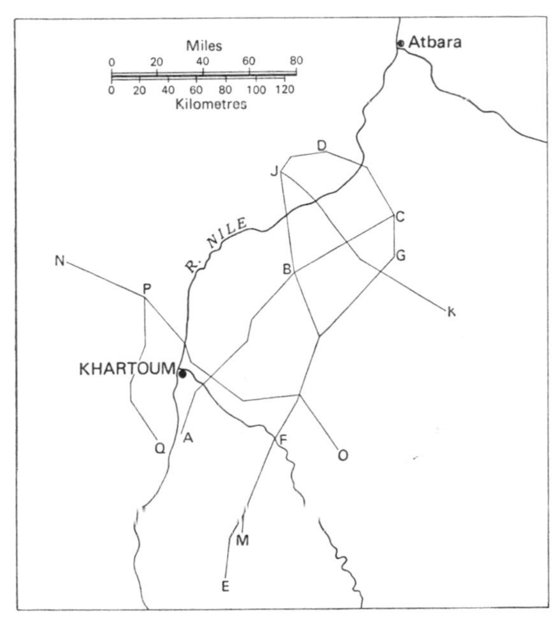

FIG. 78. Section locality map Figs. 74–77. Nile Valley Khartoum–Shendi–Atbara region

8. GROUND-WATER GEOLOGY OF GASH DELTA KASSALA PROVINCE

The Gash River rises as the Mareb in the vicinity of Asmara in Eritrea and, like the Nile and other rivers rising in Ethiopia, it floods annually, bringing down large quantities of gravel, sand, and silt. These are deposited to form the Gash delta (Plate 10). The catchment consists mainly of Basement Complex rocks, predominantly gneisses and granites, but south of Asmara some volcanic rocks occur.

Gradients are steep in Ethiopia and the coarse grained detritus swept down by the river in spate is deposited mainly upstream of Kassala, where the gradients are much less. The delta is composed mainly of clays and silts with some gravels. Kassala, situated near the head of the delta, is noted for its irrigated fruit gardens. Cotton and other irrigated crops are grown in the delta. The system of flush irrigation was described by Swann (1956). The sediments vary in thickness from 0 to 200 ft and consist of subangular quartz grains, clay minerals, granitic and gneissose fragments, and volcanic fragments. The grains are mainly subangular. As in so many delta deposits, the lithology is highly variable. The aquifers appear to be interconnected and the water appears to present under open conditions. In places the saturated thickness of the aquifer is as much as 100 ft. The best aquifers consist of sand and gravel. Depth to water varies from 20 to 60 ft below surface (Said, Sudan Geological Survey MS).

Ground-water moves towards the north-west end of the delta, the gradient of the static water surface being about 10 ft per mile. Near Kassala the static water-level stands about 1600 ft (500 m) above sea-level, but in the northern part of the delta it stands at 1552 ft (485 m). Said recorded that the annual variation in twelve wells from 1963 to 1964 was 10 ft.

Clearly the main source of recharge is water seeping into the sands and gravels from the bed of the Gash and its distributaries. The amount of water stored in 25 000 acres of the Gash delta is calculated at 375 000 000 m^3. Altogether up to 1965 there were some 400 boreholes and hand-dug wells in the delta area. Said (Geological Survey MS) cited the average yield as 4000 gal/h and noted that the wells on average are pumped 12 h/day.

9. GROUND-WATER GEOLOGY, KHOR ARBAAT, PORT SUDAN WATER SUPPLY

Khor Arbaat has a catchment area of 4000 km^2, by far the greatest in the Jubal el Bahr el Ahmer in

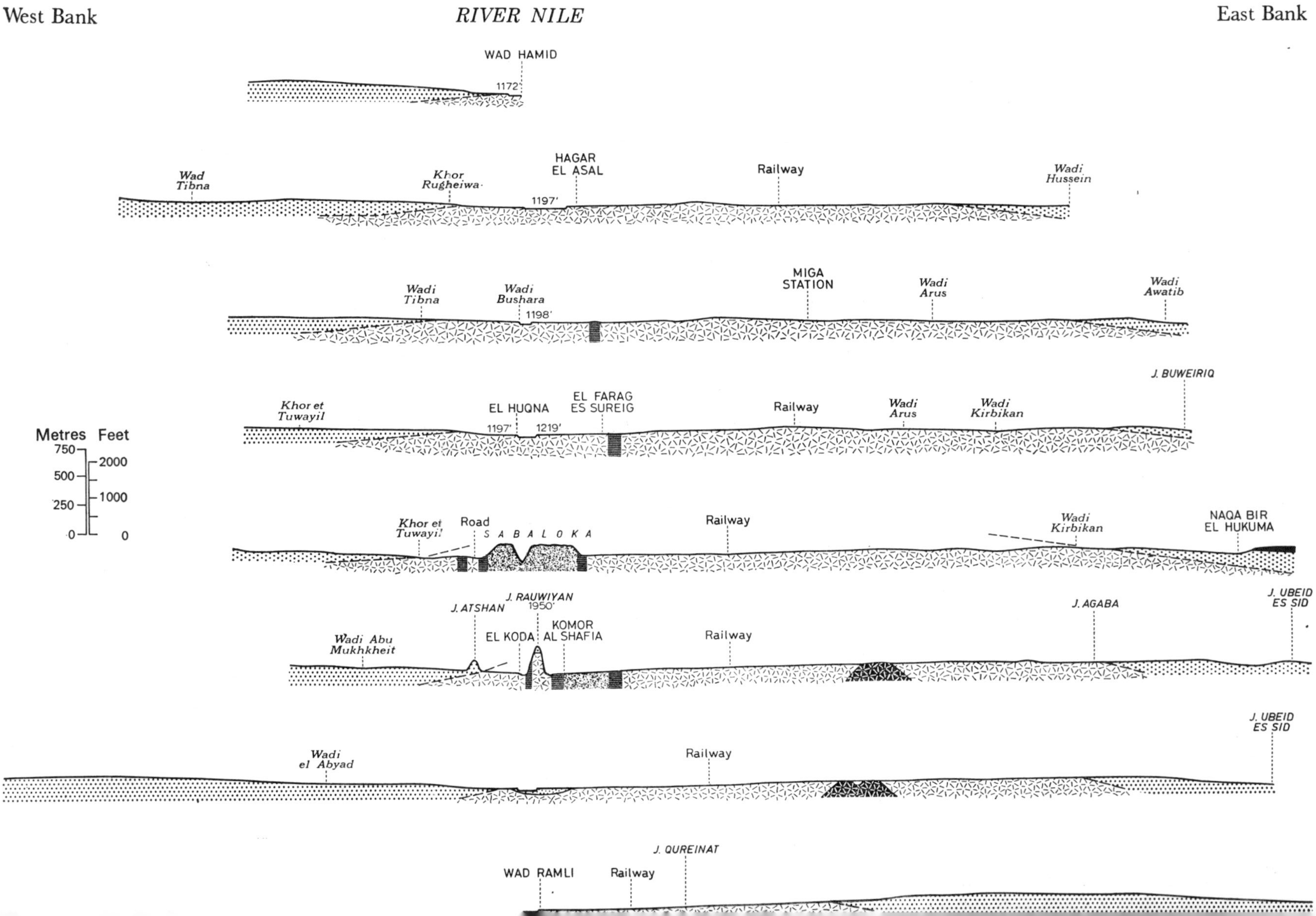

West Bank

RIVER NILE

East Bank

WAD HAMID
1172'

Wad Tibna
Khor Rugheiwa
1197'
HAGAR EL ASAL
Railway
Wadi Hussein

Wadi Tibna
Wadi Bushara
1198'
MIGA STATION
Wadi Arus
Wadi Awatib

Khor et Tuwayil
EL HUQNA
1197'
1219'
EL FARAG ES SUREIG
Railway
Wadi Arus
Wadi Kirbikan
J. BUWEIRIQ

Metres
750
500
250
0
Feet
2000
1000
0

Khor et Tuwayil
Road
S A B A L O K A
Railway
Wadi Kirbikan
NAQA BIR EL HUKUMA

Wadi Abu Mukhkheit
J. ATSHAN
J. RAUWIYAN
1950'
EL KODA
KOMOR AL SHAFIA
Railway
J. AGABA
J. UBEID ES SID

Wadi el Abyad
Railway
J. UBEID ES SID

WAD RAMLI
Railway
J. QUREINAT

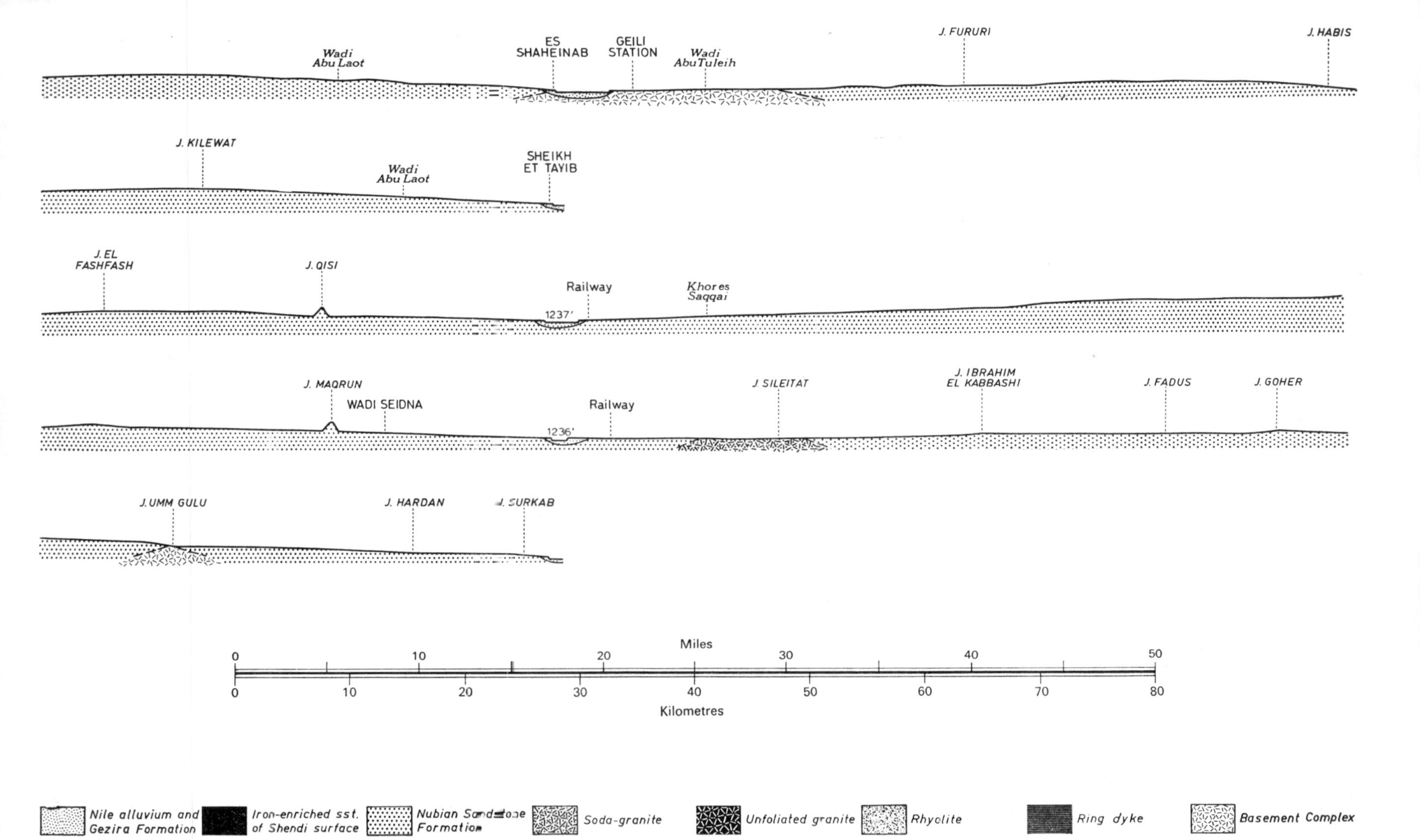

FIG. 79. East–West geological sections, Sabaloka–Khartoum region, Jebel Surkab to Wad Hamid, showing the distribution of the Basement Complex, the Sabaloka ring complex, the Nubian Formation, and superficial deposits (Based on University of Khartoum, geophysics group aneroid data, Sudan Geological Survey 1:4 000 000 maps, and geological observations by the author)

the vicinity of Port Sudan. The rainfall in the area is low and irregular and cannot be much in excess of 100 mm. The mean annual flow based on data collected over a period of 6 years is about 10 million m^3. According to Wedman *et al.* (1965) the ground-water stored in the aquifer amounts to about 48 million m^3 and the ground-water table is about 5 m below the surface.

Briefly the geology of the area is as follows. The solid geology consists of schists, greywackes, and slates with occasional bands of marble with intrusive diorites and quartz-porphyries. Xenolithic and non-xenothilic granites occur. There are parallel dyke-swarms consisting of acid, intermediate, and basic rocks. Doleritic and basaltic types are most common. The rocks are well jointed and cleavage is well developed in some of finer-grained meta-sediments. Wadi sands and gravels constitute the main aquifer and there is little water in the metamorphic and igneous rocks except joint water.

During flood periods, the greater part of the water flows to the sea and is lost. Aerial photographs available for the catchment indicate that sections of Khor Arbaat have a perennial surface flow. This occurs only where the valley is narrow and the alluvial fill is too thin to transport the discharge by underflow. Where the valley is wide and the fill thick then no dry-season run-off occurs.

The idea of utilizing Khor Arbaat as a supply for Port Sudan was advanced in 1912 but was not put into action until 1924. During this period the Khor Mog provided the town supply. In 1924, some 273 000 gal/day were abstracted from two shallow wells about 35 ft deep sunk in the khor bed. The water was piped to Port Sudan in 6-in gravity feed pipes. Tunnels were excavated in 1925 linking the two wells, but by 1941 the amount of water available was insufficient to satisfy the greatly increased consumption of Port Sudan consequent on expansion of harbour facilities, etc. An open well was dug in 1941 and all other wells connected to it by tunnels. More wells were dug in 1942 and output raised to 1 120 000 gal/day. In 1964 more wells were dug and output doubled. An electric pumping system now operates and the water is pumped to Port Sudan. Some of the wells can produce as much as 15 000 gal/h.

Wedman *et al.* (1965) concluded that 'the potential of Khor Arbaat aquifer is sufficient to meet the present and immediate future requirements of Port Sudan, provided adequate pumping and transport facilities are installed.' After a detailed consideration of the water balance of Khor Arbaat, Wedman *et al.* (1965) suggested that the maximum safe annual yield would be 10 million m^3, provided the surface and subsurface outflow were decreased by increasing infiltration by constructing low dams and wells.

An alternative source of ground-water for Port Sudan is the Khor Mog, but the possibilities here appear to be limited. Other alternatives are the ground-water of the superficial deposits and solid deposits of the littoral. Proximity to the sea and the presence of evaporites in the Tertiary sequence would necessitate close geological control if fresh water were to be produced. Virtually nothing is known about the potential of the littoral sediments. A few details and analyses are given in Grabham (1934), who also described some of the wells of the Jubal el Bahr el Ahmer at Sinkat, the Murrat wells, the Port Sudan area, etc.

VII

Carbon Fuels and Petroleum Prospects of Sudan

The Sudan is particularly poor in carbon fuels of all kinds and some 80 per cent of the surface area of the country is unlikely to yield such minerals.

1. COALS, ETC.

No workable deposits of anthracite or lignite are known in the Sudan. The (Palaeozoic) Carboniferous rocks which contain workable coals in the Colomb Bechar region of Algeria, and in Europe, occupy less than 0·5 per cent of the surface area of the Sudan and although they are of continental facies they are predominantly clastic. Rocks belonging to the Karroo Group, which in Africa south of the Equator contain large quantities of coal, do not occur. Prospects are therefore poor.

Small finds of lignite and low-grade coals have been reported from time to time but no workable deposits are known. According to Dunn (1911, p. 43) lignite was discovered by Bimbashi Hodgson in 1903 at Dongola, about 30 ft below the surface, as a carbonaceous bed in whitish clayey sandstone about 4 inches thick. Boring failed to discover any beds worth working. Dunn (1911) commented that the geology of the district does not point to there being any workable deposits and that the nearest coals are at Chelga, north of Lake Tana in Ethiopia, near Gallabat.

After reports by the District Commissioner (Merowe-Dongola), P. A. Moon (Sudan Geological Survey) was sent to investigate the presence of coals in the Hamur village and Dongola town areas (Abdullah 1955, p. 8). Two wells were dug in 1947 and a seam 2 ft 2 inches thick but of poor quality was reported. Moon noted that the samples were all of poor quality, except one from Dongola. The coal is of limited extent.

Abdullah (1966, p. 17) gave the following analysis for a coal seam (1·5 m thick) discovered in a well at Tangassi, upstream from Dongola.

Specific gravity	1·308
Moisture	12·20 per cent
Ash content	3·31 per cent
Calorific value	5460 cal/g
Calorific value	10 720 Btu/lb

A borehole was sunk for coal at Wad Kaba in 1949 in the Gedaref region. The results of this investigation are uncertain, for there are stories about the borehole having been 'salted'.

No lignite has been reported from the great swamp area of the Sudd but peat deposits exist and might be developed.

Anthracites are unknown but graphite occurs at a number of localities in the Basement Complex of the Butana and Qala en Nahl area. Dunn (1911, p. 91) reported graphite from the Bongo River district of Bahr el Ghazal Province and he noted that samples of impure graphite had been received in Khartoum from the Yambio–Meridi road. No workable deposits have been located but this may be due to inadequate prospecting.

Abdulla (1966, p. 213) recorded graphite from Qutum and Nuba Mountains. The Qutum occurrence is said to average 31 per cent C and to be amenable to concentration.

2. OIL AND GAS

(a) General

No surface indications of oil and gas are known in the Sudan, although AGIP's Durwara–2 borehole is reported to have encountered a pocket of gas. Details have not been published, however, and the hole was abandoned. The gas is said to be 20 per cent methane.

Prospects are clearly unpromising for a large part of the country; in fact the areas underlain by the Basement Complex, Nawa, Yirol, and Hudi formations, and the 'Tertiary' intrusive and extrusive rocks (some 52 per cent of the total area) may be written off from what is now known of the geology. Furthermore the great part of the remainder, underlain by the Nubian and Umm Ruwaba formations for some 47 per cent of the surface area, is a very poor prospect. The formations are thin and non-marine, and are therefore unlikely to yield oil in commercial quantities. The remainder, less than 1 per cent of the land area, is underlain by the predominantly marine formations of the Red Sea Littoral. Together with the sediments of the offshore belt of the Tokar Archipelago, these rocks constitute a potential petroleum province. Exploration in the Sudan has been limited to two areas:

(a) the Red Sea Littoral, where conditions are thought to be similar to those prevailing in the Gulf of Suez and along the Red Sea coast to Egypt; and
(b) the Jebel Uweinat region, Northern Province, where local interest was aroused because of the discovery of oil in Libya.

(b) History of exploration

(i) *Red Sea Littoral*

Lees and Wyllie (*vide* Abdullah 1955) of the former Anglo-Persian Oil Company visited the Red Sea Littoral in the 1920s with a view to assessing the petroleum prospects for the Condominium Government. They advised against further investigation because they though tthat source rocks were lacking; that the Cretaceous and Miocene strata, from which Egypt derives its oil, were also lacking; and that the few surface structures they found appeared to have been formed by basaltic intrusion and were too small and unlikely to yield oil.

Some geophysical work was done in the 1950s by specialists from the Overseas Geological Survey of Great Britain, and Abdullah (1955) presented arguments to disprove the case made out by Lees and Wyllie that the Sudan Littoral is in general a poor petroleum prospect.

Abdullah (1955) pointed out that

(i) there are possible source rocks, for example on Maghersum Island (Fig. 38), where black shales with carbonaceous plant fossils occur;
(ii) oil shows are said to occur on Dahlak Kebir, Eritrea and on the Tertiary coastal plain of Yemen;
(iii) potential reservoir rocks exist; and
(iv) the anticlinal structures of the Sudan Littoral exposed at the surface were not formed by basaltic intrusion, in fact most of them are unassociated with igneous rocks and are normal folds.

In 1959 a new petroleum law was passed with the rather unusual provision that all Government proceeds would be limited to 50 per cent of the profits of the first 10 years but after that might be increased to 70 per cent at the option of the Government (Hedberg 1959). This law eliminated such features of the Petroleum Resources Development Act 1958 as a proposed 70–30 per cent profits split in favour of the Government. The law as it then stood was 'as uninteresting as the Sudan's oil prospects' according to *World Oil* (1959). In August 1959 the Sudanese Government granted a 30-year concession along the Red Sea Littoral to the Ente Nazionale Idrocarburi's (ENI) subsidiary, Aziende Generale Italiana Petroli, and AGIP Mineraria Sudan, Ltd. was formed on 15 May 1959. The licences covered areas of 197 680 acres (800 km^2) each, a total of some 3500 square miles or 8545 km^2 (Fig. 80) (Hedberg 1960). A concession was also granted in 1959 to the General Exploration Company of California, covering 7 areas of 800 km^2 each, a total of about 2160 square miles.

Surface geological work, seismic, gravimetric, and magnetometric surveys were started. As a result of this exploration work AGIP 'spudded in' their first wildcat Durwara–1 borehole in June 1961. The hole was situated on a small island south of Port Sudan and was abandoned dry at 9521 ft.

During 1961 the ENI group proposed to construct an 8000 barrels/day refinery at Port Sudan to meet the Sudan's domestic requirements. The Sudan Government also invited Shell and British Petroleum to submit a joint proposal for the construction of a refinery. Eventually Shell–BP (Sudan), Ltd. was incorporated in 1962 in Khartoum to build and operate a 20 000 barrels/day refinery at Port Sudan. The Royal Dutch Shell Group and British Petroleum Co., Ltd. hold issued capital equally. The refinery came on stream in 1964.

The rig from Durwara–1 was subsequently moved to the Maghersum–1 location on Maghersum Island, near Mohammed Qôl (Fig. 80). This wildcat, sited on seismic evidence, was completed in 1962 at a depth of 7393 ft. Dungunab–1 drilled to 5133 ft and Abu Shagara–1 drilled to 7531 ft were completed also in 1962; all were dry.

According to Carella and Scarpa (1962) AGIP Mineraria Sudan, Ltd. carried out the following exploration work in the Red Sea Littoral from 15 May 1959 to 31 August 1962: (i) surface geology:

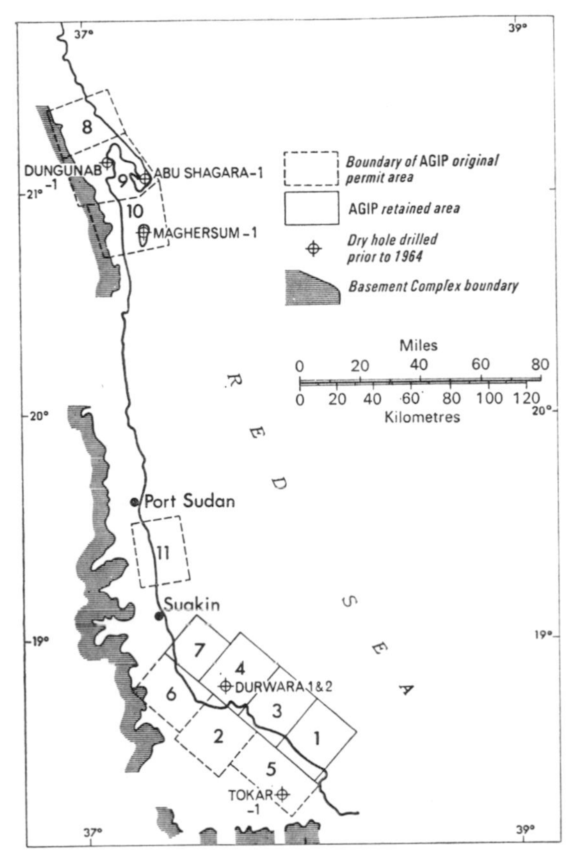

FIG. 80. Petroleum concessions, Red Sea Littoral (Based on *Bull. Am. Ass. Petrol. Geol.* 1964, p. 1253, last published concession map)

13 party months reconnaissance over 16 900 km², detail over 3710 km²; (ii) gravimetric and magnetometric surveys, 4½ party months; 2859 stations; (iii) marine seismic survey: 2 party months (total 1648 km of lines); (iv) land seismic survey: 26½ party months (total 1132 km of lines).

World Oil (1962, p. 199) reported that efforts had been made to form a joint Sudan–United Arab Republic state-owned company to search for oil in the Sudan. The Kuwait Government was also said to be interested in participating in the construction of a refinery in the Sudan.

Two dry holes were completed in 1963, Durwara-2 and Tokar-1 (Fig. 80). Durwara-2 was bottomed at 13 632 ft in basalts interbedded with sandstones. These rocks were referred to the Basement Complex.

T. Rocco (*in* Bowerman 1965) reported that no exploration or drilling was carried out during 1965 in the Sudan. However, AGIP Mineraria Sudan, Ltd. still held 1 320 000 acres in part of the Littoral.

According to local reports no exploration was carried out in 1966 but apparently attempts were made by the Sudan Government to interest the Roumanian, Kuwait, and Saudi Arabian Governments in assisting for the search for oil in the Sudan. An independent American company (Tenneco) was reported in 1967 to be interested in the off-shore area of the Suakin archipelago and the Digna Oil Company, a Sudanese–Kuwait-backed company, in association with the Continental Oil Company, obtained leases in this area. General stratigraphical information about wells drilled in the Red Sea Littoral is given in Tables 7 and 8.

Carella and Scarpa (1962) described the general results of the exploration work completed on the Red Sea Littoral by AGIP. These are discussed in detail in the section on stratigraphy and below. The author, in a short review article on the petroleum prospects of the Sudan, pointed out that the best area for investigation is the off-shore belt of the Suakin Archipelago (Whiteman 1963). Sestini (1965, pp. 1453–72) described the Tertiary and Holocene stratigraphy and structure of part of the Red Sea Littoral. Berry, Whiteman, and Bell (1966) gave a tentative Pliocene, Pleistocene, and Recent succession for the Red Sea Littoral.

(ii) *Jebel Uweinat region, Northern Province*

In 1959 the Royal Dutch–Shell Group (BIPM) and British Petroleum Company, Ltd. were requested by the Sudan Government to investigate the petroleum prospects of part of the huge uninhabited wilderness between the Nile and Jebel Uweinat. The Sudan Government granted a reconnaissance permit on 8 September 1959 for an area of some 480 000 km² west of the Nile between 16 and 22° N and 24 and 30° E. Shell–BP did not apply for leases, however.

According to the Shell–BP report (Sudan Geological Survey files) prepared by the BIPM exploration geologist A. F. Hottinger, the main purpose of the expedition was to study the stratigraphy and facies of the Palaeozoic and Mesozoic sediments in relationship to those known in southern Libya and Chad; and to study the type of structures, if any, present in the desert of the north-western Sudan.

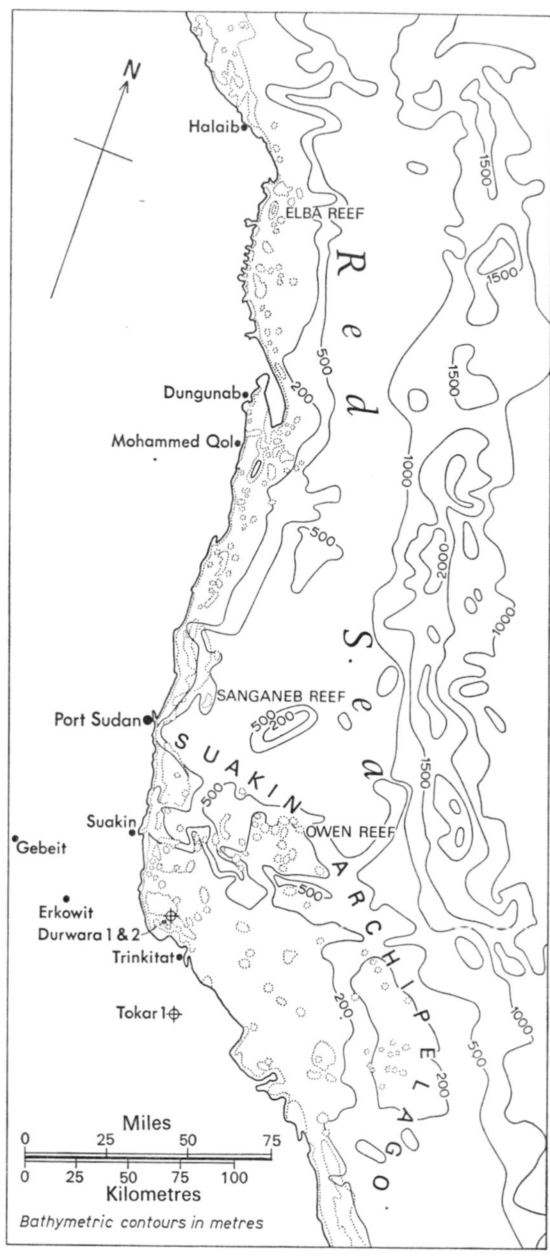

FIG. 81. Submarine topography and reefs, central Red Sea (Based on Admiralty Charts, and USGS–ARAMCO Map 1963, 1:2 000 000 scale)

The report concluded that the prospects of this region are poor, largely because much more Basement Complex is exposed than is shown on the 1:4 000 000 Sudan Geological Survey Map (second edition); the sedimentary formations are thin and largely of continental facies; and because the Palaeozoic formations pinch out within a few miles of entering the Sudan frontier (Fig. 26).

Whiteman (1964*b*) also pointed out that the petroleum prospects of the north-western Sudan are poor and that the geology is unlike the conditions prevailing in the Libyan oil fields, which are situated north of latitude 28° N.

Burollet (1963), who with a party of CORI (Italian) and CPTL, CFP, and BRP (French) geologists visited the Uweinat area, reported similarly that the petroleum prospects of this area are poor. He mentioned that the Palaeozoic sediments are intruded by granites of the Jebel Uweinat ring complex (Figs. 19, 20, and 21). Details are discussed below.

(c) Petroleum geology, Red Sea Littoral and off-shore belt

(i) *General*

The sediments of the Red Sea Littoral extend from the Egyptian frontier near Halaib to the Eritrean frontier, a distance of approximately 400 miles (Fig. 38). Their outcrop varies in width from less than a mile near Khor Tibadeb to over 20 miles in the Ras Abu Shagara region (Carella and Scarpa 1962, Sheets 1 and 2) and over 40 miles in the Baraka delta area (Sudan Geological Survey, second edition 1:4 000 000 map).

The off-shore belt extends at right angles to the coast at least 50 miles off Trinkitat (Fig. 81). The standard succession, proved mainly for the littoral north of Port Sudan, is shown in Tables 5, 7, and 8 and Fig. 39. The rocks range in age from late Cretaceous (or Palaeocene) to Recent. They consist of a series of terrigeneous clastic rocks, derived from the Basement Complex formations of the Jubal el Bahr el Ahmer, which intertongue with marine and lagoonal sediments laid down along the margin of the Red Sea Depression. Carbonates in general are subordinate to the clastic sediments and the beds are like the present coastal deposits in lithology (Plate 21). Two main thick evaporite horizons occur, separated by the predominantly calcareous and conglomeratic Middle Miocene Abu Imama Formation and A Member of the Maghersum Formation.

The contact between the Tertiary and Mesozoic sediments and the Basement Complex that occurs at the foot of the Red Sea Hills escarpment from Halaib in the north to the Eritrean frontier in the south is clearly a nonconformable overlap (Fig. 38). At least six unconformities have been recorded within the sedimentary sequence. The thickness is therefore highly variable but in general the beds thicken seawards. Basins occur but have yet to be defined accurately. The composite sequence given by Carella and Scarpa (1962) for the area north of Port Sudan totals more than 9670 ft. The Durwara–2 borehole at the edge of the off-shore belt proved 13 632 ft of sediments; and structural and seismic data indicate that the sedimentary column may be at least 20 000 ft thick on the shoulders of the submerged central trough.

(ii) *Summary of stratigraphical history*

Quite clearly the Red Sea Depression has been an area of general subsidence for a considerable part of geological time. The author has expressed the view that the great fretted escarpment that bounds the Jubal el Bahr el Ahmer on the east in the Sudan was initiated by the processes of pediplanation operating on a monoclinal flexure similar to the Lebombo monocline of South Africa, and that the Mesozoic, Tertiary, and Quaternary sediments accumulated in a megasyncline on the down-side of this structure (Whiteman 1965). The escarpment in the author's view is not a fault scarp nor a faultline scarp (Whiteman 1968).

Pre-Cambrian–Palaeozoic. The reticulate dyke pattern of the Bir Salala area may well indicate that a tension phase existed even as far back as late Pre-Cambrian times (Whiteman 1965, 1968) and evidence from the Gulf of Suez indicates that a distinct trough had already developed by Carboniferous times; its southernmost limit is unknown however.

Mesozoic. The earliest post-Basement Complex rocks encountered in the Sudan are ? late Jurassic–? early Cretaceous coarse clastics interbedded with basalts in the Khor Shinab district and the late Cretaceous–Palaeocene sediments of Maghersum–1. By late Cretaceous–Palaeocene times a zone of subsidence (limits uncertain) had developed on the down-side of the flexure, and the marine deposits of Maghersum (Carella and Scarpa 1962) and Jeddah (Karpoff 1957) indicate that the sea had already penetrated the central Red Sea area. Carella and Scarpa (1962) described the environment in which the rocks at Maghersum–1 were laid down as mariner to brackish.

Eocene. Little is known about the development of the megasyncline in Sudan in Eocene and Oligocene times. It is possible that the very thick (1070 m) Maghersum evaporites (B and C members) are older than Middle Miocene (Helvetian), the age given to the A Member by Carella and Scarpa (1962), and are lateral equivalents of the gypseous marls of the Suez region of Eocene age, if we follow the classification proposed by Said (1962) (Table 20). In the absence of fossils it is impossible to be more precise. If we accept this correlation then the Hamamit Formation may be in part the clastic continental facies of the Maghersum anhydrites, and marine conditions may have preceded the Oligocene uplift or regression.

Late Eocene and Oligocene. Evidence from Ethiopia indicates that the uplift of the Afro–Arabian swell took place during the late Eocene and that the extrusion of the gigantic mass of the Ethiopian plateau lavas (Grasty *et al.* 1963) followed during the Oligocene. The climax of Rift Valley faulting in Ethiopia is assigned to the Lower Miocene (Dainelli 1943, Mohr 1962) but how far these faults extended northwards is not known. Large-scale crustal separation took place in the Red Sea at this time.

In the Jubal el Bahr el Ahmer and the Nile Valley the Eocene seems to have been a time of erosion. The Shendi–Togni surface of the Nile Valley and Jubal el Bahr el Amer was probably formed in part during this time. At the moment, however, it is not possible to give definite ages to the higher erosion surfaces of the Jubal el Bahr el Ahmer, such as the 5000±-ft Jebel Okwer surface or the 3000±-ft Erkowit surface, because of lack of datable deposits, structural data, adequate topographic maps, etc. They may be older than the Shendi–Togni Surface, but a more probable explanation is that they have been structurally displaced. Erosion surfaces are also developed on the main escarpment of the Jubal el Bahr el Amer bearing witness to the cyclical development of the scarp, but, as so frequently happens, it is not possible to correlate erosional and depositional events, except in the broadest way (Fig. 82).

The erosion products formed during the periods when these great surfaces were cut may well be

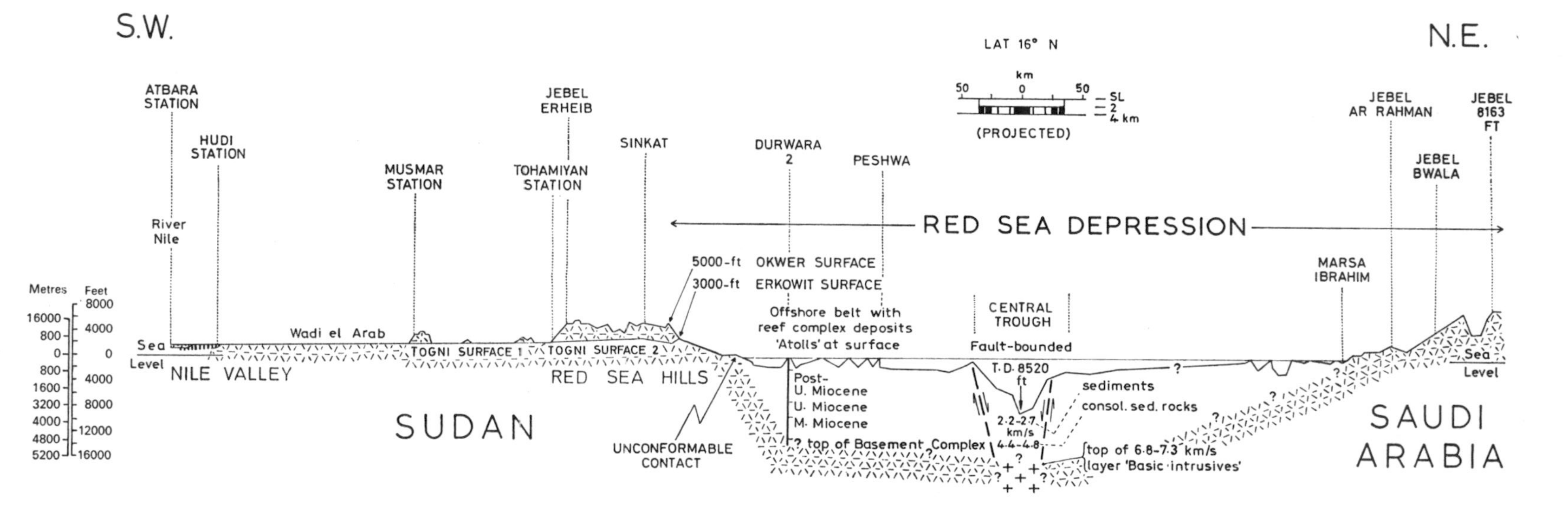

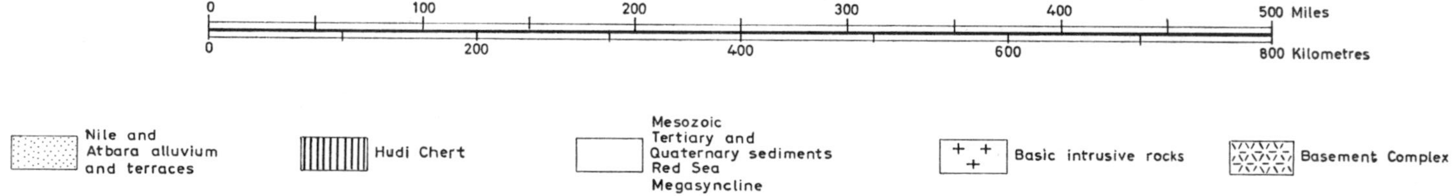

FIG. 82. Geological section, Nile Valley–Red Sea–Arabia, showing the nature of the Red Sea depression (Based on University of Khartoum geophysical expedition Red Sea Hills 1962. Drake and Girdler 1964, Vine 1967, and British Admiralty charts)

Note: Since this diagram was drawn it has been shown that the zone of separation between the African and Arabian plates is much wider than the Central Trough and is some 130–170 km wide in the Central Red Sea. The Basement Complex pattern should end and be replaced by Basic Intrusives and 'new crust' within a few kilometers of the shore.

represented in the predominantly clastic Hamamit (? Eocene–? Lower Miocene) Formation. How far these clastics extended seawards is equally uncertain but we can envisage intertonguing with marine sediments in the off-shore belt and in the zone now occupied by the shoulders of the central trough.

Oligocene time thus appears to have been a period of general uplift in the Red Sea area; and in Egypt, the Nile Valley, and the Gulf of Suez there was a great extension of continental deposits also (Tables 7 and 20).

Miocene. Throughout the Red Sea area the Miocene period opened with a transgression and general subsidence ensued. In the Gulf of Suez section of the megasyncline basalts were extruded and sands and conglomerates were laid down. The boulder beds include granitic and metamorphic rocks from the Atbai Range, and flints and basalts from the Oligocene flows. Similar deposits were probably formed to the south. Most important was the establishment of a sill at Ayun Musa (Said 1962), which subsequently developed and eventually controlled the deposition of Miocene evaporites in the Red Sea (Table 20).

The Miocene Red Sea extended southwards from the Ayun Musa sill in the north almost to the Bab el Mandeb; and the connection was clearly with the Mediterranean, as Gregory (1921) and others pointed out.† In the Sudan littoral and off-shore belts, marine conglomerates, marls, and sandstones with subordinate limestones of the Magnersum A Member were laid down.

Uplift of the rift shoulders, perhaps only local, caused the influx of the conglomerates and marls with a sandy matrix that constitutes the C Member of the Abu Imama. The pebbles are said to be predominantly igneous (Carella and Scarpa 1962). Transgression continued and the predominantly calcareous A and B members of the Middle Miocene (Tortonian) and the Abu Imama Formation were laid down. These deposits are probably equivalent to the *Globigerina* marls of the Gulf of Suez. During the deposition of the *Globigerina* marls some anhydrite was deposited. Evaporite-producing conditions became more widespread and the Evaporite Formation was laid down. In places such as Tawila–2, Gulf of Suez, over 3360 m of evaporites accumulated, but in general in Egypt the sequence is thinner and there is an alternation of lagoonal and marine deposits (Table 20).

Most probably evaporite-forming conditions intensified southwards, and rock-salt interbedded with anhydrite constituting the A Member (658 m) and gypsum interbedded with clays and sandstones constituting the B Member (54 m) were laid down. In the structurally deeper parts of the Red Sea Depression these evaporites probably attain very great thickness. On structural highs limestones were probably formed.

Pliocene. Regional subsidence ensued in the Pliocene and there was an influx of the sea from the south this time, connecting the Red Sea with the Indo-Pacific Province via the Gulf of Aden. The central Red Sea trough was formed at this time.

Gravels and grits with anhydritic bands and with occasional beds of *Ostrea gryphoides* were laid down at the north end of the Gulf of Suez. These passed laterally into marine facies south of Jebel Zeit in Egypt; in the Sudan uplift of the Rift shoulders caused an influx of coarse clastic sediments, the Upper Clastic Group of Sestini (1966). They rest with unconformity on the Dungunab and Abu Imama formations and may pass laterally into lagoonal and reef facies (Abu Shagara Formation) on the shoulders of the central trough.

Tilting and faulting took place after deposition of the Upper Clastic Group; these deposits dip seawards at 10–25°. In Egypt a regional unconformity was developed.

Pleistocene and Recent. The Pleistocene and Recent periods appear to have been a time of general stability along the Sudanese littoral and a sequence of elevated reefs was formed. Marine benches were cut into them from 16 to 2 m. The oldest of these is probably early Monastirian and the 2-m bench is probably the Recent Albrolhos stage (2100–2600 years B.P.). Most of these wave-cut features are horizontal for long distances along the littoral, so demonstrating stability (Berry, Whiteman, and Bell 1966).

The Red Sea mersas and the boat channel were excavated during the Würm–Wisconsin emergences between 75 000 and 10 000 years B.P. Terrace gravels and wadi deposits were swept in from the Red Sea Hills during the Pleistocene and Recent and these cap the elevated reefs in places.

† Heybroek (1965) mistakenly cites Montanaro (1941) as the first to draw the conclusion that the Red Sea was occupied by Mediterranean waters in the Miocene.

(iii) *Structure*

The structure of the Red Sea Littoral and the off-shore belt are described in Chapter V and adequate structures exist to provide traps for any oil that may have been generated. The thick salt sequences recorded, and the high temperature encountered in Durwara–2, point towards the possible development of halokinetic structures especially in the Suakin Archipelago (Fig. 81). None has been recognized, however, but salt piercement structures have been described by Macfadyen (1930), Lees (1931), and Wade (1931) from the coastal plains of Yemen and Asir, and recently a Middle to Upper Miocene age has been proposed for this salt (Klaus *in* Heybroek 1965).

(iv) *Petroleum prospects*

Among the essential prerequisites for an oil or gas field or a potential petroleum province we may list the following: (1) the presence in the area of a thick unmetamorphosed sequence of sedimentary and predominantly marine strata; (2) the presence of porous reservoir rocks; (3) the presence of impervious roof rocks; (4) favourable structural and stratigraphical conditions, for example anticlines, faults, unconformities, wedge belts of porosity, etc., which could act as traps; (5) evidence of source rocks; (6) actual evidence of oil or gas in or near the area in the form of tar seeps, oil shows, gas flares, producing wells, etc.

From the stratigraphical and structural data presented above, the Sudan Red Sea Littoral and the off-shore belt of the Suakin Archipelago (Fig. 81) may be classed as a potential petroleum province. This applies especially to the Suakin Archipelago, where the sedimentary column is two to three times as thick as in the littoral. The presence of thick salt-beds and other evaporites in a major rift zone enhances the petroleum possibilities of the area. In addition, the many islands form natural drilling platforms and large areas on the shoulders of the central trough are covered by only a few fathoms of water. The occurrence of gas in AGIP's Durwara–2 well is an extremely encouraging sign. AGIP's failure to bring in a major producer on the land should not be allowed to detract from the excellent prospects of the off-shore belt of the Suakin Archipelago.

(d) Petroleum geology, north-western Sudan, Jebel Uweinat area

From time to time interest has been aroused in the petroleum prospects of this area, mainly because of its proximity to Libya, which has become an important producer in recent years. Local belief is very strong in the existence of petroleum in this area.

The investigations of the Royal Dutch Shell–British Petroleum Party under the leadership of the late A. F. Hottinger and the investigations of Italian and French geologists (Burollet 1963) indicate, however, that the petroleum prospects of this region are poor.

First it should be remembered that very little oil has been discovered south of latitude 28° N in Libya and that the conditions that have enabled petroleum to accumulate in quantity in Libya do not extend into the Sudan (Fig. 26).

Secondly, as Hottinger *et al.* (1958) pointed out, there are five other factors.

(1) Compared with sections in Chad and Libya, the pre-Nubian sequence of the north-western Sudan is much thinner.
(2) In the Ennedi and Mourdi depressions the pre-Nubian sequence wedges out completely and in the Arbaaïn (20° N, 28° E) and Jebel Rahib (18° N, 27° E) areas, etc. thin continental Nubian rests unconformably at the surface. The present topography is cut into this sequence, exposing the unconformity at many points.
(3) No truly marine rocks of Palaeozoic or Mesozoic age are known from the north-western Sudan.
(4) The Basement Complex outcrops are much more extensive than was previously realized. Their distribution is shown incorrectly on the 1:4 000 000 geological maps of the Sudan.
(5) Structural features indicating down-faulting or down-warping or folding were not recognized.

It appears then from the thin sedimentary cover, and an absence of cap rocks, structures, and closed systems that the petroleum prospects of this area are low. Furthermore, Burollet (1963) reported that the Jebel Uweinat ring complex granites were intruded into the Carboniferous sediments and volcanic rocks, so reducing the petroleum prospects of these rocks still further in the environs of Jebel Uweinat.

The stratigraphical sequence in this area is shown in Table 5 and the general geology in Figs. 19, 20, and 21.

To the south in Darfur an unconfirmed report of an oil seep was made by a Sudanese Government official in 1965. No official confirmation concerning the nature of the seep has been made, however, and the report is suspect. In the author's view the petroleum prospects of this area are similar to those of Jebel Uweinat, because again the Palaeozoic sediments thin into the Sudan until ultimately the continental Nubian Formation rests on the Basement Complex.

VIII

Minerals (Excluding Carbon Fuels)

1. GENERAL

MINERALS have been mined in the Sudan since Early Dynastic times and a wide variety of minerals is known. At least 240 localities, scattered mainly throughout the Basement Complex formations, have been recorded (Fig. 84). Many of these have been mined on a small scale by local tribes but few mines are still active. Most of the localities listed below can be classed only as mineral occurrences and only about thirty have been studied by geologists. Fewer than three have so far been explored in detail.

A list of formations, etc. occurring in the Sudan is given in Table 1. Most minerals occur in the Basement Complex formations. Information on the nature and distribution of minerals is, in general, scattered and incomplete, and for many minerals only the most rudimentary statistics are available (see Table 21). Often these are unreliable and figures vary from source to source. Prior to de Kun's (1965) short article in *Mineral Resources of Africa*, the only published comprehensive account dealing with Sudan's minerals was by Dunn (1911). Information on reserves is available for only four deposits and this is incomplete.

TABLE 21

List of minerals and number of recorded occurrences in the Sudan

Minerals and non-metallic minerals	No. in Fig. 84	No. of localities
Precious metals		
Gold	1–74	74
Silver	75–76	2
Non-ferrous metals		
Copper	77–108	32
Lead	109–115	7
Zinc	109	1
Tin	116–117	2
Aluminium		
Iron and ferro-alloy metals		
Iron	118–131	14
Manganese	132–159	28
Chrome	160–166	7
Molybdenum	167–169	4
Tungsten	170–173	3
Minor metals		
Arsenic	2	1
Magnesium	174–176	3
Uranium	177–178	2
Columbium	179	1
Titanium	180	1
Barium	181–183	3
Non-metallic minerals and substances		
Clay		
Kaolin	184–185	2
Fluorspar	186	1
Talc	187–189	3
Building stones	190–195	6
Sand and gravel	196–197	2
Cement	198–199	2
Gypsum	200–211	12
Lime	212	1
Magnesite	213–214	2
Vermiculite	215–216	2
Wollastonite	217	1
Asbestos	218–219	2
Pumice	220–221	2
Diatomite	222–223	2
Mica	224–228	5
Salt	229–230	2
Natron	231–233	3
Abrasives		
Quartz	234–235	2
Glass sand	236	1
Paints and pigments	237	1
Phosphorite	238	1
Gemstones	239–240	2

At the time of writing no international mining companies are involved in ventures in the Sudan, and in the past only two major companies—Tanganyika Concession Ltd. and the Anglo-American Corporation of South Africa—have operated on a substantial scale in the country. Italian mining interests held leases prior to 1967 on the Hofrat en Nahas copper area, and a Japanese company did a limited amount of exploration work searching for high-grade deposits. International and aid-backed agencies have operated in the Sudan on a small scale only in the Shereik area, Northern Province, where a United Nations mining engineer investigated mica deposits, and in the Sufaya area, north of Jubal el Bahr el Ahmer, where the Daniel, Mann, Johnson, and Mendenhall Corporation made a feasibility study for U.S. AID of the Sufaya iron deposits. In 1967 the United Nations Development Programme and the Sudan Government initiated investigations of Hofrat en Nahas copper, Shereik mica, and Ingessana chrome.

Modern mining methods were introduced into the Sudan in the early 1900s, mainly by British companies interested primarily in finding gold. During the first 40 years of this century about eight mines were opened up, but with the exception of Gabeit Mine, northern Jubal el Bahr el Ahmer, all are now closed (Fig. 84). From 1914 to 1939 some 126 028 oz (worth £1 572 747 at the current price of gold) were extracted from Gabeit and annual production averaged about 6000 oz throughout the 1930s. After independence many local people became interested in mining and a number of companies were formed (Table 22). Little local capital was invested (excluding the cement industry) and today the number of people actively involved in the industry is small. Most of the workings can only be described as primitive.

TABLE 22

List of Sudanese mining companies (incomplete)

	Name of company or of lease	Mineral prospected for or mined	Date of registration	Expiry or liquidation
1.	Sudan Exploration Syndicate (Sudan Exploration Company) (Egyptian Sudan Exploration Company)	Au	1898	1902
2.	Sudan Gold Fields Ltd.	Au	1901	
3.	London and Sudan Development Syndicate	Au	1901	1911
4.	Hegatte Concession, Jebel Elba Concession, and Ogilvie Haig's Concession	Au	1902	1908
5.	Sudan Exploration Company	Au	1903	1906
6.	Nubia (Sudan) Development Company	Au	1903	1907
7.	Gabeit (Mining) Syndicate (The Sudan Mines Ltd.)	Au	1903	1908
8.	Dongola Concessions	Au	1903	1906
9.	Egypt and Sudan Mining Syndicate	Au	1904	1907
10.	Suakin Mining Syndicate	Au	1904	1906
11.	Tokar Prospecting Syndicate	Au	1905	1905
12.	Victoria Investment Corporation	Au	1905	1906
13.	Gabeit Mining Syndicate	Au	1903	1903
14.	Wadi Oyo (Sudan Gold Mining Syndicate Ltd.)	Au	1923	?
15.	Garabein Mining Syndicate	Au	1932	1935
16.	Gabait Gold Mines Ltd. (Sudan Gold Mines Ltd., 1903) (Gabait Mining Syndicate Ltd. 1903)	Au	1933	1942?

TABLE 22—*continued*

	Name of company or of lease	Mineral prospected for or mined	Date of registration	Expiry or liquidation
17.	Kassala Gold Mines (Sudan) Ltd.	Au	1934	1937
18.	Sudan Portland Cement Company	Limestone and clay	1946	—
19.	Atbai Gold Ltd.	Au	1947	1953
20.	Nile Cement Company	Limestone and clay	1952	—
21.	Sidki Saleh Bros. (Sufaya)	Fe	?1957	—
22.	African Mining Company Ltd.			
23.	National Mining Company (Omar Abu Amna)	Mn, marble, etc.	1950	
24.	Sudan Mining Corporation	Fe, Mica	1957	
25.	Mining and Trading Company Ltd.	Cr	1960	
26.	Al Saad Mining Company Ltd.	$CaSO_4$		
27.	Central Desert Mining Company Ltd.	Au, Fe	1953	1964 Transferred to Fodikwan Co., Ltd.
28.	Fodikwan Company, Ltd.	Fe	1964	—
29.	Abdl Moneim Samkary Co. Ltd.	Mn, Cr	1960s	

TABLE 23

Capital of companies actively engaged in mining in the Sudan

Name	Capital (Pounds Sudanese)	Percentage foreign capital	Minerals mined
Mining and Trading Co. Ltd.	100 000	100	Chrome
El Saad Mining Co. Ltd.	12 500	0	Gypsum
Fodikwan Company Ltd.	104 000	50	Iron
National Mining Co. Ltd.	5000 (Est.)	0	Manganese quartz, building materials
Abdl Moneim Samkary Co. Ltd.	5000 (Est.)	0	Manganese, chrome, etc.
Total	LS 226 500		

In Table 23 are listed the amounts of capital invested in Sudanese companies actively engaged in mining in the Sudan (excluding salt and cement companies). The role played in the Sudan's economy by the mining industry is small (Table 24). In 1965, for instance, the Sudan exported minerals and mineral products worth only £S200 242 (£S1 = £1-0-6*d* sterling) and in 1966 it is estimated that mineral exports totalled less than £S350 000. For comparison Uganda, a country much smaller than the Sudan, exported minerals worth £3 507 844 in 1960, whereas from 1960 to 1966 (April) the Sudan exported only £S560 459. Mineral production in the Sudan, according to De Kun (1965) and the Sudan Geological Survey (1965), is given in Tables 25 and 26. These figures are, however, incomplete.

TABLE 24

Minerals and mineral products exported (Data from Sudan Customs Department)

Mineral	1960	1961	1962	1963	1964	1965	1966 (Jan.–Apr.)	1960–6 (Total)
Gold	24 603	5200	8560	—	3000	—	—	
Iron ore	9975	—	3640	—	—	164 606	33 136	
Manganese	—	—	—	—	—	5688 ?	8516	
Chromite	—	—	—	2326	77 920	—	84 216	
Mica	49	—	—	—	67	—	—	
Common salt	26 911	14 815	9020	7281	5788	11 513	4602	
Lime	—	314	—	—	—	—	—	
Bricks	45	52	—	—	—	—	—	
Grinding stones	18	—	—	—	—	—	—	
Natron	—	—	—	13 688	6533	18 435	—	
Total	61 601	20 381	21 222	23 295	103 248	200 242	130 470	5 600 459

TABLE 25

Mineral production since the 1950s in the Sudan (incomplete) (Based on de Kun 1965)

Salt	60 000 tons since early 1950s
Gypsum	5000 tons in early fifties
	2000 tons per year 1956–9
Iron ore	20 000 tons in 1962
Manganese	7000–9000 tons per year 1956–8
Gold	1700 ounces per year 1951–5
	6000 ounces total 1956–8
	2500 ounces 1959–60
Mica	6 tons in 1957
	190 tons in 1958

TABLE 26

Mineral production, Sudan according to Sudan Geological Survey (Al Ayam Newspaper 1965)

Minerals extracted 'since last year': presumably some are totals and others are annual but this is not made clear (cf. de Kun 1965, Table 22).

	tons
Salt	60 253
Gypsum	2830
Iron Ore	28 534
Manganese	500
Mica	80
Chrome	30
Cement	90 722

2. PRECIOUS METALS

(a) Gold

(i) *History of development*

Gold has been mined in the Sudan for over 3000 years. A comprehensive account of the gold mines in Kush and Wawat, northern Sudan and southern Egypt, was given by Vercoutter (1959), who stated that the ancient mines were not worked before Middle Kingdom times (1900 B.C.). The distribution of ancient gold mines is shown in Fig. 83 (which is based on Vercoutter 1959, Fig. 2). Although these old workings are numerous, most of them are small and the total quantity of gold extracted, even over the centuries, must have been limited.

The modern search for gold in the Sudan may for all practical purposes be considered to have started with the expeditions of Mohammed Ali Pasha when his son Ismail Pasha raided the Fazugli area, near the Ethiopian frontier for gold and slaves in 1821 and 1822. During the next few years a number of prospectors, either employed by the Egyptian Government, or working on their own account, visited the Sudan. They included Ruppel, who prospected in Kordofan; Brocchi and Ginsberg who prospected in the Sennar area; and Linant who prospected in the Red Sea Hills (Hill 1963, p. 66). The Austrian mining engineer, von Rüssegger, and Boreani also prospected in the Sudan. Rüssegger's party, under the protection of 1000 troops, worked

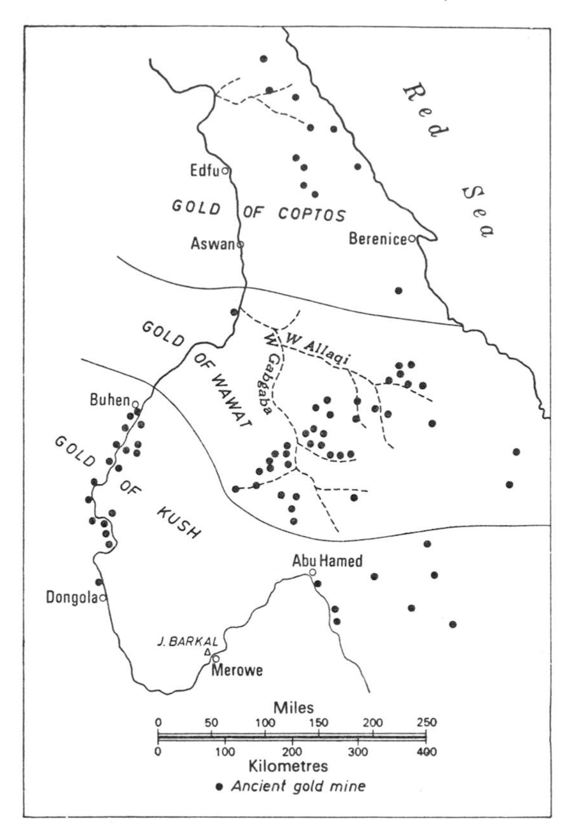

FIG. 83. Ancient gold mines of Sudan and Upper Egypt (Based on Vercoutter 1959)

with little result in the Fazughli area of the Blue Nile. Boreani's party also had little success. In 1838 Rüssegger prospected for gold in Kordofan and also inspected some of the iron ore deposits of the province. In 1838 the Pasha himself came to Khartoum and in 1839 reached Fazughli near the Ethiopian frontier, so great was the lure of gold. His prospectors found little, however. They returned to Egypt realizing that the Sudan was not an 'El Dorado' littered with gold nuggets, and that any gold present would have to be won by costly mining or the tedious process of panning. Mohammed Ali Pasha was not deterred for long, however, and in 1843 Ahmed Pasha Maniliki was sent with a body of troops to the Sudan to look for gold. During 1843–4 he visited the Sennar region but again without success. In 1845 he was replaced by Khalid Pasha Khusraw, who was charged by Mohammed Ali to search for minerals in Kordofan. Khalid visited Jebel Shaibun in Kordofan in search of gold but was unsuccessful. In desperation he tried his luck far to the east in the Beni Shangul country along the Sudan–Ethiopian frontier, where he was joined by Kovalevsky, an engineer lieutenant-colonel from the Russian Ural mines, sent by the ageing Mohammed Ali Pasha (Hill 1963, p. 83).

Kovalevsky (1849) found traces of gold and planned to extract it using modern techniques. However, soon after Kovalevsky returned Khalid reverted to the old wasteful local methods of extraction and, as Hill (1963) picturesquely puts it:

'Only a trickle of gold came from the mountains. So ended the great search which had begun with Ismail Pasha twenty-seven years before. The natives unconcernedly continued their ancient panning, wringing small profits from the unwilling sand.'

After the formation of the Condominium Government in 1899, Egyptian financial interests were once again attracted to the Sudan. Sir Ernest Cassel, who was already committed financially in Cromer's Egypt, became interested in the search for a Sudanese El Dorado. According to Hill (1963, p. 63) Cassel, Palmer (governor of the National Bank of Egypt) and other financiers, formed the Sudan Territories Exploration Syndicate (later changed to Egyptian–Sudan Exploration Company). They held prospecting licences in the Nuba Mountains, Kordofan, in the valley of the Blue Nile, and on the Red Sea coast, and were mainly interested in gold. Slatin Pasha accepted service with the company. He acted as an intermediary between the company and the Sudan Government, and as an expert on the tribes of the Sudan. He was joined by Goltlob Linck, a mineralogist from the University of Jena and an Englishman (?) named Pearless. The two scientists were engaged at £E150 per month with all found, and were to search for gold, silver, and other metals, and also oil and coal in the Nuba Mountains area (Hill 1963). The party travelled through newly pacified Kordofan in 1900 and visited among other places, Jebel Kaderu, Dilling, Jebel Tagor, and Jebel Tagali and eventually reached the White Nile, near Jebelein. Despite their more advanced prospecting techniques only traces of gold were found. Pearless visited the area again in 1902 but again without success (Hill 1965, p. 64).

During the next few years the impression that vast unexploited mineral wealth existed in the

Sudan drew many prospectors and mining companies to the country. No doubt this was founded on the knowledge of the existence of ancient Egyptian gold and fostered by travellers' tales and accounts like von Rüssegger's. Dunn (1911), an officer of the newly-founded Sudan Geological Survey, described ancient gold-mining methods and old workings and gave some details from prospectors' reports. He pointed out that Um Nabardi (Umm Nabari) Mine, Loc, 4, Fig. 84 was the only place in the Sudan where gold was then produced by mining and crushing gold-bearing quartz.

The placer workings of the Beni Shangul country, Sudan–Ethiopian frontier area, are supposed to have yielded annually some £80 000 worth of gold in addition to large quantities of gold used in Ethiopia for ornaments. The placer deposits of Beni Shangul, Tumat, and Dabus districts are said to have produced three-quarters of this gold. Dunn (1911) mentions that some £15 600 of gold was received in Addis Ababa in 1906 from these sources. At Tumat some 2000 natives were involved in the gold workings. The amount of gold produced in the Sudan at this time was probably smaller. Dunn (1911) estimated that it was unlikely that more than 200 oz were produced annually from placer deposits. He noted also that rings of pure gold from Fazughli were nearly always available in Omdurman Suq. Information on gold-prospecting in Sudan between 1900 and 1911 based on Dunn (1911) is presented in Table 27.

After Dunn's assessment (1911) little appears to have been published on this topic for more than 30 years. In the records of the Ministry of Commerce it is noted that the Gabait Mining Syndicate, a gold-mining company, was registered in 1912. The government geologist reported that the mine was closed in 1930, and the company was struck off the Register of Companies in July 1933. Beyond the fact that this company had shares in Wadi Oyo (Sudan) Gold Mining Syndicate (registered in 1923) little is known now about its activities.

TABLE 27

Gold-prospecting in the Sudan prior to 1911 (Based on Dunn 1911)

Name of company	Date prospected	Licence expired or concession abandoned	Prospected by	Mining started	Mining stopped	Remarks
Sudan Gold Fields Ltd.	Nov. 1901		A. Llewellyn	May 1902	May 1903	Mining operations at Om Nabardi (Umm Nabari) Mine, Northern Province started 1902. Thirty-two ancient workings listed. May 1908: trial crushing. Sept. 1908: continuous crushing started Value and Weight of Gold Abstracted: Up to 1910 £S 45 308 1916 £S 54 521 14 007 oz. 1917 £S 55 046 14 115 oz. 1908–1917 £S 360 482
Egyptian-Sudan Exploration (Syndicate) Company	Mar.–May 1900 Nov. 1902–Jan. 1903	Abandoned 1903	G. Linck and Pearless G. A. Wright			Prospecting in Kordofan with Slatin Pasha. Old washings of Tira Mandi visited. Financed by Cassel and Palmer (Governor of National Bank of Egypt)

TABLE 27—*continued*

Name of company	Date prospected	Licence expired or concession abandoned	Prospected by	Mining started	Mining stopped	Remarks
London and Sudan Development Syndicate	1901–2 1902–3 1903–4 1904–5	 31 Aug. 1911	C. K. Digby-Jones C. K. Digby-Jones Seal & Greaves Ackermann Ackermann	—	—	Prospecting of native washings and primary sources in Fazughli district Prospected for Messrs. Beswick and Moreing, Mining Engineers. Prospected for Mr. J. B. Means
Hegate Concession, Jebel Elba Concession and G. Ogilvy Haig's Concession	1902–3	Abandoned 1907	Capt. McCormick, J. Badge, H. Lancaster Hobbs, & G. G. Gifford	Nov. 1905	1906	Ancient workings prospected. Onib Mine 21°30′ N, 35°20′E opened up. Prospects said not to be promising. Egyptian Sudan Minerals Ltd., and Deraheib and African Syndicate listed as part owners
Sudan Exploration Company	1903–4	Surrendered 1906	A. Thomas	—	—	Prospected for Messrs. John Taylor and Sons, Mining Engineers. West of Port Sudan, Jubal el Bahr el Ahmer
Nuba (Sudan) Development Company	1903–4	Surrendered 1907	P. Wilson and A. Mackinnon	1904	1905	Ancient workings prospected in Jebel Hamra–Jebel Kuro area, Doishat development started in Jan. 1904 and continued to Mar. 1905. Moderately good vein below old workings
No details given		—	—	1904	1905	Abu Sari near Delgo, Dafaufah is an ancient mining centre with sun-dried brick kiln 50 ft high and many crucibles. Worth prospecting. Company went into liquidation
Gabeit Mining Syndicate Ltd. (The Sudan Mines Ltd.)	April–July 1903	1908	N. Griffin	1904	1908	Ancient workings. Some rich ore exploited. Gabbait. Northern Jubal el Bahr el Ahmer
Dongola Concession	1903–4	1906	G. R. Carey	—	—	Prospect situated west of the Nile in Dongola area. Carey prospected for Messrs Lake and Currie
Egyptian and Sudan Mining Syndicate	Jan.–May 1904	1907	L. Llewellyn	—	—	Ancient workings in Abu Hamed region. Quartz veins at Abu Hashim developed but values and quantities disappointing

Name of company	Date prospected	Licence expired or concession abandoned	Prospected by	Mining started	Mining stopped	Remarks
Suakin Mining Syndicate	1904 1905	Surrendered 1906	M. P. Griffiths J. Elsick	—	—	Ancient workings in Togni area of Jubal el Bahr el Ahmer. Quartz veins but with hardly any gold
Tokar Prospecting Syndicate	1905 1906		P. C. Wilson J. F. Morris	—	—	Discovered graphite and copper: for details see above
Victoria Investment Corporation	1905–6	1906	S. C. Dunn	—	—	Bayuda and Elai Jebels area prospected. No ancient gold workings or minerals recorded

The Garabein Mining Syndicate was incorporated in 1932 and included among its directors Messrs M. J. Bishop and T. M. Foley, two Irish miners who worked gold in the Sudan until the early 1950s associated with the Gabeit (Gabait) and Aberkateb (Aberkateib) mines. The syndicate had a capital of £E5000 and was wound up in 1935. Mr. M. J. Bishop of Wadi Oyo Mine acted as liquidator.

The Wadi Oyo mine was owned by Kassala Gold Mines (Sudan Ltd.). This company was registered in 1934 and up to July 1935 it had sold bullion amounting to £14 906 (uncorrected pre-1933 figures). In 1936 work was started at Wadi Oyo, Mikraff or Macruff, and Shashitaib mines. Tonnage was reduced at Wadi Oyo in 1936 owing to shortage of labour, said to have been caused by the Italo-Abyssinian War. The crushed tonnage in 1936 was about 2885 tons and 1164 ounces of gold were recovered, realizing £7681 (uncorrected pre-1933). At Mikraff mine a shaft was sunk to 200 ft and east and west drifts were opened at this level. A vein with an overall width of 12 inches was found assaying 10 oz per ton. Shashitaib No. 1 shaft encountered a quartz vein 4 ft thick, assaying 1 oz per ton at 25 ft below the ancient workings. A prospecting licence was obtained for the Onib area, considered to be the largest ancient group of mines in the Sudan. The company was wound up in 1937. Figures do not seem to be available for the total amount of gold extracted.

Gabait Gold Mines Ltd. was registered in 1933; formerly it was known as Sudan Gold Mines Ltd. (1933) and Gabait Mining Syndicate (1903). During 1933 crushing plant at Gabeit (Gabait) mine operated for most of the year and a new counter-current decantation plant was erected for cyaniding. An additional five stamps were installed, making total of ten at the mine (Plate 6B). A dam was built for water and a considerable amount of development took place before the company ceased to operate in 1942 (?). The amounts of gold extracted are given in Table 28. Unfortunately we have not been able to find complete statistics for the mine but more than 126 028 ounces are said to have been produced between 1914 and 1939. This amount was valued at £913 703 in the company report for 1938–9. In the annual report (1935–6) estimated head values are given at 10·7 dwt per ton for ore (90·56 per cent estimated extraction) and 4·13 dwt per ton for tailings. In assessing these values today it should not be forgotten that in 1933 the value of an ounce of gold rose from £4·3 to £8 and in 1950 to £12·4 and that in 1967 it was worth c.36 U.S. dollars. Work ceased at Gabeit in 1942 (?); supply problems, difficulties with spare parts, and labour problems made it impossible to continue. The chairman of the company considered that under normal conditions there could be reasonable profits for 3–4 years, but the manager considered the mine worked up. However, it should not be forgotten that as late as 5 May 1940 the company took out a licence for an area within a radius of 10 miles from the mine and that Atbai Gold Ltd. (1947–53) made the mine pay.

Andrew (MS 1946), Sudan Government geologist, mentioned that primary lodes are scattered over a wide area of the Basement Complex Group and that

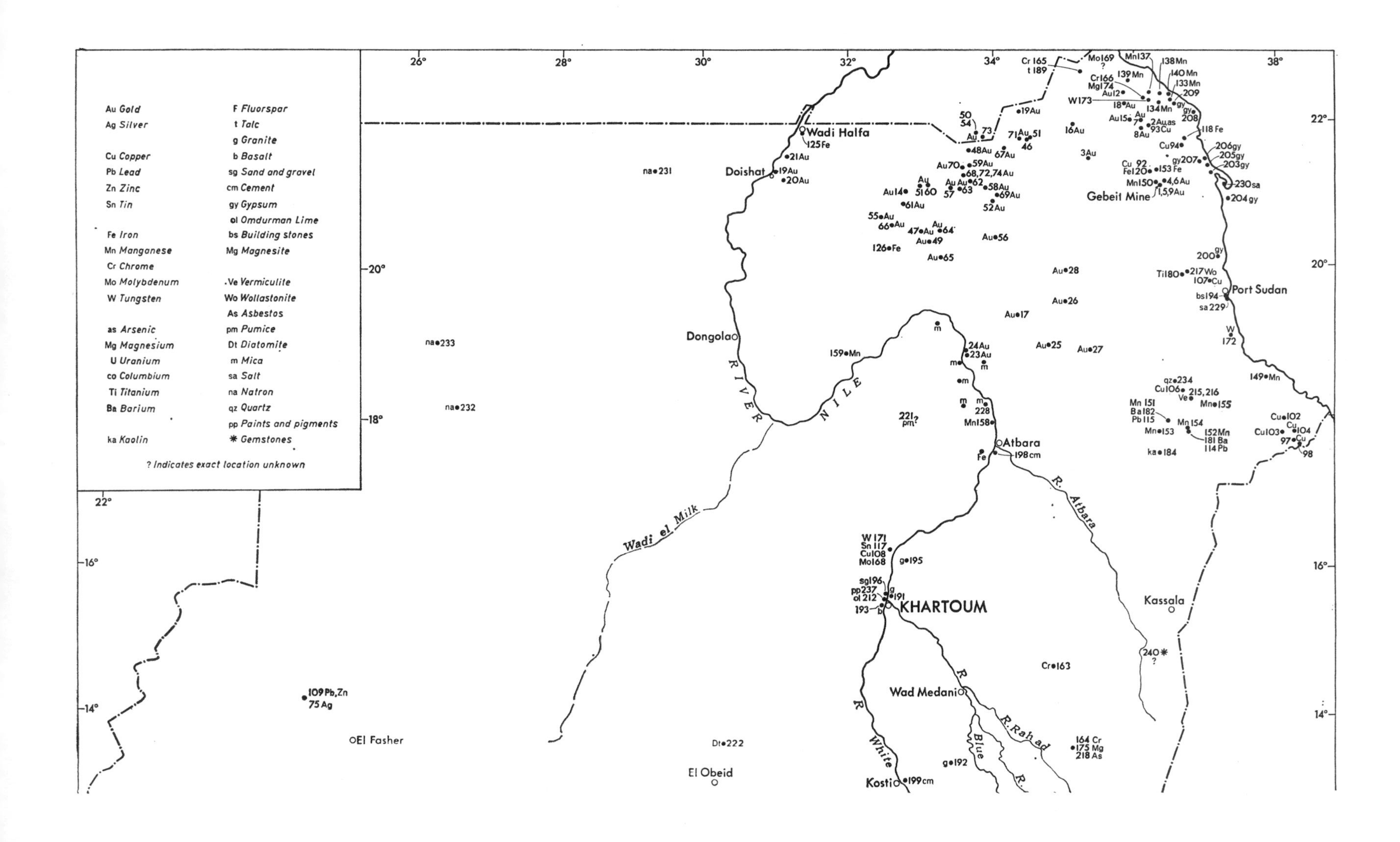

Au Gold
Ag Silver
Cu Copper
Pb Lead
Zn Zinc
Sn Tin
Fe Iron
Mn Manganese
Cr Chrome
Mo Molybdenum
W Tungsten
as Arsenic
Mg Magnesium
U Uranium
co Columbium
Ti Titanium
Ba Barium
ka Kaolin
F Fluorspar
t Talc
g Granite
b Basalt
sg Sand and gravel
cm Cement
gy Gypsum
ol Omdurman Lime
bs Building stones
Mg Magnesite
Ve Vermiculite
Wo Wollastonite
As Asbestos
pm Pumice
Dt Diatomite
m Mica
sa Salt
na Natron
qz Quartz
pp Paints and pigments
* Gemstones
? Indicates exact location unknown
Wadi Halfa
Doishat
Dongola
RIVER NILE
Wadi el Milk
Gebeit Mine
Port Sudan
Atbara
R. Atbara
KHARTOUM
Kassala
Wad Medani
R. Rahad
Blue R.
White R.
Kosti
El Obeid
El Fasher

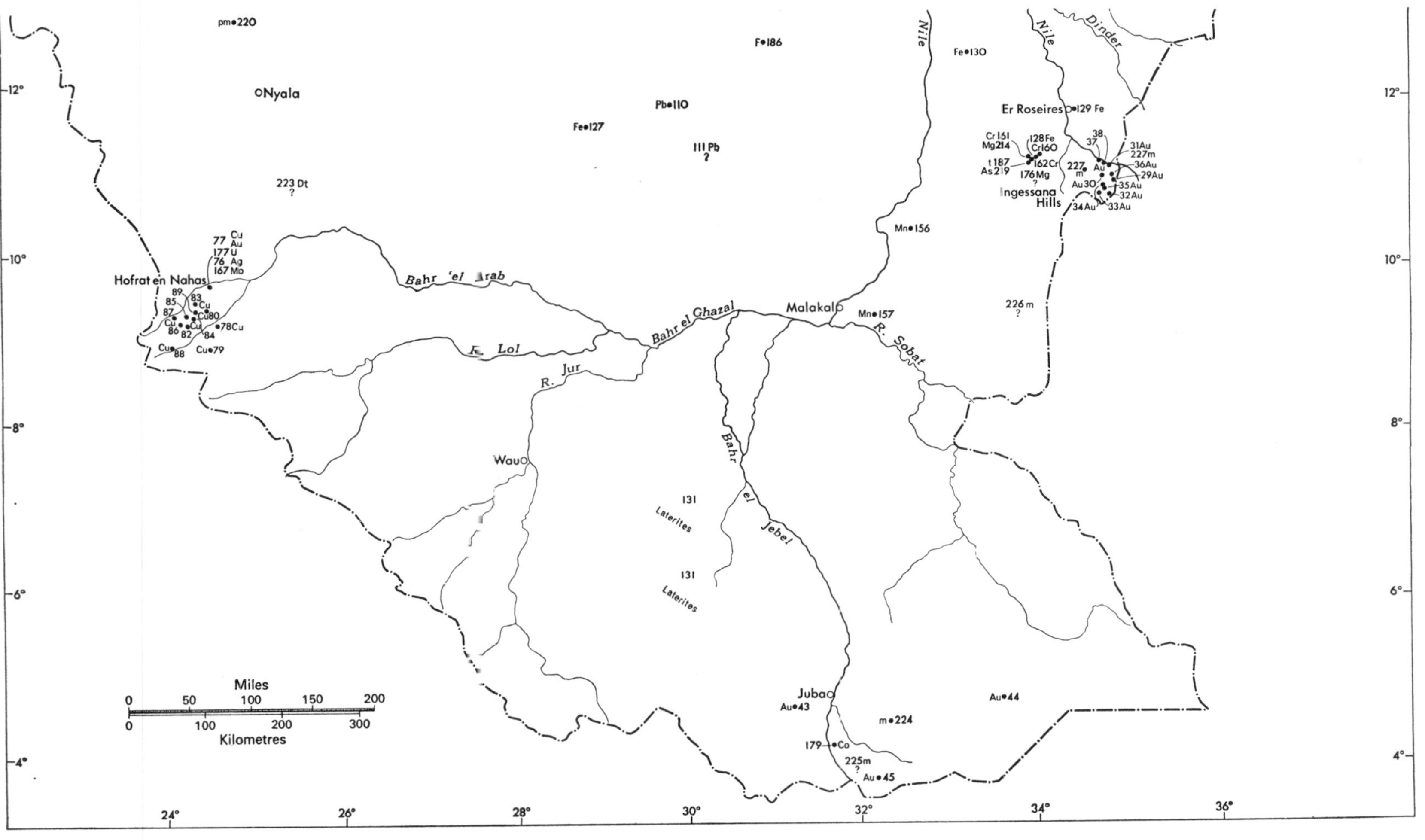

FIG. 84. Mineral Locality Map, Sudan (See Table for details)

TABLE 28

Production figures, Gabeit Gold Mines Ltd., Sudan

Year	Tons of ore crushed	Tons of tailing cyanided	Ounces extracted from tailings	Total ounces fine gold extracted	Amount realized	Remarks
1914–39	126 732			126 028·00	913 703	
1933–4	5751		(Said to contain 1800 oz but not extracted)	3621·45	23 320	No cyanide plant to process tailings
1934–5	7517		(Said to contain 1342 oz but not extracted)	4202·394	29 461	Cyanide plant started in 1934. Est. head value 10·7 dwt and 4·13 dwt tailings (averaged)
1935–6	9229	9229 (current) 13 144 (accumulated)	3531	6368·02	45 849	
1936–7	9035	9035 (C) 1414 (A)	3309·40	6212·753	43 439	
1937–8	11 690	11 690 (C) 11 890 (A)	4982	6392·00	44 630	
1938–9	—	—	—	—	—	Data not available
1939–40	11 782	14 616 (A)	—	6212·236	47 244	
1940–1	10 038	10 272	—	4368·387	35 507	(Royalty paid at 162 shillings per ounce)
1941–2	11 488	11 001	—	3963·36	28 433	(Averaged)
Totals	76 530	92 291	11 822·40	41 340·600	297 883	

(Pre-1933 values)

although many of these had been known since Middle Kingdom times, only two seem to have had a measure of success: Umm Nabari mine (closed 1945) and Gabeit (Gabait) Mine (closed 1942?). Andrew stated that gold lodes are known in the Fazughli and Kurmuk areas but according to Italian geologists, who operated during the Second World War in the area, the prospects are poor. He also stated that lodes no doubt occur in the southern Sudan but laterite cover hampers exploration.

Andrew (MS 1946) recorded that alluvial gold occurs in the Luri River, west of the Nile; in the Thingaita River, east of the Nile, Equatoria Province; and in the Tira Mandi area of the Nuba Mountains, Kordofan. Profits in these areas are said to be low and depend on very low labour costs (Fig. 84). Working was often a spare-time occupation, but one full-time gold-washing concern, run by Caley and Hagar, is said to have operated in Equatoria. The main area of operations after the Second World War were around Kapoeta, where large alluvial deposits are said to exist. Khor Luri on the Juba-Yei Road and Khor Aloto near Iwatoka were also worked by this same concern. About 100–150 oz per month are said to have been extracted. Caley and Hagar worked through the rainy seasons until the rivers dried up; this was called 'full-time'. *Reports* 77/8 and 77/11 (*Sudan Geological Survey Annual Reports* 1950–2, p. 7) deal with gold in Equatoria Province. The operations have now stopped because of the troubles.

Andrew (MS 1946) noted that alluvial deposits are unlikely to occur north of isohyet 300 mm but this idea fails to take into consideration possible Pliocene and Pleistocene concentrations when climate at times was certainly wetter than at present, even in the extreme north of the Sudan.

Atbai Gold Ltd. was registered in 1947 and a

prospecting licence taken out by M. J. Bishop to cover the Gabeit and Wadi Oyo mines. The company had a capital of £E45 000 in £E1 shares and two subscribers are listed as M. J. Bishop (mining engineer and life director) and A. J. Claxton (advocate). On 11 July 1948 Bishop sold out to the company for £E33 000 paid in 20 000 shares and £E13 000 in £1000 instalments.

In July 1951 the Sudan Government paid Bishop £S12 500 to develop Foss Mine (93 Feddans in Wadi Abu Sari, Kassala Province). He was to crush 20 tons of ore per day and after two years to form a company with £S50 000 capital. Bishop was to hold £25 000 and the Sudan Government had an option on the remainder. The mine was mortgaged in 1953 and later in the year Bishop died. Since then the mine has been worked sporadically and yields have been low. In 1956 the company was removed from the companies register.

Abdullah (1956) mentioned that in 1954 the Doishat (Dowshat) Gold Mine, 60 miles south of Wadi Halfa, was visited and that an exceptionally rich reef, shaped like an irregular cylinder and dipping north at a high angle was encountered. No assay data were given, however.

Afia and Widatalla (1961) published a report on the copper deposits of Hofrat en Nahas, which are also gold-bearing. This report contains data collected by the Nile–Congo Divide Syndicate (a subsidiary of Tanganyika Concessions Ltd.) from 1920 to 1925, and Anglo-American Corporation of South Africa staff in 1948. The occurrence of gold and copper at a number of localities between the mine and the Chad frontier is mentioned. Assay data for gold were not given.

The Central Desert Mining Company was formed in 1953 with a capital of £S50 000 incorporating the interests of the Baasher family, who own the leases of Aberkateib Gold Mine and Fodikwan Iron Mine. T. M. Foley, who had been a director of Garabein Mining Syndicate and who worked with M. J. Bishop of Atbai Gold Ltd., encouraged the Baasher family to put money into Aberkateib Mine. Foley was eventually replaced by N. Smith (an English coal-miner) but unfortunately he died of thirst while crossing from the mine to Abu Hamed on the bend of the Nile.

A list of gold occurrences in the Sudan is given in Table 29.

TABLE 29

List of gold localities, Sudan (excluding 'ancient mines' listed in Table 30). See Fig. 84 for localities

	Name	Coordinates	Sudan 1:250 000 sheet no.	Comments
Northern Jubal el Bahr el Ahmer				
1.	Gabeit (or Gabait)	21° 03½′ N 36° 19′ E	Dungunab	Working mine (details below)
2.	Wadi Oyo	21° 55′ N 36° 07½′ E	Dungunab 36–I	
3.	Onib (Alar)	21° 29′ N 35° 20′ E	Deraheib sheet 35–L	100 miles NW of Sufaya. Abandoned. Said to be largest group of ancient mines in Sudan
4.	O'haff or Haff	21° 07′ N 36° 23½′ E	Dungunab 36–I	30 miles NW of Gabeit. Abandoned
5.	Ragag Sagieb	—	Dungunab	Near Gabeit at W. Haet
6.	Garabein	21° 06′ N 36° 22′ E	Dungunab 36–L	Near Haff. Abandoned
7.	Shashitaib	21° 58′ N 36° 03′ E	Dungunab 36–I	Ancient mine
8.	Mikraff or Macruff	21° 54′ N 36° 06½′ E	Dungunab 36–I	Near Oyo. Abandoned

TABLE 29—*continued*

	Name	Coordinates	Sudan 1:250 000 sheet no.	Comments
9.	Nimra Bair	—	Dungunab 36–I	4 miles E of Gabeit. Prospected and left
10.	Foss	—	—	In Wadi Abu Sarab or Sari. 1 day's walk from Aberkateib in Nafaab country
11.	Aberkateib or Aberkateb		?	Few hours NE of Abu Hamed or 6 hours' drive by car west of Gabeit Mine
12.	Kefriai	22° 21′ N 35° 50′ E	Marsa Shaab 35–H	70–80 miles NW of Halaib, near Meisah
13.	Wali Alaqi	22° 07′ N 34° 23′ E	Haimur sheet 35–G	Gold reported in Sudan
14.	Umm Nabari or Umm Nabardi	21° 07′ N 32° 47′ E	Murat 35–J	Abandoned
15.	Amaneb	21° 57½′ N 35° 57½′ E	Deraheib sheet 35–L	
16.	Daraheib	21° 57′ N 35° 08½′ E	Deraheib sheet 35–L	Ancient mine
17.	Nigeim	19° 27½′ N 34° 18½′ E	Abu Hamed 45–C	Abandoned
18.	Mhashushenai	22° 13′ N	Deraheib sheet 35–L	Ancient mine
19.	Doishat (Dowshat)	21° 23′ N 30° 58½′ E	W. Halfa 35–I	Abandoned in 1960 (?)
20.	Om Fahm	21° 17′ N 31° 06′ E	Wadi Halfa 35–I	Nubian (Sudan) Development Co. 5 miles from Adilla Wells. Cuprous quartz veins
21.	Haisub	21° 34′ N 31° 06′ N	Wadi Halfa 35–I	Nubia (Sudan) Development Co. 12 miles east of Adilla Wells. Workings extend over 3 miles
22.	Abu Sari, near Delgo	20° 17′ N 30° 36′ E	Mosha 35–M	Nubia (Sudan) Development Co. Close to Nile
23.	Abu Hashim	18° 55′ N 33° 36′ E	Berber 45–G	Egyptian and Sudan Mining Syndicate (23–28)
24.	Marru Is.	19° 00′ N 33° 34′ E	Berber 45–G	Pegmatite with gold
25.	J. Gheiob (or J. Gaihob)	19° 05′ N 34° 45′ E	Wadi Amur 45–D	Some miles up Wadi Amur
26.	J. Kamotit	19° 35′ N 35° 00′ E	Wadi Amur 45–D	—
27.	J. Gharad (or Gurad)	18° 56′ N 35° 17′ E	Musmar 45–L	—
28.	Nawarai	C. 20° N. (?)	Wadi Amur 45–L (?)	110 miles from Abu Hamed

	Name	Coordinates	Sudan 1:250 000 sheet no.	
Blue Nile and Beni Shangul area				
29.	J. Falabut	11° 00′ N 34° 54′ E	Fazughli 66–D	Alluvial gold in khors
30.	J. Fasakuro	11° 06′ N 34° 37′ E	Fazughli 66–D	Alluvial gold in khors
31.	K. Samborung	11° 12′ N 34° 51′ E	Fazughli 66–D	Alluvial gold in khors
32.	J. Farougi	10° 52′ N 34° 54′ E	Beni Shangul 66–H	Alluvial gold in khors
33.	K. ed Dahab	10° 51′ N 34° 45′ E	Beni Shangul 66–H	Alluvial gold in khors
34.	K. Sumba	10° 57′ N 34° 48′ E	Beni Shangul 66–H	Alluvial gold in khors
35.	K. er Ragreig or Amara	10° 55′ N 34° 48′ E	Beni Shangul 66–H	Alluvial gold in khors
36.	K. Adi	11° 04′ N 34° 52′ E	Fazughli 66–D	Alluvial gold in khors
37.	K. Anis Shangani	11° 17′ N 34° 43′ E	Fazughli 66–D	Alluvial gold in khors
38.	J. Fazughli	11° 16′ N 34° 45′ E	Fazughli 66–D	Alluvial gold in khors
Kordofan				
39.	Jebel Tira Mandi or 'el Dahab'	10° 54′ N 30° 28′ E	Talodi 66–E	Alluvial
40.	J. Sheibun	11° 12′ N 30° 12′ E	Rashad 66–A	Alluvial
41.	J. Umm Gabrallah	Near J. Tira Mandi	Talodi 66–E	Alluvial
42.	Khor Eiri area	Near J. Tira Mandi	Talodi 66–E	Alluvial
Southern Sudan				
43.	Khor Luri Juba-Yei Road	4° 39′ N 31° 10′ E	Yei 78–M	Alluvial
44.	Nangolet near Kapoeta	4° 47′ N 33° 35′ E	Didinga 78–O	Alluvial
45.	Khor Allota	—	Nimule 86–B	Near Ewatoka. Alluvial

(ii) *Gabeit Mine*

(1) *General.* This is the only working gold mine in the Sudan. It has operated off and on since the Gabait Mining Syndicate was registered in 1903. Sudan Gold Mines Ltd. took over the property in 1933 and it passed to Gabeit Gold Mines Ltd. later that year. Production data provided by this company are presented in Table 26. Work creased temporarily in 1942 but Atbai Gold Ltd. took it over in 1947. Since Bishop died in 1953 it has been worked intermittently but operations at present are very small.

The mine is run now by a group of seventeen (?) merchants (Plate 6B).

(2) *Geology*. The ore bodies are quartz veins dipping at 50°–60°, bearing 045°. They cut green and red schists of the Nafirdeib Group of the Basement Complex and are thought to be three in number. The uppermost vein is said to be the richest. The schists are heavily pyritized and part of the gold occurs in pyrite and chalcopyrite. The gold is found mainly native in quartz veins as disseminations or occurs in stringers about 2 in wide and 6 ft long.

(3) *Operation*. The mine has been worked to a depth of 670 ft but at present it is worked above the 60-ft level only, the lower levels being flooded. There are three shafts, the central one being the main shaft. The drifts are small and extend only 50 ft along the ore bodies.

At present the mine employs about twenty men, who work underground with kerosene lamps and primitive equipment. The ore is crushed in an ancient stamp crusher powered by a Lister 10-h.p. engine. In 8 hours some 5–6 tons are crushed. The crushed material is passed through a mesh placed over a 2-m^2 copper plate coated with amalgam which is periodically scraped off and distilled for gold. The much more efficient cyanide process is no longer operated. About ½–1 oz of gold is extracted daily and current operations can only be described as extremely primitive and barely profitable.

(iii) *Oyo mine*

Gass (1955) briefly described the geology of Oyo mine. The reefs are said to be small and to be cut by many faults. The mine is said never to have been profitable.

(iv) *Shashitaib (Shishiteib) and Mikraff mines*

The reef at Shashitaib is extensive but contains little gold.

Mikraff mine is on a small but fairly rich reef. Work was limited, however, because the reef is faulted and could not be found again. All work stopped in 1940 (Gass 1955).

(v) *Origin of gold, northern Jubal el Bahr el Ahmer*

Gold-bearing quartz veins are common throughout the northern Jubal el Bahr el Ahmer in the metasediments and metavolcanic rocks of the Basement Complex formations. The gold-, pyrite-, and arsenopyrite-bearing hydrothermal fluids appear to have penetrated along faults and fissures as most major reefs occur along fractures. The ultimate source of gold-bearing fluids may well have been the granites that cut the metamorphic rocks. Granite masses, for instance, occur near to Oyo and Mikraff Mines. If we accept the order of superposition and the potassium–argon whole-rock date of the Jebel Hamashaweib dyke given in Chapter II, then this mineralization is older than $740 \pm 25 \times 10^6$ years (Whiteman 1965, 1968).

(vi) *Umm Nabari Mine and the ancient mines of the Sudan Mining Syndicate concession, Northern Province*

(1) *General*. Work began in this concession bounded by 20°–22° N, 24°30′ E in 1901, and the general geology and mine details were described by Llewellyn (1903) (see Chapter II). Many ancient gold mines occur in the region, which is underlain by the Basement Complex (Table 30).

TABLE 30

List of 'ancient gold mines', Sudan Mining Syndicate concession (Based on Llewellyn 1903)

Name	Lat. N	Long. E
46. Abu Dalala	21° 47′ 30″	34° 28′ 15″
47. Abu Siha	20° 35′ 25″	33° 02′ 45″
48. Abu Tundul	21° 38′ 50″	33° 40′ 20″
49. A Darawib	20° 25′ 10″	33° 06′ 00″
50. Bir Towil	21° 51′ 00″	33° 47′ 00″
	(Egyptian territory)	
51. Butana	21° 45′ 00″	34° 30′ & over
52. Dabhlakaa	21° 01′ 00″	33° 59′ 00″
53. Dayob	21° 09′ 25″	33° 00′ 00″
54. Esmat Omar	21° 49′ 20″	33° 43′ 00″
	(Egyptian territory)	
55. (UM) Fit Fit	20° 46′ 00″	32° 27′ 00″
56. Hadal Moiet	20° 30′ 45″	34° 01′ 10″
57. Idarib	21° 06′ 30″	33° 25′ 20″
58. Khabeseit	21° 07′ 45″	33° 55′ 45″
59. Lesewit	21° 25′ 10″	33° 40′ 00″
60. Mosei	21° 10′ 30″	33° 07′ 00″
61. Mundera	20° 55′ 40″	32° 46′ 10″
62. Nali	21° 13′ 20″	33° 41′ 40″
63. Nabi Tana	21° 09′ 45″	33° 35′ 30″
64. Nasb el Hosan	20° 35′ 40″	33° 17′ 25″
65. Omar Khabash	20° 13′ 00″	33° 18′ 20″

Name	Lat. N	Long. E
14. Om Nabardi	21° 07′ 30″	32° 46′ 00″
66. Rod el Ushal	20° 40′ 10″	32° 37′ 40″
67. Shashu-at-eb	21° 40′ 30″	34° 10′ 40″
68. Tana Shaib	21° 12′ 15″	33° 35′ 30″
69. Tergowi	21° 01′ 20″	34° 02′ 25″
70. Tibat Abda	21° 20′ 00″	33° 37′ 00″
71. Ufain	21° 47′ 30″	34° 23′ 30″
72. Wadi Dom	21° 15′ 00″	33° 40′ 00″
73. Wadi Howe	21° 46′ 45″	33° 42′ 20″
74. Wadi Romeit	21° 15′ 40″	33° 37′ 10″

Quartz reefs are abundant but few are auriferous. These occur mainly in 'finer trap or in schistose rocks' closely allied to diabase. Most of the ancient mines utilized these reefs but some utilized gold-bearing alluvium derived from them. In places quartz reefs form hills rising over 100 ft or more above the surrounding country. They vary in width from stringers to 6 ft thick and many extend more than several hundred yards. Richness varies inversely as thickness according to Llewellyn (1903).

The quartz is bluish-white, semi-transparent to opaque, and appears to be granular. In addition to the gold, small quantities of haematite, mispickel, and galena occur. Gold is frequently associated with 'heather red' coloured quartz, a point that was utilized by the ancient miners (Llewellyn 1903).

(2) *Umm Nabari or Umm Nabardi mine.* The country-rock is a coarsely laminated, micaceous, chloritic schist sometimes containing talc and occasionally actinolite. Altered igneous rocks also occur. The mine is situated in full desert (Plate 2). The mean strike of the foliation and of many of the reefs is north-east and they dip to the south-east. The reefs vary in thickness from stringers to 20 ft and often can be easily followed for 1500 ft. The most important reefs gave pan values ranging from 5 dwt per ton to 10 dwt per ton.

The ancient workings at Nabi occur in steep schistose rocks striking east–west and intruded by porphyritic rocks. Values of 4–6 dwt per ton were noted (Llewellyn 1903). At Nabi Tana samples from pits assayed up to 5 dwt of gold per ton.

(vii) *Nuba Mountains*

Gold has been known from the Nuba Mountains of the central Sudan for over 200 years. Brown in 1788 in his travels in Darfur mentions Sheibun as a source of gold well known to the Darfur Arabs (Dunn 1921). The main localities are Jebel Tira Mandi, Jebel Sheibun, and the Umm Gabrallah Hills.

(1) *Jebel Tira Mandi.* According to Dunn (1921), the gold occurs in khor sands and gravels; in surface soils and gravels on hill sopes; and in the subsoil, consisting of a tough reddish clay more than 8 inches thick containing abundant quartz fragments.

Mansour and Samuel (1957) described the gold centre of Mandi as a small hill at the southern end of Jebel Mandi. Numerous closely packed quartz veins forming a belt about 10 inches wide trending at 040° and dipping 50° to the west occur in the cleavage planes of the banded gneissose country-rock. Individual quartz veins vary in thickness from 8 to 60 inches, and milky white quartz, sometimes stained with iron oxide, covers the hill slopes.

Mansour and Samuel (1957) also recorded a milky quartz vein 12 ft wide cropping out for 100 yards at Mandi Village. No assay data were given by Mansour and Samuel (1957).

(2) *Jebel Sheibun.* This jebel is composed of grey-pink, fine-grained granite cut by thin quartz and pegmatite veins, particularly abundant on the northern part of the jebel (Mansour and Samuel 1957). The veins are irregular in distribution and appear to fill all available joint space. They vary in thickness between 4 and 8 inches, although there is one quartz vein over 12 ft thick.

According to Mansour and Samuels (1957) the gold workings are confined to a zone of coarse shallow soil about 3–9 ft thick and surrounding the hill for about 640 ft. These deposits were panned and yielded traces of gold.

(3) *Umm Gabrallah Hills.* Gold bearing milky white and occasional smoky quartz veins striking between 000° and 045° occur between Jebel Umm Gabrallah and Khor Eiri. Some of these are said to be tens of feet thick but commonly are only 6 ft thick. In some cases they extend for distances of over 320 ft.

No assay data were given on the three areas mentioned above. Mansour and Samuel (1957) noted that the lack of water did not encourage any further investigation.

(viii) *Current gold production, Sudan*

For the whole of the Sudan production is small and little reliance can be placed on available figures as some of the production is absorbed locally.

De Kun (1965) gave the figures shown in Table 31.

TABLE 31
Gold production, Sudan 1951–60 (According to de Kun 1965)

1951–5	1700 ounces/year
1956–8	6000 ounces (total)
1959–60	2500 ounces/year

In *Mineral Yearbook* 1964 the figures shown in Table 32 are given.

TABLE 32
Gold production, Sudan 1959–64 (*Minerals Yearbook* 1964)

1959	1419 ounces
1960	2116 ounces
1961	1226 ounces
1962	1500 ounces
1963	900 ounces
1964	877 ounces

Figures on mineral production given by the Sudan Geological Survey to *El Ayam* newspaper 18 August 1965 do not include gold. According to the Ministry of Finance, Customs Department the amounts of gold shown in Table 33 were produced in the Sudan.

TABLE 33
Gold production, Sudan 1960–6 (According to Sudan Customs Department)

	Pounds Sudanese
1960	24 603
1961	5200
1962	8560
1963	—
1964	3000
1965	No data
1966	No data

(b) Silver

Silver is known from Kutum (75), Darfur Province, where it was recorded in small traces by Wahab and Afia (1966) in analyses of lead and zinc ores; and from Hofrat en Nahas (76) where it is associated with copper ores. (Figures in brackets refer to locality map.)

3. THE NON-FERROUS METALS

(a) Copper

(i) *History of development*

Copper is known from at least thirty-two localities in the Sudan (Table 34) but only one of these, Hofrat en Nahas, has proved to be of any size and to have attracted commercial interests.

Hofrat en Nahas is situated on the Umbelasha tributary of the Bahr el Arab and has probably been worked in a small way for hundreds of years. It attracted the attention of Mohammed Ali Pasha, Viceroy of Egypt, but apparently was not visited by 'the Turks'. However, in 1846, according to Hill (1963, p. 58) the Viceroy's cabinet inquired about a letter to the Sultan of Darfur concerning the 'copper mines'. It is interesting to speculate what may have happened economically and politically if the Pasha had concentrated less on the search for gold in the Sennar region and more on copper, as these deposits are substantial and are also associated with gold.

Gray (1964, p. 65) mentioned that in the nineteenth century most of the copper produced at Hofrat en Nahas passed via Darfur into the network of Muslim trade but some was traded with the pagan tribes to the south. Barth (*Travels*, Vol. 2, p. 141) mentioned that a considerable supply of copper was imported annually into Kano, Nigeria. The size of the primitive workings at Hofrat en Nahas indicates, however, that production was small.

A right of ownership over the copper mines was claimed in 1869 by Mohammed el Hilali, reputedly from the Maghreb. He appealed to the Governor Ja'afar Pasha for military assistance in Bahr el Ghazal. This was provided on a limited scale but Hilali died in 1872 (Gray 1964, p. 121). Gessi, appointed by Gordon Pasha, was in the Bahr el Ghazal region from 1879 to 1881 and he noted that copper from Hofrat en Nahas supplied the needs of the province. During the period of the Mahdiya (1881–9) little mineral prospecting was carried on but Hofrat en Nahas figured again in Sudanese history in 1884 when Hussayn, self-styled Sultan of Darfur, surrendered the property of Hofrat en Nahas to King Leopold II and accepted the protection of the Congo Free State. An agreement was subsequently signed by Hussayn on 18 May 1884: 'with the stipulation that even if the influence of the Congo State ceased in these territories, the mines would pass in full right to the person of Leopold II and his heirs' (Collins 1964, p. 142). This clause had far-reaching consequences because between 1900 and 1906 Leopold II used his mineral rights to support his commercial and territorial claims to the whole of the province of Bahr el Ghazal.

Stockley and Morris, prospecting for the Tokar Prospecting Syndicate in 1905, discovered a number of copper-bearing quartz veins and schists in the southern Jubal el Bahr el Ahmer (for details see below). Since then nothing has been published about these occurrences except by de Kun (1965, pp. 68–70), who briefly mentioned them.

Dunn (1911) gave only a general account of the copper mines at Hofrat en Nahas, details apparently not being available. In the early 1920s interest rose again in Hofrat en Nahas. The mine had been visited by Colonel Sparkes in 1903 and samples assayed in Khartoum gave 14 per cent Cu. In 1918 Burgess Watson (inspector, Runga district) visited the area and samples were taken with a view to opening up the property. Eventually a concession of some 60 000 square miles was granted to prospect for copper and gold to the Nile–Congo Divide Syndicate (NCDS), a subsidiary of Tanganyika Concessions, Ltd. Prospecting ended in 1927 but by then reasonably extensive exploration had been carried out. Shafts had been sunk and cross-cuts made. J. G. Bower, mining engineer, estimated that at Hofrat en Nahas itself there are 20 000 tons of copper per foot (see Afia and Widatalla 1961).

The NCDS Company Report for 1927 mentioned that a large amount of development had been done, including the sinking of ten boreholes (total footage 3320 ft). The company also stated that work at Hofrat en Nahas had reached a stage when a deep shaft was required, but recommended against further development until a railway had been built. NCDS also explored south and west of Hofrat en Nahas and recorded a number of occurrences of copper and gold (Fig. 23). In fact it appears that there has been considerable mineralization in this section of Darfur, a point that does not appear to have been emphasized sufficiently hitherto, because of the concentration of interests at the mine at Hofrat en Nahas.

Andrew (MS 1946), in a list of minerals occurring in the Sudan, briefly mentions copper at Hofrat en Nahas. The ores were described as sulphides but no details were given.

Abdullah (1958) mentioned that copper had been found south of Tohamiyam, southern Jubal el Bahr el Ahmer, as malachite in chlorite-schists associated with quartz veins. Malachite, with some cuprite and chalcopyrite, was also recorded from the Khor Arbaat area near Port Sudan (Abdullah 1958).

In 1961 Afia and Widatalla published a report on the copper of Hofrat en Nahas. This contained new data obtained by the Sudan Geological Survey between 1957 and 1960; a wealth of data in the form of sections, mine plans, etc. obtained by the Nile–Congo Divide Syndicate (Tanganyika Concession Ltd.) from 1920 to 1925; and by the Anglo-American Corporation of South Africa in 1948; magnetometer and self-potential data were obtained by Messina (Transvaal) Development Company Ltd.

In recent years Italian, Japanese, and Finnish concerns have been interested in the mine, and de Kun (1965) reported that the Pera Trading Company of Livorno (Italy) and its United States affiliate, the African Mining Company, were planning to invest more than $17 million in the development of these deposits; nothing came of this however. Japanese interests investigated the deposit, searching for high-grade ore. They abandoned the area because they only found limited quantities yielding more than 7 per cent copper. The United Nations Development Programme, under the direction of Mr. A. M. Quennell and the Sudan Government were investigating the area in 1968 in an attempt to prove the reserves.

Copper-bearing localities in Sudan known to the author are listed in Table 34. Details are available for the Hofrat en Nahas area only.

(ii) *Hofrat en Nahas mine*

(1) *General.* Hofrat en Nahas (literally Hole of Copper) is an ancient mine. It was worked by the Kreish natives up to the time of the Mahdiya. Serious exploration work was done between 1921 and 1925 by the Nile–Congo Divide Syndicate who sank shafts and boreholes (Afia and Widatalla 1961).

No. 1 shaft was sunk to 64 ft in the south lode. The cross-cuts (5 × 4 ft) were driven, bearing 193°, in the direction of the ore at depths of 35 ft (short cut) and 64 ft for 101 ft. No. 2 shaft was sunk in the north lode at a depth of 76 ft. A cross-cut 5 × 3 ft was driven in a direction 318° towards the ore for 93 ft. No. 3 shaft was sunk at a point 30 ft west and 12 ft south of No. 1 shaft, and 10 boreholes were drilled.

NCDS also explored other deposits at Jebel Jangi

TABLE 34

List of copper localities, Sudan (See Fig. 84)

Name	Coordinates	Sudan 1:250 000 sheet no.	Comments
Southern Darfur and Western Bahr el Ghazal			
77. Hofrat en Nahas	09° 45′N 24° 18′E	Hofrat en Nahas 65–I	
78. J. Jangi	09° 16′N 24° 22′E	Hofrat en Nahas 65–I	
79. J. Terezol	09° 01′N 24° 18′E	Hofrat en Nahas 65–I	
80. J. Zanad	09° 27′N 24° 15′E	Hofrat en Nahas 65–I	
81. J. Nebi	?	Hofrat en Nahas 65–I	
82. Khor Sirri	09° 14′N 24° 01′E	Hofrat en Nahas 65–I	
83. J. Bishara	09° 31′N 23° 27′E	Hofrat en Nahas 65–I	
84. J. Yirongo	09° 28′N 24° 05′E	Hofrat en Nahas 65–I	
85. J. Warranga	09° 22′N 24° 02′E	Hofrat en Nahas 65–I	
86. J. Sirri	09° 16′N 23° 59′E	Bir Khadra 65–L	
87. J. Kairia	09° 28′N 23° 53′E	Bir Khadra 65–L	
88. K. between Ndongo-K. Kairia	09° 07′N 23° 52′E		
89. J. Patapan	09° 25′N 24° 08′E	Hofrat en Nahas 65–I	
Northern Jubal el Bahr el Ahmer			
90. Mohammed Qôl	?	Mohd. Qol 36–M	Two localities details uncertain
91. Halaib area	?	Halaib	One locality
92. Sufaya or Safaia	?21° 17½′N 36° 16½′E	Dungunab 36–I	
93. Shishiteib or Shashitaib Mine	21° 59′N 36° 03′E	Halaib	Traces of copper carbonate
94. Ferokit Well	21° 36′N 36° 39′E	Gungunab 36–I	Traces of copper carbonate
Southern Jubal el Bahr el Ahmer Tokar Prospecting Syndicate 17°–18½° N			
95. Karora area	?	Karora 46–J	

	Name	Coordinates	Sudan 1:250 000 sheet no.	Comments
96.	Khor Areirib		Karora 46–J	Katai Hamat or Hout area
97.	K. Amrik or Hamrik	17° 39′ N 38° 03′ E	Karora 46–J	Left bank of khor, end of camel track
98.	Agarin	17° 36′ N 38° 08′ E	Karora 46–J	Near Kalrah wells
99.	Jebel Gwerar Giobiab	?	Karora 46–J	
100.	K. Adarid	?	Karora 46–J	
101.	Abu Nakhl Wells	?	Karora 46–J	
102.	K. Ghidmut	17° 57′ N 37° 51′ E	Karora 46–J	
103.	K. Sharag	17° 47′ N 37° 53′ E	Karora 46–J	
104.	K. Mashail	17° 44′ N 38° 03′ E	Karora 46–J	
105.	K. Delisha		Karora	
106.	2 miles south of Tohamiyam			
107.	Khor Arbaat area	?	Port Sudan 46–A	Malachite occurrences near Port Sudan
108.	El Qroon el Zurug, Sabaloka	16° 19′ N 32° 37′ E	Sabaloka 45–N	55 miles north of Khartoum. Trace Cu

09° 16′ N, 24° 22′ E; Jebel Terezol 09° 1′ N, 24° 18′ E; Jebel Zanad 09° 27′ N, 24° 15′ E; Jebel Nebi; and Khor Sirri 09° 14′ N, 24° 1′ E.

In 1948 the mine was visited by T. D. Guernsey and P. E. Fairbairn on behalf of Anglo-American Corporation of South Africa. Geologists from the German Democratic Republic visited the area in 1956 and made a topographic sketch of the mine area and a geiger-counter survey. Radioactivity was noted. The Sudan Geological Survey worked for three seasons between 1957 and 1960 in the mine area. Thirteen boreholes were drilled; trenches cut; a topographic map of the mine made; a radiometric survey conducted; and ore reserves calculated. Afia and Widatalla (1961) published a report on the mine.

(2) *Geology of the Hofrat en Nahas mine area.* Because of poor exposure in the mine area a geological map was not made. However, the Nile–Congo Divide geologists recognized in the Basement Complex the Older Plains Group, the Younger Group, and intrusive rocks. (Details are given in Chapter II). The copper and gold mineralization extends as a long linear zone in a north-east–south-west direction for about 77 miles from Hofrat Mine to Jebel Nebi near the border with the Central African Republic and Sudan (Fig. 23).

Known deposits from north to south are: Hofrat en Nahas 09° 45′ N, 24° 18½′ E; Jebel Bishara 09° 31′ N, 24° 7′ E, Jebel Yirongo 09° 28′ N, 24° 5′ E; Jebel Warranja 09° 22½′ N, 24° 2½′ E; Jebel Serri 09° 16′ N, 23° 59½′ E; Jebel Kairia 09° 8′ N, 23° 53′ E; a deposit in the plain between Ndongo and K. Kairia 09° 7′ N, 23° 52′ E; and Jebel Nebi. The deposits strike north-east–south-west in general. East of the main line of mineralization are the deposits of Jebel Zanad 09° 27′ N, 24° 15′ E; Jebel Jangi

09° 16′ N, 24° 22′ E; Jebel Terezol 09° 1′ N, 24° 18′ E; J. Patapan 09° 25′ N, 24° 08′ E.

In general the mineralization is associated with a fine-grained, dark-coloured quartz–tourmaline rock. The copper content varies from stains to as much as 7–8 per cent. Usually however it is between 1 and 3 per cent. The gold content is variable but sometimes is as much as 6 dwt. Torbernite (copper–uranium phosphate), uraninite, cobalt, molybdenum, silver, lead, and zinc have also been recorded. The mineralization occurs in green schists, metamorphosed psammopelitic and ultrabasic rocks that dip moderately north-west.

The ore occurs in two lodes dipping approximately south and north at about 70°. There is a minor branch lode. The main lodes extend for 1150 and 1640 ft respectively. The north lode has a width of 50 ft and contains a rich central porphyry replacement zone some 20 ft thick. The ore is found as a network of sulphide veins in a gangue of quartz, calcite and tourmaline. Gold is associated with chalcopyrite. The zone of oxidation extends downwards for about 50 ft and consists of an upper iron pan, mainly haematite and limonite, and a lower zone rich in malachite, azurite, chalcocite, chrysocola, native copper, and cuprite. Torbernite is conspicuous in the zone of enrichment.

Pegmatitic and hydrothermal stages of mineralization have been recognized. These were separated by a period of movement when crushing and brecciation took place. Kaolinization, sericitization, and tourmalinization occur (Afia and Widatalla 1961). A. M. Quennell (personal communication 1968) thinks that Hofrat en Nahas is a porphyry-copper deposit.

(3) *Reserves at Hofrat en Nahas.* Estimates of reserves vary considerably. Without doubt the figures given by de Kun (1965) are underestimates. Afia and Widatalla's Sudan Geological Survey figures (1961) apply to the mine area only and although they are given down to 500 ft ore was not proved to this depth. The Nile–Congo Divide Syndicate's figures are probably overestimates because copper was not proved in the dimensions quoted, although there are numerous occurrences known along the 18-mile stretch mentioned. Also little information is available on the quantities of carbonate, oxide, and sulphide ores in the mine area. The estimated copper reserves at Hofrat en Nahas are shown in Table 35.

TABLE 35

Estimated copper reserves, Hofrat en Nahas (According to de Kun (1965), Afia and Widatalla (1961), and NCDS)

Source	Tons of ore	Tons of copper
1. De Kun (1965)	1 000 000	500 000
2. Afia and Widatalla (1961), Sudan Geological Survey. Estimated at mine down to 500 ft	10 000 000	278 000
3. Nile–Congo Divide Syndicate, 18 miles long zone, 50 ft wide, 1000 ft deep	400 000 000	—

(b) Lead and zinc

Lead has been recorded from the localities in the Sudan Republic shown in Table 36. Zinc is known only from Kutum. Details are not available for the Dilling, Nuba, and Jubal el Bahr el Ahmer deposits.

(i) *Kutum lead and zinc deposit: history of development*

Kutum is an ancient mine known to have been worked by the natives in the nineteenth century and perhaps earlier. Dunn (1911) noted that lead was worked at Kutum, north of Kebo, in Darfur, citing Naoum Bey Shoucair as his authority. Ali Dinar, Sultan of Darfur, is said to have run the mines until his defeat in 1916, using the lead mainly for munitions. The mines were eventually closed in 1917 (Darfur Province Annual Report).

G. V. Colchester, an officer of the Sudan Geological Survey, reported in 1923 that he had found a ton of ore and a quarter of a ton of slag at El Zeriba, near Kutum. Apparently the main part of the smelting was done there and the remainder at El Fasher. Colchester gave the following analyses (see Wahab and Afia 1966):

Lead slag	41·5 per cent Pb
Lead ore	68·1 per cent Pb
	(Pb occurs as carbonate)

The deposit seems to have aroused little interest until the late 1950s, when officers of the Sudan Geological Survey investigated the area during three

TABLE 36

List of lead localities, Sudan (See Fig. 84)

	Name	Coordinates	Sudan 1:250 000 sheet no.	Comments
109.	Kutum lead–zinc deposit	14° 14′ N	Kutum	4 miles NE of Kutum, Darfur
110.	Near Dilling	?	Abu Zabad 54–P	Details unknown
111.	Nuba Mountains	?	?	Details unknown
Jubal el Bahr el Ahmer				
112.	West of Halaib area	—	Halaib 36–E	Kabesh and Afia (1961), Kassala Province
113.	Gold-bearing quartz veins, Sudan Mining Syndicate Concession	—	—	No details
114.	Allaikaleib	17° 50′ N 36° 38′ E	Derudeb 46–I	East of Adarot Station. Traces of galena
115.	Abu Samr	17° 58′ N 36° 21′ E	Derudeb 46–I	East of Adarot Station. Traces of galena

successive seasons, 1959–61. Twenty trenches were dug and some 2668 ft of borehole drilled. Samples collected assayed up to 65·55 per cent Pb and 26·2 per cent Zn (Wahab and Afia 1966).

(ii) *General geology*

The area around Kutum is mainly composed of igneous and metamorphic rocks of the Basement Complex (Wahab and Afia 1966). There are a few patches of iron-rich conglomeratic Nubian Sandstone Formation, basalts, and volcanic rocks of (?) Tertiary age. These cap the Basement formations.

The Basement Complex consists mainly of schists and gneisses, originally of sedimentary origin, and granites associated with dykes, pegmatites, and mineral veins. According to Wahab and Afia (1966, p. 5) there are three types of granite: normal biotite granite, microgranite, and porphyritic granite. A composite north-west-trending quartz–feldspar-porphyry and dolerite dyke cut the granites and the lead deposit. The pegmatites and mineral veins are composed of similar material, said to reflect a common source.

Structurally the country-rock dips steeply south-east and forms part of a monocline striking north-east. Jointing is common in the granites. Sheared zones intruded by quartz veins form prominent north-west-trending ridges which have been shifted by as much as 130 ft by lateral faults. East–west faults that shift the metamorphic Basement are also known (Wahab and Afia 1966).

(iii) *Nature of the ore*

The veins containing ore strike mainly north-west–south-east and occur in an area of about 3280 × 5776 ft. They dip steeply to the south-west against the regional dip. Another mineralized area occurs 4 miles to the north-east at Juhr el Rusas, but the ground in between has not been proved. The zone of oxidation appears to be shallow and generally deficient in zinc. The main minerals of this zone are haematite, limonite, cerussite, anglesite, silica, pyromorphite, and mimetite. The unoxidized ore is composed of galena, sphalerite, and pyrite with minor amounts of chalcopyrite, marcasite, mispickel, and traces of silver.

The ore deposit is zoned and there is a marked increase in zinc and a decrease in lead in an easterly direction and with depth. Copper also increases with depth but the amounts are small (Wahab and Afia 1966). In both the oxidized and unoxidized zones quartz is the main gangue mineral. Chloritization and kaolinization extends several feet through the walls of the veins where small disseminations of lead and zinc sulphides occur.

Wahab and Afia (1966, p. 40–1) gave the following analyses:

	Pb	Zn	Fe	S	SiO_2	P	Cu
Oxidized zone							
Sample 3	57·12	Tr	5·03	4·08	7·4	0·05	—
Sulphide zone							
B.H.13 157–158½ ft	Tr	26·20	22·89	12·18	8·0	—	0·064

(iv) *Origin of the ore*

Wahab and Afia (1966) believe that the mineralizing solutions were derived from the granites and that there is a close association of mineralization with the pegmatites.

(v) *Reserves*

No figures are given by Wahab and Afia (1965) because of inadequate data. The presence of ore at Juhr el Rusas to the north-east may be an encouraging sign but as Wahab and Afia (1966) pointed out the ground in between has yet to be explored.

(c) Tin

Tin has been recorded from only two areas in the Sudan (Table 37).

TABLE 37

Tin localities in the Sudan (See Fig. 84)

(i) A minute crystal of cassiterite found in river sand in western Equatoria (Andrew (MS 1946) (116).

(ii) El Qroon el Zuruq stockwork, Sabaloka, near Abu Dom village, 55 miles north of Khartoum (117 refers to map Fig. 84).

(i) *Ibba–Sue basins* (116)

Andrew (MS 1943) mentioned an occurrence of cassiterite in the Ibba and Sue basins, Equatoria Province. No details were given.

(ii) *El Qroon el Zuruq, Sabaloka* (117)

General description. The presence of wolfram-bearing quartz veins in this area was first noted by the author in 1960. D. C. Almond, of the University of Khartoum, subsequently mapped the area at a scale of 1:12 500 and demonstrated that, besides containing wolfram, the quartz stockwork contains local concentrations of cassiterite and minor amounts of other sulphide minerals.

According to Almond (1967) the mineralized stockwork forms a hill, situated 1½ miles north-north-east of Abu Dom and 1900 yards north-west of the track leading from Khartoum on the west bank of the Nile. The hill rises about 65 ft above a pediment cut in Basement Complex and consists mainly of strongly foliated quartz-biotite-gneisses striking north-west and north-north-west and dipping steeply south-east (Almond 1967). Barren quartz veins and several types of granite belonging to the Sabaloka complex are intruded into the gneisses. The stockwork is associated with a belt of greissenized rock some 980 ft long, lying along the contact of a porphyritic microgranite of the ring dyke. Almond (1967) believes, however, that the mineralizing solutions may have originated from a near-by mass of biotite–muscovite granite.

The following minerals have been found within the veins: quartz, mixed haematite and black manganese oxides (after wolfram), wolfram, cassiterite, goethite, limonite, chalcedony, zinnwaldite, jarosite, powellite, galena, fluorite, molybdenite, scheelite, calcite, and malachite. Almond (1967) stated that all except the first four minerals are sparse to very sparsely distributed in veins, and that most of the minor constituents are found outside the main stockwork.

Economic potential. This is not known at present because drilling and pitting have not been undertaken and the stockwork has not been properly sampled. In addition to the prospect of the stockwork itself there are possibilities for alluvial concentrations of tin in both the Nubian Sandstone and superficial formations, since these formations rest unconformably on the Sabaloka ring complex.

(d) Aluminium

Although laterite occurs widely in the Sudan, especially in the southern provinces, there are no published references to bauxite. However, by comparison with near-by countries in the savanna belt, the presence of bauxite may be expected. The great distance from the sea of any deposits that might be proved in the southern Sudan would be against exporting bauxite, but the availability of undeveloped hydroelectric power on the Bahr el Jebel below Nimule could lead to the development of a domestic smelting industry if large reserves could be found.

4. IRON AND FERRO-ALLOY METALS

(a) Iron

(i) *History of development*

Systematic prospecting for iron started in the Sudan in the 1820s when Mohammed Ali Pasha became interested in samples of iron sent to Egypt from Kordofan Province. These deposits are laterites and zones of iron enrichment developed on the Nubian Sandstone Formation. They are small and inconsequential except on a local scale. Apparently the larger Jebel Abu Tulu deposit was not discovered by these prospectors. By 1828 Kordofan iron was used for making nails at Khurshid's shipyard at Manjara on the White Nile (Hill 1963, p. 56). Eight iron-founders came out from England and were sent to build a foundry on the White Nile. Some of them visited the ore bodies in Kordofan but the ventures eventually came to nought.

Von Rüssegger visited the iron-ore deposits, and Lambert, also in Mohammed Ali's employ, reported that smelting technique was very primitive. In an attempt to improve this, John Petherick, a Glamorganshire mining engineer and later British Consul in Khartoum, came to the Sudan in 1847 at the request of the Egyptian Government, but again the venture came to nothing (Hill 1963, p. 58).

Gessi worked in the Bahr el Ghazal region from 1879 to 1881 and proposed to establish forges using the same lateritic sources of iron as those used by the Bongo and Luo tribes (Gray 1964, p. 129). Once again these proposals to develop mineral wealth came to nothing.

Dunn (1911) described the iron ores of the Sudan, classifying them as:

(1) Solid deposits unrelated to climatic conditions.
(2) Surface deposits related to existing climatic conditions and forming two groups:
  (a) ores belonging to humid equatorial regions south of 10° N; and
  (b) ores belonging to arid regions situated north of 10° N.

The solid deposits include: specular iron ore, said to come from the Tokar Delta (Kad Idris and Khor Hamrik); 'large iron masses and iron mountains', said to exist in Bahr el Ghazal and Upper Nile provinces but details were not given; the oolitic iron ore and ferruginous sandstone in the Wadi Halfa region, described by Lyons as lenticular deposits 2–5 miles in length; and a highly ferruginous sandstone, reported by Hume of the Egyptian Survey as underlying a mass of intrusive basalt or dolerite at Jebel Alarambrah, near Kerma, Northern Province.

The surface deposits of the humid equatorial regions are the laterites and lateritic soils of Bahr el Ghazal and Upper Nile provinces. In the Rumbek and Mvolo areas in stream sections they are as much as 48 ft thick, and in Bahr el Ghazal province alone Dunn (1911) estimated that the iron-rich deposits cover 30 888 squares miles and vary from 3 to 16 ft in thickness.

The surface deposits of arid regions are mainly ferricrete crusts produced by evaporation and iron concentration, and are seldom more than a few inches thick (Dunn 1911). Dunn also mentioned that the principal tribes of the southern Sudan involved in smelting are the Jurs around Mvolo; Annuaks on the Sobat; the Aliab on the west bank of the Bahr el Jebel; and the Bongos around the Bahr el Arab. Dunn noted also that the Arab tribes smelted iron at Nahud and at several villages in Darfur. The well-known large native ironworks at Umm Semeima and Jebel Haraza in Kordofan were abandoned at the time of the Mahdiya and not re-opened. Dunn and Grabham (1911) also wrote on the iron deposits of the Sudan, mentioning much the same information as is given above.

Andrew (MS 1946) briefly mentioned iron ore, recording haematite in western Darfur, bedded iron ore near Wadi Halfa; lateritic iron ore in the southern Sudan, and magnetite at Qala en Nahl.

The discovery of the Fodikwan iron-ore deposit, northern Red Sea Hills, was announced by Andrew (1954). The deposit was discovered in 1952 by I. G. Gass, an officer of the Sudan Geological Survey engaged in regional mapping. It is a dyke-like mass some 1·2 miles in length, 66 ft wide, dipping south.

On analysis one specimen yielded 85·12 per cent Fe_2O_3, equivalent to 59·6 per cent metallic iron.

Gass (1955) recorded an iron deposit (haematite) intimately associated with the Salala ring complex. The deposit is in the form of a dyke running east–west for about 200 yards. The ore assayed 87·40 per cent Fe_2O_3.

Kabesh, Afia, and Widatalla (1958) of the Sudan Geological Survey described the Fodikwan deposit, which is small and contains only about 144 000 metric tons (70·68 per cent Fe) high grade and 95 000 tons (44·64 per cent Fe) mixed accessible ore.

Jebel Abu Tulu iron ore deposit, Dar Messeriya, Kordofan was described by Mansour and Iskander (1960). The deposit was discovered by officers of the Sudan Geological Survey in the early 1950s, and in 1957 Iskander reported it as worthy of economic consideration. At a conservative estimate there are over 35 000 000 tons of ore at Jebel Abu Tulu with an iron content of 55–65 per cent, average 61·12 per cent.

Kabesh (1960) recorded iron ore from the Ingessana Hills, Blue Nile Province. The quantities are apparently small, however, and analyses were not given.

The iron ore prospects of the Sufaya area, northern Jubal el Bahr el Ahmer were investigated in 1962 by Daniel, Mann, Johnson, and Mendenhall (MS 1962) on behalf of the Sudan Geological Survey and the United States Agency for International Development. Previously this group of deposits had been investigated by Elektrosund on behalf of a private group.

In 1964 the Fodikwan Company Ltd. was registered with £S104 000 capital to work the Fodikwan deposit and since then a limited amount of ore has been exported.

A list of main iron occurrences in the Sudan known to the author is given in Table 38.

TABLE 38

List of iron localities, Sudan (See Fig. 84)

Name	Coordinates	Sudan 1:250 000 sheet no.	Comments
Northern Jubal el Bahr el Ahmer			
118. Fodikwan	21° 44′ N 36° 42′ E	Dungunab 36–I	Working mine
119. Hashaseit		—	
120. Sufaya or Sufaia area	21° 17½′ N 36° 16½′ E	Dungunab 36–I	Working mine
121. Yaweikurar		Dungunab 36–I	
122. Ogra		Mohammed Qôl	
123. Salala	21° 17′ N 36° 11′ E	Dungunab 36–I	Associated with Salala ring complex
Nile Valley, Atbara region			
124. Jebel Hadid, Sudan Portland Cement Quarry, Atbara	17° 38′ N 33° 49′ E	Atbara 45–K	Replacement deposit in marbles in Basement Complex
Nile Valley, Wadi Halfa region			
125. Wadi Halfa area		Wadi Halfa 35–I	Sedimentary ore in Nubian Sandstone Formation
126. Station No. 7	20°–22° N 32°–34° 30′ E	J. Murrat	Irregular iron deposits associated with Basement limestones and Nubian Formation

Name	Coordinates	Sudan 1:250 000 sheet no.	Comments
Kordofan			
127. Jebel Abu Tulu	11° 41′N 28° 40′E	Kadugli 65–D	
Blue Nile Province			
128. Ingessana Hills Qala en Nahl	—	Er Roseires 66–D	Small magnetite deposits
129. Blue Nile, near Roseires	—	Er Roseires 66–D	Details unknown, minor occurrence
130. Jebel Terru	12° 33′N 33° 12′E	Kerkoj 55–O	Magnetite
Southern Sudan			
131. Various extensive laterite and lateritic deposits			Central and southern Sudan, mainly in Bahr el Ghazal, Equatoria, Upper Nile, Kordofan and Darfur Provinces

(ii) *Fodikwan iron deposits*

(1) *General geology.* The Fodikwan iron-ore deposits consist of three occurrences in the northern Jubal el Bahr el Ahmer, Kassala Province (Fig. 84). The main deposit is situated in a tributary of Khor Naakuri south of the main channel. Minor deposits are situated near by, one in the main channel of the Khor and the other to the north (Kabesh, Afia, and Widatalla 1958, p. 2).

The iron ore occurs as a dyke body in metasediments and altered volcanics of the Oyo Formation. Minor basic masses, now represented by epidorites, and some granites also occur. Ore solutions penetrated along the regional strike of the metamorphic rocks. The deposits are affected by north-west-trending major faults. Dykes and large quartz veins were intruded and these were cut by west-north-west-trending faults which acted as channels for late hydrothermal solutions. According to Kabesh *et al.* (1958) these solutions produced propylization of feldspars and caused the oxidation of magnetite into haematite in the sheared zones.

(2) *Nature of the ore.* The ore was primarily composed of magnetite and later altered either partially or totally. Minerals identified include magnetite, haematite, goethite, siderite, ilmenite, and pyrite. The sulphur and phosphorus content is negligible. The main gangue mineral is quartz but some carbonate is present in the footwall at Fodikwan. Carbonate is more abundant at Naakuri (Kabesh *et al.* 1958).

(3) *Origin of the ore.* Kabesh, Afia, and Widatalla (1958) think that the ore originated from 'ascending hydrothermal or pneumatolytic solutions' and was deposited along lines of weakness and the regional strike of metamorphic rocks. The magnetite was deposited in tabular sheets with apophyses of varying widths. Shearing then took place and the minor faults produced provided the easy way for penetration of silica-and-carbonate-rich oxidizing solutions mainly of supergene origin.

Gass (1955) gave the following analysis of a specimen from Fodikwan:

85·12 per cent	FeO_3
3·01 per cent	TiO_2
Remainder	Silica, alumina, and traces of phosphorus

Kabesh *et al.* (1958) also gave analyses of specimens from Fodikwan:

	%	%
Fe	62·5	56·9
Ti	0·87	0·68
P	0·21	n.d.
S	0·12	0·06
$Al_2O_3 + SiO_2$	7·8	—
Mn	—	0·54
Free SiO_2	—	9·16
Complex silica	—	2·44

Of twelve samples analysed from Fodikwan, values ranged from 70·86 per cent Fe in the best parts to 44·64 per cent Fe in the highly silicified parts of the ore body. Representative specimens of ore from Naakuri assayed 61·38 per cent Fe and from Naakuri North 59·14 per cent Fe.

(4) *Reserves.* Kabesh *et al.* (1958) gave the figures shown in Table 39.

TABLE 39

Iron reserves, Fodikwan, Sudan (Kabesh *et al.*)

	Tons
Fodikwan	
Total amount of accessible solid ore	143 933
Total amount of accessible mixed ore	93 644
Naakuri	
Total amount accessible ore	1157
North Naakuri	
Total ore	Less than a few hundred tons
All Deposits	
Total amount accessible ore	144 000
Total amount mixed ore	95 000
Grand total	239 000

However, according to an estimate made by Yugoslav investigators, there are at least 6 million tons of ore.

(iii) *Jebel Abu Tulu deposit*

(1) *General geology.* Jebel Abu Tulu iron-ore deposit forms part of Jebel Abu Tulu (11°41′ N, 28°40′ E), Dar Mesiriya, Kordofan, about 20 miles east-north-east of Rigl el Fula. The ore occurs in two hills, one elongated main mass trending north-east, and occupying an area of 2952 × 1312 ft (900 × 400 m) and about 262 ft (80 m) high; and another conical mass having a diameter of about 1312 ft (400 m) and rising 213 ft (65 m) above the surrounding plain (Mansour and Iskander 1960).

The iron ore occurs in Basement Complex quartz–sericite-schists associated with meta-andesites. The iron ore in the main hill mass occurs as two thick bands having sharp contacts with the metasediments. Minor bands occur and form part of the ore body. Faulting is said to have caused rotation of blocks from a north-east–south-west to an east–west trend. These east–west faults are shot with quartz veins and stringers, and silica penetrated the ore to a limited extent along cracks, joints, etc. after faulting. Small disseminations of iron ore occur in the schists, and nearer the contacts the concentration increases. Both the schists and the iron ore are banded, and foliation and cleavage are parallel to the primary structures.

(2) *Nature of the ore.* The ore is haematite, hard, dense, granular to finely crystalline. Near the contacts it becomes schistose and streaky. Haematite is the main, and magnetite the minor, constituent. Haematite commonly replaces magnetite. Pyrite is rare and the sulphur content is in general low (less than 0·38 per cent). Silica, occurring mainly as quartz, is scattered throughout. Some jasperoid silica was recorded. The iron content of the ore ranges from 55 to 65 per cent Fe. In the main mass it is 63 per cent and averages 61·12 per cent Fe. Silica ranges from 4·84 to 6·88 per cent, averaging 6·5 and 5·2 per cent on the main and conical hills respectively. A little magnesium, calcium, and manganese has been detected, but no phosphorus (Mansour and Iskander 1960).

(3) *Origin of the ore.* Mansour and Iskander (1960) believe that the ore is of sedimentary origin. If this is correct, then the question naturally arises about the regional extent of such a deposit. Unfortunately prospecting seems to have been limited to within the Jebel Abu Tulu area and nothing is known of the areas away from the jebel mass.

(4) *Reserves.* Mansour and Iskander (1960) gave the information contained in Table 40.

TABLE 40

Iron ore reserves, Jebel Abu Tulu, Kordofan (Mansour and Iskander 1960)

I.	Measured reserves	
	5 453 996	Main hill
II.	Indicated reserves	
	10 012 388 tons	Main hill
	3 732 000 tons	Conical hill
III.	Inferred reserves	
	16 533 616 tons	Main
	Total reserves	35 732 000 tons

The main problems in developing the Jebel Abu Tulu deposit are the long haul to the sea (1100 miles), lack of power at the site, and the small quantity (in international terms) of proved reserves.

(iv) *Sufaya iron deposits*

(1) *General.* The Sufaya or Sufaia iron deposits occur in Sufaya district, Kassala Province, near the Sudanese–Egyptian frontier, some 160 miles north of Port Sudan. Six deposits have been identified within an area of 62 miles diameter:

Jebel Ader Awab	(120 A)	Aquissan	(120 D)
Jebel Ankura	(120 B)	Sufaya	(120 E)
Aklat	(120 C)	Mafdeib	(120 F)

(Numbers refer to Fig. 84, the locality map)

According to Daniel, Mann, Johnson, and Mendenhall (1962) the deposits were discovered before 1957 by Sayed Mohammed Sidki Saleh and Sayed Kamal Sidki Saleh of Port Sudan, who now hold the mining lease. However, iron ore was first found in the area by Gass (1955), who mapped and prospected Dungunab sheet for the Sudan Geological Survey (Fig. 18).

Interest was aroused in the deposit by Intertrade (Yugoslavia), who arranged for a geophysical survey to be made by Elektrosond (Zagreb). The leaseholders bore all the expenses of the exploration and contracted with the Yugoslavs to supply 20 000 tons of ore from Ader Aweb deposit. Some 22 000 tons of ore was shipped from the site prior to 1962, and some 8000 tons were refused by the Yugoslavs and stockpiled near the site (Daniel *et al.* MS 1962).

The deposits of Aklat and Mafdeib are insignificant; the prospects of Aquissan and Sufaya are intermediate; whereas those of Ader Aweb and Ankura are good in local terms (Daniel *et al.* MS 1962). The iron content of ores at Ader Aweb and Ankura varies from 54 to 60 per cent.

(2) *Geology.* Little geological description is given in Daniel *et al.* (MS 1962). The country-rocks are said to be 'crypto-crystalline and reddish porphyritic rhyolites'. The rhyolites are thought to be of Cainozoic age but no proof of age is given.

(3) *Nature of the ore.* The ores occur in crushed and brecciated rocks of lamprophyric type. Contacts are sharp and the ore is found 'in layer-shaped bodies' (Daniel *et al.* MS 1962).

The following analyses are available:

	ADER AWEB	ANKURA
FeO	—	15·27
Fe_2O_3	86·46	73·17
MnO	0·10	0·19
Al_2O_3	2·60	0·03
TiO_2	0·10	0·16
P_2O_5	0·80	1·98
CaO	1·00	3·30
MgO	0·50	1·22
S	0·01	0·06
As	Tr	0·00
SiO_2	6·28	2·36
BaO	0·00	Tr
K_2O	0·10	0·12
Na_2O		0·38
Insol Res		1·81
Ign. Loss	1·62	0·04
	99·57	100·09

(4) *Reserves.* The Jebel Ader Aweb deposit is exposed at the surface over an area of 322 800 ft² and probably is at least 164 ft deep. It is estimated from visual examination that there are about 2 million tons of plus 60 per cent ore, of which about 30 000 tons have already been extracted. At Ankura there is an estimated 1 400 000 tons of plus 60 per cent ore and possibly 2 100 000 tons of 50–55 per cent ore (Daniel *et al.* MS 1962). Mr. Day, Chief Mining Engineer of Ferguson and Wild, London, estimated from visual examination that in all the Sufaya deposits there are between 7 000 000 and 8 000 000 tons of 60 per cent plus ore.

(b) Manganese

(i) *History of development*

Manganese was first reported in the Sudan by Andrew (1951) when it was discovered in drilling operations for water in Upper Nile Province at Paloich (10°27½′ N, 32°32′ E); and at Wabuit 35 miles east of Malakal (9°28′N, 32°06′ E).

In the Jubal el Bahr el Ahmer manganese was discovered in Basement Complex formations before 1955. Afia examined the Abu Samr–Tolik–Allaikaleib deposits in 1956, southern Jubal el Bahr el Ahmer (Kabesh and Afia 1961), and Abdullah (1958) reported manganese 45 miles south of Tohamiyam, central Jubal el Bahr el Ahmer, in veins and fissures in granite gneiss. Kabesh and Wahab examined the Sinkat deposits in 1958 (Kabesh and Afia 1961).

Manganese also occurs in the sediments of the Red Sea Littoral. These deposits were visited by Kabesh and Wahab in 1956. They reported on the mining operations taking place at that time (Kabesh and Afia 1961).

Kabesh and Afia (1961) reported manganese from veins in Basement Complex formations on the west bank of the Nile in the Berber district but did not give details. Daniel, Mann, Johnson, and Mendenhall, consultants for U.S. aid, visited the Sinkat area and reported on it briefly in 1963. A comprehensive account of the manganese deposits of the Sudan was presented by Kabesh and Afia (1961).

Mining of manganese began in 1955 on a small scale and production reached 7500 tons in 1956. Mining methods are extremely primitive and some of the deposits have already been exhausted. In 1965 manganese worth £S5688 was exported from the Sudan. The ores are mainly oxides and the principal commercial-grade ores consist of psilomelane, pyrolusite, and manganite. Almost nothing is known of reserves. Manganese-bearing localities known to the author are listed in Table 41.

TABLE 41

List of manganese localities, Sudan (See Fig. 84)

Name	Coordinates	Sudan 1:250 000 sheet no.	Comments
Red Sea Littoral			
132. El Amar	—	Dungunab 36–I	Lease owned by Samkary Co. Ltd.
133. Sarara	22° 16′N 36° 30′E	Halaib 36–E	Lease owned by Samkary Co. Ltd.
134. Kaleibeit	22° 13′N 36° 20′E	Halaib 36–E	Lease owned by Samkary Co. Ltd. and Egyptian interests
135. Manueib	—	Halaib 36–E	Lease owned by Samkary Co. Ltd. 30 miles west of Halaib
136. Karraf	—	Halaib 36–E	Lease owned by Samkary Co. Ltd. 45 miles north-west of Halaib
137. Arai Barar	22° 20′N 36° 12′E	Halaib 36–E	Lease owned by Samkary Co. Ltd.
138. Bashweit	22° 18′N 36° 22′E	Halaib 36–E	Lease owned by Samkary Co. Ltd. 22 miles north-west of Halaib
139. Ankalidot	22° 30′N 35° 55′E	—	Worked out
140. Almaateib	22° 19′N 36° 30′E	Halaib 36–E	Egyptian-owned 17 miles north-west of Halaib
141. Al Kashf or Belonay	—	Halaib 36–E	Egyptian-owned 17 miles north-west of Halaib
142. Kalalungab	—	Halaib 36–E	Egyptian-owned
143. Yoyder	—	Halaib 36–E	Egyptian-owned
144. Eishumhai	—	Halaib 36–E	Egyptian-owned
145. Al Hubal or El Hobal	—	Halaib 36–E	Egyptian-owned
146. Takmaniai or Takmanyai	—	Halaib 36–E	10 miles north-west of Halaib
147. Metateib	—	Halaib 36–E	4 miles north-west of Mersa Merob

Name	Coordinates	Sudan 1:250 000 sheet no.	Comments
148. Adargab	—	Halaib 35–I	2 miles north-west of Metateib
149. Tikaraeit	—	Tokar 46–F	South of Port Sudan on coastal strip
Central Jubal el Bahr el Ahmer			
150. Sinkat area	19° 07′N 36° 16′E	Port Sudan 46–A	About 40 miles north-west of Sinkat
151. Abu Samr	17° 58′N 36° 21′E	Derudeb 46–I	East of Adarot station
152. Allaikaleib	17° 50′N 36° 38′E	Derudeb 46–I	East of Adarot station
153. Tolik	17° 48′N 36° 12½′E	Derudeb 46–I	East of Adarot station
154. Bashikwan	17° 48′N 36° 38′E	Derudeb 46–E	—
155. Wurreiba	18° 10′N 37° 02′E	Sinkat 46–E	—
Upper Nile Province			
156. Paloich	10° 27′N 32° 32′E	Melut 66–F	Sedimentary manganese in the Umm Ruwaba Formation
157. Wabuit	9° 28′N 32° 06′E	Malakal 66–J	Sedimentary manganese in the Umm Ruwaba Formation
Northern Province			
158. Berber			Details not given. West bank of Nile (Kabesh and Afia 1961)
159. Umm Rahaw Station	18° 54′N 32° 02′E	Merowe 45–F	Metamorphosed deposits in marbles

(ii) *Red Sea Littoral deposits*

(1) *Geology*. Most of the deposits lie in the Halaib area and occur in moderately or steeply dipping conglomerates which form part of a sandstone, limestone, gypsiferous sequence. The ore occurs either in veins or as concretions cementing conglomerates, some composed of boulders of igneous and metamorphic rocks. Boulders are commonly encrusted with a film of manganese.

The ore veins trend mainly east–west and north–south; a few veins trend north-west, however. They dip either north-eastwards or are vertical in the southern part of the area, or dip south-west in the northern part. The veins attain a thickness of up to 6 ft, and some occur along faults (Kabesh and Afia 1961).

(2) *Nature of the ore*. The ores are mainly pyrolusite and manganite, accompanied by minor quantities of calcite and quartz, together with some barytes. Sometimes alternations of brighter crystalline and duller materials produce banding (Kabesh and Afia 1961). The manganese content varies from 23 to 53 per cent. Iron commonly occurs in the ores, 1·9–5 per cent, and silica varies from 3 to 54 per cent.

(3) *Origin of the ore*. Kabesh and Afia (1961) believe that the ore is epigenetic and was deposited from ascending manganese-rich solutions. In the author's view the origin of the manganese is sedimentary, for the deposits clearly form part of the sedimentary sequence. The stages envisaged in development of the manganiferous beds are as follows: (1) erosion supplying gravels and conglomerates from the Basement formations of Jubal el Bahr el Ahmer; (2) deposition of conglomerates

in lagoonal areas; (3) deposition of manganese from solution in films or in layers or veins together with gypsum, travertine, and aragonite; (4) supergene enrichment.

(4) *Reserves*. Because of the erratic nature of the manganese ores it is not possible to estimate reserves in the Halaib area. In conglomeratic deposits the quantity of good ore is small and there is much waste in hand picking. Some of the vein deposits are richer and sometimes yield over 50 per cent ore.

All the Halaib ore bodies lie near the surface and are worked with cheap labour by open-cast methods. Little modern exploration has been undertaken.

(iii) *Central Jubal el Bahr el Ahmer–Sinkat area*

(1) *Geology*. The deposit occurs some 40 miles north-west of Sinkat, Kassala Province (Kabesh and Afia 1961) in hornblende- and chloritic schists of the Basement Complex. The rocks are well jointed and trend north-north-east. They dip at 75°–80° to the north-west.

The ore body is a continuous vein that pinches and swells and attains a maximum width of 12 ft. Small quartz veinlets cut the vein and the country-rock.

(2) *Nature of the ore*. The ore is mainly psilomelane and pyrolusite and occurs in both massive and crystalline forms. According to Kabesh and Afia (1961) it is of epigenetic origin, the ore-bearing solutions having penetrated along fractures following the foliation.

Chemical analysis shows the ore to contain from 47 to 52 per cent Mn, 3 to 9 per cent Fe, less than 0·06 per cent P, and 0·12 per cent S (Kabesh and Afia 1961).

(3) *Reserves*. As no drilling has been done reserves cannot be estimated but they appear to be small.

(iv) *Abu Samr–Tolik–Allaikaleib area*

(1) *Geology*. Manganese is known from four localities in this area: Abu Samr, Tolik, Allaikaleib, and Bashikwan. The area is situated east of the Tehilla ring complex.

The ore bodies are found in Basement Complex interbedded with quartzites, schists, and gneisses, intruded by granites and porphyritic and quartz dykes. They consist of veins that frequently pinch and swell and are up to 860 ft long.

(2) *Nature of the ore*. The ore consists of manganese and iron, the proportions being highly variable. The main minerals are psilomelane, manganite, and some pyrolusite, rhodanite, and rhodocrosite.

Galena occurs at Allaikalaib in small amounts and there are traces at Abu Samr and Tolik. Barytes occur at Allaikalaib and Abu Samr as an important gangue mineral.

Analyses of the ore give (Kabesh and Afia 1961): Mn 7–51 per cent; Fe 44–46 per cent; P tr.–0·3 per cent; S 0–3·66 per cent; Ba 0·81–31 per cent.

(3) *Reserves*. According to Kabesh and Afia (1961) the deposits are exhausted but modern detailed exploration might well reveal new ore. Most of the workings were open-cast; all were on a small scale.

(v) *Wurreiba area*

Manganese ores also occur at Wurreiba (18° 10′ N, 37° 02′ E) in Basement Complex schists and granites. The ore is mainly rhodonite ramified with stringers of manganese oxide.

Origin of the ores in the Abu–Samr–Tolik–Allaikalaib–Wurreiba areas. Kabesh and Afia (1961) think that the manganese was originally sedimentary. The deposits were then metamorphosed on a regional scale and the manganese subsequently concentrated.

(vi) *Upper Nile Province, Paloich, and Wabuit*

(1) *Geology*. Manganese was discovered by G. Andrew in 1951 in cores from the Umm Ruwaba Formation, which consists mainly of unconsolidated sands, sandy clays, and clays.

At Paloich (10°27′N, 32°32′E) manganiferous sediments were discovered between 163 and 435 ft with concentrations of 32 per cent Mn in a 12-ft bed between 260 and 272 ft. The following chemical analyses were obtained (Kabesh and Afia 1961): Mn_3O_4 32 per cent; Fe_2O_3 38 per cent; P_2O_5 1·25 per cent.

Manganese occurs also at Wabuit (9°28′N, 32°00′E) some 35 miles east of Malakal, Upper Nile Province, at depths of 180–210 ft and 315–320 ft. Concentrations are less than 6 per cent and iron content varies from 46 to 70 per cent Fe_2O_3.

(2) *Origin of the ores*. The source of the manganese in these two areas is unknown but the intimate relationship of the sediments and the manganese point to a sedimentary origin.

(3) *Reserves*. It is not possible to estimate reserves with the borehole data available (Kabesh and Afia 1961).

(c) Chrome

(i) *History of development*

Chromite in the Sudan was first mentioned by Grabham (1930), who reported that prospectors of the Nile–Congo Divide Syndicate visited Qala en Nahl to investigate an occurrence already reported. The deposit was not thought rich enough for export. Andrew (MS 1946) mentioned this deposit again but gave no further details. Kabesh (1960) described the geology and economic prospects of the Ingessana Hills, southern Blue Nile Province where chrome deposits occur. The reserves were not estimated but are probably not large. The grade of the ore is high, however: 57·54 per cent CrO_3 (cf. 58·1 per cent for some South African ores). In 1960 the Mining and Trading Company (Sudan) Ltd. was formed to work the Ingessana deposits and mineral lease No. 193 was acquired from Messrs Saragas and Co. Ltd. (Khartoum). From 1963 to 1966 (Jan–April) chrome worth some £165 000 was exported from the Ingessana area. Abdl Moneim Corporation, Sudan commissioned a preliminary geological investigation of chrome deposits in the Ingessana Hills in 1967.

The chromite localities known to the author are listed in Table 42. Details are available for the Ingessana deposits only.

(ii) *Chrome deposits, Ingessana Hills*

(1) *Geology*. The Ingessana Hills are composed mainly of Basement Complex schists, gneisses, metaquartzites, and calcitic and dolomitic marbles cut by ultramafic bodies, which were serpentinized and heavily carbonated and in some places replaced by talc–carbonate rocks. Porphyritic and non-porphyritic biotite-granites were emplaced on a large scale and acidic and basic dykes finally intruded. Chromite is present in the serpentines and is associated with the talc–carbonate bodies and some zoned hornblende bodies (Kabesh 1960).

(2) *Nature of the ore*. The chromite bodies occur as lenses, veinlets, and disseminations in serpentines associated with talc-carbonate bodies. They range in width from a few inches to a yard. Some twenty-one chromite bodies were located but only three were classed as of importance (Kabesh 1960):

1. Bau deposit: said to be 8·5 × 6·5 × 1·5 m.
2. Soda deposit: consisting of three lenses in soft

TABLE 42

List of chrome localities, Sudan (See Fig. 84)

	Name	Coordinates	Sudan 1:250 000 sheet no.	Comments
Blue Nile Province				
160.	Ingessana Hills, Bau	11° 20′ N 34° 10′ E	Er Roseires 66–C	
161.	Ingessana Hills, Soda	11° 18′ N 33° 56′ E	Er Roseires 66–C	
162.	Ingessana Hills, Jebel Kurba	11° 17′ N 33° 57′ E	Er Roseires 66–C	
163.	Jebel Tawil, Butana	14° 54′ N 34° 40′ E	Gedaref 66–H	
164.	Qala en' Nahal	—	Gedaref 66–H	
165.	Algerif	22° 42′ N 35° 12′ E	Marsa Shash 35–H	
166.	Sol Hamid	22° 12′ N 36° 05′ E	Halaib Sheet 36–E	Samkary Co. Ltd.

TABLE 43

List of molybdenum localities, Sudan (See Fig. 84)

	Name	Coordinates	Sudan 1:250 000 sheet no.	Comments
167.	Hofrat en Nahas, Darfur	09° 45′ N 24° 19′ E		Present in small quantities. Hassan MS 1966
168.	El Qroon el Zurug, Sabaloka near Khartoum (Details see Tin at Sabaloka)	16° 19′ N 32° 37′ N	Sabaloka 45–N	Traces associated with tin and wolfram mineralization and grey mica granite. Disseminated. Almond 1967
169.	Area between Halaib and Jebel Shallal, northern Jubal el Bahr el Ahmer	—	—	Ruxton and Gass visited this area to search for reported molybdenite. Deposit not found. Andrew (1952–3); Annual Report, Sudan Geological Survey
	Jubal el Bahr el Ahmer	—	—	No details. Andrew (MS 1946)

TABLE 44

List of tungsten localities, Sudan (See Fig. 84)

	Name	Coordinates	Sudan 1:250 000 sheet no.	Comments
170.	Halaib area	—	—	Two localities, details not available; 3 per cent W according to B. A. Hassan (MS 1966)
171.	El Qroon el Zurug, Sabaloka, near Khartoum (Details see Tin at Sabaloka)	16° 19′ N 32° 37′ E	Sabaloka 45–N	A. J. Whiteman (1961) and Almond (1967)
172.	Khor Arbent, near Suakin, Jubal el Bahr el Ahmer	—	—	No details. Abdullah (1958)
173.	Gash Amer	22° 14′ N 36° 12′ E	Halaib 36–E	No details

talc–carbonate with areas varying from 15–20 × 3–8 m. The depth is unknown.

3. Jebel Kurba deposit consisting of lenses 10–30 × 2–6 m, striking north-north-west. The dip is vertical and depth unknown.

Samples of solid ore analysed gave Cr_2O_3 53–57 per cent; FeO 14–17 per cent; SiO_2 3·6–5·3 per cent; Al_2O_3 10–12 per cent; MgO 0–1·3 per cent.

The ore consists of subhedral crystals associated with other gangue minerals.

No published information is available about placer chromite in the Ingessana Hills.

(3) *Reserves.* Little information is available on the distribution of ore bodies in depth, and virtually nothing can be said about reserves.

(d) Molybdenum

Molybdenum occurs at scattered localities in the Sudan. No deposits are worked and almost nothing is known about the nature of the deposits, reserves, etc. The localities known to the author are listed in Table 43.

(e) Tungsten or wolfram

Wolframite, like molybdenum, is known from only a few localities in the Sudan. No deposits are worked and reserves are unknown. The localities known to the author are listed in Table 44.

Wolfram was discovered at El Qroon el Zurug, Sabaloka, by the author in 1961. Details based on Almond (1967) are given above under Tin.

Wolfram was originally the most abundant mineral in the veins composing the stockwork at El Qroon el Zurug, but most of the material at the surface is now altered into mixture of iron and manganese oxides. Pseudomorphs of wolfram form anhedral crystals and aggregates 0·25–2 cm and sometimes 10 cm across, moulded on quartz. According to Almond (1967) wolfram is moulded on cassiterite and is therefore the later mineral. Details are not available for the other occurrences.

5. MINOR METALS

(a) Arsenic

Arsenic is known from few localities in the Sudan. Abdullah (1956) noted the occurrence of arsenophyrite at Wadi Oyo, Jubal el Bahr el Ahmer near the gold mine, and the mineral is associated with gold elsewhere in the Jubal el Bahr el Ahmer. No details or information about reserves are available.

(b) Magnesium

Magnesium in the form of magnesite ($MgCO_3$) is known to the authors from the three localities in the Sudan given in Table 45. Details are not available for the Sol Hamed deposit but are available for Qala en Nahl and Ingessana Hills.

Wilcockson and Tyler (1933) provided chemical analyses of the Qala en Nahl ultrabasic deposits and the magnesites. Ruxton (MS 1956) also described them, and more recently the Sudan Geological Survey reinvestigated the deposits. According to Abdullah (1958) there are more than 200 000 000 tons of magnesite in this area.

In the Ingessana Hills (see under Chrome) magnesite occurs as lenses and pockets, and less commonly as veins, ramifying serpentines. It is usually associated with chromite. Altogether Kabesh (1960) recorded some thirty-two magnesite occurrences, of which five are of importance.

Chemical analyses gave the following results: $MgCO_3$ 95 per cent; $CaCO_3$ 1–3 per cent; (occasionally 35 per cent); Fe_2O_3 1·6 per cent (rarely); SiO_2 0·7 per cent (rarely). No information on reserves is available.

(c) Uranium

A search for radioactive minerals in the Sudan has so far been restricted to the Hofrat en Nahas area, Darfur (177), where torbernite (copper-uranium phosphate) and uraninite have been recorded (B. A. Hassan MS 1966). Uranium is also known from the Nuba Mountains (178) but no details are available.

TABLE 45

List of magnesite localities, Sudan (See Fig. 84)

	Name	Coordinates	Sudan 1:250 000 sheet no.	Comments
174.	Sol Hamed	22° 12′N 36° 05′E	Halaib 36–E	98 per cent Mg magnesite. Bedawi (MS 1966)
175.	Qala en Nahl		Qala en Nahl 55–L	Associated with serpentines, as asbestos and talc. Wilcockson and Tyler (1933); Ruxton (MS 1956); Bedawi (MS 1967)
176.	Ingessana Hills	—	Roseires 66–C	Associated with chromite in serpentines and talc. Kabesh (1960)

(d) Columbium

Columbite has been reported from one locality only in the Sudan between Nimule and Juba, Equatoria Province (179) (Andrew MS 1946). No details were given.

(e) Titanium

Titanium occurs as titanium oxide in gabbros and associated rocks at Dirbat Well (19° 54′N, 36° 35′E) Jubal el Bahr el Ahmer, especially in the western and central parts of the Dirbat intrusion (Kabesh and Afia) (180). The deposit, which was discovered by B. P. Ruxton of the Sudan Geological Survey in 1953, occurs in skarn zones associated with gabbroic bodies intruded into calcareous Basement Complex rocks of the Nafirdeib Formation (Ruxton 1956). The skarns contain wollastonite (see below), grossularite, schorlomite, diopside, and idocrase. The idocrase and schorlomite are titanium-bearing assaying up to 16 per cent titanium oxide. Scattered lenses contain up to 38 per cent titanium oxide.

Some 8000–1000 tons of titanium sands are said to occur in the khors of the Dirbat area. Chemical analyses of crude sands after sieving through 25-mesh gave 7·9–24 per cent TiO_2.

Magnetite–ilmenite bearing rocks occur at Onib Gorge Deraheib Sheet 35-L and containing 19 per cent TiO_2, 7·19 per cent FeO, 69 per cent Fe_2O_3, and 3·3 per cent Mn_3O_4.

(f) Barium

Barium occurs as barytes ($BaSO_4$) in small amounts as a gangue mineral at Allaikalaib (17° 50′N, 36° 38′E) (181) and Abu Samr (17° 58′N, 36° 21′E) east of Adarot Station (182), Derudeb Sheet, Central Jubal el Bahr el Ahmer; and south-west of the Halaib area (183) (B. A. Hassan 1966). No details were given.

6. NON-METALLIC MINERAL DEPOSITS

(a) Ceramic materials

(i) *Clay*

Clay is a common substance in the Sudan, especially in the riverain areas. In many localities the local clay is used for making pots, zeers (water jars), etc. It is of passing interest to note that some of the fine Meroitic pottery was made by using ground-up mudstone from the Nubian Formation mixed with local clays. Clay is used in brickmaking and 'tobh arda' for mud houses. Many of the brick clays in the Sudan contain montmorillonite, which, because it expands when wet, makes it difficult to produce good quality burnt brick.

(ii) *Kaolin and feldspar*

Kaolin and feldspar are widely distributed in the Sudan. B. A. Hassan (MS 1966) mentions that the main occurrences are in the Derudeb area (described below); on the west bank of the Nile, opposite Shereik Station, Northern Province; and in Darfur (details not given).

The Durudeb deposits occur at Chagooneen (Shagonin) (184) and Mekeikwola (185). The Chagooneen deposit lies about 10 miles east of Derudeb Station, Kassala Province (36° 12′N, 36° 14′E). According to Abbas (1967) the deposit was formed by hydrothermal and pneumatolytic alteration of felsite showing typical rhyolitic texture. Kaolinization increases with depth. Detailed chemical, thermal, and X-ray analyses were made of samples of the clay, but an estimate of reserves was not given. B. A. Hassan (1967) estimated that there are some 1 820 000 tons of kaolin with 15 per cent aluminium oxide and 70 per cent silica in the Chagooneen area.

Other ceramic materials occur in the Sudan such as bauxite (probably in southern Sudan); kyanite (Shereik area); magnesite (see above); barytes, fluorspar, and talc.

Barytes occurs in the Halaib and Derudeb areas (see above). Fluorspar occurs in veins at Semeih (186) in Kordofan (Andrew MS 1964). The deposit, which is small, was discovered by J. M. Edmonds in the 1930s. No details were published.

Large quantities of talc exist in the Sudan in the Ingessana Hills, Blue Nile Province (187); the Haiya area (188) central Jubal el Bahr el Ahmer, Kassala Province; at Gerif (189) some 70–80 miles west of Halaib (32° 12′N, 22° 42′E). The size of the last-mentioned deposit is unspecified.

(b) Structural and building materials

(i) *Building stones*

Except for public buildings, bridges, and dams, stone is used relatively little for buildings in the Sudan, despite its ready availability in many areas. Few villages possess masons and most domestic building in the northern and north central Sudan is done in 'tobh arda' composed of mud, straw, and dung, or in soft burnt brick. In the south central and southern Sudan, houses are built mainly in grass, leaves, and wood.

The main building stones of the Sudan in order of importance and use are:

(1) Sandstones of the Nubian Formation.
(2) Jebel Sileitat soda granite.
(3) Jebel Moya gneisses, granites, and charnockitic quartz-diorites.
(4) Jebel et Toriya basalt.
(5) Summit marble.
(6) Raised reef complex limestone, Port Sudan and Suakin.

(1) *Sandstones of the Nubian Formation.* This stone was chosen by the Condominium government for the construction of many official buildings in Khartoum because it could be quarried easily in near-by jebels. Its suitability as a building stone varies considerably. The Nubian has been used in the construction of Meroitic temples and pyramids, which are still well preserved, for example at Naga, Musauwarat, and Bagariwya Meroe; in Government buildings such as the Judiciary, Post Office, etc. in Khartoum; the Great Mosque and the Farouk Mosque in Khartoum; the Anglican Cathedral; the Gordon College buildings of the University of Khartoum; decorative stone in private houses in Khartoum, etc. After exposure for a number of years there is often considerable decay and scaling, especially in exposed places, or if the stone has not been carefully selected, or has been laid bed face. Cross-bedded sandstones also weather quickly and scale. This is due to the extremely high rate of chemical decay during and after the kharif or rainy season, and the ease with which water gets into the sandstone (Whiteman 1967).

Most of the stone used in large Government buildings in Khartoum was quarried at Jebel Aulia (190), 30 miles south of Khartoum on the White Nile, where a large part of the eastern face of the jebel has been excavated to expose fresh stone. Stone used for decoration in many new houses in Khartoum was quarried mainly from shallow excavations, and consequently is highly weathered. No information is available on the quantities of stone quarried.

(2) *Jebel Sileitat soda granite.* Jebel Sileitat (191) is a riebeckite-granite mass that crops out 18 miles north-north-east of Khartoum. The stone is quarried for dimension stone and is crushed for aggregate. Some of the foundation courses of buildings in Khartoum, for example the Faculty of Science, geology and zoology buildings, are of Sileitat stone. It has been used also with considerable decorative effect in the new Archaeology Museum in the Mogren, Khartoum. The central spillway section and locks at Jebel Aulia Dam are made of Sileitat soda granite. Again no figures are available concerning production.

(3) *Jebel Moya stone.* The Jebel Moya granite (192) and charnockitic quartz-diorite is quarried by Sudan Railways for dimension stone and aggregate. It was also used in the construction of the Sennar Dam.

(4) *Jebel et Toriya basalt.* The Jebel et Toriya basalt (193) outcrops 7 miles west-south-west of Omdurman and has been used together with wood to produce stylish interior walls in the perfumery of the Blue Nile Pharmacy, Khartoum. The rock is fine-grained, hard, and durable, and can easily be cut into thin slabs which can be 'hung' over cheap brick or plaster to produce pleasing effects.

(5) *Raised reef complex limestones, Port Sudan (194) and Suakin.* Reef limestones have been quarried around these two coastal towns and used in various buildings. The stone is easily worked and in Suakin it was used to produce many fine buildings and carvings, which unfortunately have now been allowed to fall into decay or have been destroyed.

(ii) *Crushed stone*

Granite from Jebel Sileitat is quarried for concrete aggregate as well as dimension stone. It was used in making the concrete for the new Shambat bridge over the Main Nile.

Jebel Rauwiyan Granite (195) from near Sabaloka, 40 miles north of Khartoum, is also used for aggregate in the Three Towns area and is of excellent quality.

Crushed basalt from Jebel et Toriya was used as top dressing for the Khartoum–Wad Medani road built by U.S. aid, and for the hard core for the new dual carriage-way from the White Nile bridge to Omdurman.

No production data are available for aggregate but

the value of crushed stone must run to many thousands of pounds per year.

(iii) *Sand and gravel*

Certain terraces of the Nile in the Khartoum area (196–197) have been extensively dug for gravel. They consist mainly of rounded quartz pebbles and cobbles derived by weathering and disintegration of conglomerate beds of the Nubian Sandstone Formation. The deposits are of two sorts: river gravels and desert lag gravels.

Exploitation has been piecemeal and the methods of working are extremely wasteful. Large quantities of gravel have been covered with waste from other workings, making reworking more expensive or impossible. As a result of uncontrolled gravel digging, workings are now over 25 miles from Khartoum on the west bank of the Nile. No production figures are available but the value of the product must run to many thousands of pounds per year.

Good quality sharp sand for building purposes is rare in many places in the Sudan. The main sources are wadis, such as the Khor Shambat, etc. and the bed of the Nile when exposed at low water. No production data are available.

(iv) *Cement*

Cement is made by the Sudan Portland Cement Company at Atbara (198) and another plant is being constructed near Rabak on the White Nile (199). Limited quantities of cement were made for construction of the Sennar Dam using limestone from Jebel Sagadi near Jebel Moya.

(1) *Sudan Portland Cement Company.* After the fall of Eritrea the cement factory of Massawa, Eritrea, was taken over from the Italians and operated by a group called Red Sea Industries, in which the Gellatly Hankey Co. Ltd. participated. Cement was made at Massawa and transported overland, but it was decided to transfer the machinery to Atbara in Sudan and the Sudan Portland Cement Company was started in 1946.

Among the reasons for choosing this locality are the occurrence of pure marble at Wadi Khurmut, on the west bank of the Nile near Atbara, and the proximity of the railway. This is a unique combination at present in the Sudan. Other limestone bodies are known, such as those in the Shereik area, Northern Province, and at Jebel Nyfr near Jebelein, Blue Nile Province. Both these deposits are dolomitic and therefore are less suitable for cement-making than the pure Khor Khurmut marble. Limestones occur at Jebel Saqadi near Sennar but most of the pure limestones seem to have been worked out and those now left appear to be mainly dolomitic.

The Sudan Portland Cement Company, Ltd. was originally financed by the Helwan Portland Cement Company (Egypt), the Societé de Ciments Portlands de Tourah (Egypt), and a number of small shareholders including Gellatly Hankey Sudan, Ltd. Both the Egyptian companies exploited the limestone deposits of the Mokattam Hills, near Cairo. The Tunnel Portland Cement Company, a British firm, held a majority interest in Helwan Portland Cement prior to sequestration in 1956.

The Atbara factory first began to operate using cement clinker transported by barge from Helwan to Wadi Halfa and then by rail to Atbara. In 1949 a kiln was started burning local limestone but clinker was still imported via Port Sudan using coal transporters. Clinker was obtained from Tunisia and Lebanon, as well as from Egypt. Lebanese clinker was much cheaper than Egyptian clinker and was paid for in Sudanese currency, which had the effect of stimulating trade with Lebanon. However, the company was eventually forced by the Sudan Government to obtain all clinker from Egypt for political reasons.

The Sudan Portland Cement Company was originally registered in London, but in 1958 the registered office was transferred to Atbara and in 1959 it was registered as a Sudanese public company. The chief shareholders are Tunnel Portland Cement; Max Schmidt Heiney and others (a Swiss company); individual Egyptian shareholders; six Sudanese shareholders; and Gellatly Hankey Co. Ltd., who also act as agents.

Total potential production per annum of clinker at Atbara is now 210 000 tons and for cement it is 100 000 tons (Table 46). The reason for the discrepancy is that the company has decided not to increase its cement-grinding capacity because of uncertainties about future development of the industry.

Lt.-Col. H. W. Dixon, General Manager of the Sudan Portland Cement Company, has estimated that the Sudan can absorb currently some 150 000 tons per annum, excluding quantities required for large construction jobs such as dam building. In

1965, some 82 611 tons of cement were imported with a value of £S600 815, while 49 082 tons were produced at Atbara. In 1964, some 337 417 tons were imported with a value of £S1 845 788. The price of a ton of cement bought by a customer in Khartoum contains approximately £S4·500 in the form of taxes, rates, rail freight, customs and import dues, and, until recently, an excise duty levied at the rate of £0·500 per ton. As is shown in Table 46 the company has never been able to operate at anything near its full capacity, which has of course precluded economic operation.

TABLE 46

Production and import data on cement, Sudan (H. W. Dixon 1966)

Year	Domestic† sales	Imported sales	Capacity	
			Cement	Clinker
1950	31 313		not known	not known
			Cement	Clinker
1951	34 917		not known	not known
			Cement	Clinker
1952	36 800		not known	not known
			Cement	Clinker
1953	39 353		not known	not known
1954	47 750		12 000	30 000
1955	64 625	23 586	120 000	60 000
1956	67 586	11 460	120 000	60 000
1957	58 672	74 129	120 000	60 000
1958	89 298	17 997	120 000	60 000
1959	94 029	19 012	120 000	60 000
1960	89 907	25 020	120 000	60 000
1961	82 026	71 953	100 000	180 000
1962	102 923	209 824	120 000	180 000
1963	116 159	278 535	150 000	180 000
1964	90 775	337 417	180 000	210 000
1965	49 082	82 611	180 000	210 000
1966	104 100	6 092	(For the first three months of the year)	

† Produced by Sudan Portland Cement Co. Ltd.

In the past Sudan Portland cement was exported to the northern Congo. The cement was handled by three firms who trucked cotton, etc. from Stanleyville to Juba and carried cement in return. Some cement was also exported by road into Chad.

(2) *The Nile Cement Company* was formed in 1957 as a public company but by the closing date for subscription not all the shares offered to the public had been taken up. These were taken up by the Sudan Government. In 1965 another 500 000 shares were offered but there were few takers; these shares were mainly taken up by the Sudan Industrial Bank. The Sudan Government therefore owns about three-quarters of the shares in this concern.

Plant was built at Rabak and a quarry area stripped Jebel Nyfr, near Jebelein. A large technical problem exists in developing the limestone because the limestone mass is both dolomitic (magnesium-bearing) and calcitic, and according to British and other international standards the acceptable magnesium content is about 3–4 per cent. In places analyses show that the limestones contain as much as 12 per cent magnesium, so if quality cement is to be made then there will have to be a careful selection of raw material. So far no cement has been made at Rabak.

(v) *Gypsum*

Limited quantities of gypsum are used in the Sudan as a stabilizer in the cement industry. Most of this is produced as a by-product in salt-making.

Enormous quantities of gypsum exist in the Red Sea Littoral in the Tertiary formations described in Chapter II. There are two main horizons: (1) the Middle Miocene Dungunab Formation, which was penetrated in AGIP's well Dungunab–1 and consists of 2125 ft of massive rock-salt interbedded with minor beds of grey anhydrite and constituting the A Member; gypsum interbedded with clays and sandstones (B Member) some 177 ft thick that occurs below the Dungunab Formation; and (2) the evaporites that occur in the Middle or Lower Miocene Maghersum Formation. The A Member (364 ft) of this formation consists of marls, sandstones, and conglomerates with subordinate limestones; gypsum is common in veins and nodules. The B Member consists of gypsum and anhydrite interbedded with sandy shales and sandstones (2818 ft thick); much of this thickness is shown on logs as gypsum and anhydrite. The underlying C Member is composed of massive rock salt with minor anhydrite and marl bands (1525 ft).

These beds crop out at various localities along the Red Sea Littoral. Known localities include Khor Eit (200); Jebel To Baham (201); Jebel Saqhum (202); Near Dungunab (203) 21°51′ N, 37°4′ E; Maghersum Island (204) 20°48′N, 37°12′E; Marsa

Shinab (205) 21°21′N, 37°02′E; Jebel Dyiba (206) 21°22′N, 37°01′E; Jebel Abu Imama (207) 21°26′N, 36°57′ E; Harboub (208) 22°08′ N, 36°47′ E; Hakurab (209) 22°14′N, 36°38′E; and near Halaib (210) 22°14′N, 36°39′E.

Gypsum also occurs at Jebel Abiad (211), west of Khandaq, near Dongola, Northern Province (B. A. Hassan MS 1966—no details). It is probably a playa deposit.

The localities given in Table 47 have been studied in detail by Ustaz Mohammed Zein Shaddad.

TABLE 47

List of gypsum and anhydrite deposits, Red Sea Littoral, Sudan. (See Fig. 84)

Name	Mineral	Location
209. Hakurab	Gypsum	2 miles west of Halaib, 22°14′N, 36°38′E
208. Harboub	Gypsum	8 miles south of Halaib and 1$\frac{1}{2}$ miles east of Halaib–Port Sudan Road, 22°08′N, 36°47′E
205. Shinab	Anhydrite	Near Marsa Shinab on Halaib–Port Sudan Road, 21°04′N, 37°02′E
207. Abu Imama	Impure gypsum	Jebel, Abu Imama, 21°26, N, 36°57′E
206. Jebel Dyiba	Impure gypsum	5 miles south of Marsa Abu Imama, 21°22′N′ 37°01′E

(1) *Hakurab deposit* (209). This deposit occurs in a khor bed 2 miles west of Halaib. The gypsum crops out for some 250 yd along the khor. It extends laterally for some 200 yards. The overburden ranges from 0 to 16 ft; much of this is unconsolidated. The gypsiferous beds dip at 20° bearing 070°.

HAKURAB

(209)

TOP	Thickness (ft)
Limestone and gypsum, bedded well laminated. This interval can be divided into an upper coarsely bedded part containing 4 beds of impure gypsum, average 5 in thick and containing about 17 laminae, average about $\frac{1}{2}$ in thick. Sample A1 was taken from this interval. Analysis 78 per cent gypsum	5
Gypsum, laminated; selenite; limestone, green. Gypsum laminae are $\frac{1}{4}$–3 in thick; 50 laminae occur. Gypsum constitutes more than 70 per cent of these beds. Beds are much disturbed due to weathering and solution but are probably less disturbed along the khor to the right. Sample A5 was taken from this part of the section. Analysis 83 per cent gypsum	4
Gypsiferous 'sand'; fine grained, recrystallized; leached and spongy	1·5
Limestone; bedded; outcrop limited by ravine coming from right bank of khor; disappears under high ground to right	6
Gypsum and limestone, green interlaminated; laminae numerous and thicker than limestone laminae. Consist mainly of gypsum which constitutes more than 80 per cent by volume of rock. Sample A4 Analysis 79·9 per cent	6
Limestone, green bedded with few conspicuous beds of gypsum ranging from $\frac{1}{4}$ to 2 inches thick. Some 17 beds were counted constituting 10 per cent volume of rock. Samples A3, A7, and A8 consist of pure selenite	10
Gypsum and limestone interbedded; gypsum laminae 1–2 inches and constitute 40 per cent by volume. Sample A2 has 74 per cent gypsum with anhydrite derived from surface weathering	6
BASE Total	38$\frac{1}{2}$

Comment: In many places gypsum is altered on the surface to a white friable powder of anydrite.

CHEMICAL ANALYSES OF SAMPLES†

Sample No.	Gypsum $CaSO_4$, $2H_2O$	Anhydrite $CaSO_4$	Total Carbonate (%)	Salt NaCl (%)	Other ingredients (%)
A1	78·4	0·0	1·6	0·0	20·0
A2	74·0	22·3	0·3	0·0	3·4
A3	76·3	13·5	0·3	0·3	9·6
A4	79·9	8·2	2·5	0·0	9·4
A5	83·4	6·4	0·0	0·2	10·0
A6	32·5	53·2	2·5	0·2	11·6
A7	93·6	6·0	0·2	0·0	0·2
A8	94·3	5·5	0·0	0·0	0·2

† Analysed by Sudan Portland Cement Company Limited.

(2) *Jebel Harboub* (208). This deposit is situated about 8 miles south of Halaib and 1½ miles east of the Port Sudan–Halaib road. The jebel consists of crescent-shaped hills, amost completely formed of gypsum. The highest of the hills does not rise more than 45 ft above the surrounding plain. The strata in general dip at an angle of 20° bearing 135°. Variations occur because of solution and slumping.

The gypsum occurs in two areas: (1) an area of 2400 × 1200 ft (Harboub 'A') and (2) nearer the sea an area of 600 × 1500 ft (Harboub 'B'). The gypsiferous beds are highly weathered and considerable recrystallization has taken place at the surface and along small khors. Powdery anhydrite is common on the surface. The hills are also covered with a skin of pebbles.

HARBOUB 'A'
(208)

TOP	Thickness (ft)
Gypsum, finely laminated; interbedded with clayey material, red, especially in lower 2 ft; gypsum recrystallized in reddish flakes. Composite sample F1 taken representative of lower 12 ft of interval	25
Limestone and gypsum, finely interbedded; laminae ⅛ to 2 in thick. Intensely weathered, recrystallized especially at surface. Gypsum constitutes about 50–70 per cent by volume of rock. Sample F2	15
BASE Total	40

Harboub 'B' (208). This constitutes a separate outcrop, situated east of Jebel Harboub. The lower part of the interval exposed at Harboub 'A' occupies an area of 1500 × 600 ft, elongated in a north–south direction. The outcrop is dissected by tributaries of Khor Harboub. Gypsum, fine-grained and laminated, is interbedded with green marl. Individual laminae average about ¾–½ in but some reach 1½ in. More than 10 ft is exposed in the Harboub 'B' area.

As elsewhere the deposit is recrystallized at the surface to selenite. This crust ranges in thickness from 6 in to 5 ft. Anhydrite occurs as an alteration product on the surface. The selenite crust, however, covers about 70 per cent of the surface outcrop.

(3) *Jebel Abu Imama deposit.* Jebel Abu Imama (207) lies west of the Halaib–Port Sudan road, 5 miles south-west of Mersa Abu Imama and about 15 miles south of Mersa Oseif (Fodikwan Iron Company's dock). The jebel extends north–south for about 3 miles and east–west for about 2 miles.

The section along Khor Hibit was studied. At the base red sandstone occurs, followed by green sandstone with occasional conglomerates and isolated gypsum laminae (¼ inch). The sandstones are at least 300 ft thick. Bedded calcareous gypsum, some 30 ft thick, lies on top of the sandstone. The section is capped by resistant recrystallized fossiliferous limestone. All beds dip uniformly at 20° bearing 070°. A bed of calcareous gypsum forms a continuous band along the landward side of the north–south limb. The outcrop is mainly marked by talus. At the surface the gypsum is altered to white powdery anhydrite. Immediately below the surface selenite occurs.

Near the north-east corner of the jebel, three spot samples were taken about 150 ft apart. These have a high carbonate content, gypsum being less than 62 per cent.

CHEMICAL ANALYSES OF SAMPLES

Sample No.	Gypsum $CaSO_4$, $2H_2O$ (%)	Anhydrite $CaSO_4$ (%)	Total carbonate (%)	Salt NaCl (%)	Other ingredients (%)
B1	61·6	11·8	26·0	0·0	0·6
B2	51·7	5·9	41·6	0·0	0·8
B3	31·3	7·0	60·6	0·0	1·1

(4) *Jebel Dyiba* (206). This deposit lies at the north-west corner of Jebel Dyiba a few hundred feet from Marsa Halaka, near the Halaib–Port Sudan road, about 20 miles south of Marsa Oseif.

Between the north-west corner of the jebel and the marsa the ground is much dissected. The rocks are bedded green limestones and sandstones capped by thin calcareous gypsum. This is highly weathered in the lower part and there are numerous vertical and oblique veins. Parts of the underlying limestone are permeated with gypsum in the form of pods up to 3 ft in diameter.

South of this dissected area gypsiferous limestone occur. Samples were taken but all showed low gypsum content.

JEBEL DYIBA
(206)

TOP	Thickness (ft)
Limestone	9
Limestone, gypsiferous, white Sample D10	3
Limestone, green	1
Limestone, gypsiferous white. Sample D9	3
Limestone, nodules of gypsum, green	3
Limestone, fossiliferous with gypsum. Samples D6, D7, D8	2
Limestone, green	1
Limestone, highly fossiliferous with gypsum. Surface leached and pitted. Samples D+, D5	2
Limestone, abundant gypsum in nodules and veins, Sample D3	2
Limestone, green	1
Limestone, gypsiferous. Samples D1, D2	4
BASE Total	31

CHEMICAL ANALYSIS OF SAMPLES

Sample No.	Gypsum $CaSO_4$, $2H_2O$ (%)	Anhydrite $CaSO_4$ (%)	Total carbonate (%)	Salt NaCl (%)	Other ingredients (%)
D1	32·4	10·9	54·1	0·0	2·6
D2	46·2	0·3	52·9	0·2	0·4
D3	20·4	22·6	56·6	0·0	0·4
D4	35·2	0·0	51·4	0·0	13·4
D5	3·5	33·7	60·5	0·0	0·3
D6	19·2	21·6	58·4	0·0	0·8
D7	38·6	3·8	56·1	0·0	1·5
D8	—	—	—	—	—
D9	35·2	8·6	55·1	0·2	0·9
D10	36·0	2·7	56·8	0·0	4·5

(5) *Marsa Shinab* (205). This deposit occurs on the south end of Jebel Dyiba and extends from the western top of Marsa Shinab for about 3300 ft to the west. The thickness of the beds is about 30 ft. Anhydrite, white and chalklike, occurs in the eastern part interbedded with red sandstones and limestone. South of the jebel there are many patches anhydrite resting on green limestone. The beds dip at 20° bearing 300°.

MARSA SHINAB
(205)

TOP	Thickness (ft)
Limestone, fossiliferous	24
Sandstone, fine grained red	18
Anhydrite, coarsely bedded, white. Sample E5 (channel 6 ft long)	18
Sandstone red and limestone green	15
Anhydrite, coarse bedded beds up to 10 in thick. Samples E2, E3, E4	12
Sandstone red, and limestone green resting unconformably on igneous (dunite) rocks	21
BASE Total	108

An outcrop of the same anhydrite beds occurs at the foot of the jebel a few yards to west of Marsa Shinab along the Halaib–Port Sudan road.

Anhydrite occurs as isolated tabular bodies in the limestone, overlain by anhydrite. The anhydrite ranges from a few inches to 6 ft thick. Sample E 6 is representative of the lowermost 5 ft of this formation. Recrystallized gypsum, a few inches to a foot thick, covers the surface outcrops.

CHEMICAL ANALYSIS OF SAMPLES

Sample No.	Gypsum $CaSO_4$ $2H_2O$ (%)	Anhydrite $CaSO_4$ (%)	Total carbonate (%)	Salt NaCl (%)	Other ingredients (%)
E1	0·0	57·0	41·1	1·0	0·9
E2	11·2	76·3	0·5	0·7	11·3
E3	4·3	93·9	1·3	0·3	0·2
E4	2·8	92·4	0·5	0·5	3·8
E5	1·6	73·8	2·5	0·2	21·9
E6	6·0	83·1	5·7	1·5	3·7

Among the deposits investigated, only Hakurab and Harboub deposits are suitable as sources of gypsum for the cement industry. Certainly other deposits exist in the area visited by the writers but could not be investigated because of shortage of time.

Without doubt millions of tons of gypsum and anhydrite exist along the Red Sea Littoral but owing to the low price of gypsum it would have to be used for local consumption. Development would entail open-cast mining and shipment by sea and by rail.

(vi) *Lime*

Only small amounts of lime are used in the Sudan for mortar and very little limestone or marble is burnt for lime. 'Omdurman lime' is a ground-up form of caliche, a mixture of calcium carbonate and sulphate, etc., which is formed as a result of capillary action and alternate wetting and drying of sediments along the banks of the present day Nile (212). It also occurs in terrace deposits. It is used ground up to a powder which is added to water and used as a lime wash for painting buildings.

(vii) *Magnesite*

Magnesite is used for refractory bricks, cements, and flooring in the building industry. It occurs in great quantities associated with ultrabasic rocks in the Qala en Nahl area (213), Ingessana Hills (214), and a 9 per cent magnesite deposit occurs in the Halaib area.

No estimates of reserves are available except for Qala en Nahl, where Abdullah (1958) estimated that there are over 200 million tons of talc–magnesite-schist in quarriable form.

(viii) *Heat and sound insulators*

Various nonmetallic materials are used for this purpose. Such substances occurring in Sudan include

(1) Vermiculite.
(2) Wollastonite.
(3) Asbestos.
(4) Pumice.
(5) Diatomite.

(1) *Vermiculite.* According to B. A. Hassan (MS 1966) vermiculite occurs west of Tohamiyam Station (216), Jubal el Bahr el Ahmer. U.S. aid Report on Minerals prepared by Daniel Mann, Johnson, and Mendenhall (MS 1962) gave another locality (216) '20 miles south-east of the railroad town of Tohamiyam'. At this last locality an area 200 × 1000 ft was trenched and small strips of vermiculite, 1–3 ft wide were exposed. A large amount of waste would have to be excavated to yield small amounts of vermiculite, and according to Daniel *et al.* (MS 1962) it is extremely doubtful if this deposit could be mined profitably.

Daniel *et al.* (MS 1962) also mentioned another vermiculite deposit, similar to the one mentioned above, some 10 miles south of Tohamiyam.

(2) *Wollastonite.* Wollastonite was discovered by B. P. Ruxton of the Sudan Geological Survey in 1953 at Dirbat well (217), Jubal el Bahr el Ahmer, 19° 53′ N, 36° 31′ E. It occurs in skarn zones associated with gabbroic bodies intruded into calcareous Basement Complex rocks. The skarns contain wollastonite, grossularite, schorlomite, diopside, and idocrase. The idocrase and schorlomite are titanium-bearing with up to 16 per cent titanium oxide (Kabesh and Afia 1959). Wollastonite samples contain about 71 per cent wollastonite, and according to Kabesh and Afia (1959) the deposit is of good quality and could be worked open-cast.

(3) *Asbestos.* Asbestos is known from Qala en Nahl (218) and the Ingessana Hills (219) where it is associated with chromite, magnesite, and talc. The deposits were studied by Wilcockson and Tyler (1933), Ruxton (MS 1956) and other officers of the Sudan Geological Survey. Daniel, Mann, Johnson, and Mendenhall (MS 1962) also reported on the deposit for U.S. aid and the Sudan Government.

There are a number of serpentine chrysotile asbestos deposits in the Qala en Nahl area, Kassala Province. Four deposits have been explored by trenching and pitting.

No. 1 deposit. There are many outcrops of asbestos in an area of 50 × 200 ft on the side of a hill rising about 100 ft high above the surrounding plains. Talus and weathered rock covers much of the slope. Asbestos with fibre lengths of $\frac{1}{16}$ inch of good clean grades was found in pits, and Daniel *et al.* (MS 1962) estimated that there is a possible 10 000 tons of ore with a 5–10 per cent asbestos content which could easily be mined.

No. 2 deposit. This deposit has yielded fibres up to $\frac{3}{4}$ inch and exploratory work showed 5–12 per cent asbestos content. The surface area under which the ore lies is about 300 × 500 ft. Possibly in the area there is more than 100 000 tons of 5–12 per cent asbestos-bearing ore.

No. 3 deposit. Fibre lengths of $\frac{3}{4}$ inch were noted in ore with a content of 5–10 per cent outcropping in a mineralized area of 100 × 300 ft. Daniel *et al.* (1962) estimate that there is about 30 000 tons or more of asbestos ore in this area.

No. 4 deposit. This deposit is the largest recorded and many veins have ½-inch fibres. The mineralized zone is about 800 × 400 ft and there are a possible 200 000 tons of ore in this deposit with up to 20 per cent asbestos.

All the deposits consist of chrysotile filling fractures in serpentine and all can be worked by open cut. Altogether in the region there are several hundred thousand tons of ore with a grade of 5 per cent asbestos or better. Asbestos is also known from the Sol Hamid area 22° 12′ 30″ N, 36° 05′ E, Halaib sheet. No details are available.

(4) *Pumice.* Pumice occurs in Darfur in the Jebel Marra volcanic range (220) and in the Bayuda volcanic field (221) but it is not worked.

(5) *Diatomite.* Diatomaceous earth was recorded by Andrew (1946 MS) at Bara in Kordofan (222) and in central Darfur (223). It is said to occur near the base of the Kordofan sands. The deposit is not worked.

(ix) *Miscellaneous materials*

(1) *Mica.* Mica has been recorded in commercial quantities from the localities given in Table 48.

In 1957 the African Mining Company was incorporated to take over from Sayed Abdl el Rofhail Gereis all mica-mining leases held in the Rubatab or Shereik area, Northern Province.

The mica area of Shereik was described by Kabesh (1960) and subsequently investigated by an English mining group headed by D. Sutton, who prospected the whole of the Rubatab area. Sutton thinks that it would be extremely difficult to build up a mica industry based on these deposits that would be competitive in the well-established world market (Sutton, personal communication 1967).

The Rubatab mica belt extends from Jebel Nakhara, near Berber in the south mainly along the west bank of the Nile to Wadi El Koro in the north. A few occurrences occur east and south-east of Shereik. The rocks are composed of metamorphosed sediments injected by gabbros, granites, and pegmatites (for details see Chapter II).

The pegmatite bodies are developed on the west side of the Nile mainly (Kabesh 1960). The bodies are lenticular and irregularly shaped and are usually concordant with the foliation of the country-rock. Some are up to 1300 ft long and over 30 ft wide. The majority are 6–12 ft wide. They pinch and swell and show boudinage structure.

The pegmatites are of the acid type with quartz and microcline feldspar forming graphic intergrowths. The cores are usually composed of quartz and sometimes contain microperthite. Some pegmatites show late albitization and contain small quantities tourmaline (crystals up to 5 cm), spessar-

TABLE 48

List of mica localities, Sudan (See Fig. 84)

Name	Coordinates	Sudan	Comments
Southern Sudan			
224. Longairo, Mongala district	4° 29′ N 32° 16′ E	Juba 78–N	
225. Near Lufir	—	Juba 78–N	On Juba–Nimule Road
Southern Fung			
226. Khor Budini, south of Yabus River	—		
227. Near Bikori	11° 12′ N 34° 40′ E	Fazughli 66–D	On Roseires–Qeissan Road
Northern Province			
228. Shereik area	—		Details given below

Details are available for the Shereik area only.

tite garnet (up to 1 cm diameter), pale-green whitish-beryl (6 × 10 cm) and apatite (up to 8 cm) Kabesh (1960). Kyanite is common and well developed in the country-rocks.

Muscovite is scattered throughout most of the pegmatites in the area and occurs in books of industrial sizes (20 × 20 cm or 10 × 10 cm). Muscovite occasionally encloses tourmaline, garnet, and apatite crystals. Books range from 2·5 to 5 cm and some attain 10 cm thickness. The muscovite occurs mainly in the side walls and appears to be more abundant on hanging walls.

The development of mica in the pegmatite is irregular and therefore estimation of reserves is difficult. The yield of mica to rock mined is low (approximately 5 per cent) and only about 5–10 per cent of this is actually marketed. The mica is classed as 'fairly stained ruby mica'. Working has been on a small scale and can only be described as primitive.

A list of coordinates for mica localities in the Rubatab area is given below.

N	E
18° 11′	33° 32′ 30″
18° 14′	03° 50′
18° 32′	33° 34′
18° 33′	33° 30′
18° 47′ 30″	33° 30′
18° 48′ 30″	33° 50′
19° 16′	33° 12′ 30″

(2) *Salt and natron.* No salt is mined in the Sudan although large quantities exist in the Tertiary formations of the Red Sea Littoral (see Gypsum).

Salt is produced by evaporation of sea water in artificial lagoons south of Port Sudan (229). Production is small but some is exported. Salt was made at a British-owned salt works in the early years of this century on the west side of Ras Abu Shagara (230), utilizing water from Dungunab Bay.

Old salt works are recorded between Khartoum and Rufaa on the western edge of the Butana. The source and nature of the salt here is not clear. It may have been derived from kankar deposits in the Nile terraces or from deeper ? volcanic sources; or from old lake deposits. There are small sporadic salt fields in Kordofan and Darfur, for example in Dar Hamid (L. Hill, personal communication).

Natron occurs in the desert west of the Nile in Northern Province (231) and northern Darfur (232 and 233). It is used locally as a laxative and figures in minerals exported from the Sudan (Table 8).

Grabham (1928), describing rocks and well water collected by Messrs Newbold and Shaw on a journey through Bir Natrun (232), Nakheila (232) to Selima (231), gave the following analysis for Nukheila water: NaCl 10·09 g/100 cm^3 water; Na_2SO_4 2·88 g/100 cm^3 water; Na_2CO_3 2·46 g/100 cm^3 water; Total 15·45 g/100 cm^3 water. Calcium and magnesium are absent, and potassium, although not estimated, is said to be small.

(3) *Abrasive materials.* Spessartite garnet is the only abrasive, other than diatomite (mentioned above) recorded from the Sudan.

(4) *Quartz and glass sand.* A limited amount of quartz is mined and crushed for glass-making in the Sudan. According to Daniel, Mann, Johnson, and Mendenhall (MS 1962) about 450 tons per month of quartz was quarried from a conical mound of high grade quartz 35 ft high and 150 ft base diameter, situated about 15 miles south-west of Tohamiya (234), Jubal el Bahr el Ahmer. This was sold to Khartoum Bottle Company at £S3·100 per ton delivered in Khartoum. The quartz was transported by road and rail to Khartoum and was a profitable operation only because of the low labour costs. The quartz was quarried by the National Mining Company, Sayed Omer Abu Amna, Gebeit.

Quartz veins of considerable purity occur 40 miles north of Khartoum in the Sabaloka inlier (235) but are not utilized for glass-making.

Glass sands may exist in Kordofan but have not been prospected in detail (236).

(5) *Paints and pigments.* Ochre derived both from iron-rich Nubian Sandstone Formation and Basement Complex rocks is common in the Sudan, especially in Kordofan and Darfur provinces. The multi-coloured materials are crushed, and used largely for colour wash for houses (237).

(6) *Phosphorite.* Phosphorite is said to occur in the Halaib district (238). No details are known. Apatite occurs in small quantities at Shereik, Northern Province (see under Mica above).

(x) *Gemstones*

Minerals of gemstone quality are apparently rare in the Sudan. Only one small diamond has been recorded so far. No locality was given by Andrew (MS 1946); it is said to have been found in river sand in Equatoria Province (239).

Agate, chalcedony, and carnelian occur in the gravels of the Atbara River (240) and are common in the Khashm el Girba region for instance. These and other minerals were used by the Neolithic and Meroitic peoples for making jewellery, and at the Bagrawiya Merowe site, near Kabushiya some of the buildings, now ruined, must have been jewellers' shops, judging from the quantity and nature of mineral fragments to be found there now.

Numerous minerals exist in the Sudan suitable for the establishment of a semi-precious stone or a 'chunky jewellery' industry as in east and central Africa.

Bibliography

ABDL SALAAM, Y. (1966) The ground-water geology of the Gezira. M.Sc. Thesis, University of Khartoum.

ABDL WAHAB, O. (1963) Geology of Shambat bridge site. *Sudan Geological Survey Bulletin No.* 13. Ministry of Mineral Resources, Khartoum.

ABDULLAH, M. A. (1955*a*) *Report 1953–5, Sudan Geological Survey*. Sudan Government, Khartoum.

—— (1955*b*) Progress of oil prospecting in the Sudan. In *Colloque sur la Géologie Appliquée dans de Proche-Orient*, pp. 165–7. UNESCO, Ankara.

—— (1958) *Annual Report 1955–7, Sudan Geological Survey*. Ministry of Mineral Resources, Khartoum.

—— (1960) *Annual Report 1959–60, Sudan Geological Survey*. Ministry of Mineral Resources, Khartoum.

—— (1965)

ADDISON, F. (1949) *Jebel Moya, the Wellcome excavations in the Sudan*, Vols. 1 and 2. Oxford University Press.

AFIA, M. S., and WIDATALLA, A. L. (1961) An investigation of the Hofrat en Nahas copper deposit. *Sudan Geological Survey, Bulletin No.* 10. Ministry of Mineral Resources, Khartoum.

AGIP MINERARIA LTD. (1963) Petroleum development in Africa 1962: Sudan. *Bull. Am. Ass. Petrol. Geol.* 47, 1392.

AGIP MINERARIA SUDAN LTD. (1965) Petroleum development in North Africa. *Bull. Am. Ass. Petrol. Geol.* 49, 1253.

ALEXANDER, B. (1907) From the Niger, by Lake Chad to the Nile. *Geogrl J.* 30, 119–52.

ALLAN, T. D. (1964) A preliminary magnetic survey in the Red Sea and Gulf of Aden. *Boll. Geofis. teor. appl.* 6.

ALLEN, P. E. T. (1963) Photogrametry in the Sudan. Eighth Annual Conference 1960, Phil. Soc. Sudan.

ALMOND, D. C. (1962) *Explanation of the geology of sheet 15 Kitgum*. Geological Survey, Uganda.

—— (1964) Preliminary account of the geology of Sabaloka. *Annual Report, Research Committee, University of Khartoum.*

—— (1967) Discovery of a tin–tungsten mineralization in northern Khartoum Province, Sudan. *Geol. Mag.* 104, (1) 1–12.

—— (1967) Petrology of the basalt at Jebel et Toriya, Khartoum. *Sudan Notes Rec.* 48, 141.

ANDREW, G. (1934) The structure of the Esh Mellaha Range (Eastern Desert of Egypt). *Bull. Inst. Égypte* 16, 47–9.

—— (1936) Sur les roches hyperalcalines d'Egypte. *Bull. Soc. fr. Minér. Cristallogr.* 59, 338.

—— (1937*a*) On the Nubian Sandstone of the Eastern Desert of Egypt. *Bull. Inst. Egypte* 19, 93–115.

—— (1937*b*) The late Tertiary igneous rocks of Egypt. *Bull. Fac. Sci. Egypt. Univ.* 10.

—— (1943) MS. The geology of the Sudan. University of Khartoum Geology Library. Original MS for 'The geology of the Sudan' in TOTHILL (1948). Contains regional descriptions not in Andrew (1948).

—— (1944*a*) Notes on Quaternary climates in Sudan. In *Soil Conservation Report*, Appendix 25. Sudan Government.

—— (1944*b*) Rural water supplies. In *Soil Conservation Report*, Appendix IX. Sudan Government.

—— (1945) Sources of information on the geology of the Anglo-Egyptian Sudan. *Sudan Geological Survey Bulletin No.* 3. Khartoum.

—— (1946) MS. Mineral resources of Sudan. Incomplete. University of Khartoum Geology Library.

—— (1947) The development of the Sudan plain in the Quaternary. *Proceedings of the first Pan-Africa Congress on Prehistory*, pp. 73–5.

—— (1948) The geology of the Sudan. In TOTHILL, J. D. (1948) *Agriculture in the Sudan*, pp. 84–128. Oxford University Press, London

—— (1952*a*) Contribution to discussion on laterite, East Africa High Commission Paper No. 5. *Proceedings of the fifth Inter-territorial Geological Conference* (Dodoma), pp. 58–62.

—— (1952*b*) Iron ores in the Anglo-Egyptian Sudan. *Proc. nineteenth int. geol. Congr.*, Algiers, fasc. 21, 390.

—— (1953*a*) Late Tertiary movements in the Anglo-Egyptian Sudan. *Colon. Geol. Miner. Resour.* 3, 248.

—— (1953*b*) *Report 1950–2, Sudan Geological Survey*. Sudan Government, Khartoum.

—— (1954) *Report 1952–3, Sudan Geological Survey*, pp. 1–20. Sudan Government, Khartoum.

—— and ARKELL, A. J. (1943) A middle Pleistocene discovery in the Anglo-Egyptian Sudan. *Nature, Lond.* 151, 226.

ANDREW, G. and KARKANIS, G. Y. (1945) Stratigraphical notes, Anglo-Egyptian Sudan. *Sudan Notes Rec.* **26**, 157–66.

ANDREWS, C. W. (1912) Notes on the molar tooth of an elephant from the Nile near Khartoum. *Geol. Mag.* 49, 110–13.

ANDREWS, F. W. (1948) Vegetation of the Sudan. In TOTHILL, J. D. (1948) *Agriculture in the Sudan.* Oxford University Press, London.

ANON. (1922) *Monograph on mineral resources, Egypt and the Sudan.* Imperial Institute.

—— (1944) *Soil Conservation Committee's Report.* Sudan Government, Khartoum.

—— (1955*a*) Jonglei investigation team, equatorial Nile project. *The Jonglei Report.* Sudan Government.

—— (1955*b*) Food and society in the Sudan. Third Annual Conference, Phil. Soc. Sudan.

—— (1955*c*) Resultats scientifiques des compagnes de la 'Calypso'. I. Compagnes 1951–2 en Mer Rouge. *Annals Inst. océanogr., Monaco* **30**, 1–204.

—— (1956*a*) *Catalogue of topographical maps.* Sudan Survey Department.

—— (1956*b*) Resultats scientifiques des compagnes de la 'Calypso'. II. Compagne en Mer Rouge (suite). *Annls Inst. océanogr., Monaco* **32**.

—— (1957*a*) *Sudan irrigation.* Ministry of Irrigation and Hydroelectric Power, Sudan Republic.

—— (1957*b*) Fifth Annual Conference on Transport and Communications in Sudan, Phil. Soc. Sudan.

—— (1956–7) *Fourth Annual Report, Hydrobiological Research Unit. University of Khartoum.*

—— (1956–8) *Fifth Annual Report, Hydrobiological Research Unit. University of Khartoum.*

—— (1958–9) *Sixth Annual Report, Hydrobiological Research Unit, University of Khartoum.*

—— (1960–1) *Eighth Annual Report, Hydrobiological Research Unit, University of Khartoum.*

—— (1961–3) *Tenth Annual Report, Hydrobiological Research Unit, University of Khartoum.*

—— (1961) Effect of rural water development on agricultural expansion. *Sudan Daily* 27 August 1961.

—— (1962*a*) A gravity–magnetic survey in Sudan. Rural water development. *Final Report Contract No. ICAC-2185, Project No.* 650-52-021.

—— (1962*b*) *Sudan in pictures.* Central Office of Information, Ministry of Information and Labour, Sudan Government.

—— (1963) *Sudan almanac.* An official handbook. Sudan Republic.

—— (1964) Research in the Sudan. Twelfth Annual Conference, Phil. Soc. Sudan.

—— (1965*a*) Agricultural development in the Sudan. Thirteenth Annual Conference, Phil. Soc. Sudan.

—— (1965*b*) Geological Survey extracts big quantities of minerals. *El Ayam* newspaper 18 August 1965, p. 6.

—— (1965*c*) *Sudan research information Bulletin No.* 1. Sudan Unit, University of Khartoum.

—— (1965*d*) MS. Geology of Sudan Portland Cement Company quarry. F. L. Smidth & Co.

—— (1966) *Sudan almanac.* An official handbook. Sudan Republic.

ARAMBOURG, C. (1933) Les formations prétertiares de la bordure occidentale du lac Rudolf. *C. r. hebd. Séanc. Acad. Sci., Paris* **197**, 1663.

—— (1935) Mission scientifique de l'Omo 1932–33. I. Geologie, Anthropologie, fasc. 1. Esquisse géologique de la bordure occidentale du lac Rudolf. *Mém. Mus. natr. Hist. nat., Paris* **1**, 1–59.

ARKELL, A. J. (1949*a*) The Old Stone Age in the Anglo-Egyptian Sudan. *Sudan Antiquities Service Pamphlet No.* 1, pp. 1–52.

—— (1949*b*) *Early Khartoum.* Oxford University Press.

—— (1953) *Shaheinab.* Oxford University Press.

—— (1957) Khartoum's part in the development of the Neolithic. *Kush* 6, 8–12.

—— (1962*a*) The Aterian and Great Wanyanga (Ounianga Kebir). In MORTELMANS, G. (editor) (1962) pp. 233–42.

—— (1962*b*) The distribution in Central Africa of one Early Neolithic Wave (Dotted Wavy line pottery) and its possible connection with the Beginning Pottery. In MORTELMANS, G. (editor) (1962).

—— (1964) *Wanyanga and an archaeological reconnaissance of the south-western Libyan desert.* The British Ennedi Expedition 1957. Oxford University Press.

—— (1966) *A history of the Sudan to 1821*, 2nd edn 1961, reprinted 1966.

ARKELL, W. J. (1928) Aspects of the ecology of certain fossil coral reefs. *J. Ecol.* **16**, 135–49.

—— (1951) Origin of the Red Sea Graben. *Geol. Mag.* 88, 70.

—— (1956) *The Jurassic geology of the world.* Oliver and Boyd.

ARLDT, T. (1918) Zur Palaeogeographie das Nillandes in Kreide und Tertiar. *Geol. Rdsch.* **9**, 47–56, 104–24.

ARSANDAUX, H. (1906) Contributions a l'étude des roches alcalines de l'est Afrique. *C. r. Séanc. Mission Duschesne-Fournot.*

ATHILL, L. F. T. (1920) Through south-western Abyssinia to the Nile. *Geogrl J.* **56**, 347–67.

ATTIA, M. I. and MURRAY, G. W. (1952) Lower Cretaceous Ammonites in the marine intercalation in the 'Nubian Sandstone' of the eastern desert of Egypt. *Q. Jl geol. Soc. Lond.* (1907) pp. 442.

AUDEN, J. B. (1958) In QUENNELL, A. M. The structural and geomorphic evolution of the Dead Sea Rift. *Q. Jl geol. Soc. Lond.* **114**, 1–24.

AUSTEN, H. H. (1901) A survey of the Sobat region. *Geogrl J.* **17**, 475–512.

—— (1902) A journey from Omdurman to Mombasa via Lake Rudolf. *Geogrl J.* **19**, 669–90.

AWAD, M. (1928) Some stages in the evolution of the River Nile. *Rep. int. geogr. Congr.* 1928, pp. 268–87.

BABET, V. (1941) Note preliminaire sur l'orographie de la colonie du Chad. *Gouv. Gen. Afr. Equat. Franc. Libre Services des Mines.*

BADR, A. M. and CROSSLAND, C. (1939) Topography of the Red Sea floor. *Publ. mar. biol. Stn Ghardaqa* **1**, 13–20.

BADR EL DIN KHALIL (1966) MS. Ring complexes and newer granites. Department of Geology, University of Khartoum, Geology Library.

BAGNOLD, R. A. (1931) Journeys in the Libyan desert 1929 and 1931. *Geogrl J.* **78**, 13–39, 524–53.

—— (1933) A further journey through the Libyan desert. *Geogrl J.* **82**, 103–29, 211–35.

BAKER, B. H. (1965*a*) An outline of the geology of the Kenya Rift Valley. *Rep. UMC/UNESCO Seminar on the East African Rift System*, pp. 1–19. Nairobi.

—— (1965*b*) The Rift System in Kenya. *Rep. UMC/UNESCO Seminar on the East African Rift System*, pp. 82–4. Nairobi.

BALDACCI, L. (1910) Studi dui giacimenti minerari nello Colonia Eritrea. *Min. degli. Affari esteri.*

BALFOUR-PAUL, H. G. (1955) History and antiquities of Darfur. *Sudan Antiquities Service Museum Pamphlet No.* 3, pp. 1–28.

BALL, J. (1903) The Semna Cataract or rapid of the Nile, a study of river erosion. *Q. Jl geol. Soc. Lond.* **59**, 65–79.

—— (1907) *A description of the first or Aswan Cataract of the Nile.*

—— (1910) On the origin of the Nile Valley and the Gulf of Suez. *Geol. Mag.* **7**, (5) 71–6.

—— (1911) Gulf of Suez. *Geol. Mag.* **8**, (5) 1–10.

—— (1912) The geography and geology of south-eastern Egypt. *Mem. geol. Surv. Dep. Egypt.*

—— (1927) Problems of the Libyan Desert. *Geogrl J.* **70**, 21.

—— (1939) *Contributions to the geography of Egypt*, pp. 1–308. Cairo.

—— (1960) Kharga Oasis: its topography and geology. *Mem. geol. Surv. Dep. Egypt.*

BARBOUR, K. M. (1950) The Wadi Azum from Zalingei to Murnei. *Sudan Notes Rec.* **31**, pt. 1, 105.

—— (1961) *The Republic of the Sudan.* University of London Press.

BARTHOUX, J. C. (1922) Chronologie et description des roches ignées du Désert arabique. *Mem. Inst. Egypte* **5**.

BARRON, T. (1907) *The topography and geology of the peninsula of Sinai (western portion).* Survey Department, Egypt.

—— and HUME, W. F. (1902) *Topography and geology of the eastern desert of Egypt (central portion).* Geological Survey, Egypt.

BATE, D. M. A. (1947) An extinct reed rat from the Sudan (*Thryonomys arkelli*). *Ann. Mag. nat. Hist.* **14**, 65–71.

—— (1949) A new African fossil long-horned Buffalo. *Ann. Mag. nat. Hist.* **3**, 981–5.

—— (1951) The mammals from Singa and Abu Huggar. The Pleistocene fauna of two Blue Nile sites. *Fossil Mammals of Africa No.* 2. British Museum (Natural History), London.

BAYOUMI, A. R. (1965) U.N. special fund project of land and water use, survey of Kordofan Province. Phil. Soc. Sudan.

BEADNELL, H. J. L. (1905) The relations of the Eocene and Cretaceous Systems in the Esna-Aswan reach of the Nile Valley. *Q. Jl geol. Soc. Lond.* **61**, 667–78.

—— (1909*a*) The relation of the Nubian Sandstone and the crystalline rocks south of the oasis of Kharga, Egypt. *Q. Jl geol. Soc. Lond.* **65**, 41–54.

—— (1909*b*) *An Egyptian oasis: an account of the oasis of Kharga in the Libyan desert*, pp. 1–248. Murray, London.

—— (1924) Report on the geology of the Red Sea coast between Qoseir and Wadi Ranga. *Petrol. Res. Bull., Cairo No.* 13.

—— (1934) Libyan desert dunes. *Geogrl J.* **84**, 337.

BEAM, W. (1906) Composition of Nile waters. *Wellcome trop. Res. Lab. second Rep.*, pp. 206–14.

—— (1908) Chemical composition of Nile waters. *Wellcome trop. Res. Lab., third Rep.*

—— (1911) The mechanical analysis of arid soils. *Wellcome trop. Res. Lab. fourth Rep.*

—— (1911) Sobat river water. *Wellcome trop. Res. Lab. fourth Rep.*, pp. 32–3.

BEATON, A. C. (1948) *Equatoria Province Handbook*, Vol. 2. Khartoum.

BELL, S. V. (1964) *First annual report arid zone research unit*, 1962–4. University of Khartoum.

DE BELLFONDS, L. M. L. (1868) *L'Etbaye—mines d'or.* Atlas folio.

BERRY, E. W. (1910) Epidermal characters of *Frenelopsis ramosissima. Bot. Gaz.* **1**, 305–9.

BERRY, L. (1959) Physical features of the Nile between Sabaloka and Malakal. *Sixth Annual Report, Hydrobiological Research Unit, University of Khartoum*, 1958–9, pp. 4–11.

—— (1962*a*) Alluvial islands in the Nile. *Revue Géomorph. dyn.* **12**.

—— (1962*b*) The characteristics and mode of formation of Nile islands between Malakal and Sabaloka. *Eighth Annual Report, Hydrobiological Research Unit, University of Khartoum*, pp. 7–13.

—— (1962*c*) The physical history of the White Nile. *Eighth Annual Report, Hydrobiological Research Unit, University of Khartoum*, pp. 14–19.

—— (1963*a*) A note on the coast line of the Sudan. *Geogrl Mag. Khartoum* **1**, 16–19.

—— (1963*b*) A morphological map of the Khor Eit region. *Proc. int. geogr Union Congr.*, Strasbourg, 1962.

—— (1963*c*) Maps and development in the Sudan. *Proceedings of the ninth annual conference on surveying for development in Sudan*, 1960, pp. 100–4.

—— (1963*d*) A note on the Nile between Khartoum and Atbara. *Tenth Annual Report, Hydrobiological Research Unit, University of Khartoum.*

—— (1964*a*) Some erosional features of the semi-arid Sudan. *Proc. int. geogr. Union Congr.*, London, 1963.

—— (1964*b*) Geographical research in Sudan. Twelfth Annual Conference, Phil. Soc. Sudan.

—— (1964*c*) Some notes on the relationships between geomorphology and soils in the central Sudan. *Proc. Int. geogr. Union Congr.*, Liverpool, 1964.

—— and CLOUDSLEY THOMPSON, J. L. (1960) Autumn temperatures in the Red Sea Hills. *Nature, Lond.* **188**, 843.

—— and RUXTON, B. P. (1960*a*) Notes on weathering zones and soils in two tropical regions. *J. Soil Sci.* **10**, 54–63.

—— —— (1960*b*) The Butana grass patters. *J. Soil Sci.* **11**, 61–2.

—— and SESTINI, J. (1963) Mapcomento geomorphologico so longo da Mare Vermalho. *Bull. Bahearo Geogr. Brazil* **3**, 1.

——, WHITEMAN, A. J. and BELL, S. V. (1966) Some radiocarbon dates and their geomorphological significance, emerged reef complex of the Sudan. *Z. Geomorph.* **10**, 119–43.

—— and WHITEMAN, A. J. (1968) The Nile in the Sudan. *Geogrl J.* **134**, 1–37.

BICKELL, R. S. and PETERS, C. M. F. (1962) MS. Geological study of the Coastal Province, Sudan between 20° 10′N and 21° 52′N and 21° 52′E. General Exploration Company of California.

BISSET, C. B. (1935) Notes on the volcanic rocks of central Karamoja. *Uganda geol. Surv. Bull. No.* **2**, pp. 40–3.

BISHOP, W. W. (1966) In DRURY, G. H., *Essays in geomorphology. Stratigraphical geomorphology*, pp. 139–76.

BLANCKENHORN, M. (1902) Die Geschichte Nil-Stromes in der Tertiar and Quartar Periode. *Z. Geo. Erdk. Berlin*, pp. 694–722, 753–62.

—— (1912) Aegypton. *Handbuch der regionalen Geologie*, Bd. VII, Abt. 9, Heft. 23. Heidelberg.

BLANDFORD, W. T. (1869) On the geology of a portion of Abyssinia. *Q. Jl geol. Soc. Lond.* **25**, 401–6.

—— (1870) *Observations on the geology and zoology of Abyssinia made during the progress of the British expedition to that country in 1867–68.* London.

BLOKHUIS, W. A. (1963) *Soil survey report on Khashm el Girba Scheme: Khashm el Girba South*, Pt. I, No. 6/KA/1. Ministry of Agriculture, Khartoum.

BLUNDELL, H. W. (1900) A journey through Abyssinia to the Nile. *Geogrl J.* **15**, 97–118, 264–7.

—— (1906) Exploration in the Abbai Basin, Abyssinia. *Geogr. J.* **37**, 529–51.

BOULNOIS, P. K. (1924) On the western frontiers of the Sudan. *Geogrl J.* **63**, 465–77.

BOUTET, G. and NEVILLE, H. (1930) Recherches sur le *Hylochoerus. Ann. Mag. nat. Hist. Paris* **5**, 215–307.

BROOKS, C. E. P. (1947) A long term plan for the Nile Basin. *Nature, Lond.* **160**, 215–16.

BUIST, G. (1854) On the physical geography of the Red Sea. *Geogrl J.* **24**, 227–38.

BULLARD, E. C. (1936) Gravity measurements in East Africa. *Phil. Trans. R. Soc.* **A235**, 445–531.

BUNTING, A. H. and LEA, J. D. (1962) The soils and vegetation of the Fung, East Central Sudan. Agricultural Research Division, Ministry of Agriculture. *J. Ecol.* **50**, 529–58.

BUROLLET, P. F. (1963) Reconnaissance geologique dans le Sud-Est du Bassin du Kafra. First Saharan Symposium Tripoli 1963. *Revue I. F.P.* **18**, 219–27.

BURTON, A. N. and WICKENS, G. E. (1966) Jebel Marra volcano, Sudan. *Nature, Lond.* **210**, 1146–7.

BUTZER, K. W. (1959) Contributions to the Pleistocene geology of the Nile Valley. *Erkunde*, Bd. 13, pp. 46–7.

CAHEN, L. and SNELLING, N. J. (1966) *The geochronology of equatorial Africa*. North Holland.

CARELLA, R. and ACARPA, N. (1962) Geological results of exploration in Sudan by AGIP Mineraria Ltd. Fourth Arab Petroleum Congress, Beirut, 1962.

CAVENDISH, H. S. H. (1898) Through Somaliland and around and south of Lake Rudolf. *Geogrl J.* **11**, 372–94.

CHAMPION, A. M. (1935) Teleki's volcano and the lava fields at the southern end of Lake Rudolf. *Geogrl J.* **85**, 323–36.

—— (1937) Physiography of the region west and south-west of Lake Rudolf. *Geogrl J.* **89**, 97–118.

CHEESEMAN, R. E. (1928) The upper waters of the Blue Nile. *Geogrl J.* **71**, 358–74.

—— (1936) *Lake Tana and the Blue Nile*.

CHRISTY, C. (1917) The Nile-Congo watershed. *Geogrl J.* **50**, 199–216.

—— (1933) The Bahr el Ghazal and its waterways. *Geogrl J.* **61**, 313–35.

CHURCHILL, W. S. (1899) *The River War*. Eyre and Spottiswoode.

CLARK, J. D. (1965) The later Pleistocene cultures of Africa. *Science* **150**, 833–47.

CLAYTON, P. A. (1933) The western side of the Gilf Kebir. *Geogrl J.* **81**, 254–9.

CLOUDSLEY-THOMPSON, J. L. (1964*a*) Desert invertebrates of Khartoum Province. *Sudan Notes Rec.* **45**, 1.

—— (1964*b*) *Guide to Natural History of Khartoum Province*, Pt. I. Arid Zone Research Unit, University of Khartoum.

COLCHESTER, G. V. (1927) Malha crater, Darfur. *Sudan Notes Rec.* **10**, 233–5.

COLE, S. (1964) *The Pre-history of East Africa*. Weidenfeld and Nicholson.

COLLINS, R. O. (1962) *The southern Sudan 1883–1898. A struggle for control*. Yale University Press.

COLSTON, R. E. (1887) La géologie de la region Berenice et Berber. *Bull. Soc. khéd. Geogr. Cairo* **2**.

COMYN, D. (1907) Western sources of the Nile. *Geogrl J.* **30**, 524–30.

—— (1908*a*) The desert west of Wadi Halfa. *Geogrl J.* **31**, 442–4.

—— (1908*b*) Lt. Comyn's survey of the Pibor River. *Geogrl J.* **31**, 304–7.

CONANT, L. C. and GOUDARZI, G. M. (1967) Stratigraphic and tectonic framework of Libya. *Bull. Am. Ass. Petrol. Geol.* **51**, 719–30.

CORBYN, E. N. (1945) Soil conservation in the Sudan. *Nature, Lond.* **155**, 70–1.

COUSTEAU, J. Y., NESTEROFF, W., and TAZIEFF, H. (1952) Coupes transversales de la Mer Rouge. *C. r. int. geol. Congr.*, Sec. VI, pp. 75–7.

COX, L. R. (1929) Notes on the post-Miocene Ostreidae and Pectinidae of the Red Sea region with remarks on the geological significance of their distribution. *Proc. malac. Soc. Lond.* **18**, 165–209.

—— (1932) On fossiliferous siliceous boulders from the Anglo-Egyptian Sudan. *Abstr. Proc. geol. Soc. Lond.*, No. 1254, pp. 17–18.

—— (1932–3) A lower Tertiary siliceous rock from the Anglo-Egyptian Sudan. *Bull. Inst. Geol. Egypt* **15**, 315–48.

—— (1956) Lamellibranchiata from the Nubian Sandstone Series of Egypt. *Bull. Inst. Geol. Egypt* **37**, 465–80.

CRAWFORD, O. G. S. (1949) Some mediaeval theories about the Nile. *Geogrl J.* **114**, 6–29.

—— and ADDISON, F. (1951) *Abu Geili. Wellcome excavations in the Sudan*, Vol. III.

—— and MACADAM, M. F. L. (1961) Castles and churches in the Middle Nile Region. *Sudan Antiquities Service Occ. Papers No. 2*.

CROWSHAW, H. B. (1914) An epidote from the Sudan. *Geol. Mag.* **1**, 381; *Nature, Lond.* **93**, 472.

CROSSLAND, C. (1907) Recent history of the coral reefs of the Sudan. *J. Linn. Soc.* **31**, 14.

—— (1911*a*) Reports on the marine biology of the Sudanese Red Sea. *J. Linn. Soc.* **31** C, 1–515.

—— (1911*b*) A physical description of Khor Dongonab, Red Sea. *in* Crossland 1911*a*, pp. 256–8.

—— 1913) *Desert and water gardens of the Red Sea*. Cambridge University Press.

CROWFOOT, J. W. and GRIFFITH, F. L. (1911) The island of Meroe. *Archaeological Survey of Egypt Memoir* 19.

CURRY, P. A. (1913) The value of gravity at eight stations in Egypt and the Sudan. *Survey Department Paper* 18. Cairo.

CUVILLIER, J. (1941) Le passage du Cretacée à l'Éocene dans la region de Suez. *C. r. Rebd. Séanc. Acad. Sci., Paris* **212**, 710–12.

DAINELLI, G. (1943) *Geologia dell'Africa orientale* (3 Vols. text and 1 Vol. maps). R. Acc. Italia, Rome.

DALLONI, M. (1934) Mission au Tibesti. *Mém. Inst. France* **61**.

DANIEL, MANN, JOHNSON, and MENDENHALL, (1962*a*) *Report on minerals in Sudan*. Report made for U.S. AID.

—— —— —— —— (1962*b*) Safaya iron deposits, Red Sea Hills, Kassala Province. Report made for U.S. AID.

DARWIN, C. (1842*a*) *The structure and distribution of coral reefs*. London.

—— (1842*b*) (Reprint 1962) *The structure and distribution of coral reefs*. University of California Press.

DAUOUD, A. S. (1968) MS. A geological and geophysical study of the Area, south-east of Qerri Station, Sabaloka area, Nile Valley, Central Sudan. M.Sc. thesis, University of Khartoum.

DAVIDSON, B. (1959) *Old Africa rediscovered.*

—— (1964) *African past*. Longmans.

DAVIES, K. A. (1935) A contribution to the study of the geology of Karamoja. *Uganda Geological Survey Bulletin No.* 2, pp. 30–6.

—— (1951) The Uganda section of the Western Rift. *Geol. Mag.* 88, 377–85.

DEBENHAM, F. (1954) The water resources of Africa. *Sudan Notes Rec.* 35 pt. 2, 69.

DELANY, F. M. (1952) Recent contributions to the geology of the Anglo-Egyptian Sudan. *Chron. Mines colon.* 197.

—— (1953) Report on a visit to Eastern District, Equatoria. Open file, No. 78/13, Sudan Geological Survey, Khartoum.

—— (1954) Recent contributions to the geology of the Sudan. *Nineteenth int. geol. Congr.* Algiers, Vol. 20, p. 11–18.

—— (1955) Ring structures in the northern Sudan. *Eclog. geol. Helv.* **48**, (1) 133–48.

—— (1958) Observations in the Sabaloka Series of the Sudan. *Trans. geol. Soc. S. Afr.* **61**, 111–24.

—— (1966) *Lexique stratigraphique international Sudan*, Vol. 4, fasc. 46, pp. 77–94.

DELORME, J. and DELANY, F. M. (1958) Etude preliminaire de la Série Argilo-greseuse de la Région Diamantifère de l'Ouest-Oubangui, A.E.F. *Proc. twentieth int. geol. Congr.*, Mexico City, 1956, pp. 65–72.

DENAEYER, M. E. (1924) L'Ouadai Oriental et les regions voisines. *Bull. Soc. geol. Fr.* **24**, 538–76.

—— and CARRIER, A. (1924) Les principaux résultats géologiques et lithologiques de la Mission de délimitation Ouadai-Darfur. *C. r. hebd. Séanc. Acad. Sci., Paris* **178**, 1197.

DESIO, A. (1935) *Missione Scientifica a Cufra* (1931). *Studi geologici sulla Cirenaica sul Deserto Libico, sulla Tripolitania e sul Fessan orientale.* R. Acc. d'Italia Viaggio de studi ed esplor.

—— (1939) Le nostre conoscenza geologiche sulla Libia sino al 1938. *Pubbl. Ist. Geol. Paleont. Geogr. fis. R. Univ. Milano*, Ser. G, **10**.

DIXEY, F. (1952) African geomorphology. *Geogrl J.* **118**, 75–7.

—— (1956) The east African Rift System. *Colon. Geol. Miner. Resour.* Bull. Suppl. 1. H.M.S.O., London.

DOXIADIS ASSOCIATES (1964*a*) *Land use and water survey in Kordofan Province, Sudan Republic, Bulletin No.* 71.

—— —— (1964*b*) Details of bench marks and datum points for the general hydrometric observations. *Land and water survey, Kordofan Province, Sudan, Dox-Sud-A* 29.

—— —— (1964*c*) General hydrometric records for the 1962–64 hydrological years. *Land use and water survey, Kordofan Province, Sudan, Dox-Sud-A* 32.

—— —— (1964*d*) First report on Hafir hydrology, hydrological year 1963–64. *Land use and water survey, Kordofan Province, Sudan, Dox-Sud-A* 31.

—— —— (1964*e*) First report on well-field localities investigation, hydrological year 1963–64. *Land use and water survey, Kordofan Province, Sudan, Dox-Sud-A* 33.

—— —— (1965*a*) Rainy seasons diagrams. *Land use and water survey, Kordofan Province, Sudan, Dox-Sud-A* 34.

—— —— (1965*b*) Second report on rainy season diagrams. *Land use and water survey, Kordofan Province, Sudan, Dox-Sud-A* 38.

—— —— (1965*c*) General hydrometric records, hydrobiological years 1962–65. *Land use and water survey, Kordofan Province, Sudan. Dox-Sud-A* 39.

—— —— (1965*d*) Second report on well field localities investigations, hydrological year 1964–65. *Land use and water survey, Kordofan Province, Sudan. Dox-Sud-A* 40.

—— —— (1965*e*) Second report on the Rahad Lake: season 1964–65. *Land use and water survey, Kordofan Province, Sudan. Dox-Sud-A* 4.

—— —— (1966*a*) Third and concluding report on Rahad Lake. Compilation of data collected and recommendations. *Land Use and Water Survey, Kordofan Province, Sudan. Dox-Sud-A* 52.

—— —— (1966*b*) Third and concluding report on Hafir hydrology. Compilation of data collected and recommendations. *Land use and water survey, Kordofan Province, Sudan, Dox-Sud-A* 53.

DRAKE, C. L. and GIRDLER, R. W. (1964) A geophysical study of the Red Sea. *Q. Jl R. Astr. Soc.* 8, (5) 173–95.

—— —— and LANDISMAN, M. (1959) Geophysical measurements in the Red Sea. Reprint *Proc. int. oceanogr. Congr.*, p. 21.

DRIBERG, J. H. (1927) The Didinga Mountains. *Geogrl J.* 89, 507–21.

DUNHAM, D. (1957) Royal tombs at Merowe and Barkal. *The royal cemetries of Kush*, Vol. 4. Museum of Fine Arts, Boston.

DUNLOP, A. (1937) The Dadessa Valley. *Geogrl J.* 89, 507.

DUNN, S. C. (1911*a*) Notes on the mineral deposits of the Anglo-Egyptian Sudan. *Geological Survey Bulletin No.* 1. Khartoum.

—— (1911*b*) Ancient gold mining in the Sudan. *Rep. Wellcome trop. Res. Lab.* 4, 207–15.

—— (1921) Native gold washings in the Nuba Mountains Province. *Sudan Notes Rec.* 4, 138–45.

—— and GRABHAM, G. C. (1910) The iron ore deposits of the Anglo-Egyptian Sudan. *C. r. int. geol. Congr.*, Vol. 2, pp. 1021–5.

EDMONDS, J. M. (1942) The distribution of Kordofan Sands, Anglo-Egyptian Sudan. *Geol. Mag.* 79, 18–30.

EDWARDS, W. N. (1926) Fossil plants from the Nubian Sandstone of eastern Darfur. *Q. Jl geol. Soc. Lond.* **82**, 94–100.

FABUNI, L. A. (1960) *The Sudan in Anglo-Egyptian relations. A case study in power politics 1800–1956*. Longmans.

FAIRBRIDGE, R. W. (1958) Dating the latest movements of Quaternary sea-level. *Trans. N.Y. Acad. Sci.* **20**, 471–82.

—— (1962) New radiocarbon dates of Nile sediments. *Nature, Lond.* **196**, 108–10.

—— (1963) Nile sedimentation above Wadi Halfa during the last 20 000 years. *Kush* **11**, 96–107.

FAIRSERVIS, W. A. (1962) *The Ancient Kingdoms of the Nile*. Mentor, New York.

FAROUK, A. (1968) MS. The geology of the Jebel Geili, Butana, and Jebel Sileitat–Es Sufr. Igneous complexes, Nile Valley, central Sudan. M.Sc. Thesis, University of Khartoum.

FLOYER, E. A. (1892) The mines of the northern Etbai and of northern Ethiopia. *Jl R. Asiat. Soc.* **24**, 811–33.

FOLEY, E. J. (1941) Geological survey between Ras Beras and the Sudan frontier, West Coast, Red Sea. F.O. Egypt 50, N.Y.O. Egypt 59.

FORTAU, R. (1902) Sur les grès nubien. *C. r. hebd. Séanc. Acad. Sci., Paris* **135**, 803–4.

FREULON, J. M. (1964) Étude geologique des Series Primaires due Sahara Central. *C.R.N.S.* Ser. Geol. 3.

FRITEL, P. H. and CARRIER, A. (1924) Sur les vestiges de plantes devonniennes et carboniferienne recueillis en Quadai par la mission du lieutenant-colonel Grossard. *C. r. hebd. Séanc. Acad. Sci., Paris* **178**, 505.

FOLSTER, H. (1964) Morphogenese der sudsudanesischen Pediplane. *Z. geomorph.* 8, 393–423.

FUCHS, V. E. (1935) The Lake Rudolf Rift Valley expedition 1934. *Geogrl J.* 86, 114–42.

—— (1939) Geological history of the Lake Rudolf Basin, Kenya colony. *Phil. Trans. R. Soc.* **B229**, 219–74.

FURON, R. (1963) *Geology of Africa*. Oliver and Boyd.

GABERT, G. (1958) Report on mapping the north-eastern part of sheet Deraheib. Open file, Sudan Geol. Surv., Khartoum.

——, RUXTON, B. P., and VENZLAFF, H. (1960) Uber Untersuchungen im Kristallin der nordlichen Red Sea Hills im Sudan. *Geol. Jber.* 77, 241–67.

GAITSKELL, A. (1959) *Gezira. A study of development in Sudan*. Faber.

GALLITELLI, P. (1935) Studi lithologici sull Cirenaica e sulla Tripolitanica orientale; contributo alla conocenza dalla rocce metamorfiche de erruptive della Libia. *Missione scientifica a Cufra* (1931), Vol. 3, 323–450.

—— (1939) Ritrovimento di gen. *Pachyseris* (*P. imperalis* sp. nov.) nelle terri pleistoceniche de Mar Rosso. *Riv. ital. Paleont. Stratigr.* 45, f 4/4, 51–6.

GARDENER, E. W. (1932) Some problems of Pleistocene hydrography of Kharga Oasis, Egypt. *Geol. Mag.* **69**, 386.

—— (1935) The Pleistocene fauna and flora of Kharga Oasis, Egypt. *Q. Jl geol. Soc. Lond.* **91**, 479–518.

GASS, J. G. (1955) Geology of Dungunab, Sudan. M.Sc. thesis, University of Leeds.

—— (1960) Unpublished petroleum report, General Exploration Company of California.

GAUTIER, E. F. (1928) *Le Sahara*. Paris.

GAY, P. A. (1957–8) Riverain swamps in the Sudan and Uganda. *Fifth Annual Report, Hydrobiological Research Unit, University of Khartoum*, 7–15.

—— and BERRY, L. (1959) The water hyacinth. A new problem on the Nile. *Geogrl J.* **125**, 89–91.

GIBB, SIR ALEXANDER and PARTNERS. (1954) *Estimation of irrigable areas in Sudan, 1951–53*. Sudan Government, London.

GILLIAN, J. A. (1918) Jebel Marra and the Deriba lakes. *Sudan Notes Rec.* **1**, 263.

GIRDLER, R. W. (1958) The relationship of the Red Sea to the east African Rift System. *Q. Jl geol. Soc. Lond.* **114**, 79–105.

—— (1964) Geophysical studies of rift valleys. In AHRENS, L. N. *et al.* (1964) *Physics and chemistry of the earth*, Vol. 5. Pergamon Press.

—— (1965) The formation of new oceanic crust. In BLACKETT, P. M. S. *et al.* (1965) A symposium on continental drift. *Phil. Trans. R. Soc.* **108**, (8) 123–36.

GIRDLER, (1968) Drifting and rifting in Africa. *Nature, Lond.* **217**, 1102–06.

—— and HARRISON, J. C. (1957) Submarine gravity measurement in the Atlantic Ocean, Indian Ocean, Red Sea and Mediterranean Sea. *Proc. R. Soc.* A239, 202–13.

GLEICHEN, A. E. W. (1898) *Handbook of the Sudan.*

—— (1899) *Handbook of the Sudan, Supplement.*

—— (1905) *The Anglo-Egyptian Sudan.* 2 Vols.

GRABHAM, G. W. (1906) Well-water supply of the north-eastern Sudan. *Rep. Br. Ass. Advmt Sci.*, p. 712.

—— (1909) Wells in the north-eastern Sudan. *Geol. Mag.* 6, 265–71, 311–18.

—— (1910) Note on some recent contributions to the study of desert water supplies. *Cairo scient. J.* 4, (46) 166–74.

—— (1919) Climatic change. *Sudan Notes Rec.* **2**, 154.

—— (1920*a*) Desiccation. *Sudan Notes Rec.* **3**, 130–6.

—— (1920*b*) The Bayuda. Volcanic fields. *Sudan Notes Rec.* 8, 133–6.

—— (1920*c*) Volcanic rocks in the Anglo-Egyptian Sudan. *Nature, Lond.* **105**, 199–200.

—— (1920*d*) Fossil bones. *Sudan Notes Rec.* **3**, 131–3.

—— (1926) Note on red colouration under climatic influence in the Sudan. *Geol. Mag.* 73, 280–2.

—— (1927) *Report of the Geological Survey of the Anglo-Egyptian Sudan for 1926.* Khartoum.

—— (1928) Notes on rocks collected by Messrs. Newbold and Shaw during a journey through Bir Natrun, Nukheila to Selima. *Sudan Notes Rec.* **11**, 151–6.

—— (1929) Gold in the Anglo-Egyptian Sudan and Abyssinia. *C. r. int. geol. Congr.*, Vol. 2, pp. 279–80.

—— (1930*a*) *The Gold resources of the world*, pp. 33–4. Pretoria.

—— (1930*b*) *Report of the Geological Survey of the Anglo-Egyptian Sudan for 1930.* Sudan Government, Khartoum.

—— (1934) Water supplies in the Anglo-Egyptian Sudan. *Sudan Geological Survey Bulletin No.* 2, pp. 1–42.

—— (1935) *The Anglo-Egyptian Sudan from within*, pp. 257–81. London.

—— (1936) Water supplies in the Anglo-Egyptian, Sudan. *Water* 38, 9–17.

—— (1937) *Geological note on the Boma plateau.* Khartoum.

—— (1938) Note on the geology of the Singa district of the Blue Nile. *Antiquity* **12**, 193.

—— (1941) Problems of Egyptian Geography. *Geogrl J.* 96, 207–8.

—— and BLACK, R. P. (1925) *Report of the mission to Lake Tana, 1920–1921.* Cairo.

GRAY, R. (1964) *A history of the southern Sudan 1839–1899.* Oxford University Press.

GREGORY, J. W. (1894) Contributions to the physical geography of British East Africa. *Geogrl J.* 4, 289–315, 408–24.

—— (1896) *The Great Rift Valley.*

—— (1999) The geological history of the Rift Valley. *Jl E. Africa Uganda nat. Hist. Soc.*, pp. 429–40.

—— (1920) The African Rift Valleys. *Geogrl J.* 56, 14–47.

—— (1921*a*) The *Times* Africa flight; a discovery of a new volcanic field. *Nature, Lond.* **104**, 667–9.

—— (1921*b*) *The Rift Valleys and geology of East Africa.* London.

—— (1921*c*) The African Rift Valleys. *Geogrl J.* 57 238–9.

GREENE, H. (1920) Soil profiles in the eastern Gezira. *J. agric. Sci.* **17**, 3.

—— (1935) Soil problems in the Sudan. *Proc. third int. Congr. Soil Sci.* Vol. 1, pp. 350–3.

GROSSARD, G. (1925) *Mission de delimitation de l'Afrique equatoriale francais et du Sudan anglo-egyptien.* Paris.

GROVES, A. W. (1932) Petrology and the western rift of central Africa. *Geol. Mag.* 69, 497–510.

—— (1935) The charnockite series of Uganda. *Q. Jl geol. Soc. Lond.* **91**, 150–206.

GUILCHER, A. (1955) Geomorphologie de l'extremité Septrionale du banc Farasan (Mer Rouge). *Annls Inst. océanogr. Monaco* **30**, 55–97.

GWYNN, C. W. (1901) Surveys on the proposed Sudan–Abyssinian frontier. *Geogrl J.* **18**, 562–3.

HALWAGY, R. (1961) The vegetation of the semi-desert north-east of Khartoum, Sudan. *Oikos* **12**, (1) 87–110.

HAMMERTON, D. (1964) Hydrobiological research in the Sudan. Twelfth Annual Conference, Phil. Soc. Sudan.

—— (1966) Hot springs and fumaroles at Jebel Marra, Sudan. *Nature, Lond.* **209**, 1290.

HAMILTON, J. S. (1920) The Deriba country east of the Bahr el Jebel. *Geogrl J.* 56, 398–401.

HARRIS, N., PALLISTER, J. W. and BROWN, J. M. (1956) Oil in Uganda. *Memoir No.* 9. Geological Survey, Uganda.

HARRISON, M. N. and JACKSON, J. K. (1958) Ecological classification of the vegetation of the Sudan. *Ministry of Agriculture, Forests Department, Bulletin No.* 2.

HASSAN, B. A. (1966) Mineral deposits of the Sudan. Open file report, Sudan Geological Survey, Uganda

HASSAN, I. H. (1959) The utilization of Nile waters. Ph.D. thesis, University of London.

HASSANEIN, A. M. (1924) Through Kufra to Darfur. *Geogrl J.* 64, 273–91.

—— (1925) *The lost Oases.*

HAUGHTON, S. H. (1963) *The stratigraphical history of Africa south of the Sahara.* Oliver and Boyd.

HEBBERT, H. E. (1935) The Port Sudan water supply. *Sudan Notes Rec.* 18, 89–101.

HEDBERG, H. D. *et al.* (1959) Petroleum development in Africa. Sudan. *Bull. Am. Ass. Petrol. Geol.* 43, 1672.

—— —— (1960) Petroleum development in Africa Sudan. *Bull. Am. Ass. Petrol. Geol.* 44, 1141.

—— —— (1961) Petroleum development in Africa. *Bull. Am. Ass. Petrol. Geol.* 45, 1183.

HEINZELIN, J. DE and PAEPE, R. (1964) The geological history of the Nile Valley in Sudanese Nubia. Preliminary results. *Contribution to the Prehistory of Nubia,* assembled by F. Wendorf.

HEPWORTH, J. V. (1964) *Explanation of the geology of sheets* 19, 20, 28, *and* 29. Southern west Nile. Geological Survey, Uganda.

HEYBROEK, F. (1965) *Salt basins around Africa.* Institute of Petroleum, London.

HILL, R. (1956) The search for the White Nile's source: two explorers who failed. *Geogrl J.* 122, 247–9.

—— (1963) *Egypt in the Sudan 1820–1881.* Oxford University Press.

—— (1965*a*) *Sudan transport: a history of railway, marine, and river services in the Republic of the Sudan.* Oxford University Press.

—— (1965*b*) *Slatin Pasha.* Oxford University Press.

HOBBS, W. H. (1918) The peculiar weathering processes of desert regions with illustrations from Egypt and the Sudan. *Twentieth Michigan Academy Report,* pp. 93–9.

HOBBS, H. F. (1918) Notes on Jebel Marra, Darfur. *Geogrl J.* 52, 357–63.

HODGKIN, R. A. (1961) *Sudan geography.* Longmans.

HOLLAND, W. P. (1926) Volcanic action north of Lake Rudolf in 1918. *Geogrl J.* 68, 488–91.

HOLMES, A. (1916) Notes on the structure of the Nile–Tanganyika rift valley. *Geogrl J.* 48, 149–56.

—— (1965) *Principles of physical geology.* Nelson.

HOLT, H. L. (1954) Rural water supply and land use development in the semi-arid regions of the Sudan. *Soil Conservation Report.* Sudan Government.

HOPWOOD, A. T. (1929) *Hylochoerus grabhami,* a new species of fossil pig from the White Nile at Kosti. *Ann. Mag. nat. Hist.* 10, (4) 289–90.

HOTTINGER, A. F. (1959) Geological reconnaissance of the north-western Sudan by Royal Dutch Shell and British Petroleum survey party. MS. The Hague 1959. Open file, Sudan Geological Survey Library.

HOWELL, E. J. (1956) Water supply and development in Kordofan, Sudan. *Proc. int. geogr. Congr.* Washington, 1952, pp. 89–92.

HOWELL, P. P. (1953) The equatorial Nile project and its effects in the Sudan. *Geogrl J.* 119, 33–48.

HULL, E. (1896) Observations on the geology of the Nile Valley, and on the evidence of greater volume of water at a former period. *Q. Jl geol. Soc. Lond.* 22, 303–19.

HUME, F. W. (1906) Notes on the history of the Nile and its valley. *Geogrl J.* 27, 52–9.

—— (1937) *Geology of Egypt.* Vol. II, *The fundamental Pre-Cambrian rocks of Egypt and the Sudan.*

HUNT, J. M., HAYS, E. H., DEGENS, E. T., and ROSS, D. A. (1967) Red Sea: detailed survey of hot brine areas. *Science* 156, 514–16.

HUNTING TECHNICAL SERVICES LTD. (1958) *Jebel Marra investigations, report on phase* I *studies.* Ministry of Irrigation and Hydroelectric Power, Khartoum.

HURST, H. E. (1927*a*) Progress in the study of the hydrology of the Nile in the last twenty years. *Geogrl J.* 70, 440–58.

—— (1927*b*) The lake plateau basin of the Nile. *Physical Department Paper* 23, 1 and 2. Cairo.

—— (1932) The Sudd region of the Nile. *Jl R. Soc. Arts* 81, 720–36.

—— (1944) A short account of the Nile basin. *Physical Department Paper* 45. Cairo.

—— (1950) *The Nile basin.* Vol. 8, *The hydrology of Sobat and White Nile and the topography of the Blue Nile and Atbara.* Cairo.

—— and PHILLIPS, P. (1931) The Nile basin. General description of the basin. *Physical Department Paper,* Vol. 1. Cairo.

—— —— (1938) *The Nile Basin. The hydrology of the lake plateau and Bahr el Jebel.* Cairo.

HUSSEIN, KEMAL EL-DINE (1928) L'exploration du desert libique. *Geographie* 50, 171–83, 320–36.

HYLAND, J. S. (1890) On some specimens from Wadi Halfa, Upper Egypt. *Proc. R. Soc. Dublin* 6, 438–47; *Miner. Mag.* 10, 41.

IRVING, E. and TARLING, D. H. (1961) Palaeomagnetism of the Aden volcanics. *J. Geophys. Res.* 66, 549–56.

IRWIN, H. T. and WHEAT, J. B. (1965) Report of the Palaeolithic Section, University of Colorado, Nubian Expedition. *Kush* 13, 17–23.

ISKANDER, W. (1957) File No. 55G, Sudan Geological Survey, Khartoum.

JACKSON, H. G. (1926) A trek in the Abu Hamed district. *Sudan Notes Rec.* 9, 1–35.

JACKSON, J. K. (1957) Changes in the climate and vegetation of the Sudan. *Sudan Notes Rec.* 28, 47–66.

JEFFERSON, H. K. (1952) *Soil conservation in the Sudan.* Ministry of Agriculture, Khartoum.

JEWITT, T. N. (1955) The soils of the Sudan. Typescript.

JONES, F. (1954) The geology of the Damazin area. In: Sir Alexander Gibb and Partners, *Roseires Dam project.* Sudan Government.

JOHNSON, R. J. (1960) *Explanation of the geology of sheet* 69, *Lake Wamala.* Geological Survey, Uganda.

JONGLEI INVESTIGATION TEAM (1954) *The equatorial Nile project and its effects on the Anglo-Egyptian Sudan.* Sudan Government.

JOSEPH, A. F. (1924) The composition of some Sudan soils. *J. agric. Sci.* **14**, 490–7.

—— (1927) *Report of the Government chemist.* Sudan Government.

—— (1935) *The Anglo-Egyptian Sudan from within*, pp. 257–81.

KABESH, M. L. (1960*a*) The mica deposits of the northern Sudan. *Sudan Geological Survey Bulletin No.* 7.

—— (1960*b*) The geology and economic minerals and rocks of the Ingessana Hills. *Sudan Geological Survey Bulletin No.* 11.

—— (1962) The geology of Mohammed Qôl sheet. *Sudan Geological Survey Memoir No.* 3.

—— and AFIA, M. S. (1959) The wollastonite deposit of Dirbat well. *Sudan Geological Survey Bulletin No.* 5.

—— —— (1961) Manganese ore deposits of the Sudan. *Sudan Geological Survey Bulletin No.* 9.

—— —— and WIDATALLA, A. L. (1958) Fodikwan iron ore deposits. *Sudan Geological Survey Bulletin No.* 4.

KARPOFF, R. (1957) Sur l'existence du Maestrichtien au de Djeddah (Arabie Seoudite). *C. r. hebd. Séanc. Acad. Sci., Paris* **245**, 1322–4.

KARKANIS, B. G. (1965) Hydrochemical facies of ground-water in the western Provinces of Sudan. M.Sc. thesis, University of Arizona.

KELLY, H. H. (1912) The Pibor River: II, the Beir country. *Geogrl J.* **40**, 497–501.

KENNEDY, W. Q. (1965) The influence of Basement Structure on the evolution of coastal basins. In *Salt basins around Africa*, Institute of Petroleum.

KHEIRALLA, M. K. (1966) A study of the Nubian Sandstone Formation of the Nile Valley between 14° N and 17°42′ N with reference to the ground-water geology. M.Sc. Thesis, University of Khartoum.

KHOGALI, M. M. (1964) The significance of the railway to the economic development of the Republic of the Sudan with special reference to the western Provinces. M.A. thesis.

KLEINSORGE, H. and ZSCHEKED, J. G. (1958) *Geologic and hydrologic research in the arid and semi-arid zone of western Sudan.* Amt. fur Bordenforschung, Hanover.

KLITZSCH, E. (1963) Geology of the north-east flank of the Murzuk basin (Djebel Ben Ghnema-Dor el Gussa area). First Saharan Symposium Tripoli 1963, *Review I.F.P.*, Vol. 18, No. 10, pp. 97–113.

KOETTLITZ, R. (1900) A journey through Somaliland and southern Abyssinia to the Shangalla and Buta country on the Blue Nile, and through the Sudan to Egypt. *Scott. geogr. Mag.* **10**; *Geogrl J.* **16**, 264–72.

KOWALEWESKI, J. R. (1849) Expedition dirigée vers les sources du Nil a la recherche de sables aurifères. *Annls Voyage* **4**.

KRENKEL, E. (1925) *Geologie Afrikas*, Vol. 1.

—— (1926) *Handbuch der Regionalen Geologie*, Vol. 7, 8a, Abessomalien.

DE KUN, N. (1965) *The mineral resources of Africa.* Elsevier.

LACAILLE, A. D. (1951) *The stone industry of Singa, Abu Hugar. The Pleistocene fauna of two Blue Nile sites. Fossil mammals of Africa No.* 2. British Museum (Natural History), London.

LACROIX, A. and TILHO, J. A. M. (1919) Les volcans de Tibesti. *C. r. hebd. Séanc. Acad. Sci., Paris* **168**, 1237–40.

DE LAPPARENT, A. F. (1954) État actuel de nos connaissances sur la stratigraphie, la palaeontologie, et la tectonique des 'Gres de Nubie' du Sahara central. *C. r. int. geol. Congr.*, Alger, 1952, Vol. 21, pp. 113–27.

LAWSON, A. C. (1927) The valley of the Nile. *University of California Chronicle, No.* 29.

LEACH, T. A. (1926) The Selima oases. *Sudan Notes Rec.* **9**, 37–46.

LEBON, J. H. G. (1965) Land use in Sudan. *World Land Use Survey Memoir No.* 4.

—— and ROBERTSON, V. C. (1961) The Jebel Marra, Darfur, and its Region. *Geogrl J.* **127**, 30–49.

LEFEVRE, A. (1839) Sur la géologie de la vallée du Nil jusq'au Chartoum. *Bull. Soc. geol. Fr.* **10**, 144–8.

LEPERSONNE, D. (1956) Les surfaces d'érosion des hauts plateaux de l'interieur de l'Afrique centrale. *Bull. Acad. r. Sci. Coll.* **2**, 4.

LINCK, G. (1901) Bericht uber seine Reise nach Kordofan. *Verh. cl. Ges. fur Erdk. Jg.* 1901, pp. 217–25.

—— (1903) Beitrage zur Geologie und Petrographie von Kordofan. *Neues Jb. Miner Geol. Palaont.* **17**, 391–463.

LLEWELLYN, A. (1903) *Report on a mining concession in the Egyptian Sudan.*

LOEBLICH, A. R. and TAPPAN, H. (1964) *Treatise on invertebrate palaeontology.* Part C, Protista 2, Sarcodina, Vols. 1 and 2. Geological Society of America and University of Kansas Press.

LOMBARDINI, E. (1865) *Essai sur l'hydrologie du Nil.* Paris and Milan.

LOTFI, M. and KABESH, M. L. (1964) On a new classification of the Basement Complex. Rocks of the Red Sea Hills, Sudan. *Bull. Soc. Géogr. Egypte* 38, 91–9.

LYDEKKER, R. (1887) On a molar of a Pliocene type of *Equus* from Nubia. *Q. Jl geol. Soc. Lond.* 43, 161–4.

LYNES, H. (1921) Notes on the natural history of Jebel Marra. *Sudan Notes Rec.* 4, 119–37.

—— and CAMPBELL SMITH, W. (1921) Preliminary note on the rocks of Darfur. *Geol. Mag.* 58, 206–15.

LYONS, H. G. (1894) On the stratigraphy and physiography of the Libyan desert. *Q. Jl geol. Soc. Lond.* 50, 531–47.

—— (1897) Note on a portion of the Nubian desert, south-east of Korosko. *Q. Jl geol. Soc. Lond.* 53, 360–73.

—— (1905) On the Nile Flood and its variation. *Geogrl J.* 26, 259–72.

—— (1906) *Physiography of the River Nile and its basin.* Cairo.

—— (1908) Some geographical aspects of the Nile. *Geogrl J.* 32, 449–75.

—— (1909) The longitudinal section of the Nile. *Geogrl J.* 34, 35–51.

MACDONALD, SIR M. (1932) General account of the projects now contemplated for controlling the Nile. *Nature, Lond.* 130, 731.

MACFADYEN, W. A. (1930) The geology of the Farsan Islands, Gizan, and Kamaran Island, Red Sea. Pt. I, General geology. *Geol. Mag.* 67, 310–15.

—— (1932) On the volcanic Zebayir Islands, Red Sea. *Geol. Mag.* 69, 63–7.

MACMICHAEL, M. A. (1920) The Kheiran. *Sudan Notes Rec.* 3, 231–44.

—— (1927) Notes on Jebel Haraza. *Sudan Notes Rec.* 10, 61–7.

MCCONNEL, R. B. (1955) The erosion surfaces of Uganda. *Colon. Geol. Miner. Resour.* 5, 425–8.

MANSOUR, A. O. (1961) Sources of water supply as related to geology in Kordofan Province. *Sudan Geological Survey Bulletin No.* 8.

—— (1963) Sources of water supply as related to geology in Kordofan Province. *Proc. ninth annual Conference on Surveying for Development in Sudan*, pp. 93–9. Phil. Soc. Sudan.

—— and ISKANDER, W. (1960) An iron ore deposit at Jebel Abu Tulu, Dar Messeriya, southern Kordofan. *Sudan Geological Survey Bulletin No.* 6.

—— and SAMUEL, A. (1957) Geology and hydrogeology sheet 66 A, (Rashad) and sheet 66 E, (Talodi) *Sudan Geological Survey, Regional Geology, Memoir No.* 1, 1–48.

MASON, A. (1880) Dar For. *Petermann Mitt.* B. 26.

MAYDON, H. C. (1923) North Kordofan to south Dongola. *Geogrl J.* 61, 34–41.

MENCHIKOFF, N. (1926) Observations géologiques faites au cours de l'expedition de S.A.S. le Prince Kemal-el-Dine Hussein dans le désert de Libye (1925–26). *C. r. hebd. Séanc. Acad. Sci., Paris* 183, 1047.

—— (1927*a*) Les roches cristallines et volcaniques du centru du désert de Libye. *C. r. hebd. Séanc. Acad. Sci., Paris* 184, 215–17.

—— (1927*b*) Etude petrographique des roches cristallines et volcaniques de la région d'Ouenat. *Bull. Soc. geol. Fr.* 27, 337–54.

MILLER, A. R., DENSMORE, C. D., DEGENS, E. T., HATHAWAY, J. C., MAWHEIM, E. T., MACFARLIN, P. F., POCKLINGTON, R., and JOKEILA, A. (1966) Hot brines and recent iron deposits in deeps of the Red Sea. *Geochim. cosmochim. Acta* 30, 341–59.

MINERALS YEARBOOK (1964) Vol. 4. Area Reports. International.

MOHR, P. (1962) *The geology of Ethiopia.* University College of Addis Ababa Press.

—— (1967) Major volcano-tectonic lineament in the Ethiopian Rift System. *Nature, Lond.* 213, 664–5.

MOON, F. W. (1924) Notes on the geology of Hassanein Bey's expedition Sollum Darfur. *Geogrl J.* 64, 388–93.

MOON, P. A. (1954) The static water surface in the sedimentary formations of the north-western Sudan. Open file Rep., Sudan Geological Survey.

MUNSEY, D. F. (1959) First-order and second-order triangulation 1943–52. *Sudan Topographic Survey Records No.* 2.

—— and BULLARD, E. C. (1937) Gravity measurements in the Anglo-Egyptian Sudan. *Mon. Not. R. astr. Soc. geophys. Suppl.* 4, 114–21.

NESTEROFF, W. D. (1955) Les récifs coralliens du Banc Farsan Nord (Mer Rouge) *Annls Inst. océanogr., Monaco* 30, 7–55.

—— (1959) Ages des derniers mouvements du graben de la méthode du C 14 applique aux récifs fossils. *Bull. geol. Soc. Fr.* 7, 415–18.

NEWBOLD, D. (1924*a*) Two journeys to Kufra Oasis. *Sudan Notes Rec.* 7, 104–7.

—— (1924*b*) A desert odyssey of a thousand miles. *Sudan Notes Rec.* 7, 43–92.

—— (1928*a*) More oases of the Libyan Desert. *Geogrl J.* 72, 547–54.

—— (1928*b*) Rock pictures and archaeology in the Libyan Desert. *Antiquity* 2, 261.

—— and SHAW, W. B. K. (1928) An exploration of the south Libyan desert. *Sudan Notes Rec.* 11, 103.

NEWBOLD, L. (1848) On the geological position of the silicified wood of the Egyptian and Libyan deserts, with the description of a petrified forest. *Q. Jl geol. Soc. Lond.* 4, 349.

NEWELL, N. D., RIGBY, J. K., FISCHER, A. G., WHITEMAN, A. J., HICKOX, J. E., and BRADLEY, J. S. (1953) *The Permian reef complex of the Guadalupe Mountains region, Texas and New Mexico.* Freeman, San Francisco.

—— ——, WHITEMAN, A. J., and BRADLEY, J. S. (1951) Shoal water geology and environments, eastern Andros Island, Bahamas. *Bull. Am. Mus. nat. Hist.* 97, 7–29.

NEWHOUSE, F. L. (1928) *The problem of the Upper Nile.* Cairo.

NEWTON, R. S. (1909) On some fossils from the Nubian Series of Egypt. *Geol. Mag.* 6, 352–9, 288–97.

NILSSON, E. (1938) Pluvial lakes in east Africa. *Geol. För. Stockh. Förh.* 60, 423–33.

—— (1940) Ancient changes of climate in British East Africa and Abyssinia. *Geogr. Annlr* 22, 1–29.

—— (1953) Contribution to the history of the Blue Nile. *Bull. Soc. Géogr. Égypte* 25, 29–47.

OAKLEY, K. P. (1963) *Man the tool maker*, 5th edn. British Museum (Natural History), London.

OCHTMAN, L. H. J. (1965) *Soil Survey Report Khashm el Girba Scheme. Khashm el Girba North. Pt. 2, No. 12/KA/3.* Ministry of Agriculture, Khartoum.

OLIVER, J. (1965) Guide to the natural history of Khartoum Province. Part II, the climate of Khartoum Province. *Sudan Notes Rec.* 46, 1–40.

PALLISTER, J. N. (1959) The geology of southern Mengo. *Geological Survey of Uganda, Report* I, pp. 1–124.

PALLME, I. (1844) *Travels in Kordofan*, pp. 293–4. Ref. gold at J. Sheibun.

PARKER, G. and MACKINTOSH, W. P. (1934) The Veveno-Pibor Canal project survey. *Geogrl J.* 84, 53.

PEARSON, H. D. (1912) The Pibor River. I, the upper sources. *Geogrl J.* 40, 486–97.

PERCIVAL, C. (1907) Captain Percival's surveys in Bahr el Ghazal Province. *Geogrl J.* 30, 604–87.

PERKINS, D. (1965) Three faunal assemblages from Sudanese Nubia *Kush* 13, 56–61.

PICARD, L. (1937) On the structure of the Arabian Peninsula. *Bull. Hebrew Univ. Jerusalem Geol. Dept.* Vol. 1, Bull. 3.

—— (1943) *Structure and evolution of Palestine.* Jerusalem.

PLAUCHUT, B. (1960) Notice explicative sur la carte géologique du Bassin du Djado (feuilles Djado et Toummo). *B.R.G.M.* Dakar.

POMEYROL, R. (1968) Nubian Sandstone. *Bull. Am. Ass. Petrol. geol.* 52, (4) 589–60.

PREUMONT, G. E. (1905) Notes on the geological aspects of some of the north-eastern territories of the Congo Free State. *Q. Jl geol. Soc. Lond.* 61, 641–66.

PULFREY, W. (1960) Shape of the Sub-Miocene erosion bevel in Kenya. *Geological Survey of Kenya Bulletin No.* 3.

QUENNEL, A. M. (1958) The structural and geomorphic evolution of the Dead Sea Rift. *Q. Jl geol. Soc. Lond.* 114, 1–24.

QURESHI, I. R., ALMOND, D. C., and SADIG, A. A. (1966) On unusually shaped basalt intrusion in the Sudan. *Bull. Geofis. téor. appl.* 3, (30) 151–60.

—— and SADIG, A. A. (1967) Earthquakes and associated faulting in central Sudan. *Nature, Lond.* 215, 263–5.

RAEBURN, C. and JONES, B. (1934) The Chad basin, geology and water supply. *Nigerian Geological Survey Bulletin No.* 15. (Kaduna).

RAISIN, C. A. (1897) Note on a portion of the Nubian desert south-east of Korosko. II, Petrology. *Q. Jl geol. Soc. Lond.* 53, 364–73.

REED, F. R. C. (1921) *The geology of the British Empire.*

RIGASSI, D. (1969) Nubian Sandstone. Discussion. *Bull. Am. Ass. Petrol. Geol.* 53, (1) 183–4.

RITTMAN, A. (1954) Remarks on the eruption mechanism of the Tertiary volcanoes of Egypt. *Bull. volcan.* ser. 2, 15, 109–17.

ROBERTS, F. W. C. (1923) *Report on Tokar.* Cairo.

RODIS, H. G., HASSAN, A., and WAHADAN, L. (1963) Availability of ground water in Kordofan Province, Sudan. *Sudan Geological Survey Bulletin No.* 12.

—— and ISKANDER, W. (1963) Ground water in the Nahud outlier the Nubian Series, Kordofan Province, Sudan. *U.S.G.S. Prof. Paper* 475-B, Art, 49, 179–81.

RUHE, R. V. (1954) Erosion surfaces of central African interior high plateaus. *INEAC*, ser. scient., *Publ.* 59.

—— (1956) Landscape evolution in the High Ituri, Belgian Congo. *INEAC*, ser. scient., *Publ.* 66.

RÜSSEGGER, J. (1837*a*) Kreids and Sandstein: Enfluss var Granit auf letzteren. *Neues Jb. Miner. Geol. Palaont.*, pp. 665–9.

—— (1837*b*) Porphyre gesteine in Egypten, Nubien bis nach Sennar. *Neues Jl Miner. Geol. Palaont.* pp. 665–9.

—— (1838) Ueber das Vorkommen und die verabeitung des Raseneisensteins des nordliches Kordofan u. uber das Vorkommen des Goldes um Gebet Tira in Lande Nuba. *Arch. Miner.* 2, 315–31.

—— (1848) *Reisen in Europa, Asien und Afrika*, Vols 2 und 3. Stuttgart.

RUXTON, B. P. (1956*a*) MS. An account of the geology of the Sennar and Gedaref area. MS. University of Khartoum Geology Library.

—— (1956*b*) The major rock groups of the northern Red Sea Hills. *Geol. Mag.* **93**, 314–30.

—— (1958) Weathering and subsurface erosion in granite at the piedmont angle, Balos, Sudan. *Geol. Mag.* **95**, 353–77.

—— and BERRY, L. (1960) The Butana grass patterns. *J. Soil Sci.* **2**, (1) 61–2.

—— —— (1961*a*) Weathering profiles and geomorphic position on granite in two tropical regions. *Revue Geomorph. dyn.* **12**, 16–31.

—— —— (1961*b*) Notes on faceted slopes, rock fans, and domes on granite in the east-central Sudan. *Am. J. Sci.* **259**, 194–206.

RZOSKA, J. (1961) Observations on tropical rainpools and general remarks on temporary waters. *Hydrobiologica* **17**, (4) 265–86.

SABET, A. H. (1958) Geology of some dolerite flows on the Red Sea coast south of el Qoseir. *Egypt. J. Geol.* **2**, 45–8.

SADIG, A. A. (1968) MS. A gravity and magnetic survey of the Sabaloka area with details on the ring complex. M.Sc. Thesis, University of Khartoum.

SAGGERSON, E. P. (1962) The physiography and geology of east Africa. *Natural resources of east Africa*. Nairobi.

SAID, R. (1962) *The geology of Egypt*. Elsevier.

—— and SHUKRI, N. M. (1955) Ancient shore lines of Egypt. Pt. 1, The Palaeozoic. *Bull. Soc. geogr. Egypt*, **28**, 41–9.

SANDFORD, K. S. (1932) Recent developments in the study of Palaeolithic Man in Egypt. *Am. J. semit. Lang. Lit.* **48**, 3.

—— (1933*a*) Lower Tertiary rocks within the Province of Berber, Anglo-Egyptian Sudan. *Geol. Mag.* **70**, 301–4.

—— (1933*b*) Volcanic craters in the Libyan Desert. *Nature, Lond.* **131**, 46.

—— (1933*c*) The geology and geomorphology of the southern Libyan desert. *Geogrl J.* **82**, 213–23.

—— (1934) Palaeolithic Man and the Nile Valley in Upper and Middle Egypt. *Publs orient. Inst. Univ. Chicago No.* 18.

—— (1935*a*) Sources of water in the north-west Sudan. *Geogrl J.* 412–31.

—— (1935*b*) Extinct volcanoes and associated intrusions in the Libyan Desert. *Trans. R. geol. Soc. Corn.* **16**, 332–58.

—— (1935*c*) Sources of water in the north-west Sudan. *Geogrl J.* **85**, 412.

—— (1935*d*) Geological observations on the north-west frontiers of the Anglo-Egyptian Sudan and the adjoining part of the southern Libyan desert. *Q. Jl geol. Soc. Lond.* **91**, 323–81.

—— (1936*a*) Observation on the distribution of land and freshwater Mollusca in the southern Libyan Desert. *Q. Jl geol. Soc. Lond.* **92**, 201–20.

—— (1936*b*) Notes on geological data from the southern Libyan desert. *Geogrl J.* **87**, 211–14.

—— (1936*c*) Problems of the Nile Valley. *Geogrl Rev.* **26**, 67.

—— (1936*d*) Geology and geomorphology of the southern Libyan Desert. *Proc. int. geol. Congr.*, Washington, 1933, 779.

—— (1936*e*) Man and Pleistocene climate changes in the north-western Sudan. *Proc. int. geol. Congr.*, Washington, 1933, 812.

—— (1937) Observations on the geology of northern central Africa. *Q. Jl geol. Soc. Lond.* **93**, 534–80.

—— (1946) The geology of French Equatorial Africa. *Geogrl J.* **107**, 114–49.

—— (1949) Notes on the Nile Valley in Berber and Dongola. *Geol. Mag.* **86**, (2) 97–109.

—— (1951) The stratigraphical position of the Nubian Series. *Proc. int. geol. Congr.*, London, Vol. 14, p. 180.

—— and ARKELL, W. J. (1929) On the relation of Palaeolithic man to the history and geology of the Nile Valley in Egypt. *Man* **29**, 65.

—— —— (1933) Palaeolithic man and the Nile Valley in Nubia and Upper Egypt. *Publs orient. Inst. Chicago No.* 17.

SESTINI, J. (1965) Cenozoic stratigraphy and depositional history, Red Sea coast, Sudan. *Bull. Am. Ass. Petrol. Geol.* 9, 1453–72.

SHAW, W. B. K. (1934) The mountain of Uweinat. *Antiquity* 3, 63

—— (1935) International boundaries in Libya. *Geogrl J.* 85, 50–3.

—— (1936*a*) An expedition to the southern Libyan desert. *Geogrl J.* 87, 193–221.

—— (1936*b*) Archaeological investigations in the Libyan Desert. *Nature, Lond.*, **137**, 159.

SHAWA, M. S. (1969) Nubian Sandstone. Discussion. *Bull. Am. Ass. Petrol. geol.* **53**, (1) 182.

SHINNIE, P. L. (1967) *Meroe—a civilization of the Sudan*. Thames and Hudson, London.

SHUKRI, N. M. (1945) The geology of the Nubian Sandstone. *Nature, Lond.* **156**, 3952.

SHUKRI, N. M. (1950) The mineralogy of some Nile sediments. *Q. Jl geol. Soc. Lond.* **105**, 511–34.

—— and AYOUTY, M. K. (1953) The mineralogy of the Nubian Sandstone in Aswan. *Bull. Inst. Égypte* **3**, (2) 65–88.

—— and SAID, R. (1944) Contribution to the geology of the Nubian Sandstone. Pt. 1, Field observations and mechanical analyses. *Bull. Fac. Sci. Egypt. Univ.* **25**, 149–72.

—— —— (1946) Contributions to the geology of Nubian Sandstone. Pt. 2. Mineral analyses. *Bull. Inst. Égypte* **27**, 229–64.

SMITH, W. C. (1920) Riebeckite-rhyolite from Kordofan. *Nature, Lond.* **104**, 693.

SOLECKI, R. S. (editor) (1963) Preliminary statement of the prehistoric investigations of the Columbia University Nubian Expedition in Sudan 1961–62. *Kush* **11**, 70–92.

STEBBING, E. P. (1935) The encroaching Sahara. *Geogrl J.* **85**, 506–19.

—— (1953) *The creeping desert in the Sudan and elsewhere in Africa*. McCorquodale, Sudan.

SUTCLIFFE, J. H. (1944) *Report on Soil Conservation Committee*. Sudan Government, Khartoum.

—— The hydrology of the Sudd region of the upper Nile. University of Cambridge Ph.D. thesis.

SWAN, C. H. (1956) *The recorded behaviour of the River Gash*. Ministry of Irrigation and Hydroelectric Power, Khartoum.

SWARTZ, D. H. and ARDEN, D. D., Jr. (1960) Geological history of the Red Sea area. *Bull. Am. Ass. Petrol. Geol.* **44**, 1621–37.

TALLING, J. F. (1954–5) Characteristics of Nile waters in relation to phytoplankton production. *Second Annual Report, Hydrobiological Research Unit, University of Khartoum.*

—— (1957) The longitudinal succession of water characteristics in the White Nile. *Hydrobiologia* **11**, (1) 73–89.

TAYLOR, J. H. (1952) Geological structures in the Ethiopian region of east Africa. *Trans. N.Y. Acad. Sci.*

TAZIEFF, H. (1952) Une recent compagne oceanographique dans la mer Rouge. *Bull. Soc. belge Geol. Paléont. Hydrol.* **61**, 84–90.

THEOBALD, A. B. (1965) *Ali Dinar : last Sultan of Darfur 1898–1916*. Longmans.

THORBURN, O. H. (1922) The Pibor River. *Geogrl J.* **60**, 210–17.

TILHO, J. (1916–17) Exploration du Cdt. Tilho en Afrique centrale: Borkou, Ennedi, Tibesti, Darfur (1912–1917). *Geographie* **31**.

—— (1919) Une mission scientifique de l'Institut de France en Afrique centrale (Tibesti, Borkou, Ennedi). *C. r. hebd. Séanc. Acad. Sci., Paris* **168**, 984.

—— (1920) The exploration of Tibesti Erdi, Borkou, and Ennedi in 1912–1917. *Geogrl J.* **56**, 81.

—— (1921) La frontière franco-anglo-egyptienne et la ligne de partage des eaux entre les bassins du Nil et du lac Tchad. *C. r. hebd. Séanc. Acad. Sci., Paris* **173**, 563.

TOTHILL, J. D. (1946) The origin of the Sudan Gezira Clay. *Sudan Notes Rec.* **27**, 153–83.

—— (1948*a*) A note on the origin of the soils of the Sudan. In *Agriculture in the Sudan* (editor J. D. Tothill), pp. 129–43. Oxford University Press.

—— (editor) (1948*b*) *Agriculture in the Sudan*. Oxford University Press.

—— *et al.* (1944) *Report of the Soil Conservation Committee*. Sudan Government, Khartoum.

TRENDALL, A. E. (1959) The topography under the northern part of the Kadam volcanics. *Rec. geol. Surv. Uganda* 1955–6, 1–8.

TYLER, W. H. (1932) Chromite in the Sudan. *Min. Mag. Lond.* **47**, 83–8.

—— (1934) An investigation into the commercial utility of a deposit of magnesite-bearing rocks in the Anglo-Egyptian Sudan. *Trans. Ceram. Soc.* **33**, 104–27.

VACEK, M. (1876) Ueber einen forilen Buffelschadel aus Kordofan. *Verh. geol. Reichsanst.* (*St. Anst./ Landesanst.*), *Wein* pp. 141–4.

VAGELER, P. (1933) *Introduction to tropical soils*, Trans. H. Greene. London.

VERCOUTTER, J. (1959) The gold of Kush. *Kush* **7**, 120–53.

VENZLAFF, H. (1958) Report on mapping the south-eastern part of sheet Deraheib. Open file, Sudan Geological Survey, Khartoum.

VINE, F. J. (1967) Spreading of the ocean floor. *Science* **154**, 1405–15.

VITTEMBERGA, P. and CARDELLO, R. (1963) Sedimentologie et Petrographie du Palaeozoique de Bassin du Kufra. First Saharan Symposium, Tripoli, 1963. *Revue I.F.P.* Vol. 18, No. 10, pp. 228–39.

WAKEFIELD, R. C. and MUNSEY, D. F. (1950) The arc of thirtieth meridian between the Egyptian frontier and latitude 13°45′, 1935–1940. *Sudan Topographic Survey Department Records* 1. Ministry of Mineral Resources Sudan Republic.

WALTHER, J. K. (1888) Die Korallenriffe de Sinaihalbinsel. Geologische und Biologische Beobachtungen. *Abh. sächs. Akad. Wiss.* Math–Naturw. Kl. **14**, 439–505.

WARREN, A. (1944) The dunes of Kordofan. *Hunting Group Review No.* 3, pp. 5–9.

—— (1964) The fixed sand dunes of Kordofan in the Republic of Sudan. British Association for the Advancement of Science, Southampton, Section E.

WATSON, G. M. (1894) The Suakin Berber route to the Sudan. *J. Manchr geogr. Soc.* **10**.

WAYLAND, E. J. (1921) Some account of the geology of the Lake Albert Rift Valley. *Geogrl J.* 58, 344–59.

—— (1926) The geology of the Kaiso Bone Beds. *Uganda Geological Survey Paper* 2, pt. 1, pp. 5–11.

—— (1943) A middle Pleistocene discovery in the Anglo-Egyptian Sudan. *Nature, Lond.* **151**, 334.

WEDMANN, E. J., VAN WERINGEN, and ABDL RAHMAN, M. (1965) MS. Khor Arbaat geohydrology and summary. Shell–British Petroleum Report, Khartoum.

WENDORF, F., SHINER, J. L., MARKS, A. E., HEINZELIN, J. DE and CHMIELEWSKI, W. (1965). The combined prehistoric expedition: summary of 1963–64 field season. *Kush* 13, 28–55.

WHITEHOUSE, G. T. (1931) The Largia–Acholi mountain region of the Sudan. *Geogrl J.* 77, 140–79.

WHITEMAN, A. J. (1962) Report on arid zone research in the University of Khartoum. University of Khartoum, Geology Library.

—— (1963) MS. Foundation survey, Khor Abu Anga Bridge, Omdurman. Engineering Geology Report, University of Khartoum, Geology Library.

—— (1964*a*) Occurrence of the Karroo System in Ethiopia. *Nature, Lond.* **261**, 1311.

—— (1964*b*) Geological research in Sudan. *Proc. twelfth annual Conference Phil. Soc. Sudan* 1964, *University of Khartoum*, pp. 219–39.

—— (1965*a*) Essay review. The Geological Map of Africa, scale 1:5 000 000, new edn 1963, *Geol. Mag.* **102**, (1) 80–6.

—— (1965*b*) Summary of present knowledge of the Rift Valley and associated structures in Sudan. *Report UMC/UNESCO Seminar on East African Rift System*, pp. 34–6. Nairobi.

—— (1965*c*) Salt basins around Africa. Written contribution dealing with Red Sea Littoral, pp. 119–20. Institute of Petrology and Geological Society, London.

—— (1967*a*) Red Sea Depression. *New Scient.* 34, 554.

—— (1967*b*) Geology and the builder in Sudan. *Archit. J.*, University of Khartoum.

—— Bibliography of Sudan geology. *Sudan Notes Rec.* (in the press).

—— (1968) MS. The Pleistocene and Recent stratigraphy of Sudan Republic.

—— (1968) Formation of Red Sea Depression. *Geol. Mag.* **105**, 231–46.

WHITEMAN, A. J. (1970) Nubian Group: origin and status. *Bull Am. Ass. Petrol. Geol.* 54, 3, 522–8.

—— and KHEIRALLA, K. M. (1967) MS. The ground-water geology of the Nile Valley between the latitudes of Sennar and Atbara.

—— and SALAAM, Y. A. (1967) MS. The ground-water geology of the Gezira, Sudan.

—— and SHADDAD, M. Z. (1967) The mineral resources of Sudan Republic. *Sudan Notes Rec.* (in the press).

WIDATALLA, A. L. (1963) Ground-water resources of the Sudan. *Proc. ninth annual Conference on Surveying for Development, Phil. Soc. Sudan*, pp. 88–92.

WILCOCKS, W. (1904) *The Nile*. London.

WILCOCKSON, W. H. and TYLER, W. H. (1933) On an area of ultrabasic Rocks in Kassala Province of the Anglo-Egyptian Sudan. *Geol. Mag.* **70**, 306–20.

WILLIAMS, M. A. J. (1966) Age of alluvial clays in the western Gezira, Sudan Republic. *Nature, Lond.* **211**, 270–1.

WILLIMOT, S. G. (1957) Soils and vegetation of the Boma plateau and eastern district of Equatoria. *Sudan Notes Rec.* 38, 10–20.

—— Report on some soils and vegetation of the Boma plateau and eastern district, Equatoria. *YEF/9-A-4*. Yambio. Sudan Government.

WILLIS, R., BICKEL, R. S., JOHNSON, G. H., WILLIS, T. H., and BALLANTYNE, R. S., Jr. (1961) MS. Geologic study of the Coastal Province, Sudan between 20° 10′N, and 21° 52′N. General Exploration Company of California.

WOODWARD, A. S. (1938) A fossil skull of an anustral Bushman from the Anglo-Egyptian Sudan. *Antiquity* pp. 190–5.

WORRALL, G. A. (1956) A pedological study of the Khartoum district. M.Sc. thesis, University of Khartoum.

—— (1957) A simple introduction to the geology of the Sudan. *Sudan Notes Rec.* 38, 2–9.

—— (1958) Deposition of silt by irrigation waters of the Nile Kheti. *Geogrl J.* **124**, 219–22.

—— (1959) The Butana grass patterns. *J. Soil Sci.* **10**, 34–53.

WRIGHT, E. P. (1955) Geological Report No. 9/14/G. Eastern Equatoria. Open file, Sudan Geological Survey, Khartoum.

WRIGHT, H. P. (1917) The Imatong Agora Mountains. *Geogrl J.* 50, 283–7.

WRIGHT, J. W. (1949) Some characteristics of the White Nile and Sobat flood plains. *Nature, Lond.* **164**, 926.

World Oil August 1959, p. 172. Sudan.

World Oil August 1960, pp. 176–77. Sudan.

World Oil August 1961, p. 140. Sudan.

World Oil August 1962, p. 199. Sudan.

World Oil August 1963, p. 167. Sudan.

World Oil August 1964, p. 144. Sudan.

Index

PLATES

PLATE 1. Rauwiyan Island, Sabaloka or Sixth Cataract, showing Nubian Sandstone Formation unconformable on Sabaloka ring complex. (Vertical aerial photograph, Sudan Topographic Survey.)

PLATE 2. Jebel Nahogaret Group, looking east. Situated about 5 miles east of prominent bend in Abu Hamed–Wadi Halfa railway, near Station No. 6. Metamorphic schists, gneisses, and igneous Basement Complex. Note the tail dunes and wadi systems, the latter being evidence of rain as far north as lat. 21° N. (Sudan Topographic Survey print of U.S.A.A.F. Trimetrogon photograph No. 29/100R/1060.)

PLATE 3. Folded metamorphic Basement Complex formations. Looking north towards Nile Valley from a point about 15 miles east of Kumna, sheet 35–I Northern Province. (Sudan Topographic Survey print of U.S.A.A.F. Trimetrogon photograph 161/88R/1060.)

PLATE 4. (A) Boiler-plate and onion-skin weathering, granite jebels near Geri Station, Sabaloka, north of Khartoum. Note the sharply defined pediment and free face with its boulder cascade.

(B) Wadi Derudeb drainage, Kassala Province, looking northwards Basement Complex rocks, Odi schists, and Older gneissose formations.

PLATE 5. Second Cataract from near Abu Sir Rock, looking east towards east bank of Nile Cataract Zone composed of diorites overlain unconformably by the Nubian Sandstone Formation, which forms the back ground hills.

PLATE 6. (A) Erosion surfaces cut across Basement Complex formations. Sinkat area, central Jubal el Bahr el Ahmer. This view is typical of the central section of the Jubal.

(B) General view of Gebeit Mine, northern Jubal el Bahr el Ahmer. (photograph by M. Z. Shaddad, 1967.)

PLATE 7. Tehilla ring complex, Derudeb, Central Jubal el Bahr el Ahmer, showing central gabbro mass (dark) with granite walls of ring behind.

PLATE 8. El Sabaloka or 'waterspout' exit of Sabaloka Gorge, Sixth Cataract, Main Nile. Basal part of the Sabaloka complex, agglomerates, rhyolites, and ignimbrites. Aerial photograph by Sudan Topographic Survey.

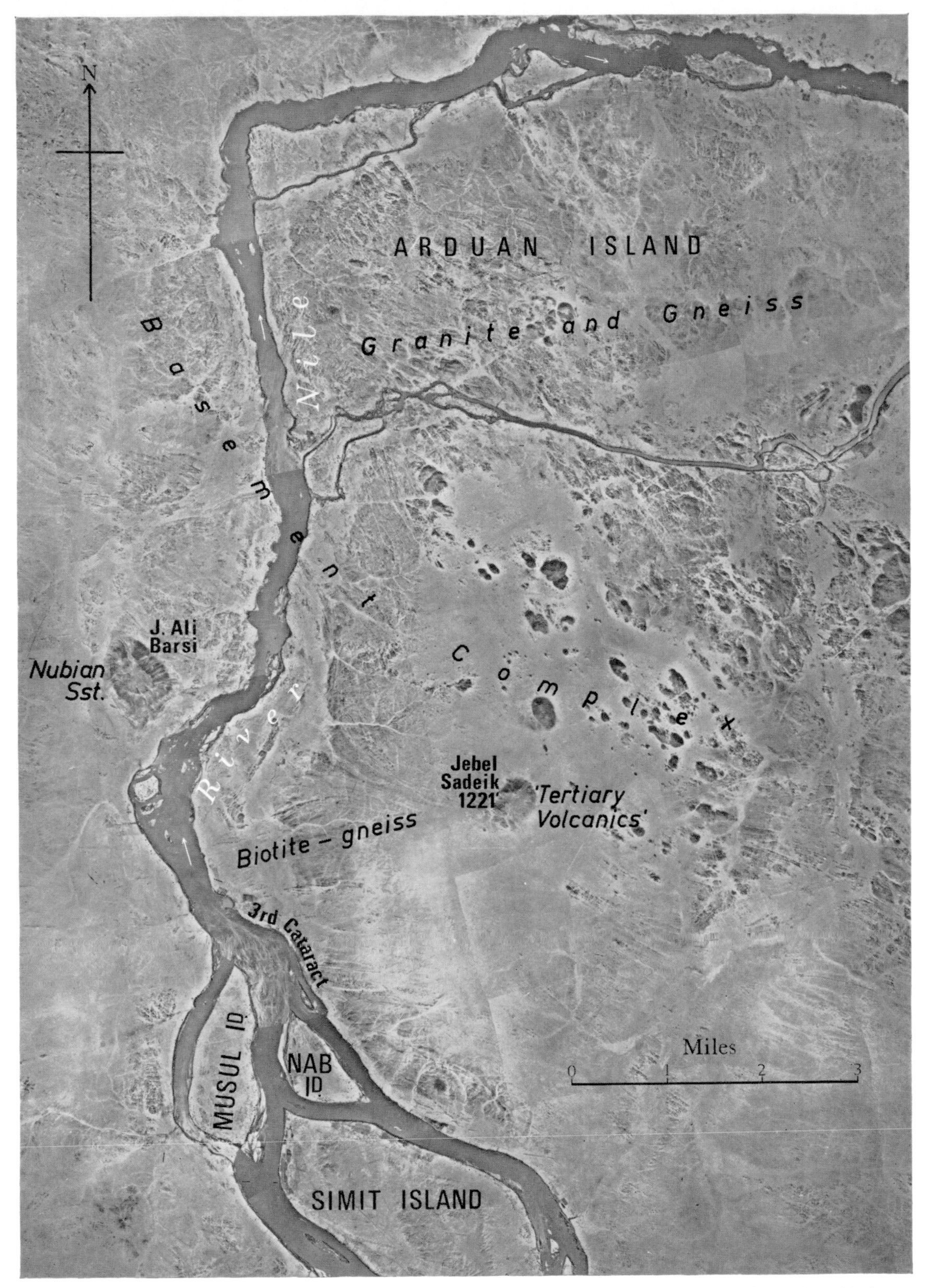

PLATE 9. Third Cataract and Arduan Island north of Kerwa, Northern Province. Mosaic proposed by Fairey Air Surveys, England. Photograph from Sudan Topographic Survey.

PLATE 10. Jebel Kassala, River Gash and the Gash Desert delta from the north. Jebel Kassala is composed of jointed gneissose rocks. Note the prominent 'weathering aureoles' round the jebels.

PLATE 11. Nubian Sandstone jebels, east bank of Nile, south of Abu Simbl. Note narrow belt of 'Nile silt' and the well-developed pediments.

PLATE 12. Nile Valley, looking east from over Kermakol, south of Dongola. Nubian Sandstone Formation largely sand-covered. Note well-developed barchans and tail dunes. The Nile is flowing towards the observer. (Sudan Topographic Survey print of U.S.A.A.F. Trimetrogon photograph No. 560/88R/1060.)

PLATE 13. Dissected, flat-lying Nubian Sandstone Formation, probably iron-capped, north of Elai jebels, east of Wadi el Milk sheet 451. The dark colour of the outcrop is due to iron enrichment. (Sudan Topographic Survey print of U.S.A.A.F. Trimetrogon photograph No. 36/88L/1060.)

PLATE 14. Eastern margin of Ethiopian plateau near Adi Arcai, Ethiopia: looking south-east towards Ras Dascan. Highly dissected plateau lavas and plugs resting on Antalo Limestone, Adigrat Sandstone, and Basement Complex.

PLATE 15. Western slopes of Jebel Marra near Kronga ash beds and lava flows. Note the extensive artificial cultivation terraces. Stream water in this area is saline. (Photograph by Hunting Surveys Ltd., No. F.5848.)

PLATE 16. Jebel Marra Caldera, highest point about 10 000 ft above sea-level, showing walls (1500 ft high and 3 miles wide), central ash cone inside which is a small crater lake (A) some 357 ft deep, and a shallow lake (B) some 38 ft deep occupying a shallow depression caldera floor. (Photograph by D. Hammerton.)

PLATE 17. Wadi Azum, western Darfur. Gorge cut in Basement Complex rocks near Dereisa. (Photograph by Hunting Surveys Ltd., No. F.5724.)

PLATE 18. Wind-blasted lag gravels composed mainly of metamorphic and quartz fragments, resting on coarse sand from which most of the finer grades have been blown away, resting on coarse to fine sand-mixed clay.

PLATE 19. Late Acheulian implement from quartz gravels and sands, Khor Umar, between Omdurman and Wadi Seidna, west bank of Nile.

PLATE 20. 'Karib' Lands, west bank of Atbara River, Butana bridge on Khashim el Girba–Kassala track: cut in Pleistocene deltaic deposits of River Gash–Atbara System.

PLATE 21. Suakin and the Khor Arbent fan, lagoonal, raised reef, and reef deposits, looking east. Sudan Topographic Survey print of U.S.A.A.F. Trimetrogon photograph.

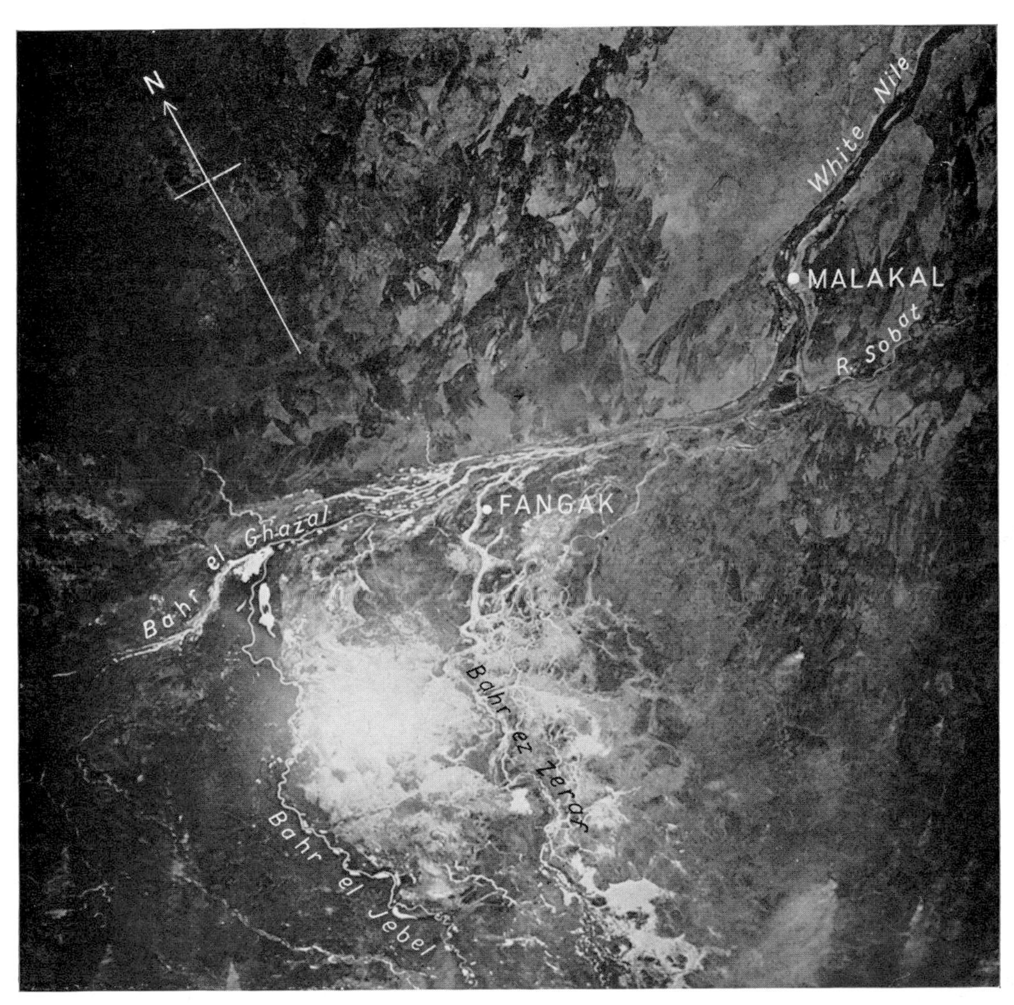

PLATE 22. The northern Sudd, Bahr el Ghazal Province, southern Sudan. (Gemini satellite photograph Gt. 6, 15 December 1965, NASA.)

PRINTED IN GREAT BRITAIN
BY WILLIAM CLOWES & SONS LIMITED
LONDON, COLCHESTER AND BECCLES